MME - Mobility Mana~
S-GW- Serving Gateway

GSM /WCDMA - voice evolution
HSPA /LTE - data evolution

LTE for UMTS

Evolution to LTE-Advanced
Second Edition

LTE for UMTS

Evolution to LTE-Advanced
Second Edition

Edited by

Harri Holma and Antti Toskala
Nokia Siemens Networks, Finland

A John Wiley and Sons, Ltd., Publication

This edition first published 2011
© 2011 John Wiley & Sons, Ltd

Registered office
John Wiley & Sons Ltd, The Atrium, Southern Gate, Chichester, West Sussex, PO19 8SQ, United Kingdom

For details of our global editorial offices, for customer services and for information about how to apply for permission to reuse the copyright material in this book please see our website at www.wiley.com.

Library of Congress Cataloging-in-Publication Data

LTE for UMTS : Evolution to LTE-Advanced / edited by Harri Holma, Antti Toskala. – Second Edition.
  p. cm
  Includes bibliographical references and index.
  ISBN 978-0-470-66000-3 (hardback)
  1. Universal Mobile Telecommunications System. 2. Wireless communication systems – Standards. 3. Mobile communication systems – Standards. 4. Global system for mobile communications. 5. Long-Term Evolution (Telecommunications) I. Holma, Harri (Harri Kalevi), 1970-II. Toskala, Antti. III. Title: Long Term Evolution for Universal Mobile Telecommunications Systems.
  TK5103.4883.L78 2011
  621.3845′6 – dc22

2010050375

A catalogue record for this book is available from the British Library.

Print ISBN: 9780470660003 (H/B)
ePDF ISBN: 9781119992950
oBook ISBN: 9781119992943
ePub ISBN: 9781119992936

Typeset in 10/12 Times by Laserwords Private Limited, Chennai, India.

3   2012

To Kiira and Eevi

– Harri Holma

To Lotta-Maria, Maija-Kerttu and Olli-Ville

– Antti Toskala

Contents

Preface

The number of mobile subscribers has increased tremendously in recent years. Voice communication has become mobile in a massive way and the mobile is the preferred method of voice communication. At the same time data usage has grown quickly in networks where 3GPP High Speed Packet Access (HSPA) was introduced, indicating that the users find broadband wireless data valuable. Average data consumption exceeds hundreds of megabytes and even a few gigabytes per subscriber per month. End users expect data performance similar to fixed lines. Operators request high data capacity with low cost of data delivery. 3GPP Long Term Evolution (LTE) is designed to meet those targets. The first commercial LTE networks have shown attractive performance in the field with data rates of several tens of mbps. This book presents 3GPP LTE standard in Release 8 and describes its expected performance.

The book is structured as follows. Chapter 1 presents the introduction. The standardization background and process is described in Chapter 2. System architecture evolution

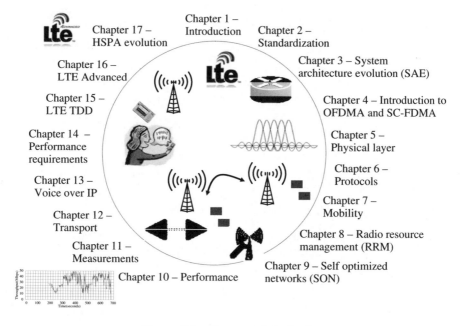

Figure 0.1   Contents of the book

(SAE) is presented in Chapter 3 and the basics of the air interface in Chapter 4. Chapter 5 describes 3GPP LTE physical layer solutions and Chapter 6 protocols. Mobility aspects are addressed in Chapter 7 and the radio resource management in Chapter 8. Self-optimized Network (SON) algorithms are presented in Chapter 9. Radio and end-to-end performance is illustrated in Chapter 10 followed by the measurement results in Chapter 11. The back-haul network is described in Chapter 12. Voice solutions are presented in Chapter 13. Chapter 14 explains the 3GPP performance requirements. Chapter 15 presents the LTE Time Division Duplex (TDD). Chapter 16 describes LTE-Advanced evolution and Chapter 17 HSPA evolution in 3GPP Releases 7 to 10.

LTE can access a very large global market – not only GSM/UMTS operators but also CDMA and WiMAX operators and potentially also fixed network service providers. The large potential market can attract a large number of companies to the market place pushing the economies of scale that enable wide-scale LTE adoption with lower cost. This book is particularly designed for chip set and mobile vendors, network vendors, network operators, application developers, technology managers and regulators who would like to gain a deeper understanding of LTE technology and its capabilities.

The second edition of the book includes enhanced coverage of 3GPP Release 8 content, LTE Release 9 and 10 updates, introduces the main concepts in LTE-Advanced, presents transport network protocols and dimensioning, discusses Self Optimized Networks (SON) solutions and benefits, and illustrates LTE measurement methods and results.

Acknowledgements

The editors would like to acknowledge the hard work of the contributors from Nokia Siemens Networks, Nokia, Renesas Mobile, ST-Ericsson and Nomor Research: Andrea Ancora, Iwajlo Angelow, Dominique Brunel, Chris Callender, Mieszko Chmiel, Mihai Enescu, Marilynn Green, Kari Hooli, Woonhee Hwang, Seppo Hämäläinen, Juha Kallio, Pasi Kinnunen, Tommi Koivisto, Troels Kolding, Krzysztof Kordybach, Juha Korhonen, Jarkko Koskela, István Z. Kovács, Markku Kuusela, Daniela Laselva, Petteri Lunden, Timo Lunttila, Atte Länsisalmi, Esa Malkamäki, Earl McCune, Torsten Musiol, Peter Muszynski, Laurent Noël, Jussi Ojala, Kari Pajukoski, Klaus Pedersen, Karri Ranta-aho, Jussi Reunanen, Timo Roman, Claudio Rosa, Cinzia Sartori, Peter Skov, Esa Tiirola, Ingo Viering, Haiming Wang, Colin Willcock, Che Xiangguang and Yan Yuyu.

We would also like to thank the following colleagues for their valuable comments: Asbjörn Grovlen, Kari Heiska, Jorma Kaikkonen, Michael Koonert, Peter Merz, Preben Mogensen, Sari Nielsen, Gunnar Nitsche, Miikka Poikselkä, Nathan Rader, Sabine Rössel, Benoist Sebire, Mikko Simanainen, Issam Toufik and Helen Waite.

The editors appreciate the fast and smooth editing process provided by Wiley-Blackwell and especially Susan Barclay, Sarah Tilley, Sophia Travis, Jasmine Chang, Michael David, Sangeetha Parthasarathy and Mark Hammond.

We are grateful to our families, as well as the families of all the authors, for their patience during the late-night and weekend editing sessions.

The editors and authors welcome any comments and suggestions for improvements or changes that could be implemented in forthcoming editions of this book. Feedback may be sent to the editors' email addresses: harri.holma@nsn.com and antti.toskala@nsn.com.

List of Abbreviations

1×RTT	1 times Radio Transmission Technology
3GPP	Third Generation Partnership Project
AAA	Authentication, Authorization and Accounting
ABS	Almost Blank Subframes
ACF	Analog Channel Filter
ACIR	Adjacent Channel Interference Rejection
ACK	Acknowledgement
ACLR	Adjacent Channel Leakage Ratio
ACS	Adjacent Channel Selectivity
ADC	Analog-to Digital Conversion
ADSL	Asymmetric Digital Subscriber Line
AKA	Authentication and Key Agreement
AM	Acknowledged Mode
AM/AM	Amplitude Modulation to Amplitude Modulation conversion
AMBR	Aggregate Maximum Bit Rate
AMD	Acknowledged Mode Data
AM/PM	Amplitude Modulation to Phase Modulation conversion
AMR	Adaptive Multi-Rate
AMR-NB	Adaptive Multi-Rate Narrowband
AMR-WB	Adaptive Multi-Rate Wideband
AP	Antenna Port
ARCF	Automatic Radio Configuration Function
ARP	Allocation Retention Priority
ASN	Abstract Syntax Notation
ASN.1	Abstract Syntax Notation One
ATM	Adaptive Transmission Bandwidth
AWGN	Additive White Gaussian Noise
BB	Baseband
BCCH	Broadcast Control Channel
BCH	Broadcast Channel
BE	Best Effort
BEM	Block Edge Mask
BICC	Bearer Independent Call Control Protocol
BiCMOS	Bipolar CMOS
BLER	Block Error Rate

BO	Backoff
BOM	Bill of Material
BPF	Band Pass Filter
BPSK	Binary Phase Shift Keying
BS	Base Station
BSC	Base Station Controller
BSR	Buffer Status Report
BT	Bluetooth
BTS	Base Station
BW	Bandwidth
CA	Carrier Aggregation
CAC	Connection Admission Control
CAZAC	Constant Amplitude Zero Autocorrelation Codes
CBR	Constant Bit Rate
CBS	Committed Burst Size
CC	Component Carrier
CCCH	Common Control Channel
CCE	Control Channel Element
CCO	Coverage and Capacity Optimization
CDD	Cyclic Delay Diversity
CDF	Cumulative Density Function
CDM	Code Division Multiplexing
CDMA	Code Division Multiple Access
CDN	Content Distribution Network
CGID	Cell Global Cell Identity
CIF	Carrier Information Field
CIR	Carrier-to-Interference Ratio
CIR	Committed Information Rate
CLM	Closed Loop Mode
CM	Cubic Metric
CMOS	Complementary Metal Oxide Semiconductor
CoMP	Coordinated Multiple Point
CoMP	Coordinated Multipoint Transmission
CP	Cyclic Prefix
CPE	Common Phase Error
CPE	Customer Premises Equipment
CPICH	Common Pilot Channel
C-Plane	Control Plane
CQI	Channel Quality Information
CRC	Cyclic Redundancy Check
C-RNTI	Cell Radio Network Temporary Identifier
CRS	Cell-specific Reference Symbol
CRS	Common Reference Symbol
CS	Circuit Switched
CSCF	Call Session Control Function
CSFB	Circuit Switched Fallback
CSI	Channel State Information

CT	Core and Terminals
CTL	Control
CW	Continuous Wave
DAC	Digital to Analog Conversion
DARP	Downlink Advanced Receiver Performance
D-BCH	Dynamic Broadcast Channel
DC	Direct Current
DCCH	Dedicated Control Channel
DCH	Dedicated Channel
DC-HSDPA	Dual Cell HSDPA
DC-HSPA	Dual Cell HSPA
DC-HSUPA	Dual Cell HSUPA
DCI	Downlink Control Information
DCR	Direct Conversion Receiver
DCXO	Digitally-Compensated Crystal Oscillator
DD	Duplex Distance
DeNB	Donor eNodeB
DFCA	Dynamic Frequency and Channel Allocation
DFT	Discrete Fourier Transform
DG	Duplex Gap
DHCP	Dynamic Host Configuration Protocol
DL	Downlink
DL-SCH	Downlink Shared Channel
DPCCH	Dedicated Physical Control Channel
DR	Dynamic Range
DRX	Discontinuous Reception
DSCP	DiffServ Code Point
DSL	Digital Subscriber Line
DSP	Digital Signal Processing
DTCH	Dedicated Traffic Channel
DTM	Dual Transfer Mode
DTX	Discontinuous Transmission
DVB-H	Digital Video Broadcast – Handheld
DwPTS	Downlink Pilot Time Slot
EBS	Excess Burst Size
E-DCH	Enhanced DCH
EDGE	Enhanced Data Rates for GSM Evolution
EFL	Effective Frequency Load
EFR	Enhanced Full Rate
EGPRS	Enhanced GPRS
E-HRDP	Evolved HRPD (High Rate Packet Data) network
eICIC	Enhanced Inter-Cell Interference Coordination
EIR	Excess Information Rate
EIRP	Equivalent Isotropic Radiated Power
EMI	Electromagnetic Interference
EMS	Element Management System
EPA	Extended Pedestrian A

EPC	Evolved Packet Core
EPDG	Evolved Packet Data Gateway
ETU	Extended Typical Urban
E-UTRA	Evolved Universal Terrestrial Radio Access
EVA	Extended Vehicular A
EVC	Ethernet Virtual Connection
EVDO	Evolution Data Only
EVM	Error Vector Magnitude
EVS	Error Vector Spectrum
FACH	Forward Access Channel
FCC	Federal Communications Commission
FD	Frame Delay
FD	Frequency Domain
FDD	Frequency Division Duplex
FDE	Frequency Domain Equalizer
FDM	Frequency Division Multiplexing
FDPS	Frequency Domain Packet Scheduling
FDV	Frame Delay Variation
FE	Fast Ethernet
FE	Front End
FFT	Fast Fourier Transform
FLR	Frame Loss Ratio
FM	Frequency Modulated
FNS	Frequency Non-Selective
FR	Full Rate
FRC	Fixed Reference Channel
FS	Frequency Selective
GB	Gigabyte
GBF	Guaranteed Bit Rate
GBR	Guaranteed Bit Rate
GDD	Group Delay Distortion
GE	Gigabit Ethernet
GERAN	GSM/EDGE Radio Access Network
GF	G-Factor
GGSN	Gateway GPRS Support Node
GMSK	Gaussian Minimum Shift Keying
GP	Guard Period
GPON	Gigabit Passive Optical Network
GPRS	General packet radio service
GPS	Global Positioning System
GRE	Generic Routing Encapsulation
GSM	Global System for Mobile Communications
GTP	GPRS Tunneling Protocol
GTP-C	GPRS Tunneling Protocol, Control Plane
GUTI	Globally Unique Temporary Identity
GW	Gateway
HARQ	Hybrid Adaptive Repeat and Request

HB	High Band
HD-FDD	Half-duplex Frequency Division Duplex
HFN	Hyper Frame Number
HII	High Interference Indicator
HO	Handover
HPBW	Half Power Beam Width
HPF	High Pass Filter
HPSK	Hybrid Phase Shift Keying
HRPD	High Rate Packet Data
HSDPA	High Speed Downlink Packet Access
HS-DSCH	High Speed Downlink Shared Channel
HSGW	HRPD Serving Gateway
HSPA	High Speed Packet Access
HS-PDSCH	High Speed Physical Downlink Shared Channel
HSS	Home Subscriber Server
HS-SCCH	High Speed Shared Control Channel
HSUPA	High Speed Uplink Packet Access
IC	Integrated Circuit
IC	Interference Cancellation
ICI	Inter-carrier Interference
ICIC	Inter-cell Interference Control
ICS	IMS Centralized Service
ID	Identity
IDU	Indoor Unit
IEEE	Institute of Electrical and Electronics Engineers
IETF	Internet Engineering Task Force
IFFT	Inverse Fast Fourier Transform
IL	Insertion Loss
iLBC	Internet Lob Bit Rate Codec
IM	Implementation Margin
IMD	Intermodulation
IMS	IP Multimedia Subsystem
IMT	International Mobile Telecommunications
IMT-A	IMT-Advanced
IoT	Interference over Thermal
IOT	Inter-Operability Testing
IP	Internet Protocol
IR	Image Rejection
IRC	Interference Rejection Combining
ISD	Inter-site Distance
ISDN	Integrated Services Digital Network
ISI	Inter-system Interference
ISTO	Industry Standards and Technology Organization
ISUP	ISDN User Part
ITU	International Telecommunication Union
IWF	Interworking Function
L2VPN	Layer 2 VPN

L3VPN	Layer 3 VPN
LAI	Location Area Identity
LB	Low Band
LCID	Logical Channel Identification
LCS	Location Services
LMA	Local Mobility Anchor
LMMSE	Linear Minimum Mean Square Error
LNA	Low Noise Amplifier
LO	Local Oscillator
LOS	Line of Sight
LTE	Long Term Evolution
LTE-A	LTE-Advanced
M2M	Machine-to-Machine
MAC	Medium Access Control
MAP	Maximum *a posteriori*
MAP	Mobile Application Part
MBMS	Multimedia Broadcast/Multicast Service
MBMS	Multimedia Broadcast Multicast System
MBR	Maximum Bit Rate
MCH	Multicast Channel
MCL	Minimum Coupling Loss
MCS	Modulation and Coding Scheme
MDT	Minimization of Drive Testing
MEF	Metro Ethernet Forum
MGW	Media Gateway
MIB	Master Information Block
MIMO	Multiple Input Multiple Output
MIP	Mobile IP
MIPI	Mobile Industry Processor Interface
MIPS	Million Instructions Per Second
MLB	Mobility Load Balancing
MM	Mobility Management
MME	Mobility Management Entity
MMSE	Minimum Mean Square Error
M-Plane	Management Plane
MPLS	Multiprotocol Label Switching
MPR	Maximum Power Reduction
MRC	Maximal Ratio Combining
MRO	Mobility Robustness
MSC	Mobile Switching Center
MSC-S	Mobile Switching Center Server
MSD	Maximum Sensitivity Degradation
MSS	Maximum Segment Size
MTU	Maximum Transmission Unit
MU	Multiuser
MU-MIMO	Multiuser MIMO
MWR	Microwave Radio

NACC	Network Assisted Cell Change
NACK	Negative Acknowledgement
NAS	Non-access Stratum
NAT	Network Address Table
NB	Narrowband
NBAP	Node B Application Part
NDS	Network Domain Security
NF	Noise Figure
NGMN	Next Generation Mobile Networks
NMO	Network Mode of Operation
NMS	Network Management System
NRT	Non-real Time
NTP	Network Time Protocol
OAM	Operation Administration Maintenance
OCC	Orthogonal Cover Codes
OFDM	Orthogonal Frequency Division Multiplexing
OFDMA	Orthogonal Frequency Division Multiple Access
OI	Overload Indicator
OLLA	Outer Loop Link Adaptation
O&M	Operation and Maintenance
OOB	Out of Band
OOBN	Out-of-Band Noise
PA	Power Amplifier
PAPR	Peak to Average Power Ratio
PAR	Peak-to-Average Ratio
PBR	Prioritized Bit Rate
PC	Personal Computer
PC	Power Control
PCB	Printed Circuit Board
PCC	Policy and Charging Control
PCC	Primary Component Carrier
PCCC	Parallel Concatenated Convolution Coding
PCCPCH	Primary Common Control Physical Channel
PCell	Primary Serving Cell
PCFICH	Physical Control Format Indicator Channel
PCH	Paging Channel
PCI	Physical Cell Identity
PCM	Pulse Code Modulation
PCRF	Policy and Charging Resource Function
PCS	Personal Communication Services
PD	Packet Delay
PDCCH	Physical Downlink Control Channel
PDCP	Packet Data Convergence Protocol
PDF	Probability Density Function
PDN	Packet Data Network
PDSCH	Physical Downlink Shared Channel
PDU	Payload Data Unit

PDU	Protocol Data Unit
PDV	Packet Delay Variation
PER	Packed Encoding Rules
PF	Proportional Fair
P-GW	Packet Data Network Gateway
PHICH	Physical HARQ Indicator Channel
PHR	Power Headroom Report
PHS	Personal Handyphone System
PHY	Physical Layer
PKI	Public Key Infrastructure
PLL	Phase Locked Loop
PLMN	Public Land Mobile Network
PLR	Packet Loss Ratio
PMI	Precoding Matrix Index
PMIP	Proxy Mobile IP
PN	Phase Noise
PRACH	Physical Random Access Channel
PRB	Physical Resource Block
PRC	Primary Reference Clock
PS	Packet Switched
PSD	Power Spectral Density
PSS	Primary Synchronization Signal
PTP	Precision Time Protocol
PUCCH	Physical Uplink Control Channel
PUSCH	Physical Uplink Shared Channel
QAM	Quadrature Amplitude Modulation
QCI	QoS Class Identifier
QD	Quasi Dynamic
QN	Quantization Noise
QoS	Quality of Service
QPSK	Quadrature Phase Shift Keying
RACH	Random Access Channel
RAD	Required Activity Detection
RAN	Radio Access Network
RAR	Random Access Response
RAT	Radio Access Technology
RB	Resource Block
RBG	Radio Bearer Group
RF	Radio Frequency
RI	Rank Indicator
RLC	Radio Link Control
RLF	Radio Link Failure
RN	Relay Node
RNC	Radio Network Controller
RNL	Radio Network Layer
RNTP	Relative Narrowband Transmit Power
ROHC	Robust Header Compression

RR	Round Robin
RRC	Radio Resource Control
RRM	Radio Resource Management
RS	Reference Signal
RSCP	Received Symbol Code Power
RSRP	Reference Symbol Received Power
RSRQ	Reference Symbol Received Quality
RSSI	Received Signal Strength Indicator
RT	Real Time
RTT	Round-Trip Time
RV	Redundancy Version
S1AP	S1 Application Protocol
SA	Services and System Aspects
SAE	System Architecture Evolution
SAIC	Single Antenna Interference Cancellation
SCC	Secondary Component Carrier
S-CCPCH	Secondary Common Control Physical Channel
SC-FDMA	Single Carrier Frequency Division Multiple Access
SCH	Shared Channel
SCH	Synchronization Channel
SCM	Spatial Channel Model
SCTP	Stream Control Transmission Protocol
SDQNR	Signal to Distortion Quantization Noise Ratio
SDU	Service Data Unit
SE	Spectral Efficiency
SEG	Security Gateway
SEM	Spectrum Emission Mask
SF	Spreading Factor
SFBC	Space Frequency Block Coding
SFN	Single Frequency Network
SFN	System Frame Number
SGSN	Serving GPRS Support Node
S-GW	Serving Gateway
SIB	System Information Block
SID	Silence Indicator Frame
SIM	Subscriber Identity Module
SIMO	Single Input Multiple Output
SINR	Signal to Interference and Noise Ratio
SLA	Service Level Agreement
SLS	Service Level Specification
SMS	Short Message Service
SNR	Signal to Noise Ratio
SON	Self Organizing Networks
SORTD	Space-Orthogonal Resource Transmit Diversity
S-Plane	Synchronization Plane
SR	Scheduling Request
S-RACH	Short Random Access Channel

SRB	Signaling Radio Bearer
S-RNC	Serving RNC
SRS	Sounding Reference Signals
SR-VCC	Single Radio Voice Call Continuity
SSS	Secondary Synchronization Signal
S-TMSI	S-Temporary Mobile Subscriber Identity
SU-MIMO	Single User Multiple Input Multiple Output
SyncE	Synchronous Ethernet
TA	Tracking Area
TBS	Transport Block Size
TD	Time Domain
TDD	Time Division Duplex
TD-LTE	Time Division Long Term Evolution
TD-SCDMA	Time Division Synchronous Code Division Multiple Access
TM	Transparent Mode
TNL	Transport Network Layer
TPC	Transmit Power Control
TRX	Transceiver
TSG	Technical Specification Group
TTI	Transmission Time Interval
TU	Typical Urban
UDP	Unit Data Protocol
UE	User Equipment
UHF	Ultra High Frequency
UICC	Universal Integrated Circuit Card
UL	Uplink
UL-SCH	Uplink Shared Channel
UM	Unacknowledged Mode
UMD	Unacknowledged Mode Data
UMTS	Universal Mobile Telecommunications System
UNI	User Network Interface
U-Plane	User Plane
UpPTS	Uplink Pilot Time Slot
USB	Universal Serial Bus
USIM	Universal Subscriber Identity Module
USSD	Unstructured Supplementary Service Data
UTRA	Universal Terrestrial Radio Access
UTRAN	Universal Terrestrial Radio Access Network
VCC	Voice Call Continuity
VCO	Voltage Controlled Oscillator
VDSL	Very High Data Rate Subscriber Line
VLAN	Virtual LAN
VLR	Visitor Location Register
V-MIMO	Virtual MIMO
VoIP	Voice over IP
VPN	Virtual Private Network
VRB	Virtual Resource Blocks

WCDMA	Wideband Code Division Multiple Access
WG	Working Group
WLAN	Wireless Local Area Network
WRC	World Radio Conference
X1AP	X1 Application Protocol
ZF	Zero Forcing

1

Introduction

Harry Holma and Antti Toskala

1.1 Mobile Voice Subscriber Growth

The number of mobile subscribers increased tremendously from 2000 to 2010. The first billion landmark was passed in 2002, the second billion in 2005, the third billion 2007, the fourth billion by the end of 2008 and the fifth billion in the middle of 2010. More than a million new subscribers per day have been added globally – that is more than ten subscribers on average every second. This growth is illustrated in Figure 1.1. Worldwide mobile phone penetration is 75%[1]. Voice communication has become mobile in a massive way and the mobile is the preferred method of voice communication, with mobile networks covering over 90% of the world's population. This growth has been fueled by low-cost mobile phones and efficient network coverage and capacity, which is enabled by standardized solutions, and by an open ecosystem leading to economies of scale. Mobile voice is not the privilege of the rich; it has become affordable for users with a very low income.

1.2 Mobile Data Usage Growth

Second-generation mobile networks – like the Global System for Mobile Communications (GSM) – were originally designed to carry voice traffic; data capability was added later. Data use has increased but the traffic volume in second-generation networks is clearly dominated by voice traffic. The introduction of third-generation networks with High Speed Downlink Packet Access (HSDPA) boosted data use considerably.

Data traffic volume has in many cases already exceeded voice traffic volume when voice traffic is converted into terabytes by assuming a voice data rate of 12 kbps. As an example, a European country with three operators (Finland) is illustrated in Figure 1.2. The HSDPA service was launched during 2007; data volume exceeded voice volume during 2008 and the data volume was already ten times that of voice by 2009. More than 90% of the bits in the radio network are caused by HSDPA connections and less than 10% by voice calls. High Speed Downlink Packet Access data growth is driven by

---

[1] The actual user penetration can be different since some users have multiple subscriptions and some subscriptions are shared by multiple users.

LTE for UMTS: Evolution to LTE-Advanced, Second Edition. Edited by Harri Holma and Antti Toskala.
© 2011 John Wiley & Sons, Ltd. Published 2011 by John Wiley & Sons, Ltd.

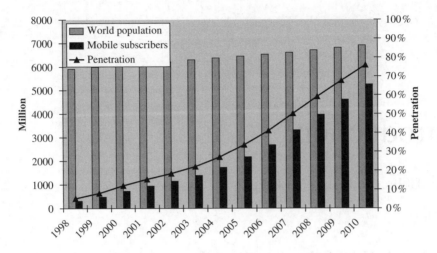

Figure 1.1   Growth of mobile subscribers

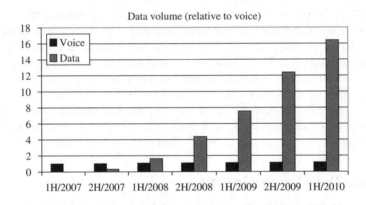

Figure 1.2   HSDPA data volume exceeds voice volume (voice traffic 2007 is scaled to one)

high-speed radio capability, flat-rate pricing schemes and simple device installation. In short, the introduction of HSDPA has turned mobile networks from voice-dominated to packet-data-dominated networks.

Data use is driven by a number of bandwidth-hungry laptop applications, including internet and intranet access, file sharing, streaming services to distribute video content and mobile TV, and interactive gaming. Service bundles of video, data and voice – known also as triple play – are also entering the mobile market, causing traditional fixed-line voice and broadband data services to be replaced by mobile services, both at home and in the office.

A typical voice subscriber uses 300 minutes per month, which is equal to approximately 30 megabytes of data with the voice data rate of 12.2 kbps. A broadband data user can easily consume more than 1000 megabytes (1 gigabyte) of data. The heavy broadband data use takes between ten and 100 times more capacity than voice usage, which sets high requirements for the capacity and efficiency of data networks.

It is expected that by 2015, five billion people will be connected to the internet. Broadband internet connections will be available practically anywhere in the world. Already, existing wireline installations can reach approximately one billion households and mobile networks connect more than three billion subscribers. These installations need to evolve into broadband internet access. Further extensive use of wireless access, as well as new wireline installations with enhanced capabilities, is required to offer true broadband connectivity to the five billion customers.

1.3 Evolution of Wireline Technologies

Wide-area wireless networks have experienced rapid evolution in terms of data rates but wireline networks are still able to provide the highest data rates. Figure 1.3 illustrates the evolution of peak user data rates in wireless and wireline networks. Interestingly, the shape of the evolution curve is similar in both domains with a relative difference of approximately 30 times. Moore's law predicts that the data rates should double every 18 months. Currently, copper-based wireline solutions with Very-High-Data-Rate Digital Subscriber Line (VDSL2) can offer bit rates of tens of Mbps and the passive optical-fiber-based solution provides rates in excess of 100 Mbps. Both copper and fiber based solutions will continue to evolve in the near future, increasing the data rate offerings to the Gbps range.

Wireless networks must push data rates higher to match the user experience that wireline networks provide. Customers are used to wireline performance and they expect the wireless networks to offer comparable performance. Applications designed for wireline networks drive the evolution of the wireless data rates. Wireless solutions also have an important role in providing the transport connections for the wireless base stations.

Wireless technologies, on the other hand, have the huge advantage of being able to offer personal broadband access independent of the user's location – in other words, they provide mobility in nomadic or full mobile use cases. The wireless solution can also

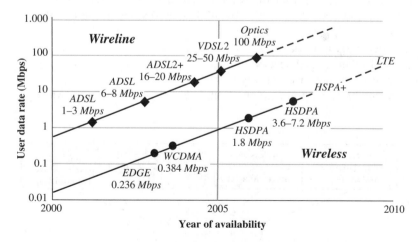

Figure 1.3 Evolution of wireless and wireline user data rates GPON = Gigabit Passive Optical Network. VDSL = Very High Data Rate Subscriber Line. ADSL = Asymmetric Digital Subscriber Line

provide low-cost broadband coverage compared to new wireline installations if there is no existing wireline infrastructure. Wireless broadband access is therefore an attractive option, especially in new growth markets in urban areas as well as in rural areas in other markets.

1.4   Motivation and Targets for LTE

Work towards 3GPP Long Term Evolution (LTE) started in 2004 with the definition of the targets. Even though High-Speed Downlink Packet Access (HSDPA) was not yet deployed, it was evident that work for the next radio system should be started. It takes more than five years from system target settings to commercial deployment using interoperable standards, so system standardization must start early enough to be ready in time. Several factors can be identified driving LTE development: wireline capability evolution, need for more wireless capacity, need for lower cost wireless data delivery and competition from other wireless technologies. As wireline technology improves, similar evolution is required in the wireless domain to ensure that applications work fluently in that domain. There are also other wireless technologies – including IEEE 802.16 – which promised high data capabilities. 3GPP technologies must match and exceed the competition. More capacity is needed to benefit maximally from the available spectrum and base station sites. The driving forces for LTE development are summarized in Figure 1.4.

LTE must be able to deliver performance superior to that of existing 3GPP networks based on HSPA technology. The performance targets in 3GPP are defined relative to HSPA in Release 6. The peak user throughput should be a minimum of 100 Mbps in the downlink and 50 Mbps in the uplink, which is ten times more than HSPA Release 6. Latency must also be reduced to improve performance for the end user. Terminal power consumption must be minimized to enable more use of multimedia applications without recharging the battery. The main performance targets are listed below and are shown in Figure 1.5:

- spectral efficiency two to four times more than with HSPA Release 6;
- peak rates exceed 100 Mbps in the downlink and 50 Mbps in the uplink;
- enables a round trip time of <10 ms;
- packet switched optimized;
- high level of mobility and security;
- optimized terminal power efficiency;
- frequency flexibility with allocations from below 1.5 MHz up to 20 MHz.

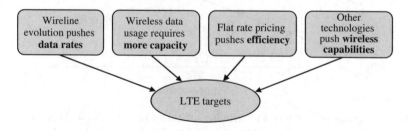

Figure 1.4   Driving forces for LTE development

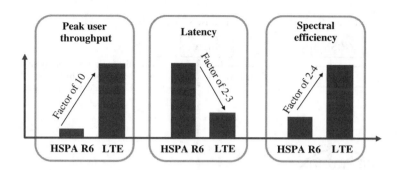

Figure 1.5 Main LTE performance targets compared to HSPA Release 6

1.5 Overview of LTE

The multiple-access scheme in the LTE downlink uses Orthogonal Frequency Division Multiple Access (OFDMA). The uplink uses Single Carrier Frequency Division Multiple Access (SC-FDMA). Those multiple-access solutions provide orthogonality between the users, reducing interference and improving network capacity. Resource allocation in the frequency domain takes place with the resolution of 180 kHz resource blocks both in uplink and in downlink. The frequency dimension in the packet scheduling is one reason for the high LTE capacity. The uplink user specific allocation is continuous to enable single-carrier transmission, whereas the downlink can use resource blocks freely from different parts of the spectrum. The uplink single-carrier solution is also designed to allow efficient terminal power amplifier design, which is relevant for terminal battery life. The LTE solution enables spectrum flexibility. The transmission bandwidth can be selected between 1.4 MHz and 20 MHz depending on the available spectrum. The 20 MHz bandwidth can provide up to 150 Mbps downlink user data rate with 2×2 MIMO and 300 Mbps with 4×4 MIMO. The uplink peak data rate is 75 Mbps. The multiple access schemes are illustrated in Figure 1.6.

High network capacity requires efficient network architecture in addition to advanced radio features. The aim of 3GPP Release 8 is to improve network scalability for increased traffic and to minimize end-to-end latency by reducing the number of network elements. All radio protocols, mobility management, header compression and packet retransmissions are located in the base stations called eNodeB. These stations include all those algorithms that

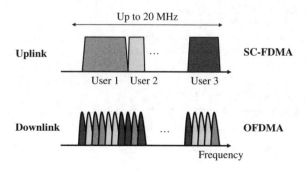

Figure 1.6 LTE multiple access schemes

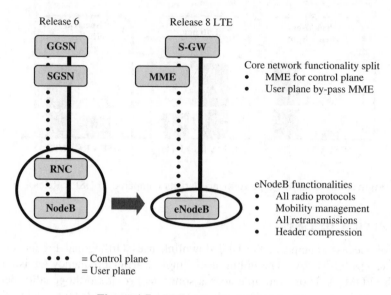

Figure 1.7   LTE network architecture

are located in Radio Network Controller (RNC) in 3GPP Release 6 architecture. The core network is streamlined by separating the user and the control planes. The Mobility Management Entity (MME) is just a control plane element and the user plane bypasses MME directly to Serving Gateway (S-GW). The architecture evolution is illustrated in Figure 1.7.

1.6   3GPP Family of Technologies

3GPP technologies – GSM/EDGE and WCDMA/HSPA – are currently serving 90% of global mobile subscribers. The market share development of 3GPP technologies is illustrated in Figure 1.8. A number of major CDMA operators have already turned to, or

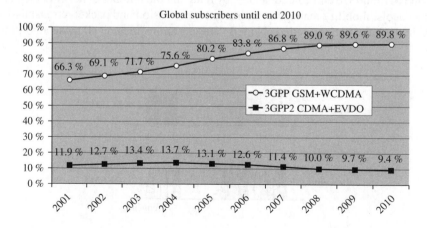

Figure 1.8   Global market share of 3GPP and 3GPP2 technologies

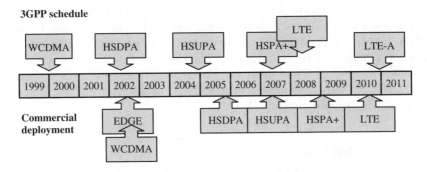

Figure 1.9   Schedule of 3GPP standard and their commercial deployments

will soon be turning to, GSM/WCDMA for voice evolution and to HSPA/LTE for data evolution to access the benefits of the large and open 3GPP ecosystem and for economies of scale for low-cost mobile devices. The number of subscribers using 3GPP-based technologies is currently more than 4.5 billion. The 3GPP Long Term Evolution (LTE) will be built on this large base of 3GPP technologies.

The time schedules of 3GPP specifications and the commercial deployments are illustrated in Figure 1.9. The 3GPP dates refer to the approval of the specifications. WCDMA Release 99 specification work was completed at the end of 1999 and was followed by the first commercial deployments during 2002. The HSDPA and HSUPA standards were completed in March 2002 and December 2004 and the commercial deployments followed in 2005 and 2007. The first phase of HSPA evolution, also known as HSPA+, was completed in June 2007 and the deployments started during 2009. The LTE standard was approved at the end of 2007, backwards compatibility started in March 2009 and the first commercial networks started during 2010. The next step is LTE-Advanced (LTE-A) and the specification was approved in December 2010.

The new generations of technologies push the data rates higher. The evolution of the peak user data rates is illustrated in Figure 1.10. The first WCDMA deployments 2002 offered 384 kbps, first HSDPA networks 3.6–14 Mbps, HSPA evolution 21–168 Mbps, LTE 150–300 Mbps and LTE-Advanced 1 Gbps, which is a more than 2000 times higher data rate over a period of ten years.

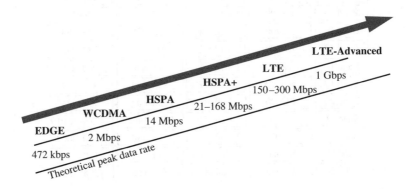

Figure 1.10   Peak data rate evolution of 3GPP technologies

The 3GPP technologies are designed for smooth interworking and coexistence. The LTE will support bi-directional handovers between LTE and GSM and between LTE and UMTS. GSM, UMTS and LTE can share a number of network elements including core network elements. It is also expected that some of the 3G network elements can be upgraded to support LTE and there will be single network platforms supporting both HSPA and LTE. The subscriber management and SIM (Subscriber Identity Module)-based authentication will be used also in LTE.

1.7 Wireless Spectrum

The LTE frequency bands in 3GPP specifications are shown in Figure 1.11 for paired bands and in Figure 1.12 for unpaired bands. Currently 22 paired bands and nine unpaired bands have been defined and more bands will be added during the standardization process. Some of the bands are currently used by other technologies and LTE can coexist with the legacy technologies. In the best case in Europe there is over 600 MHz of spectrum available for the mobile operators when including the 800, 900, 1800, 2100 and 2600 MHz Frequency Division Duplex (FDD) and Time Division Duplex (TDD) bands. In the USA the LTE

Operating band	3GPP name	Total spectrum	Uplink (MHz)	Downlink (MHz)
Band 1	2100	2 × 60 MHz	1920–1980	2110–2170
Band 2	1900	2 × 60 MHz	1850–1910	1930–1990
Band 3	1800	2 × 75 MHz	1710–1785	1805–1880
Band 4	1700/2100	2 × 45 MHz	1710–1755	2110–2155
Band 5	850	2 × 25 MHz	824–849	869–894
Band 6	800	2 × 10 MHz	830–840	875–885
Band 7	2600	2 × 70 MHz	2500–2570	2620–2690
Band 8	900	2 × 35 MHz	880–915	925–960
Band 9	1700	2 × 35 MHz	1750–1785	1845–1880
Band 10	1700/2100	2 × 60 MHz	1710–1770	2110–2170
Band 11	1500	2 × 25 MHz	1427.9–1452.9	1475.9–1500.9
Band 12	US700	2 × 18 MHz	698–716	728–746
Band 13	US700	2 × 10 MHz	777–787	746–756
Band 14	US700	2 × 10 MHz	788–798	758–768
Band 17	US700	2 × 12 MHz	704–716	734–746
Band 18	Japan800	2 × 15 MHz	815–830	860–875
Band 19	Japan800	2 × 15 MHz	830–845	875–890
Band 20	EU800	2 × 30 MHz	832–862	791–821
Band 21	1500	2 × 15 MHz	1447.9–1462.9	1495.9–1510.9
Band 22	3500	2 × 90 MHz	3410–3500	3510–3600
Band 23	S-band	2 × 20 MHz	2000–2020	2180–2200
Band 24	L-band	2 × 34 MHz	1626.5–1660.5	1525–1559

Figure 1.11 Frequency bands for paired bands in 3GPP specifications

Operating band	3GPP name	Total spectrum	Uplink and downlink (MHz)
Band 33	UMTS TDD1	1 × 20 MHz	1900–1920
Band 34	UMTS TDD2	1 × 15 MHz	2010–2025
Band 35	US1900 UL	1 × 60 MHz	1850–1910
Band 36	US1900 DL	1 × 60 MHz	1930–1990
Band 37	US1900	1 × 20 MHz	1910–1930
Band 38	2600	1 × 50 MHz	2570–2620
Band 39	UMTS TDD	1 × 40 MHz	1880–1920
Band 40	2300	1 × 100 MHz	2300–2400
Band 41	2600 US	1 × 194 MHz	2496–2690

Figure 1.12 Frequency bands for unpaired bands in 3GPP specifications

networks will initially be built on 700 and 1700/2100 MHz frequencies. In Japan the LTE deployments start using the 2100 band followed later by 800, 1500 and 1700 bands.

Flexible bandwidth is desirable to take advantage of the diverse spectrum assets: refarming typically requires a narrowband option below 5 MHz while the new spectrum allocations could take advantage of a wideband option of data rates of 20 MHz and higher. It is also evident that both FDD and TDD modes are required to take full advantage of the available paired and unpaired spectrum. These requirements are taken into account in the LTE system specification.

1.8   New Spectrum Identified by WRC-07

The ITU-R World Radiocommunication Conference (WRC-07) worked in October and November 2007 to identify the new spectrum for IMT. The objective was to identify low bands for coverage and high bands for capacity.

The following bands were identified for IMT and are illustrated in Figure 1.13. The main LTE band will be in the 470–806/862 MHz UHF frequencies, which are currently

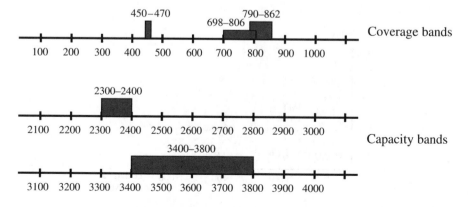

Figure 1.13 Main new frequencies identified for IMT in WRC-07

used for terrestrial TV broadcasting. The 790–862 MHz sub-band was identified in Europe and Asia-Pacific. The availability of the band depends on the national time schedules of the analogue to digital TV switchover. The first auction for that band was conducted in Germany in May 2010 and the corresponding frequency variant is Band 20. The band allows three operators, each running 10 MHz LTE FDD.

The 698–806 MHz sub-band was identified for IMT in Americas. In the US part of the band has already been auctioned. In Asia, the band plan for 698–806 MHz is expected to cover 2 × 45 MHz FDD operation.

The main capacity band will be 3.4–4.2 GHz (C-band). A total of 200 MHz was identified in the 3.4–3.8 GHz sub-band for IMT in Europe and in Asia-Pacific. This spectrum can facilitate the deployment of larger bandwidth of IMT-Advanced to provide the highest bit rates and capacity.

The 2.3–2.4 GHz band was also identified for IMT but this band is not expected to be available in Europe or in the Americas. This band was identified for IMT-2000 in China at the WRC-2000. The 450–470 MHz sub-band was identified for IMT globally, but it is not expected to be widely available in Europe. This spectrum will be narrow with maximum 2 × 5 MHz deployment. Further spectrums for IMT systems are expected to be allocated in the WRC-2016 meeting.

1.9 LTE-Advanced

International Mobile Telecommunications – Advanced (IMT-Advanced) is a concept for mobile systems with capabilities beyond IMT-2000. IMT-Advanced was previously known as 'Systems beyond IMT-2000'. The candidate proposals for IMT-Advanced were submitted to ITU in 2009. Only two candidates were submitted: LTE-Advanced from 3GPP and IEEE 802.16m.

It is envisaged that the new capabilities of these IMT-Advanced systems will support a wide range of data rates in multi-user environments with target peak data rates of up to approximately 100 Mbps for high mobility requirements and up to 1 Gbps for low mobility requirements such as nomadic/local wireless access. IMT-Advanced work within 3GPP is called LTE-Advanced (LTE-A) and it is part of Release 10. 3GPP submitted an LTE-Advanced proposal to ITU in October 2009 and more detailed work was done during 2010. The content was frozen in December 2010 and the backwards compatibility is expected

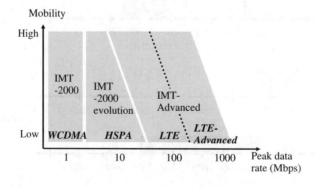

Figure 1.14 Bit rate and mobility evolution to IMT-Advanced

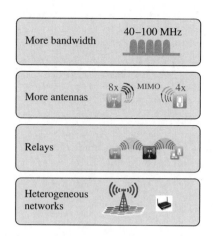

Figure 1.15   LTE-Advanced includes a toolbox of features

in June 2011. The high-level evolution of 3GPP technologies to meet IMT requirements is shown in Figure 1.14.

The main technology components in Release 10 LTE-Advanced include:

- carrier aggregation up to 40 MHz total band, and later potentially up to 100 MHz;
- MIMO evolution up to 8×8 in downlink and 4×4 in uplink;
- relay nodes for providing simple transmission solution;
- heterogeneous networks for optimized interworking between cell layers including macro, micro, pico and femto cells.

LTE-Advanced features are designed in a backwards-compatible way where LTE Release 8 terminals can be used on the same carrier where new LTE-Advanced Release 10 features are activated. LTE-Advanced can be considered as a toolbox of features that can be flexibly implemented on top of LTE Release 8. The main features of LTE-Advanced are summarized in Figure 1.15.

2

LTE Standardization

Antti Toskala

2.1 Introduction

Long-Term Evolution (LTE) standardization is being carried out in the Third Generation Partnership Project (3GPP), as was the case for Wideband CDMA (WCDMA) and the later phase of GSM evolution. This chapter introduces the 3GPP LTE release schedule and the 3GPP standardization process. The requirements set for LTE by the 3GPP community are reviewed and the anticipated steps for later LTE Releases, including LTE-Advanced work for the IMT-Advanced process, are covered. This chapter concludes by introducing LTE specifications and 3GPP structure.

2.2 Overview of 3GPP Releases and Process

The development of 3GPP dates from 1998. The first WCDMA release, Release 99, was published in December 1999. This contained basic WCDMA features with theoretical data rates of up to 2 Mbps, with different multiple access for Frequency Division Duplex (FDD) mode and Time Division Duplex (TDD). After that, 3GPP abandoned the yearly release principle and the release naming was also changed, continuing from Release 4 (including TD-SCDMA), completed in March 2001. Release 5 followed with High Speed Downlink Packet Access (HSDPA) in March 2002 and Release 6 with High Speed Uplink Packet Access (HSUPA) in December 2004 for WCDMA. Release 7 was completed in June 2007 with the introduction of several HSDPA and HSUPA enhancements. In 2008 3GPP finalized Release 8 (with a few issues pending for March 2009, including RRC ASN.1 freezing), which brought further HSDPA/HSUPA improvements, often referred to jointly as High Speed Packet Access (HSPA) evolution, as well as the first LTE Release. A more detailed description of the WCDMA/HSPA Release content can be found in Chapter 17 covering Release 8, 9 and 10 and in [1] for the earlier releases. The feature content for Release 8 was completed in December 2008 and then work continued with further LTE releases, as shown in Figure 2.1, with Release 9 completed at end of 2009 and Releases 10 and 11 scheduled to be finalized in March 2011 and the second half of 2012 respectively. Three months' additional time was allowed for the ASN.1 freeze.

LTE for UMTS: Evolution to LTE-Advanced, Second Edition. Edited by Harri Holma and Antti Toskala.
© 2011 John Wiley & Sons, Ltd. Published 2011 by John Wiley & Sons, Ltd.

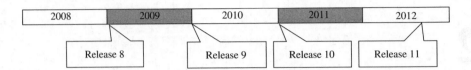

Figure 2.1 3GPP LTE Release schedule up to Release 11

The earlier 3GPP releases are related to the LTE in Release 8. Several novel features, especially features adopted in HSDPA and HSUPA, are also used in LTE. These include base station scheduling with physical layer feedback, physical layer retransmissions and link adaptation. The LTE specifications also reuse the WCDMA design in the areas where this could be done without compromising performance, thus facilitating reuse of the design and platforms developed for WCDMA. The first LTE release, Release 8, supports data rates up to 300 Mbps in the downlink and up to 75 Mbps in the uplink with low latency and flat radio architecture. Release 8 also facilitates the radio level interworking with GSM, WCDMA and cdma2000.

Currently 3GPP is introducing new work items and study items for Release 11, some of them related to features postponed from earlier releases and some of them related to new features. Release 9 content was finalized at the end of 2009 with a few small additions in early 2010. Release 10 contains further radio capability enhancement in the form of LTE-Advanced, submitted to the ITU-R IMT-Advanced process with data rate capabilities foreseen to range up to 1 Gbps. Release 10 specifications were ready at the end of 2010 with some fixes during the first half of 2011.

It is in the nature of the 3GPP process that more projects are started than eventually end up in the specifications. Often a study is carried out first for more complicated issues, as was the case with LTE. Typically, during a study, several alternatives are examined and only some of these might eventually enter a specification. Sometimes a study results in the conclusion that there is not enough gain to justify the added complexity in the system. Sometimes a change request from the work-item phase could be rejected for the same reason. The 3GPP process is shown in Figure 2.2.

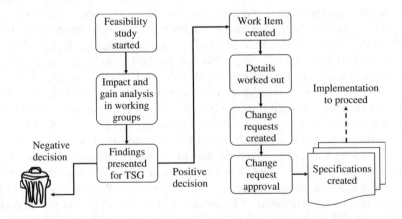

Figure 2.2 3GPP process for moving from study towards work item and specification creation

2.3 LTE Targets

At the start of work, during the first half of 2005, 3GPP defined the requirements for LTE development. The key elements included in the target setting for LTE feasibility study work, as defined in [2], were as follows:

- The LTE system should be packet-switched domain optimized. This means that circuit switched elements are not really considered but everything is assumed to be based on the packet type of operation. The system was required to support IP Multimedia Sub-system (IMS) and further evolved 3GPP packet core.
- As the data rates increase, the latency also needs to come down in order for the data rates to be improved. Thus the requirement for LTE radio round trip time was set to be below 10 ms and access delay below 300 ms.
- The requirements for the data rates were defined to ensure sufficient step in terms of data rates in contrast to HSPA. The peak rate requirements for uplink and downlink were set to 50 Mbps and 100 Mbps respectively.
- As the 3GPP community was used to a good level of security and mobility with earlier systems, starting from GSM, it was also a natural requirement to maintain a good level of mobility and security. This included inter-system mobility with GSM and WCMA, as well as cdma2000, as there was (and is) major interest in the cdma2000 community to evolve to LTE for next generation networks.
- With WCDMA, terminal power consumption was one of the topics that presented challenges, especially in the beginning, so it was necessary to improve terminal power efficiency.
- In the 3GPP technology family there was both a narrowband system (GSM with 200 kHz) and wideband system (WCDMA with 5 MHz), so it was necessary for the new system to facilitate frequency allocation flexibility with 1.25/2.5, 5, 10, 15 and 20 MHz allocations. Later during the course of work, the actual bandwidth values were slightly adjusted for the two smallest bandwidths (to use 1.4 and 3 MHz bandwidths) to match both GSM and cdma2000 refarming cases. It was also required to be able to use LTE in a deployment with WCDMA or GSM as the system on the adjacent band.
- The 'standard' requirement for any new system is to have higher capacity. The benchmark level chosen was 3GPP Release 6, which had a stable specification and known performance level at the time. Thus Release 6 was a stable comparison level for running the LTE performance simulations during the feasibility study phase. Depending on the case, 2–4 times higher capacity than provided with the Release 6 HSDPA/HSUPA reference case, was required.
- One of the drivers for the work was cost – to ensure that the new system could facilitate lower investment and operating costs compared to the earlier system. This was the natural result of the flat-rate charging model for data use and created pressure on the price the data volume level.

It was also expected that the further development of WCDMA would continue in parallel with LTE activity. This was done with Release 8 HSPA improvements, as covered in Chapter 14.

2.4 LTE Standardization Phases

The LTE work was started as a study in 3GPP, with the first workshop held in November 2004 in Canada. In the workshop the first presentations were given both on the expected requirements for the work and on the expected technologies to be adopted. Contributions were made both from the operator and vendor sides.

Following the workshop, 3GPP TSG RAN approved the start of the study for LTE in December 2004, with work first running at the RAN plenary level to define the requirements, and then moving to working groups for detailed technical discussions for multiple access, protocol solutions and architecture. The first key issues to be resolved were the requirements, as discussed above – these were mainly settled during the first half of 2005, with the first approved version in June 2005. Then work focused on solving two key questions:

- What should the LTE radio technology be in terms of multiple access?
- What should the system architecture be?

The multiple access discussion was concluded rather quickly with the decision that something new was needed instead of just an extension to WCDMA. This conclusion was due to the need to cover different bandwidths and data rates in a reasonably complex way. It was obvious that Orthogonal Frequency Division Multiple Access (OFDMA) would be used in the downlink (this had already been reflected in many of the presentations in the original LTE workshop in 2004). For uplink multiple access, the Single Carrier Frequency Division Multiple Access (SC-FDMA) soon emerged as the most favorable choice. It was supported by a large number of key vendors and operators, as could be seen, for example, in [3]. A noticeable improvement from WCDMA was that both FDD and TDD modes were receiving the same multiple access solution, and this is addressed in Chapter 15. Chapter 4 covers OFDMA and SC-FDMA principles and motivational aspects further. The multiple-access decision was officially endorsed at the end of 2005 and, after that, LTE radio work focused on those technologies chosen, with the LTE milestones shown in Figure 2.3. The FDD/TDD alignment refers to the agreement on the adjustment of the frame structure to minimize the differences between FDD and TDD modes of operation.

With regard to LTE architecture, it was decided, after some debate, to aim for a single-node RAN, with the result that all radio-related functionality was to be placed in the base station. This time the term used in 3GPP was 'eNodeB' with 'e' standing for 'evolved'.

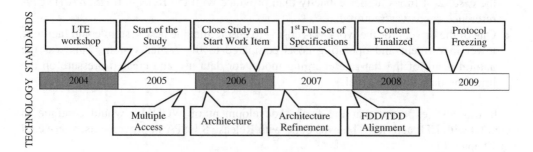

Figure 2.3 LTE Release 8 milestones in 3GPP

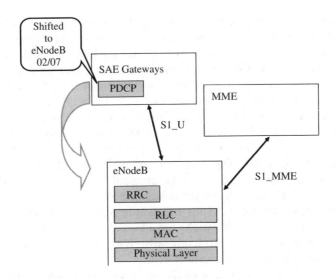

Figure 2.4   Original network architecture for LTE radio protocols

The original architecture split, as shown in Figure 2.4, was endorsed in March 2006 with a slight adjustment made in early 2007 (with the Packet Data Convergence Protocol (PDCP) shifted from the core network side to eNodeB). The fundamental difference with the WCDMA network was the lack of the Radio Network Controller (RNC) element. The architecture is described further in Chapter 3.

The study also evaluated the resulting LTE capacity. The studies reported in [4], and more refined studies summarized in [5], show that the requirements could be reached.

The study part of the process was closed formally in September 2006 and detailed work was started to make the LTE part of 3GPP Release 8 specifications.

The LTE specification work produced the first set of approved physical-layer specifications in September 2007 and the first full set of approved LTE specifications in December 2007. Clearly, there were open issues in the specifications at that point in time, especially in the protocol specifications and in the area of performance requirements. The remaining specification freezing process could be divided into three different steps:

1   Freezing the functional content of the LTE specifications in terms of what would be finalized in the Release 8. This meant leaving out some of the originally planned functionality like support for broadcast use (point-to-multipoint data broadcasting). Functional freeze thus means that no new functionality can be introduced but the agreed content will be finalized. In LTE, the introduction of new functionality was basically over after June 2008 and during the rest of 2008 the work was focusing on completing the missing pieces (and correcting the errors detected) especially in the protocol specifications, which were mostly completed by December 2008.

2   Once all the content is expected to be ready for a particular release, the next step is to freeze the protocol specifications in terms of starting backwards compatibility. The backwards compatibility defines for a protocol the first version that can be the commercial implementation baseline. Until backwards compatibility is started in the protocol specifications, they are corrected by deleting information elements

that do not work as intended and replacing them with new ones. Once the start of backwards compatibility is reached, the older information elements are no longer removed but extensions are used. This allows equipment based on the older version to work based on the old information elements (though not necessary 100% optimally) while equipment with newer software can read the improved/corrected information element after noticing the extension bit being set. Obviously, the core functionality needs to work properly before start of the backwards compatibility makes sense, because if something is totally wrong, fixing it with a backwards compatible correction does not help older software versions if the functionality is not operational at all. This step was reached with 3GPP Release 8 protocol specifications in March 2009 when the protocol language used – Abstract Syntax Notation One (ASN.1) – review for debugging all the errors was completed. With Release 9 specifications the ASN.1 backwards compatibility was started in March 2010 while, for Release 10, 3GPP has scheduled ASN.1 backward compatibility to be started from June 2011 onwards. With further work in Release 11, the corresponding milestone is December 2012.

3   The last phase is a 'deep' freeze of the specifications, when no further changes to specifications will be allowed. This is something that is valid for a release that has already been rolled out in the field, like Release 5 with HSDPA and Release 6 with HSUPA. With the devices out in the field the core functionality has been tested and proven – there is no point in changing those releases any more. Improvements would need to be made in a later release. Problems may arise in cases where a feature has not been implemented (and thus no testing with network has been possible) and the problem is only detected later. Then it could be corrected in a later release and a recommendation could be made to use it only for devices that are based on this later release. For LTE Release 8 specifications this phase was achieved more-or-less at the end of 2010. Changes that were made during 2009 and 2010 still allowed backwards compatibility to be maintained.

Thus, from a 3GPP perspective, Release 8 has reached a very stable state. The last topics to be covered were UE-related performance requirements in different areas. The amount of changes requested concerning physical layers and key radio-related protocols decreased sharply after March 2009, as shown in Figure 2.5, allowing RRC backwards compatibility from March 2009 to be maintained. With the internal interfaces (S1/X2) there was one round of non-backwards compatible corrections still in May 2009 after which backwards compatibility there was also retained.

2.5 Evolution Beyond Release 8

The work of 3GPP during 2008 focused on finalizing Release 8, but work was also started for issues beyond Release 8, including the first Release 9 projects and LTE-Advanced for IMT-Advanced. The following projects have been addressed in 3GPP during Release 9 and 10 work:

• LTE MBMS, which covers operations related to broadcast-type data for both for dedicated MBMS carriers and for shared carriers. When synchronized properly, OFDMA-based broadcast signals can be sent in the same resource space from different base stations (with identical content) and then signals from multiple base stations can be combined in the devices. This principle is already in use in, for example, Digital Video Broadcasting for Handheld (DVB-H) devices in the market. DVB-H is also an OFDMA-based system but is only intended for broadcast use. Release 9 supports the carrier with shared MBMS and point-to-point. No specific MBMS-only carrier is defined. There are no changes in that aspect in Release 10.

- Self-Optimized Network (SON) enhancements. 3GPP has worked on the self-optimization/configuration aspects of LTE and that work continued in Releases 9 and 10. It is covered in more detail in Chapter 9.
- Minimization of Drive Test (MDT). This is intended to reduce the need to collect data with an actual drive test by obtaining the necessary information from the devices in the field, as elaborated in more detail in Chapter 6.
- Requirements for multi-bandwidth and multi-radio access technology base stations. The scope of this work is to define the requirements for in cases where the same Radio Frequency (RF) part is used for transmitting, for example, LTE and GSM or LTE and WCDMA signals. Currently requirements for emissions on the adjacent frequencies, for example, take only a single Radio Access Technology (RAT) into account; requirements will now be developed for different combinations including running multiple LTE bandwidths in parallel in addition to the multi-RAT case. This was completed in Release 10 for the operation of contiguous frequency allocations. From mid-2010 onwards, work will be carried out for non-contiguous spectrum allocations (for an operator, for example, in 900 MHz band having spectrums in different parts of the band).
- Enhancing support for emergency calls, both in terms of enabling prioritization of the emergency calls as well as adding (in addition to the GPS-based methods) support for position location in the LTE network itself with the inclusion of OTDOA measurements in the UEs, completed in Release 9. There is also ongoing work to add uplink-based solutions as part of LTE specifications in the Release 11 based on Uplink TDOA (UTDOA).

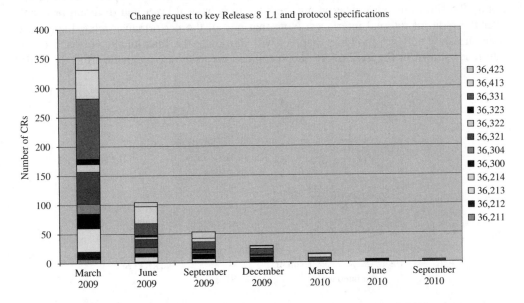

Figure 2.5 Release 8 Change Request statistics from March 2009

The next release, then, is Release 11 with work commencing in early 2011. It is scheduled to be finalized by the end of 2012. The content of Release 11 has not yet been decided in 3GPP but it seems obvious that there will be plenty of new projects started because, during the work for Release 10, a large number of topics were identified that could not be initiated in order to keep to the Release 10 schedule.

2.6   LTE-Advanced for IMT-Advanced

In parallel to the work for LTE corrections and further optimization in Releases 9 and 10, 3GPP is also creating input for the IMT-Advanced process in ITU-R, as covered in Chapter 16 in more detail. The ITU-R is developing the framework for next-generation wireless networks with up to 1 Gbps for nomadic (low mobility) and 100 Mbps for high-mobility data rates being part of the requirements for IMT-Advanced technology. 3GPP is aiming to meet ITU-R specifications in 2011, so 3GPP will submit the first full set of specification around the end of 2010. The work for LTE-Advanced includes multiple work items, improving both downlink and uplink capacity and peak data rates.

2.7   LTE Specifications and 3GPP Structure

The LTE specifications mostly follow a similar notation to the WCDMA specifications but using the 36-series numbering. For example when WCDMA RRC is 25.331, the corresponding LTE spec is 36.331. The LTE specifications use the term 'Evolved Universal Terrestrial Radio Access' (E-UTRA) whereas WCDMA specification use the UTRA term (and UTRAN, with 'N' standing for 'Network'). There are some differences in the physical layer – for example, specifications on spreading and modulation, like WCDMA specification 25.213, were not needed. Now, due to use of the same multiple access, the FDD and TDD modes are covered in the same physical layer specification series. Figure 2.6 shows the specification numbers for the physical layer and different protocols over the radio or internal interfaces. Note that not all the performance-related specifications are shown. The following chapters will introduce the functionality in each of the interfaces shown in Figure 2.6. All the specifications listed are available from the 3GPP

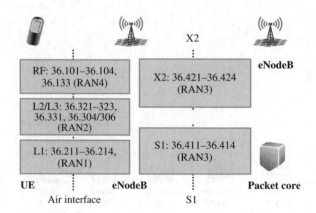

Figure 2.6   Specifications with responsible working groups for different LTE interfaces

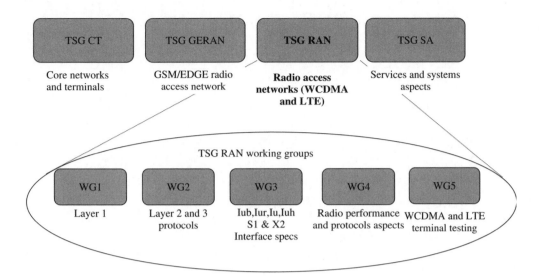

Figure 2.7   3GPP structure

website [6]. When using a 3GPP specification it is recommended that the latest version for the release in question should be used. For example, version 8.0.0 is always the first approved version and versions with numbers 8.4.0 (or higher) are normally more stable, with fewer errors.

Inside 3GPP, the 3GPP TSG RAN is responsible for LTE specification development. The detailed specification work is covered in the Working Groups (WGs) under each TSG. TSG RAN has a total of five working groups, as shown in Figure 2.7. Working groups under other TSGs are not shown. Specifications for the Evolved Packet Core (EPC) are covered in TSA SA and in TSG CT. They are also needed for an end-to-end functioning system. The TSG GERAN is responsible for the Release 8 changes in GSM/EDGE specifications to facilitate the LTE-GERAN interworking from a GERAN perspective.

The 36.2 series physical layer specifications are developed by WG1 as shown in Figure 2.7. The Layer 2 (L2) and Layer 3 (L3) specifications are in the 36.3 series from WG2; internal interfaces are in the 36.4 series from WG3, and radio performance requirements are in the 36.1 series from WG4. The LTE terminal test specifications are from WG5. All groups also cover their respective areas for further WCDMA/HSPA releases.

References

[1] H. Holma, A. Toskala, 'WCDMA for UMTS', 5th edition, Wiley, 2010.
[2] 3GPP Technical Report, TR 25.913, 'Requirements for Evolved UTRA (E-UTRA) and Evolved UTRAN (E-UTRAN)' version 7.0.0, June 2005.
[3] 3GPP Tdoc, RP-050758, LS on UTRAN LTE Multiple Access Selection, 3GPP TSG RAN WG1, November 2005.
[4] 3GPP Technical Report, TR 25.814, 'Physical Layer Aspects for Evolved Universal Terrestrial Radio Access (UTRA)', 3GPP TSG RAN, September 2006.
[5] 3GPP Tdoc, RP-060535, LS on LTE SI Conclusions, 3GPP TSG RAN WG1, September 2006.
[6] www.3gpp.org.

3

System Architecture Based on 3GPP SAE

Atte Länsisalmi and Antti Toskala

3.1   System Architecture Evolution in 3GPP

When the evolution of the radio interface started, it soon became clear that the system architecture would also need to be evolved. The general drive towards optimizing the system only for packet switched services is one reason that alone would have set the need for evolution, but some of the radio interface design goals – such as removal of soft handover – opened up new opportunities in the architecture design. Also, since it had been shown by High Speed Packet Access (HSPA) that all radio functionality can be efficiently co-located in the NodeB, the door was left open for discussions of flatter overall architecture.

Discussions for System Architecture Evolution (SAE) then soon followed the radio interface development, and it was agreed to schedule the completion of the work in Release 8. There had been several reasons for starting this work, and there were also many targets. The following lists some of the targets that possibly shaped the outcome the most:

- optimization for packet switched services in general, when there is no longer a need to support the circuit switched mode of operation;
- optimized support for higher throughput required for higher end user bit rates;
- improvement in the response times for activation and bearer set-up;
- improvement in the packet delivery delays;
- overall simplification of the system compared to the existing 3GPP and other cellular systems;
- optimized inter-working with other 3GPP access networks;
- optimized inter-working with other wireless access networks.

Many of the targets implied that a flat architecture would need to be developed. Flat architecture with less involved nodes reduces latencies and improves performance.

LTE for UMTS: Evolution to LTE-Advanced, Second Edition. Edited by Harri Holma and Antti Toskala.
© 2011 John Wiley & Sons, Ltd. Published 2011 by John Wiley & Sons, Ltd.

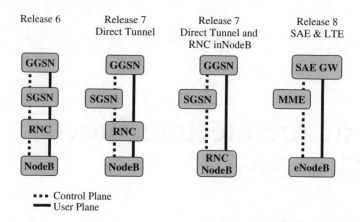

Figure 3.1 3GPP architecture evolution towards flat architecture

Development towards this direction had already started in Release 7 where the Direct Tunnel concept allows User Plane (UP) to bypass the SGSN, and the placement of RNC functions to HSPA NodeB was made possible. Figure 3.1 shows these evolution steps and how this aspect was captured at a high level in SAE architecture.

Some of the targets seem to drive the architecture development in completely different directions. For example, optimized inter-working with several wireless access networks (ANs) indicates the need to introduce a set of new functions and maybe even new interfaces to support specific protocols separately for each one of them. This works against the target of keeping the architecture simple. Therefore, since it is likely that that none of the actual deployments of the architecture would need to support all of the potential inter-working scenarios, the 3GPP architecture specifications were split into two tracks:

- GPRS enhancements for E-UTRAN access [1]: This document describes the architecture and its functions in its native 3GPP environment with E-UTRAN and all the other 3GPP ANs, and defines the inter-working procedures between them. The common nominator for these ANs is the use of GTP (GPRS Tunnelling Protocol) as the network mobility protocol.
- Architecture enhancements for non-3GPP accesses [2]: This document describes the architecture and functions when inter-working with non-3GPP ANs, such as cdma2000® High Rate Packet Data (HRPD), is needed. The mobility functionality in this document is based on IETF protocols, such as MIP (Mobile Internet Protocol) and PMIP (Proxy MIP), and the document also describes E-UTRAN in that protocol environment.

This chapter further describes the 3GPP system architecture in some likely deployment scenarios: basic scenario with only E-UTRAN, legacy 3GPP operator scenario with existing 3GPP ANs and E-UTRAN, and finally E-UTRAN with non-3GPP ANs, where inter-working with cdma2000® is shown as a specific example.

3.2   Basic System Architecture Configuration with only E-UTRAN Access Network

3.2.1   Overview of Basic System Architecture Configuration

Figure 3.2 describes the architecture and network elements in the architecture configuration where only the E-UTRAN AN is involved. The logical nodes and connections
shown in this figure represent the basic system architecture configuration. These elements
and functions are needed in all cases when E-UTRAN is involved. The other system
architecture configurations described in the next sections also include some additional
functions.

This figure also shows the division of the architecture into four main high level domains:
User Equipment (UE), Evolved UTRAN (E-UTRAN), Evolved Packet Core Network
(EPC), and the Services domain.

The high level architectural domains are functionally equivalent to those in the existing
3GPP systems. The new architectural development is limited to Radio Access and Core
Networks, the E-UTRAN and the EPC respectively. UE and Services domains remain
architecturally intact, but functional evolution has also continued in those areas.

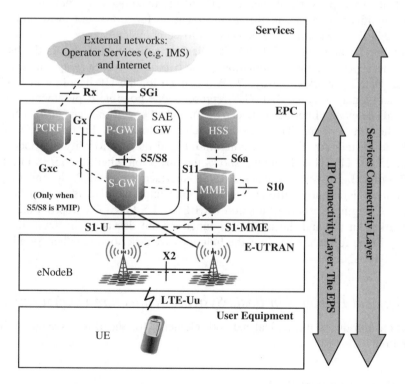

Figure 3.2   System architecture for E-UTRAN only network

UE, E-UTRAN and EPC together represent the Internet Protocol (IP) Connectivity Layer. This part of the system is also called the Evolved Packet System (EPS). The main function of this layer is to provide IP based connectivity, and it is highly optimized for that purpose only. All services will be offered on top of IP, and circuit switched nodes and interfaces seen in earlier 3GPP architectures are not present in E-UTRAN and EPC at all. IP technologies are also dominant in the transport, where everything is designed to be operated on top of IP transport.

The IP Multimedia Sub-System (IMS) [3] is a good example of service machinery that can be used in the Services Connectivity Layer to provide services on top of the IP connectivity provided by the lower layers. For example, to support the voice service, IMS can provide Voice over IP (VoIP) and interconnectivity to legacy circuit switched networks PSTN and ISDN through Media Gateways it controls.

The development in E-UTRAN is concentrated on one node, the evolved Node B (eNodeB). All radio functionality is collapsed there, i.e. the eNodeB is the termination point for all radio related protocols. As a network, E-UTRAN is simply a mesh of eNodeBs connected to neighbouring eNodeBs with the X2 interface.

One of the big architectural changes in the core network area is that the EPC does not contain a circuit switched domain, and no direct connectivity to traditional circuit switched networks such as ISDN or PSTN is needed in this layer. Functionally the EPC is equivalent to the packet switched domain of the existing 3GPP networks. There are, however, significant changes in the arrangement of functions and most nodes and the architecture in this part should be considered to be completely new.

Both Figure 3.1 and Figure 3.2 show an element called SAE GW. As the latter figure indicates, this represents the combination of the two gateways, Serving Gateway (S-GW) and Packet Data Network Gateway (P-GW) defined for the UP handling in EPC. Implementing them together as the SAE GW represents one possible deployment scenario, but the standards define the interface between them, and all operations have also been specified for when they are separate. The same approach is followed in this chapter of the book.

The Basic System Architecture Configuration and its functionality are documented in 3GPP TS 23.401 [1]. This document shows the operation when the S5/S8 interface uses the GTP protocol. However, when the S5/S8 interface uses PMIP, the functionality for these interfaces is slightly different, and the Gxc interface also is needed between the Policy and Charging Resource Function (PCRF) and S-GW. The appropriate places are clearly marked in [1] and the additional functions are described in detail in 3GPP TS 23.402 [2]. In the following sections the functions are described together for all cases that involve E-UTRAN.

3.2.2   Logical Elements in Basic System Architecture Configuration

This section introduces the logical network elements for the Basic System Architecture configuration.

3.2.2.1   User Equipment (UE)

UE is the device that the end user uses for communication. Typically it is a hand held device such as a smart phone or a data card such as those used currently in 2G and 3G, or it could be embedded, e.g. to a laptop. UE also contains the Universal Subscriber

Identity Module (USIM) that is a separate module from the rest of the UE, which is often called the Terminal Equipment (TE). USIM is an application placed into a removable smart card called the Universal Integrated Circuit Card (UICC). USIM is used to identify and authenticate the user and to derive security keys for protecting the radio interface transmission.

Functionally the UE is a platform for communication applications, which signal with the network for setting up, maintaining and removing the communication links the end user needs. This includes mobility management functions such as handovers and reporting the terminals location, and in these the UE performs as instructed by the network. Maybe most importantly, the UE provides the user interface to the end user so that applications such as a VoIP client can be used to set up a voice call.

3.2.2.2 E-UTRAN Node B (eNodeB)

The only node in the E-UTRAN is the E-UTRAN Node B (eNodeB). Simply put, the eNodeB is a radio base station that is in control of all radio related functions in the fixed part of the system. Base stations such as eNodeB are typically distributed throughout the networks coverage area, each eNodeB residing near the actual radio antennas.

Functionally eNodeB acts as a layer 2 bridge between UE and the EPC, by being the termination point of all the radio protocols towards the UE, and relaying data between the radio connection and the corresponding IP based connectivity towards the EPC. In this role, the eNodeB performs ciphering/deciphering of the UP data, and also IP header compression/decompression, which means avoiding repeatedly sending the same or sequential data in IP header.

The eNodeB is also responsible for many Control Plane (CP) functions. The eNodeB is responsible for the Radio Resource Management (RRM), i.e. controlling the usage of the radio interface, which includes, for example, allocating resources based on requests, prioritizing and scheduling traffic according to required Quality of Service (QoS), and constant monitoring of the resource usage situation.

In addition, the eNodeB has an important role in Mobility Management (MM). The eNodeB controls and analyses radio signal level measurements carried out by the UE, makes similar measurements itself, and based on those makes decisions to handover UEs between cells. This includes exchanging handover signalling between other eNodeBs and the MME. When a new UE activates under eNodeB and requests connection to the network, the eNodeB is also responsible for routing this request to the MME that previously served that UE, or selecting a new MME, if a route to the previous MME is not available or routing information is absent.

Details of these and other E-UTRAN radio interface functions are described extensively elsewhere in this book. The eNodeB has a central role in many of these functions.

Figure 3.3 shows the connections that eNodeB has to the surrounding logical nodes, and summarizes the main functions in these interfaces. In all the connections the eNodeB may be in a one-to-many or a many-to-many relationship. The eNodeB may be serving multiple UEs at its coverage area, but each UE is connected to only one eNodeB at a time. The eNodeB will need to be connected to those of its neighbouring eNodeBs with which a handover may need to be made.

Both MMEs and S-GWs may be pooled, which means that a set of those nodes is assigned to serve a particular set of eNodeBs. From a single eNodeB perspective this

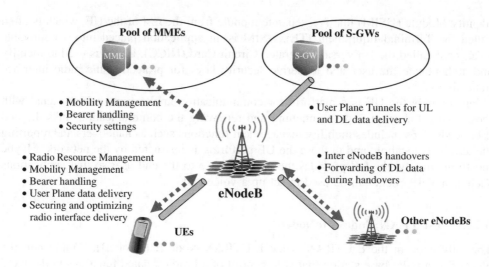

Figure 3.3   eNodeB connections to other logical nodes and main functions

means that it may need to connect to many MMEs and S-GWs. However, each UE will be served by only one MME and S-GW at a time, and the eNodeB has to keep track of this association. This association will never change from a single eNodeB point of view, because MME or S-GW can only change in association with inter-eNodeB handover.

3.2.2.3   Mobility Management Entity (MME)

Mobility Management Entity (MME) is the main control element in the EPC. Typically the MME would be a server in a secure location in the operator's premises. It operates only in the CP, and is not involved in the path of UP data.

In addition to interfaces that terminate to MME in the architecture as shown in Figure 3.2, the MME also has a logically direct CP connection to the UE, and this connection is used as the primary control channel between the UE and the network. The following lists the main MME functions in the basic System Architecture Configuration:

- Authentication and Security: When a UE registers to the network for the first time, the MME initiates the authentication, by performing the following: it finds out the UE's permanent identity either from the previously visited network or the UE itself; requests from the Home Subscription Server (HSS) in UE's home network the authentication vectors which contain the authentication challenge – response parameter pairs; sends the challenge to the UE; and compares the response received from the UE to the one received from the home network. This function is needed to assure that the UE is who it claims to be. The details of EPS-AKA authentication are defined in [4]. The MME may repeat authentication when needed or periodically. The MME will calculate UEs ciphering and integrity protection keys from the master key received in the authentication vector from the home network, and it controls the related settings in E-UTRAN for UP and CP separately. These functions are used to protect the communication from eavesdropping and from alteration by unauthorized third parties respectively. To protect the UE privacy, MME also allocates each UE a temporary identity called the Globally Unique Temporary Identity (GUTI), so that the need to send the permanent UE identity – International Mobile Subscriber Identity (IMSI) – over the radio interface is minimized. The GUTI may be re-allocated, e.g. periodically to prevent unauthorized UE tracking.

- Mobility Management: The MME keeps track of the location of all UEs in its service area. When a UE makes its first registration to the network, the MME will create an entry for the UE, and signal the location to the HSS in the UE's home network. The MME requests the appropriate resources to be set up in the eNodeB, as well as in the S-GW which it selects for the UE. The MME will then keep tracking the UE's location either on the level of eNodeB, if the UE remains connected, i.e. is in active communication, or at the level of Tracking Area (TA), which is a group of eNodeBs in case the UE goes to idle mode, and maintaining a through connected data path is not needed. The MME controls the setting up and releasing of resources based on the UE's activity mode changes. The MME also participates in control signalling for handover of an active mode UE between eNodeBs, S-GWs or MMEs. MME is involved in every eNodeB change, since there is no separate Radio Network Controller to hide most of these events. An idle UE will report its location either periodically, or when it moves to another Tracking Area. If data are received from the external networks for an idle UE, the MME will be notified, and it requests the eNodeBs in the TA that is stored for the UE to page the UE.
- Managing Subscription Profile and Service Connectivity: At the time of a UE registering to the network, the MME will be responsible for retrieving its subscription profile from the home network. The MME will store this information for the duration it is serving the UE. This profile determines what Packet Data Network connections should be allocated to the UE at network attachment. The MME will automatically set up the default bearer, which gives the UE the basic IP connectivity. This includes CP signalling with the eNodeB, and the S-GW. At any point later on, the MME may need to be involved in setting up dedicated bearers for services that benefit from higher treatment. The MME may receive the request to set up a dedicated bearer either from the S-GW if the request originates from the operator service domain, or directly from the UE, if the UE requires a connection for a service that is not known by the operator service domain, and therefore cannot be initiated from there.

Figure 3.4 shows the connections MME has to the surrounding logical nodes, and summarizes the main functions in these interfaces. In principle the MME may be connected to any other MME in the system, but typically the connectivity is limited to one operator network only. The remote connectivity between MMEs may be used when a UE that has travelled far away while powered down registers to a new MME, which then retrieves the UE's permanent identity, the International Mobile Subscriber Identity (IMSI), from the previously visited MME. The inter-MME connection with neighbouring MMEs is used in handovers.

Connectivity to a number of HSSs will also need to be supported. The HSS is located in each user's home network, and a route to that can be found based on the IMSI. Each MME will be configured to control a set of S-GWs and eNodeBs. Both the S-GWs and eNodeBs may also be connected to other MMEs. The MME may serve a number of UEs at the same time, while each UE will only connect to one MME at a time.

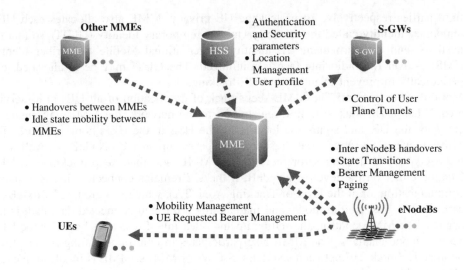

Figure 3.4   MME connections to other logical nodes and main functions

3.2.2.4   Serving Gateway (S-GW)

In the Basic System Architecture configuration, the high level function of S-GW is UP tunnel management and switching. The S-GW is part of the network infrastructure maintained centrally in operation premises.

When the S5/S8 interface is based on GTP, the S-GW will have GTP tunnels on all its UP interfaces. Mapping between IP service flows and GTP tunnels is done in P-GW, and the S-GW does not need to be connected to PCRF. All control is related to the GTP tunnels, and comes from either MME or P-GW. When the S5/S8 interface uses PMIP, the S-GW will perform the mapping between IP service flows in S5/S8 and GTP tunnels in S1-U interfaces, and will connect to PCRF to receive the mapping information.

The S-GW has a very minor role in control functions. It is only responsible for its own resources, and it allocates them based on requests from MME, P-GW or PCRF, which in turn are acting on the need to set up, modify or clear bearers for the UE. If the request was received from P-GW or PCRF, the S-GW will also relay the command on to the MME so that it can control the tunnel to eNodeB. Similarly, when the MME initiated the request, the S-GW will signal on to either the P-GW or the PCRF, depending on whether S5/S8 is based on GTP or PMIP respectively. If the S5/S8 interface is based on PMIP, the data in that interface will be IP flows in one GRE tunnel for each UE, whereas in the GTP based S5/S8 interface each bearer will have its own GTP tunnel. Therefore S-GW supporting PMIP S5/S8 is responsible for bearer binding, i.e. mapping the IP flows in S5/S8 interface to bearers in the S1 interface. This function in S-GW is called Bearer Binding and Event Reporting Function (BBERF). Irrespective of where the bearer signalling started, the BBERF always receives the bearer binding information from PCRF.

During mobility between eNodeBs, the S-GW acts as the local mobility anchor. The MME commands the S-GW to switch the tunnel from one eNodeB to another. The MME may also request the S-GW to provide tunnelling resources for data forwarding, when there is a need to forward data from source eNodeB to target eNodeB during the time UE makes the radio handover. The mobility scenarios also include changing from one S-GW to another, and the MME controls this change accordingly, by removing tunnels in the old S-GW and setting them up in a new S-GW.

For all data flows belonging to a UE in connected mode, the S-GW relays the data between eNodeB and P-GW. However, when a UE is in idle mode, the resources in eNodeB are released, and the data path terminates in the S-GW. If S-GW receives data packets from P-GW on any such tunnel, it will buffer the packets, and request the MME to initiate paging of the UE. Paging will cause the UE to re-connect, and when the tunnels are re-connected, the buffered packets will be sent on. The S-GW will monitor data in the tunnels, and may also collect data needed for accounting and user charging. The S-GW also includes functionality for Lawful Interception, which means the capability to deliver the monitored user's data to authorities for further inspection.

Figure 3.5 shows how the S-GW is connected to other logical nodes, and lists the main functions in these interfaces. All interfaces have to be configured in a one-to-many fashion from the S-GW point of view. One S-GW may be serving only a particular geographical area with a limited set of eNodeBs, and likewise there may be a limited set of MMEs that control that area. The S-GW should be able to connect to any P-GW in the whole network, because P-GW will not change during mobility, while the S-GW may be relocated, when the UE moves. For connections related to one UE, the S-GW will always signal with only one MME, and the UP points to one eNodeB at a time (indirect data forwarding is the exception, see next paragraph). If one UE is allowed to connect to multiple PDNs through different P-GWs, then the S-GW needs to connect to those separately. If the S5/S8 interface is based on PMIP, the S-GW connects to one PCRF for each separate P-GW the UE is using.

Figure 3.5 also shows the indirect data forwarding case where UP data is forwarded between eNodeBs through the S-GWs. There is no specific interface name associated to the interface between S-GWs, since the format is exactly the same as in the S1-U

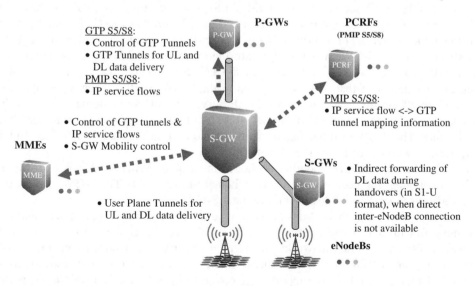

Figure 3.5   S-GW connections to other logical nodes and main functions

interface, and the involved S-GWs may consider that they are communicating directly with an eNodeB. This would be the case if indirect data forwarding takes place via only one S-GW, i.e. both eNodeBs can be connected to the same S-GW.

3.2.2.5 Packet Data Network Gateway (P-GW)

Packet Data Network Gateway (P-GW, also often abbreviated as PDN-GW) is the edge router between the EPS and external packet data networks. It is the highest level mobility anchor in the system, and usually it acts as the IP point of attachment for the UE. It performs traffic gating and filtering functions as required by the service in question. Similarly to the S-GW, the P-GWs are maintained in operator premises in a centralized location.

Typically the P-GW allocates the IP address to the UE, and the UE uses that to communicate with other IP hosts in external networks, e.g. the internet. It is also possible that the external PDN to which the UE is connected allocates the address that is to be used by the UE, and the P-GW tunnels all traffic to that network. The IP address is always allocated when the UE requests a PDN connection, which happens at least when the UE attaches to the network, and it may happen subsequently when a new PDN connectivity is needed. The P-GW performs the required Dynamic Host Configuration Protocol (DHCP) functionality, or queries an external DHCP server, and delivers the address to the UE. Also dynamic auto-configuration is supported by the standards. Only IPv4, only IPv6 or both addresses may be allocated depending on the need, and the UE may signal whether it wants to receive the address(es) in the Attach signalling, or if it wishes to perform address configuration after the link layer is connected.

The P-GW includes the PCEF, which means that it performs gating and filtering functions as required by the policies set for the UE and the service in question, and it collects and reports the related charging information.

The UP traffic between P-GW and external networks is in the form of IP packets that belong to various IP service flows. If the S5/S8 interface towards S-GW is based on GTP, the P-GW performs the mapping between the IP data flows to GTP tunnels, which represent the bearers. The P-GW sets up bearers based on request either through the PCRF or from the S-GW, which relays information from the MME. In the latter case, the P-GW may also need to interact with the PCRF to receive the appropriate policy control information, if that is not configured in the P-GW locally. If the S5/S8 interface is based on PMIP, the P-GW maps all the IP Service flows from external networks that belong to one UE to a single GRE tunnel, and all control information is exchanged with PCRF only. The P-GW also has functionality for monitoring the data flow for accounting purposes, as well as for Lawful Interception.

P-GW is the highest level mobility anchor in the system. When a UE moves from one S-GW to another, the bearers have to be switched in the P-GW. The P-GW will receive an indication to switch the flows from the new S-GW.

Figure 3.6 shows the connections P-GW has to the surrounding logical nodes, and lists the main functions in these interfaces. Each P-GW may be connected to one or more PCRF, S-GW and external network. For a given UE that is associated with the P-GW, there is only one S-GW, but connections to many external networks and respectively to many PCRFs may need to be supported, if connectivity to multiple PDNs is supported through one P-GW.

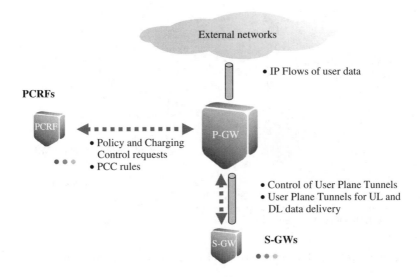

Figure 3.6   P-GW connections to other logical nodes and main functions

3.2.2.6   Policy and Charging Resource Function (PCRF)

Policy and Charging Resource Function (PCRF) is the network element that is responsible for Policy and Charging Control (PCC). It makes decisions on how to handle the services in terms of QoS, and provides information to the PCEF located in the P-GW, and if applicable also to the BBERF located in the S-GW, so that appropriate bearers and policing can be set up. PCRF is part of the PCC framework defined in [5]. PCRF is a server usually located with other CN elements in operator switching centres.

The information the PCRF provides to the PCEF is called the PCC rules. The PCRF will send the PCC rules whenever a new bearer is to be set up. Bearer set-up is required, for example, when the UE initially attaches to the network and the default bearer will be set up, and subsequently when one or more dedicated bearers are set up. The PCRF will be able to provide PCC rules based on request either from the P-GW and also the S-GW in PMIP case, like in the attach case, and also based on request from the Application Function (AF) that resides in the Services Domain. In this scenario the UE has signalled directly with the Services Domain, e.g. with the IMS, and the AF pushes the service QoS information to PCRF, which makes a PCC decision, and pushes the PCC rules to the P-GW, and bearer mapping information to S-GW in PMIP S5/S8 case. The EPC bearers are then set up based on those.

The connections between the PCRF and the other nodes are shown in Figure 3.7. Each PCRF may be associated with one or more AF, P-GW and S-GW. There is only one PCRF associated with each PDN connection that a single UE has.

3.2.2.7   Home Subscription Server (HSS)

Home Subscription Server (HSS) is the subscription data repository for all permanent user data. It also records the location of the user in the level of visited network control node, such as MME. It is a database server maintained centrally in the home operator's premises.

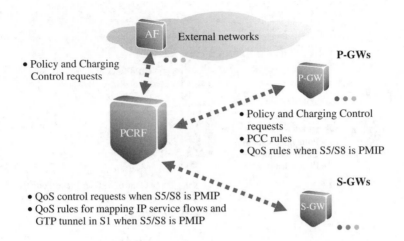

Figure 3.7   PCRF connections to other logical nodes and main functions

The HSS stores the master copy of the subscriber profile, which contains information about the services that are applicable to the user, including information about the allowed PDN connections, and whether roaming to a particular visited network is allowed or not. For supporting mobility between non-3GPP ANs, the HSS also stores the Identities of those P-GWs that are in use. The permanent key, which is used to calculate the authentication vectors that are sent to a visited network for user authentication and deriving subsequent keys for encryption and integrity protection, is stored in the Authentication Center (AuC), which is typically part of the HSS. In all signalling related to these functions, the HSS interacts with the MME. The HSS will need to be able to connect with every MME in the whole network, where its UEs are allowed to move. For each UE, the HSS records will point to one serving MME at a time, and as soon as a new MME reports that it is serving the UE, the HSS will cancel the location from the previous MME.

3.2.2.8   Services Domain

The Services domain may include various sub-systems, which in turn may contain several logical nodes. The following is a categorization of the types of services that will be made available, and a short description of what kind of infrastructure would be needed to provide them:

- IMS based operator services: The IP Multimedia Sub-system (IMS) is service machinery that the operator may use to provide services using the Session Initiation Protocol (SIP). IMS has 3GPP defined architecture of its own, and is described in section 3.6, and more thoroughly, e.g. in [3].
- Non-IMS based operator services: The architecture for non-IMS based operator services is not defined in the standards. The operator may simply place a server into their network, and the UEs connect to that via some agreed protocol that is supported by an application in the UE. A video streaming service provided from a streaming server is one such example.

- Other services not provided by the mobile network operator, e.g. services provided through the internet: This architecture is not addressed by the 3GPP standards, and the architecture depends on the service in question. The typical configuration would be that the UE connects to a server in the internet, e.g. to a web-server for web browsing services, or to a SIP server for internet telephony service (i.e. VoIP).

3.2.3 Self-configuration of S1-MME and X2 Interfaces

In 3GPP Release 8 development it has been agreed to define the support for self-configuration of the S1-MME and X2 interfaces. The basic process is as presented in Figure 3.8, where the eNodeB once turned on (and given that the IP connection exists) will connect to the O&M (based on the known IP address) to obtain then further parameters in terms of which other network elements to connect (and also for eNodeB software download) as well as initial parameters for the operation, such as in which part of the frequency band to operate and what kind of parameters to include for the broadcast channels.

This is expected to include setting the S1-MME connection by first setting up the SCTP association with at least one MME, and once that is connected to continue with application level information exchange to make S1-MME interface operational. Once the link to MME exists, there needs to be then association with S-GW created for UP data transfer.

To enable functionalities such as mobility and inter-cell interference control, the X2 interface configuration follows similar principles to the S1-MME interface. The difference here is that initially the eNodeB will set up the X2 connection for those eNodeBs indicated from the O&M and it may then later adapt more to the environment based on the Automatic Neighbour Relationship (ANR) functionality – as covered in Chapter 7 – to further optimize the X2 connectivity domain based on actual handover needs. The parameters that are exchanged over the X2 interface include:

- global eNodeB ID;
- information of the cell specific parameters such as Physical Cell ID (PCI), uplink/downlink frequency used, bandwidth in use;
- MMEs connected (MME Pool).

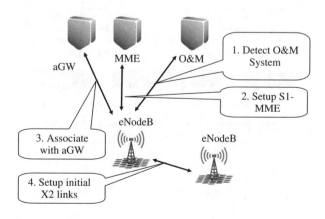

Figure 3.8   eNodeB self-configuration step

For the PCI there is also support for auto-configuration in the Release 8 specifications as covered in Chapter 5, other parameters then coming from the O&M direction with procedures that can be automated to limit the need for on-site configuration by installation personnel.

3.2.4 Interfaces and Protocols in Basic System Architecture Configuration

Figure 3.9 shows the CP protocols related to a UE's connection to a PDN. The interfaces from a single MME are shown in two parts, the one on top showing protocols towards the E-UTRAN and UE, and the bottom one showing protocols towards the gateways. Those protocols that are shown in white background are developed by 3GPP, while the protocols with light grey background are developed in IETF, and represent standard internet technologies that are used for transport in EPS. 3GPP has only defined the specific ways of how these protocols are used.

The topmost layer in the CP is the Non-Access Stratum (NAS), which consists of two separate protocols that are carried on direct signalling transport between the UE and the MME. The content of the NAS layer protocols is not visible to the eNodeB, and the eNodeB is not involved in these transactions by any other means, besides transporting the messages, and providing some additional transport layer indications along with the messages in some cases. The NAS layer protocols are:

- EPS Mobility Management (EMM): The EMM protocol is responsible for handling the UE mobility within the system. It includes functions for attaching to and detaching from

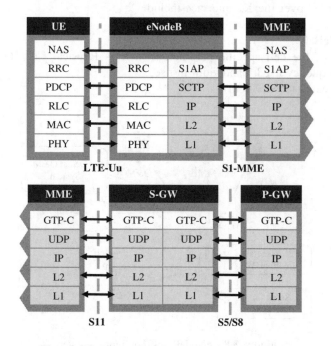

Figure 3.9   Control plane protocol stack in EPS

the network, and performing location updating in between. This is called Tracking Area Updating (TAU), and it happens in idle mode. Note that the handovers in connected mode are handled by the lower layer protocols, but the EMM layer does include functions for re-activating the UE from idle mode. The UE initiated case is called Service Request, while Paging represents the network initiated case. Authentication and protecting the UE identity, i.e. allocating the temporary identity GUTI to the UE are also part of the EMM layer, as well as the control of NAS layer security functions, encryption and integrity protection.

- EPS Session Management (ESM): This protocol may be used to handle the bearer management between the UE and MME, and it is used in addition for E-UTRAN bearer management procedures. Note that the intention is not to use the ESM procedures if the bearer contexts are already available in the network and E-UTRAN procedures can be run immediately. This would be the case, for example, when the UE has already signalled with an operator affiliated Application Function in the network, and the relevant information has been made available through the PCRF.

The radio interface protocols are (only short descriptions are included here, since these functions are described extensively in other sections of this book):

- Radio Resource Control (RRC): This protocol is in control of the radio resource usage. It manages UE's signalling and data connections, and includes functions for handover.
- Packet Data Convergence Protocol (PDCP): The main functions of PDCP are IP header compression (UP), encryption and integrity protection (CP only).
- Radio Link Control (RLC): The RLC protocol is responsible for segmenting and concatenation of the PDCP-PDUs for radio interface transmission. It also performs error correction with the Automatic Repeat Request (ARQ) method.
- Medium Access Control (MAC): The MAC layer is responsible for scheduling the data according to priorities, and multiplexing data to Layer 1 transport blocks. The MAC layer also provides error correction with Hybrid ARQ.
- Physical Layer (PHY): This is the Layer 1 of LTE-Uu radio interface that takes care of OFDMA and SC-FDMA Layer functions.

The S1 interface connects the E-UTRAN to the EPC, and involves the following protocols:

- S1 Application Protocol (S1AP): S1AP handles the UE's CP and UP connections between the E-UTRAN and EPC, including participating in the handover when EPC is involved.
- SCTP/IP signalling transport: The Stream Control Transmission Protocol (SCTP) and Internet Protocol (IP) represent standard IP transport suitable for signalling messages. SCTP provides the reliable transport and sequenced delivery functions. IP itself can be run on a variety of data link and physical layer technologies (L2 and L1), which may be selected based on availability.

In the EPC, there are two alternative protocols for the S5/S8 interface. The following protocols are involved, when GTP is used in S5/S8:

- GPRS Tunnelling Protocol, Control Plane (GTP-C): It manages the UP connections in the EPC. This includes signalling the QoS and other parameters. If GTP is used in the

S5/S8 interface it also manages the GTP-U tunnels. GTP-C also performs the mobility management functions within the EPC, e.g. when the GTP-U tunnels of a UE need to be switched from one node to the other.

- UDP/IP transport. The Unit Data Protocol (UDP) and IP are used as the standard and basic IP transport. UDP is used instead of Transmission Control Protocol (TCP) because the higher layers already provide reliable transport with error recovery and re-transmission. IP packets in EPC may be transported on top of a variety of L2 and L1 technologies. Ethernet and ATM are some examples.

The following protocols are used, when S5/S8 is based on PMIP:

- Proxy Mobile IP (PMIP): PMIP is the alternative protocol for the S5/S8 interface. It takes care of mobility management, but does not include bearer management functions as such. All traffic belonging to a UE's connection to a particular PDN is handled together.
- IP: PMIP runs directly on top of IP, and it is used as the standard IP transport.

Figure 3.10 illustrates the UP protocol structure for UE connecting to P-GW.

The UP shown in Figure 3.10 includes the layers below the end user IP, i.e. these protocols form the Layer 2 used for carrying the end user IP packets. The protocol structure is very similar to the CP. This highlights the fact that the whole system is designed for generic packet data transport, and both CP signalling and UP data are ultimately packet data. Only the volumes are different. Most of the protocols have been introduced already above, with the exception of the following two that follow the selection of protocol suite in S5/S8 interface:

- GPRS Tunnelling Protocol, User Plane (GTP-U): GTP-U is used when S5/S8 is GTP based. GTP-U forms the GTP-U tunnel that is used to send End user IP packets belonging to one EPS bearer. It is used in S1-U interface, and is used in S5/S8 if the CP uses GTP-C.
- Generic Routing Encapsulation (GRE): GRE is used in the S5/S8 interface in conjunction with PMIP. GRE forms an IP in IP tunnel for transporting all data belonging to one UE's connection to a particular PDN. GRE is directly on top of IP, and UDP is not used.

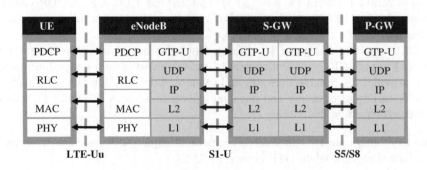

Figure 3.10   User plane protocol stack in EPS

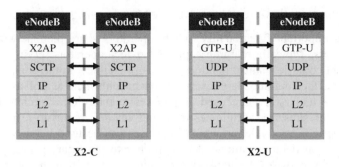

Figure 3.11 Control and user plane protocol stacks for X2 interface

Figure 3.11 illustrates the X2 interface protocol structure, which resembles that of the S1 interface. Only the CP Application Protocol is different. X2 interface is used in mobility between the eNodeBs, and the X2AP includes functions for handover preparation, and overall maintenance of the relation between neighbouring eNodeBs. The UP in the X2 interface is used for forwarding data in a transient state during handover, when the radio interface is already disconnected on the source side, and has not yet resumed on the target side. Data forwarding is done for the DL data, since the UL data can be throttled effectively by the UE.

Table 3.1 summarizes the protocols and interfaces in Basic System Architecture configuration.

Table 3.1 Summary of interfaces and protocols in Basic System Architecture configuration

Interface	Protocols	Specification
LTE-Uu	CP: RRC/PDCP/RLC/MAC/PHY	36.300 [6]
	UP: PDCP/RLC/MAC/PHY	(stage 2)
X2	CP: X2AP/SCTP/IP	36.423 [7]
	UP: GTP-U/UDP/IP	29.274 [8]
S1-MME	S1AP/SCTP/UDP/IP	36.413 [9]
S1-U	GTP-U/UDP/IP	29.274 [8]
S10	GTP-C/UDP/IP	29.274 [8]
S11	GTP-C/UDP/IP	29.274 [8]
S5/S8 (GTP)	GTP/UDP/IP	29.274 [8]
S5/S8 (PMIP)	CP: PMIP/IP	29.275 [10]
	UP: GRE/IP	
SGi	IP (also Diameter & Radius)	29.061 [11]
S6a	Diameter/SCTP/IP	29.272 [12]
Gx	Diameter/SCTP/IP	29.212 [13]
Gxc	Diameter/SCTP/IP	29.212 [13]
Rx	Diameter/SCTP/IP	29.214 [14]
UE – MME	EMM, ESM	24.301 [15]

3.2.5 Roaming in Basic System Architecture Configuration

Roaming is an important functionality, where operators share their networks with each other's subscribers. Typically roaming happens between operators serving different areas, such as different countries, since this does not cause conflicts in the competition between the operators, and the combined larger service area benefits them as well as the subscribers. The words *home* and *visited* are used as prefixes to many other architectural terms to describe where the subscriber originates from and where it roams to respectively.

3GPP SAE specifications define which interfaces can be used between operators, and what additional considerations are needed if an operator boundary is crossed. In addition to the connectivity between the networks, roaming requires that the operators agree on many things at the service level, e.g. what services are available, how they are realized, and how accounting and charging is handled. This agreement is called the *Roaming Agreement*, and it can be made directly between the operators, or through a broker. The 3GPP specifications do not cover these items, and operators using 3GPP technologies discuss roaming related general questions in a private forum called the GSM Association, which has published recommendations to cover these additional requirements.

Roaming defined for SAE follows quite similar principles to the earlier 3GPP architectures. The E-UTRAN is always locally in the visited network, but the data may be routed either to the home network, or can break out to external networks directly from the visited network. This aspect differentiates the two roaming models supported for SAE, which are defined as follows:

- Home Routed model: The P-GW, HSS and PCRF reside in the home operator network, and the S-GW, MME and the radio networks reside in the visited operator network. In this roaming configuration the interface between P-GW and S-GW is called S8, whereas the same interface is called S5 in the non-roaming case when S-GW and P-GW are in the same operator's network. S5 and S8 are technically equivalent. When the S8 interface is based in GTP, the roaming architecture is as shown in Figure 3.2

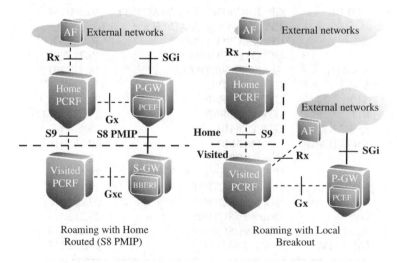

Figure 3.12 Home routed and local breakout roaming

(Gxa does not apply with GTP). When the S8 interface uses PMIP, the PCRF will also be divided into home and visited nodes with the S9 interface between them. This is the scenario shown in Figure 3.12 on the left, and is explained with more detail in section 3.7.1. The Home Routed roaming model applies to legacy 3GPP ANs in the same way, the additional detail being that the SGSN introduced in the next chapter and shown in Figure 3.12 resides in the visited network.

● Local Breakout model: In this model, shown in the right side of Figure 3.12, the P-GW will be located in the visited network, and the HSS is in the home network. If dynamic policy control is used, there will again be two PCRFs involved, one in the home network, and the other in the visited network. Depending on which operator's services are used, the PCRF in that operator's network is also connected to the AF. Also this scenario is explained with more detail in section 3.7.1. With these constraints the Local Breakout model also works with the legacy 3GPP ANs.

3.3 System Architecture with E-UTRAN and Legacy 3GPP Access Networks

3.3.1 Overview of 3GPP Inter-working System Architecture Configuration

Figure 3.13 describes the architecture and network elements in the architecture configuration where all 3GPP defined ANs, E-UTRAN, UTRAN and GERAN, are connected to

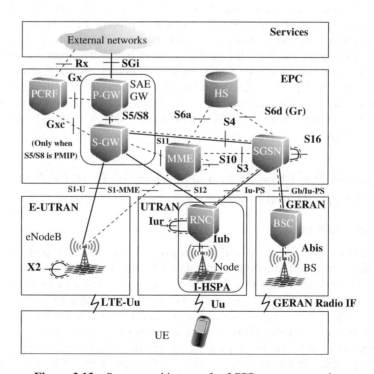

Figure 3.13 System architecture for 3GPP access networks

the EPC. This is called here the 3GPP Inter-working System Architecture Configuration, and it allows optimized inter-working between the mentioned accesses.

Functionally the E-UTRAN, UTRAN and GERAN all provide very similar connectivity services, especially when looking at the situation from the end user point of view, where the only difference may be the different data rates and improved performance, but architecturally these ANs are quite different, and many things are carried out differently. There are, for example, big differences in how the bearers are managed in the EPS compared to the existing networks with UTRAN or GERAN access. However, when UTRAN or GERAN is connected to EPC, they may still operate as before from this perspective, and for this purpose the S-GW simply assumes the role of the Gateway GPRS Support Node (GGSN). Also in optimized inter-working with the E-UTRAN, the GERAN and UTRAN ANs behave almost the same way as they behave when inter-working between themselves. The differences become more visible in the EPC, because what used to be the fixed GGSN is now the S-GW that may be changed along with the SGSN change during UE mobility.

All nodes and functions described in the previous section for the Basic System Architecture Configuration are needed here also. The EPC needs the addition of a few new interfaces and functions to connect and inter-work with UTRAN and GERAN. The corresponding functions will also be required from GERAN and UTRAN. The new interfaces are S3, S4 and S12 as shown in Figure 3.12. The interface from SGSN to HSS can also be updated to Diameter based S6d, but the use of the legacy MAP based Gr is also possible.

Keeping E-UTRAN, i.e. the eNodeB design as focused to, and as optimized for the requirements of the new OFDMA radio interface, and as clean of inter-working functionality as possible, was an important guideline for the inter-working design. Consequently, the eNodeB does not interface directly with the other 3GPP ANs, and the interaction towards the EPC is the same as in other mobility cases that involve EPC. However, optimized inter-working means that the network is in control of mobility events, such as handovers, and provides functionality to hand the communication over with minimum interruption to services. This means that an eNodeB must be able to coordinate UE measuring UTRAN and GERAN cells, and perform handover decisions based on measurement results, and thus E-UTRAN radio interface protocols have been appended to support the corresponding new functions. Similar additions will be required from UTRAN and GERAN to support handover to E-UTRAN.

3.3.2 Additional and Updated Logical Elements in 3GPP Inter-working System Architecture Configuration

3.3.2.1 User Equipment

From the UE point of view, inter-working means that it needs to support the radio technologies in question, and the mobility operations defined for moving between them. The optimized inter-working means that the network controls the usage of radio transmitter and receiver in the UE in a way that only one set of them needs to be operating at the same time. This is called single radio operation, and allows UE implementations where only one pair of physical radio transmitter and receiver is implemented.

The standard does not preclude implementing multiple radio transmitters and receivers, and operating them simultaneously in dual radio operation. However, single radio operation is an important mode, because the different ANs often operate in frequencies that are so close to each other that dual radio operation would cause too much interference within the terminal. That, together with the additional power consumption, will decrease the overall performance.

3.3.2.2   E-UTRAN

The only addition to E-UTRAN eNodeB compared to the Basic System Architecture Configuration is the mobility to and from other 3GPP ANs. From the eNodeB perspective the functions are very similar irrespective of whether the other 3GPP AN is UTRAN or GERAN.

For the purpose of handover from E-UTRAN to UTRAN or GERAN, the neighbouring cells from those networks need to be configured into the eNodeB. The eNodeB may then consider handover for those UEs that indicate corresponding radio capability. The eNodeB requests the UE to measure the signal level of the UTRAN or GERAN cells, and analyses the measurement reports. If the eNodeB decides to start the handover, it signals the need to the MME in the same way that it would signal inter-eNodeB handover when the X2 interface is not available. Subsequently, the eNodeB will receive the information needed for the Handover Command from the target Access System via the MME. The eNodeB will send the Handover Command to the UE without the need for interpreting the content of this information.

In the case of handover from UTRAN or GERAN to E-UTRAN, the eNodeB does not need to make any specific preparations compared to other handovers where the handover preparation request comes through the MME. The eNodeB will allocate the requested resources, and prepare the information for handover command, which it sends to the MME, from where it is delivered to the UE through the other 3GPP Access System that originated the handover.

3.3.2.3   UTRAN

In UTRAN, the radio control functionality is handled by the Radio Network Controller (RNC), and under its control the Node B performs Layer 2 bridging between the Uu and Iub interfaces. UTRAN functionality is described extensively in [16].

UTRAN has evolved from its initial introduction in Release 99 in many ways, including the evolution of architectural aspects. The first such item is Iu flex, where the RNC may be connected to many Serving GPRS Support Nodes (SGSNs) instead of just one. Another such concept is I-HSPA, where the essential set of packet data related RNC functions is included with the Node B, and that connects to Iu-PS as a single node. Figure 3.13 also shows the direct UP connection from RNC to S-GW, which is introduced to 3G CN by the Direct Tunnel concept, where the SGSN is bypassed in UP.

Inter-working with E-UTRAN requires that UTRAN performs the same measurement control and analysis functions as well as the transparent handover information delivery in Handover Command that were described for eNodeB in the earlier section. Also the UTRAN performs similar logic that it already uses with Relocation between RNCs, when the Iur interface is not used.

3.3.2.4 GERAN

GSM EDGE Radio AN (GERAN) is the evolved version of GSM AN, which can also be connected to 3G Core Network. It consists of the Base Station Controller (BSC) and the Base Station (BS), and the radio interface functionalities are divided between them. An overview of GERAN functionality and the whole GSM system can be found in [17].

The GERAN is always connected to the SGSN in both Control and UPs, and this connection is used for all the inter-working functionality. Also the GERAN uses logic similar to that described above for E-UTRAN and UTRAN for inter-working handover.

3.3.2.5 EPC

The EPC has a central role for the inter-working system architecture by anchoring the ANs together. In addition to what has been described earlier, the MME and S-GW will support connectivity and functions for inter-working. Also the SGSN, which supports the UTRAN and GERAN access networks, will need to support these functions, and when these additions are supported, it can be considered to belong to the EPC.

The S-GW is the mobility anchor for all 3GPP access systems. In the basic bearer operations and mobility between SGSNs, it behaves like a GGSN towards the SGSN, and also towards the RNC if UP tunnels are set up in Direct Tunnel fashion bypassing the SGSN. Many of the GGSN functions are actually performed in the P-GW, but this is not visible to the SGSN. The S-GW retains its role as a UP Gateway, which is controlled by either the MME or the SGSN depending on which AN the UE is being served by.

To support the inter-working mobility, the MME will need to signal with the SGSN. These operations are essentially the same as between those two MMEs, and have been described earlier in section 3.2. An additional aspect of the MME is that it may need to combine the change of S-GW and the inter-working mobility with SGSN.

The SGSN maintains its role as the controlling node in core network for both UTRAN and GERAN. These functions are defined in [18]. The SGSN has a role very similar to that of the MME. The SGSN needs to be updated to support for S-GW change during mobility between SGSNs or RNCs, because from the legacy SGSN point of view this case looks like GGSN changing, which is not supported. As discussed earlier, the SGSN may direct the UP to be routed directly between the S-GW and UTRAN RNC, or it may remain involved in the UP handling. From the S-GW point of view this does not really make a difference, since it does not need to know which type of node terminates the far end of the UP tunnel.

3.3.3   Interfaces and Protocols in 3GPP Inter-working System Architecture Configuration

Table 3.2 summarizes the interfaces in the 3GPP Inter-working System Architecture Configuration and the protocols used in them. Interfaces and protocols in legacy 3GPP networks are not listed. Interfaces and protocols listed for Basic System Architecture Configuration are needed in addition to these.

Table 3.2 Summary of additional interfaces and protocols in 3GPP Inter-working System Architecture configuration

Interface	Protocols	Specification
S3	GTP-C/UDP/IP	29.274 [8]
S4	GTP/UDP/IP	29.274 [8]
S12	GTP-U/UDP/IP	29.274 [8]
S16	GTP/UDP/IP	29.274 [8]
S6d	Diameter/SCTP/IP	29.272 [12]

3.3.4 Inter-working with Legacy 3GPP CS Infrastructure

While the EPS is purely a Packet Switched (PS) only system without a specific Circuit Switched (CS) domain with support for VoIP, the legacy 3GPP systems treat CS services such as voice calls with a specific CS infrastructure. IMS VoIP may not be ubiquitously available, and therefore the SAE design includes two special solutions that address inter-working with circuit switched voice. A description of how inter-working between E-UTRAN and the legacy 3GPP CS domain can be arranged is given in Chapter 10 on VoIP. Two specific functions have been defined for that purpose, Circuit Switched Fall Back (CSFB) and Single Radio Voice Call Continuity (SR-VCC).

CSFB [19] is a solution for networks that do not have support for IMS VoIP. Instead, the voice calls are handled by the CS domain, and the UE is handed over there at the time of a voice call. The SGs interface between the MME and MSC Server is used for related control signalling, as shown with more detail in Chapter 10.

SR-VCC [20] is a solution for converting and handing over an IMS VoIP call to a CS voice call in the legacy CS domain. This functionality would be needed when the coverage of an IMS VoIP capable network is smaller than that of the legacy CS networks. SR-VCC allows a UE entering the edge of the VoIP coverage area with an ongoing VoIP call to be handed over to the CS network without interrupting the call. SR-VCC is a one way handover from the PS network with VoIP to the CS network. If E-UTRAN coverage becomes available again, the UE may return there when the call ends and the UE becomes idle. The solution relies on running only one radio at a time, i.e. the UE does not need to communicate simultaneously with both systems. In this solution the MME is connected to the MSC Server in the CS domain via a Sv interface, which is used for control signalling in the SR-VCC handover. The details of the solution are presented in Chapter 10. A summary of additional interfaces and protocols for inter-working with legacy 3GPP CS infrastructure is given in Table 3.3.

Table 3.3 Summary of additional interfaces and protocols for inter-working with legacy 3GPP CS infrastructure

Interface	Protocols	Specification
SGs	SGsAP/SCTP/IP	29.118 [21]
Sv	GTP-C(subset)/UDP/IP	29.280 [22]

3.4   System Architecture with E-UTRAN and Non-3GPP Access Networks

3.4.1   Overview of 3GPP and Non-3GPP Inter-working System Architecture Configuration

Inter-working with non-3GPP ANs was one of the key design goals for SAE, and to support it, a completely separate architecture specification [2] was developed in 3GPP. The non-3GPP Inter-working System Architecture includes a set of solutions in two categories. The first category contains a set of generic and loose inter-working solutions that can be used with any other non-3GPP AN. Mobility solutions defined in this category are also called Handovers without Optimizations, and the same procedures are applicable in both connected and idle mode. The second category includes a specific and tighter inter-working solution with one selected AN, the cdma2000® HRPD. This solution category is also called Handovers with Optimizations, and it specifies separate procedures for connected and idle mode.

The generic non-3GPP Inter-working System Architecture is shown in Figure 3.14. The specific application of the architecture for cdma2000® HRPD inter-working and the required additional interfaces are described with more detail in section 3.5.

Figure 3.14 describes the generic inter-working solution that relies only on loose coupling with generic interfacing means, and without AN level interfaces. Since there are so many different kinds of ANs, they have been categorized to two groups, the trusted and untrusted non-3GPP ANs, depending on whether it can be safely assumed that 3GPP defined

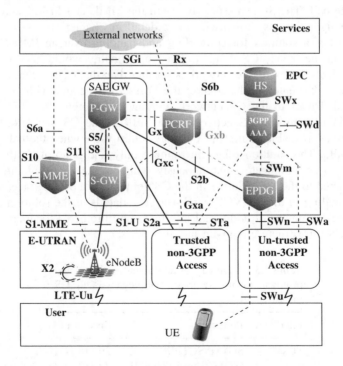

Figure 3.14   System architecture for 3GPP and non-3GPP access networks

authentication can be run by the network, which makes it trusted, or if authentication has to be done in overlay fashion and the AN is un-trusted. The P-GW will maintain the role of mobility anchor, and the non-3GPP ANs are connected to it either via the S2a or the S2b interface, depending on whether the non-3GPP AN functions as a Trusted or Un-trusted non-3GPP AN. Both use network controlled IP layer mobility with the PMIP protocol. For networks that do not support PMIP, Client MIPv4 Foreign Agent mode is available as an option in S2a. In addition to mobility functions, the architecture includes interfaces for authenticating the UE within and through the non-3GPP ANs, and also allows PCC functionality in them via the Gxa and Gxb interfaces. Note that the detailed functions and protocols for Gxb are not specified in Release 8.

In addition to the network controlled mobility solutions, a completely UE centric solution with DSMIPv6 is also included in the inter-working solutions. This scenario is depicted in Figure 3.15.

In this configuration the UE may register in any non-3GPP AN, receive an IP address from there, and register that to the Home Agent in P-GW. This solution addresses the mobility as an overlay function. While the UE is served by one of the 3GPP ANs, the UE is considered to be in home link, and thus the overhead caused by additional MIP headers is avoided.

Another inter-working scenario that brings additional flexibility is called the chained S8 and S2a/S2b scenario. In that scenario the non-3GPP AN is connected to S-GW in the visited Public Land Mobile Network (PLMN) through the S2a or S2b interface, while the P-GW is in the home PLMN. This enables the visited network to offer a roaming subscriber the use of non-3GPP ANs that might not be associated with the home operator at all, even in the case where P-GW is in the home PLMN. This scenario requires that S-GW performs functions that normally belong to P-GW in order to behave as the termination point for the S2a or S2b interfaces. In Release 8, this scenario does not support dynamic policies through the PCC infrastructure, i.e. the Gxc interface will not be used. Also, chaining with GTP based S5/S8 is not supported. All other interfaces related to non-3GPP ANs are used normally as shown in Figure 3.14.

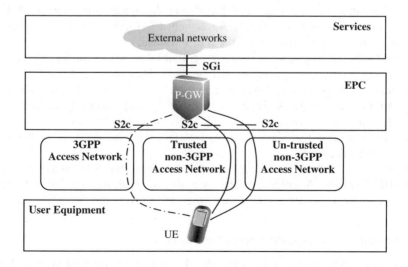

Figure 3.15 Simplified system architecture showing only S2c

3.4.2   Additional and Updated Logical Elements in 3GPP Inter-working System Architecture Configuration

3.4.2.1   User Equipment

Inter-working between the non-3GPP ANs requires that the UE supports the corresponding radio technologies, and the specified mobility procedures. The mobility procedures and required radio capabilities vary depending on whether optimizations are in place or not. The procedures defined for Handovers without Optimizations do not make any assumption about the UE's capability to use the radio transmitters and receivers simultaneously, and both single radio and dual radio configurations can use the procedures. However, the handover gap time is expected to be shorter, if preparing the connections towards the target side can start already while data are still flowing through the source side. This is caused by the fact that Handovers without Optimizations do not have procedures in the network side to assist in handover preparations, and the procedures follow the principle where UE registers to the target network according to the method defined for that network, and then the network switches the flow to the target network. This may be time consuming, since it normally includes procedures such as authentication. Also, the decision to make these handovers is the responsibility of the UE.

The Handovers with Optimizations, i.e. inter-working with cdma2000® HRPD, assume that they do include network control for connected mode, so the handovers are decided by the network, while the idle mode mobility relies on UE decision making, which may use cdma2000® HRPD related information in the LTE-Uu broadcast. Furthermore, the procedures are designed with the assumption that single radio configuration is enough for the UE.

3.4.2.2   Trusted Non-3GPP Access Networks

The term trusted non-3GPP AN refers to networks that can be trusted to run 3GPP defined authentication. 3GPP Release 8 security architecture specification for non-3GPP ANs [23] mandates that the Improved Extensible Authentication Protocol Method for 3rd Generation Authentication and Key Agreement (EAP-AKA') [24] is performed. The related procedures are performed over the STa interface.

The trusted non-3GPP ANs are typically other mobile networks, such as the cdma2000® HRPD. The STa interface supports also delivery of subscription profile information from Authentication, Authorization and Accounting (AAA)/HSS to the AN, and charging information from the AN to AAA Server, which are typical functions needed in mobile networks. It can also be assumed that such ANs may benefit from connecting to the PCC infrastructure, and therefore the Gxc interface may be used to exchange related information with the PCRF.

The trusted non-3GPP AN connects to the P-GW with the S2a interface, with either PMIP or MIPv4 Foreign Agent mode. The switching of UP flows in P-GW is therefore the responsibility of the trusted non-3GPP AN when UE moves into the AN's service area.

3.4.2.3   Un-trusted Non-3GPP Access Networks

To a large extent, the architectural concepts that apply for un-trusted non-3GPP ANs are inherited from the Wireless Local Area Network Inter-Working (WLAN IW) defined originally in Release 6 [25]. The Release 8 functionality for connecting un-trusted non-3GPP ANs to EPC is specified fully in [2] with references to the earlier WLAN IW specifications when applicable.

The main principle is that the AN is not assumed to perform any other functions besides delivery of packets. A secure tunnel is established between UE and a special node called the Enhanced Packet Data Gateway (EPDG) via the SWu interface, and the data delivery takes place through that tunnel. Furthermore, the P-GW has a trust relationship with the EPDG connected to it via the S2b interface, and neither node needs to have secure association with the un-trusted non-3GPP AN itself.

As an optional feature, the un-trusted non-3GPP AN may be connected to the AAA Server with the SWa interface, and this interface may be used to authenticate the UE already in the non-3GPP AN level. This can be done only in addition to authentication and authorization with the EPDG.

3.4.2.4 EPC

The EPC includes quite a few additional functions for the support of non-3GPP ANs, when compared to the previously introduced architecture configurations. The main changes are in the P-GW, PCRF and HSS, and also in S-GW for the chained S8 and S2a/S2b scenario. In addition, completely new elements, such as the EPDG (Evolved Packet Data Gateway) and the AAA are introduced. The AAA infrastructure contains the AAA Server, and it may also contain separate AAA proxies in roaming situations. Figure 3.16 highlights the AAA connections and functions for non-3GPP ANs.

The P-GW is the mobility anchor for the non-3GPP ANs. For PMIP based S2a and S2b interfaces, the P-GW hosts the Local Mobility Anchor (LMA) function in a manner similar to that for the S5/S8 PMIP interfaces. Also the Home Agent (HA) function for the Client MIPv4 Foreign Agent mode in S2a is located in P-GW. The relation between

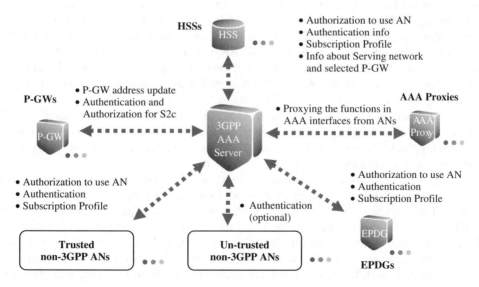

Figure 3.16 3GPP AAA server interfaces and main functions

P-GWs and non-3GPP ANs is many to many. The P-GW will also interface with the AAA Server, which subsequently connects to HSS. This interface is used for reporting the selected P-GW to the HSS so that it is available in mobility between non-3GPP ANs, and to authenticate and authorize users connecting with S2c mode. Each P-GW may connect to more than one AAA server.

The PCRF supports PCC interfaces for non-3GPP ANs. The Gxa is used towards trusted non-3GPP ANs, and Gxb towards un-trusted non-3GPP ANs. Only Gxa is specified in detail level in Release 8. The Gxa interface functions in a fashion similar to that of the Gxc interface. In this case the BBERF function will be in the non-3GPP AN, and it will receive instructions from the PCRF on how to handle the bearer level functions for the IP flows in the S2a interface. The further bearer functions internal to non-3GPP ANs are not addressed in 3GPP specifications.

The EPDG is a dedicated node for controlling the UE and inter-network connection, when an un-trusted non-3GPP AN is connected to EPC. Since the AN is not trusted, the main function is to secure the connection, as defined in [23]. The EPDG establishes an IPsec tunnel to the UE through the un-trusted non-3GPP AN with IKEv2 signalling [26] over the SWu interface. During the same signalling transaction the EAP-AKA authentication is run, and for that the EPDG signals with the AAA Server through the SWm interface. While the SWm interface is logically between UE and the EPDG, the SWn interface represents the interface on a lower layer between the EPDG and the un-trusted non-3GPP AN. The Release 8 specifications do not assume that EPDG would signal with PCRF for any PCC functions, but the architecture already contains the Gxb interface for that purpose.

The 3GPP AAA Server, and possibly a AAA Proxy in the visited network, performs a 3GPP defined set of AAA functions. These functions are a subset of what the standard IETF defined AAA infrastructure includes, and do not necessarily map with the way other networks use AAA infrastructure. The AAA Server acts between the ANs and the HSS, and in doing so it creates a context for the UEs it serves, and may store some of their information for further use. Thus, the 3GPP AAA Server consolidates the signalling from different types of ANs into a single SWx interface towards the HSS, and terminates the access specific interfaces S6b, STa, SWm and SWa. Most importantly the AAA Server performs as the authenticator for the EAP-AKA authentication through the non-3GPP ANs. It checks the authenticity of the user, and informs the AN about the outcome. The authorization to use the AN in question will also be performed during this step. Depending on the AN type in question, the AAA Server may also relay subscription profile information to the AN, which the AN may further use to better serve the UE. When the UE is no longer served by a given non-3GPP AN, the AAA Server participates in removing the UE's association from the HSS. Figure 3.16 summarizes the AAA Server main functions in relation to other nodes.

The HSS performs functions similar to those for the 3GPP ANs. It stores the main copy of the subscription profile as well as the secret security key in the AuC portion of it, and when requested, it provides the profile data and authentication vectors to be used in UEs connecting through non-3GPP ANs. One addition compared to 3GPP ANs is that since the non-3GPP ANs do not interface on the AN level, the selected P-GW needs to be stored in the HSS, and retrieved from there when the UE mobility involves a non-3GPP AN. The variety of different AN types are mostly hidden from the HSS, since the AAA

Server terminates the interfaces that are specific to them, and HSS only sees a single SWx interface. On the other hand, the subscription profile stored in the HSS must reflect the needs of all the different types of ANs that are valid for that operator.

3.4.3 Interfaces and Protocols in Non-3GPP Inter-working System Architecture Configuration

Connecting the non-3GPP ANs to EPC and operating them with it requires additional interfaces to those introduced in earlier sections. Table 3.4 lists the new interfaces.

3.4.3.1 Roaming in Non-3GPP Inter-working System Architecture Configuration

The principles for roaming with non-3GPP accesses are equivalent to those described in section 3.2.4 for 3GPP ANs. Both home routed and local breakout scenarios are supported and the main variations in the architecture relate to the PCC arrangement, which depends on where the services are consumed. This aspect is highlighted more in section 3.7.1.

The additional consideration that non-3GPP ANs bring to roaming is related to the case where the user is roaming to a visited 3GPP network in Home Routed model, and it would be beneficial to use a local non-3GPP AN that is affiliated with the visited network, but there is no association between that network and the home operator. For this scenario, the 3GPP Release 8 includes a so-called *chained case*, where the S-GW may behave as the anchor for the non-3GPP ANs also, i.e. it terminates the S2a or S2b interface, and routes the traffic via the S8 interface to the P-GW in the home network.

Table 3.4 Summary of additional interfaces and protocols in non-3GPP Inter-working System Architecture configuration

Interface	Protocols	Specification
S2a	PMIP/IP, or MIPv4/UDP/IP	29.275 [10]
S2b	PMIP/IP	29.275 [10]
S2c	DSMIPv6, IKEv2	24.303 [27]
S6b	Diameter/SCTP/IP	29.273 [28]
Gxa	Diameter/SCTP/IP	29.212 [13]
Gxb	*Not defined in Release 8*	N.A.
STa	Diameter/SCTP/IP	29.273 [28]
SWa	Diameter/SCTP/IP	29.273 [28]
SWd	Diameter/SCTP/IP	29.273 [28]
SWm	Diameter/SCTP/IP	29.273 [28]
SWn	PMIP	29.275 [10]
SWu	IKEv2, MOBIKE	24.302 [29]
SWx	Diameter/SCTP/IP	29.273 [28]
UE – foreign agent in trusted non-3GPP Access	MIPv4	24.304 [30]
UE – Trusted or Un-trusted non-3GPP access	EAP-AKA	24.302 [29]

3.5   Inter-working with cdma2000® Access Networks

3.5.1   Architecture for cdma2000® HRPD Inter-working

The best inter-working performance in terms of handover gap time is achieved by specifying the networks to inter-operate very tightly to exchange critical information. This creates a specific solution that is valid for only the ANs in question. With the limited time and resources available for specification work, the number of such solutions in 3GPP Release 8 could only be limited. A tight inter-working solution also requires changes in the other ANs, and by definition the development of non-3GPP ANs is not within the control of 3GPP. Achieving a well designed solution requires special attention to coordination between the developments in different standardization bodies. With these difficulties at hand, 3GPP Release 8 only includes an optimized inter-working solution with cdma2000® HRPD AN.

Figure 3.17 highlights the architecture for cdma2000® HRPD inter-working. It shows the Evolved HRPD (E-HRPD) network, where a number of modifications have been applied to make it suitable for connecting to the EPC. Due to these modifications it will be called E-HRPD in this chapter to distinguish it from legacy HRPD systems that do not support these functions. The radio interface and the Radio Access Network have been kept as similar as possible, but the HRPD Serving Gateway (HSGW) is a completely new node inheriting many of its functions from S-GW.

The E-HRPD is generally treated as a trusted non-3GPP AN, and it is therefore connected to the EPC via S2a, Gxa and STa interfaces. These interfaces operate as described

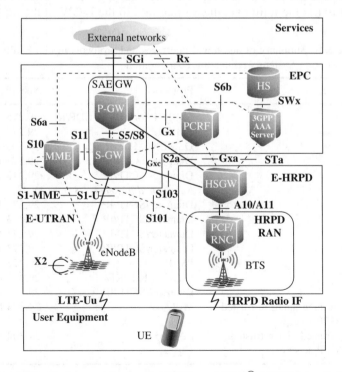

Figure 3.17   System architecture for 3GPP and 2000® HRPD inter-working

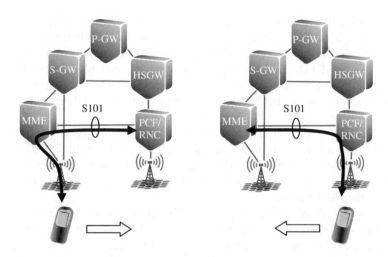

Figure 3.18 Tunnelled pre-registration to eHRPD and to E-UTRAN

earlier. Since the inter-working solution is optimized, and does not rely on UE performing the attach procedure directly to the target network, two new interfaces, S101 and S103, were defined for the CP and UP interactions respectively.

3GPP ANs and the 3GPP2-defined cdma2000® ANs share many things in common, but many things are also different. Both systems use a prepared handover, where the source system signals to the target system to give it essential parameters to be able to serve the terminal there, and the target system gives the source system parameters that can be further given to the terminal to guide it to make the access to the target radio. While there are similarities in these methods, the parameters themselves do not match well at all, and this method could not be used by applying a simple protocol conversion. To ease up on the need to align every information element that would need to be exchanged in handover, it was decided to use a transparent signalling transport method.

Figure 3.18 shows how the transparent tunnel is used in mobility from E-UTRAN to E-HRPD on the left, and the opposite direction is shown on the right. The thick black arrow indicates the signalling which is carried transparently through the source access system and over the S101 interface to the target system. In this method the source access system gives guidance to the UE to register to the target system through the tunnel. This creates the UE context in the target system without the source system having to convert its information to the target system format. This is called pre-registration, and the purpose is to take the time consuming registration/attach function away from the time critical path of handover. The transparent tunnel may also be used to build the bearer context in the target system so that when the time to make the handover is at hand, everything will be ready and waiting at the target side. The actual handover is decided based on radio interface conditions, and this solution requires that both systems are able to handle measurements from the other system. The following inter-working scenarios are supported between E-UTRAN and E-HRPD:

- E-UTRAN → E-HRPD handover: The pre-registration may be performed well before the actual handover takes place, and also all bearers are set up in the E-HRPD side.

The UE remains in a dormant state (equal to idle mode) from the E-HRPD system point of view before handover, and this state may be long lived. When the radio conditions indicate the need for handover, the eNodeB commands the UE to start requesting traffic channel from E-HRPD. This takes place through the transparent tunnel, and once it is completed, the eNodeB commands the UE to make the actual handover. The S103 interface is used only in this handover scenario to forward DL data during the time when the UE is making the switch between the radios.

- E-UTRAN → E-HRPD idle mode mobility: The pre-registration state works as described above for the handover. The UE is in idle mode in the E-UTRAN also, and it moves within the system, selecting the cells on its own, and when it selects an E-HRPD cell, it briefly connects to the E-HRPD network to get the mobility pointers updated to the E-HRPD side.

- E-HRPD → E-UTRAN handover: The E-HRPD AN will request the UE to make tunnelled pre-registration (attach) only at the time the handover is needed, and the UE will immediately proceed to requesting connection directly from the E-UTRAN cell after the registration is complete. The bearers are set up in embedded fashion with the registration and connection request procedures.

- E-HRPD → E-UTRAN idle mode mobility: The idle mode procedure follows the same guidelines as the handover for the tunnelled registration (attach), but the UE accesses the E-UTRAN radio only by reporting its new location (Tracking Area), since there is no need to set up bearers in E-UTRAN for UE in idle mode.

3.5.2   Additional and Updated Logical Elements for cdma2000®
HRPD Inter-working

Inter-working with eHRPD in an optimized manner brings a lot of new features in the basic SAE network elements, and introduces few totally new elements in the HRPD side. The UE, eNodeB, MME and S-GW will all be modified to support new functions, and MME and S-GW will also deal with new interfaces. The eHRPD is a new network of its own, and it consists of elements such as Base Station, Radio Network Controller (RNC), Packet Control Function (PCF) and HRPD Serving Gateway (HSGW).

The UE will need to support both radio interfaces. The design of the procedure assumes that UE is capable of single mode operation only. On the other hand, the integration is kept loose enough so that it would be possible to implement terminal with separate chip sets for E-UTRAN and E-HRPD. This means that the UE is not required to make measurements of cells in the other technology in as tightly a timewise controlled manner as is normally seen within a single radio technology. The UE will also need to support the tunnelled signalling operation. The tunnelled signalling itself is the same signalling as the UE would use directly with the other system.

The main new requirement for the eNodeB is that it also needs to be able to control mobility towards the eHRPD access. From the radio interface perspective it does this much in the same manner as with the other 3GPP accesses, by instructing the UE to make measurements of the neighbouring eHRPD cells, and making the handover decision based on this information. On the other hand, the eNodeB does not signal the handover preparation towards the eHRPD, like it would for other handovers in S1 interface. Instead the handover preparation works so that the UE sends traffic channel requests to the eHRPD AN through the transparent tunnel, and the eNodeB is only responsible for marking the uplink messages with appropriate routing information, so that the MME can select the right node in the eHRPD AN, and noting the progress of the handover from the headers of the S1 messages carrying the eHRPD signalling.

The MME implements the new S101 interface towards the eHRPD RAN. For UE originated messages, it needs to be able to route them to the right eHRPD RAN node based on a reference given by the eNodeB. In the reverse direction the messages are identified by the IMSI of the UE, and the basis that the MME can route them to the right S1 signalling connection. The MME does not need to interpret the eHRPD signalling message contents, but the status of the HO progress is indicated along with those messages that require special action from the MME. For example, at a given point during E-UTRAN → E-HRPD handover, the MME will set up the data forwarding tunnels in the S-GW. The MME also needs to memorize the identity of the E-HRPD AN node that a UE has been signalling with, so that if MME change takes place, the MME can update the S101 context in the HRPD AN node.

The S-GW supports the new S103 interface, which is used for forwarding DL data during the time in handover, when the radio link cannot be used. The forwarding function is similar to the function S-GW has for the E-UTRAN handovers. The difference is that S103 is based on a GRE tunnel, and there will be only one tunnel for each UE in handover, so the S-GW needs to map all GTP tunnels from the S1-U interface to a single GRE tunnel in the S103 interface.

The E-HRPD network is a completely new way to use the existing HRPD radio technology with the SAE, by connecting it to the EPC. Compared to the original HRPD, many changes are caused by the inter-working, and connecting to the EPD requires some new functions, e.g. the support of EAP-AKA authentication. The HSGW is taking the S-GW role for E-HRPD access, and performs much like a S-GW towards the P-GW. The HSGW also includes many CP functions. Towards the eHRPD AN, it behaves like the Packet Data Serving Node (PDSN) in a legacy HRPD network. It also signals with the 3GPP AAA Server to authenticate the UE, and to receive its service profile. The CN aspects of the E-HRPD are specified in [31] and the evolved RAN is documented in [32].

3.5.3 Protocols and Interfaces in cdma2000® HRPD Inter-working

The optimized inter-working introduces two new interfaces – S101 and S103 – to the architecture (see Table 3.5). The S2a, Gxc and STa are as described earlier. The following summarizes the new interfaces:

- S101 is a CP interface that in principle forms a signalling tunnel for the eHRPD messages. The CP protocol is S101AP, which is specified in [33]. The S101AP uses the same message structure and coding as the newest version of GTP. The main function is to carry the signalling messages, with the IMSI as a reference and with an additional handover status parameter that is set by either the UE or either one of the networks it signals with. In addition, when the data forwarding tunnel needs to be set up, the address information is also included in S101AP. S101AP also includes a procedure to switch the interface from one MME to another if handover in E-UTRAN causes MME change.
- S103 is a simple GRE tunnel for UP data forwarding in handover. It is only used for DL data in handover from E-UTRAN to E-HRPD. S103 is a UP interface only, and

Table 3.5 Additional interfaces and protocols
for inter-working with cdma2000® eHRPD

Interface	Protocols	Specification
S101	S101AP/UDP/IP	29.276 [33]
S103	GRE/IP	29.276 [33]

all control information to set up the GRE tunnel is carried in other interfaces. It is
specified with S101AP in [33].

3.5.4   Inter-working with cdma2000® 1xRTT

The cdma2000® 1xRTT is a system supporting CS bearers, and is primarily used for voice
calls. In this respect it is functionally equivalent to the legacy 3GPP CS infrastructure
such as the MSC and the CS bearer capabilities of GERAN and UTRAN. As described in
Chapter 10, the 3GPP standard includes two functions to support inter-working between
the E-UTRAN and the legacy CS infrastructure. These are the CSFB [19] and SR-VCC
[20]. These functions have been extended to cover inter-working with cdma2000® 1xRTT
also, and at a high level they work in the same way as described in Chapter 10. In the
1xRTT case, the interface between MME and the cdma2000® 1xRTT infrastructure is
called S102. S102 carries a protocol specified in 3GPP2 for the A21 interface, which is
used in cdma2000® systems for voice call continuity (see Table 3.6).

3.6   IMS Architecture

3.6.1   Overview

The IP Multimedia Services Sub-System (IMS) is the preferred service machinery for
LTE/SAE. IMS was first introduced in Release 5, and with the well defined inter-working
with existing networks and services that have been introduced since, the Rel-8 IMS
can now be used to provide services over fixed and wireless accesses alike. The IMS
architecture is defined in [36], and the functionality is defined in [37]. A comprehensive
description of IMS can also be found in [3]. For the purpose of this book, the functional
architecture of IMS is presented in Figure 3.19, and a short description of the main
functions follows below.

IMS is an overlay service layer on top of the IP connectivity layer that the EPS
provides. Figure 3.19 shows a thick grey line from UE to P-GW that represents the UE's

Table 3.6 Additional interfaces and protocols
for inter-working with cdma2000® 1xRTT

Interface	Protocols	Specification
S102	S102 protocol	29.277 [34]
A21	A21 protocol	A.S0008-C [35]

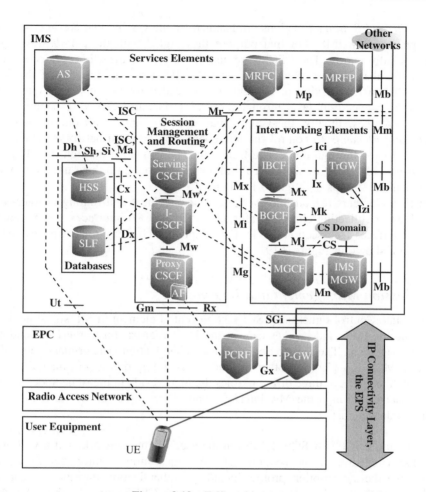

Figure 3.19 IMS architecture

IP connectivity to IMS and other external networks through the RAN and EPC. The signalling interfaces Gm and Ut run on top of this connection, which typically use the default bearer a UE will always have in LTE/SAE. The services may further require that dedicated bearers are set up through EPC, and the service data flows may need to be handled by one of the Inter-working or Services Elements.

In principle the IMS is independent of the connectivity layer, which requires its own registration and session management procedures, but it has also been specifically designed to operate over the 3GPP-defined ANs, and it works seamlessly with the PCC described in section 3.7. IMS uses SIP protocol for registration and for controlling the service sessions. SIP is used both between the terminal and the IMS (Gm Interface) and between various IMS nodes (ISC, Mw, Mg, Mr, Mi, Mj, Mx, Mk and Mm Interfaces). The SIP usage in IMS is defined in [38]. Diameter (Cx, Dx, Dh and Sh Interfaces) and H.248 (Mp) are the other protocols used in IMS.

The UE primarily signals with the CSCFs for the services it wishes to use, and in addition some service specific signalling may be run directly with the Application Servers. The signalling may also be network originated for terminating services. The *Session*

Management and Routing functions are handled by the CSCFs that are in control of the UE's registration in IMS. For that purpose they signal with the *Databases* to get the appropriate information. The CSCFs are also in control of the UE's service sessions in IMS, and for that purpose they may need to signal with one or more of the *Services Elements* to know what kind of connectivity is needed for the service in question, and then with the connectivity layer through the Rx interface to make corresponding requests to bearer resources. Finally, the CSCFs may need to signal with one or more of the *Inter-working Elements* to control the interconnection between networks. Whenever the UP flow is routed through one of the IMS elements, it is done through the Mb interface that connects IMS to IP networks. The following sections introduce the main functions in the functional groups highlighted in Figure 3.19.

Most IMS elements responsible for session management and routing or inter-working are involved in collecting charging information. Rf and Ro interfaces are the main IMS charging interfaces (see section 3.7.1). For simplicity, charging related nodes and interfaces are not shown in Figure 3.19.

3.6.2   Session Management and Routing

The Call State Control Function (CSCF) is the central element in SIP signalling between the UE and the IMS, and it takes care of the UE's registration to the IMS, and service session management. The registration includes authentication. The primary authentication method is IMS-AKA [39], but other methods such as http digest [40] may also be used. CSCF is defined to have three different roles that may reside in the same node, or separate nodes connected through the Mw interface, and all are involved in the UE-related SIP signalling transactions:

- The Serving CSCF (S-CSCF) locates in the user's home network, and it will maintain the user's registration and session state. At registration, it interfaces with the HSS to receive the subscription profile, including authentication information, and it will authenticate the UE. For the service sessions, the S-CSCF signals with the UE through the other CSCFs, and may also interact with the Application Servers (ASs) or the MRFCs for setting up the service session properly. It also carries the main responsibility for controlling the Inter-working Elements. The S-CSCF may also need to interact with MGCF for inter-working with CS networks, or with other multimedia networks for UE requested services.
- The Interrogating CSCF (I-SCSF) is located at the border of the home network, and it is responsible for finding out the UE's registration status, and either assigning a new S-CSCF or routing to the right existing S-CSCF. The request may come from Proxy CSCF (P-CSCF), from other multimedia networks, or from CS networks through the Media Gateway Control Function (MGCF). Also I-CSCF may need to interact with the ASs for service handling. The Ma interface is used for this when Public Service Identity (PSI) is used to identify the service, and the I-CSCF can route the request directly to the proper AS.
- The (P-CSCF) is the closest IMS node the UE interacts with, and it is responsible for all functions related to controlling the IP connectivity layer, i.e. the EPS. For this purpose the P-CSCF contains the Application Function (AF) that is a logical element for the PCC concept, which is described in section 3.7.1. The P-CSCF is typically located in the same network as the EPS, but the Rel-8 includes a so-called Local Breakout

concept that allows P-CSCF to remain in the home network, while PCRF in the visited network may still be used.

In addition to the above-mentioned three CSCF roles, a fourth role, the Emergency CSCF (E-CSCF), has been defined. As the name indicates, the E-CSCF is dedicated to handling the emergency call service in IMS. The E-CSCF connects to the P-CSCF via the Mw interface, and these nodes must always be in the same network. In addition, the E-CSCF is also connected to a Location Retrieval Function (LRF) through the Mi Interface. The LRF can provide the location of the UE, and routing information to route the emergency call appropriately. The E-CSCF and LRF are not shown in Figure 3.19 for simplicity

The CSCFs are connected to each other with the Mw interface, and to other multimedia networks through the Mm interface. Interconnection between CSCFs in different operators' networks may be routed through a common point called the Interconnection Border Control Function (IBCF). See section 3.6.5.

3.6.3 Databases

The Home Subscriber Server (HSS) is the main database used by the IMS. The HSS contains the master copy of subscription data, and it is used in much the same way as with the IP connectivity layer. It provides the location and authentication information based on requests from the I- or S-CSCF, or the AS. The interface between the HSS and the Services Elements will be either Sh or Si depending on the type of services elements. The Sh interface is used in case of SIP or OSA service capability server and the Si when CAMEL based AS is in question.

When there is more than one addressable HSS, another database called the Subscription Locator Function (SLF) may be used to find the right HSS.

3.6.4 Services Elements

The actual service logic is located in the Application Servers (AS). A variety of different services may be provided with different ASs, and the standards do not aim to cover all possible services. Some of the main services are covered in order to facilitate easier interworking with operators in roaming, and to provide for consistent user experience. One example of a standardized AS is the Telephony Application Server (TAS), which may be used to provide the IMS VoIP service.

The media component of the service can be handled by the Multimedia Resource Function (MRF), which is defined as a separate controller (MRFC) and processor (MRFP). The UP may be routed through MRFP for playing announcements as well as for conferencing and transcoding. For coordination purposes, the MRFC may also be connected to the related AS.

3.6.5 Inter-working Elements

The Inter-working Elements are needed when the IMS interoperates with other networks, such as other IMS networks, or CS networks. The following are the main functions of the standardized inter-working elements:

- The Breakout Gateway Control Function (BGCF) is used when inter-working with CS networks is needed, and it is responsible for selecting where the interconnection will take place. It may select the Media Gateway Control Function (MGCF) if the breakout is to happen in the same network, or it may forward the request to another BGCF in another network. This interaction may be routed through the Interconnection Border Control Function (IBCF).

- The Interconnection Border Control Function (IBCF) is used when interconnection between operators is desired to be routed through defined points, which hide the topology inside the network. The IBCF may be used in interconnection between CSCFs or BGCFs and it is in control of Transition Gateway (TrGW), which is used for the same function in the UP. Note that the IBCF–TrGW interface is not fully specified in Release 8. The IBCFs and the TrGWs in different operators' networks may be interconnected to each other via the Ici and Izi interfaces respectively, and together they comprise the Inter IMS Network to Network Interface (II-NNI).
- The Media Gateway Control Function (MGCF) and IMS-Media Gateway (IMS-MGW) are the CP and UP nodes for inter-working with the CS networks such as the legacy 3GPP networks with CS domain for GERAN or UTRAN, or for PSTN/ISDN. Both incoming and outgoing IMS VoIP calls are supported with the required signalling inter-working and transcoding between different voice coding schemes. The MGCF works in the control of either the CSCF or BGCF.

3.7   PCC and QoS

3.7.1   PCC

Policy and Charging Control (PCC) has a key role in the way users' services are handled in the Release 8 LTE/SAE system. It provides a way to manage the service related connections in a consistent and controlled way. It determines how bearer resources are allocated for a given service, including how the service flows are partitioned to bearers, what QoS characteristics those bearers will have, and finally, what kind of accounting and charging will be applied. If an operator uses only a very simple QoS model, then a static configuration of these parameters may be sufficient, but Release 8 PCC allows the operator to set these parameters dynamically for each service and even each user separately.

The PCC functions are defined in [5] and the PCC signalling transactions as well as the QoS parameter mapping are defined in [41]. Figure 3.20 shows the PCC functions and interfaces in the basic configuration when PCC is applied in one operator's network.

The primary way to set up service flows in Release 8 is one where the UE first signals the request for the service in the Service Layer, and the Application Function (AF) residing in that layer contacts the Policy and Charging Resource Function (PCRF) for appropriate bearer resources. The PCRF is in charge for making the decisions on what PCC to use for the service in question. If subscriber specific policies are used, then the PCRF may enquire subscription related policies from the Subscription Profile Repository (SPR). Further details about SPR structure and, for example, its relation to HSS, and the Sp interface are not specified in Release 8. Based on the decision, the PCRF creates the appropriate PCC rules that determine the handling in the EPS.

If the interface from P-GW to the S-GW is based on GTP, the PCRF pushes the PCC rules to the Policy and Charging Enforcement Function (PCEF) residing in the P-GW, and it alone will be responsible for enforcing the PCC rules, e.g. setting up the corresponding

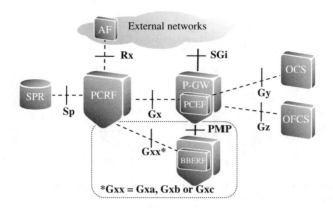

Figure 3.20   Basic PCC functions

dedicated bearers, or modifying the existing bearers so that the new IP service flows can be mapped to them, and by ensuring that only authorized service flows are allowed and QoS limits are not exceeded. In this case the Gxx interface shown in Figure 3.20 does not apply.

If the interface from P-GW towards the AN is based on PMIP, i.e. if it is S5 PMIP, S2a or S2b, there is no means to signal the bearer level information onwards from the P-GW, and the PCRF will create a separate set of QoS rules, and those are first sent to the BBERF, which will handle the mapping between IP service flows and bearers over the AN. Depending on the AN type, the BBERF may reside in S-GW (S5 PMIP), trusted non-3GPP AN, e.g. in HSGW (S2a), or in the EPDG (S2b) for the un-trusted non-3GPP AN (S2b is not supported in Release 8). Also in this case the PCC rules are also sent to PCEF in the P-GW, and it performs the service flow and QoS enforcement.

Release 8 also supports UE initiated bearer activation within the EPS, which is applicable to the case when there is no defined service that both the UE and the serving network could address. In this case the UE signals with the AN and the BBERF requests the service resources from the PCRF. The PCRF makes the PCC decision, and the logic then continues as described above.

The PCC standard [5] defines two charging interfaces, Gy and Gz, which are used for online and offline charging respectively. The Gy interface connects the PCEF to the Online Charging System (OCS), which is used for flow based charging information transfer and control in an online fashion. The Gz interface is used between the P-GW and the Offline Charging System (OFCS), and it is applied when charging records are consolidated in an offline fashion. The charging specifications [42] and [43] further define that the Gy interface is functionally equivalent to the Ro interface that uses Diameter Credit-Control Application as defined in [44]. The Gz interface may be based on either the Rf interface, which relies on the mentioned Diameter Credit-Control Application, or the Ga interface, which uses the 3GPP defined GTP protocol. The Ro and Rf interfaces are also used for charging in IMS, and were originally specified for that purpose.

The PCRF in control of the PCEF/P-GW and the BBERF typically reside in the same operator's network. In the case of roaming, they may reside in different networks, and the S9 interface between PCRFs is used to enable the use of a local PCRF. The S9 interface

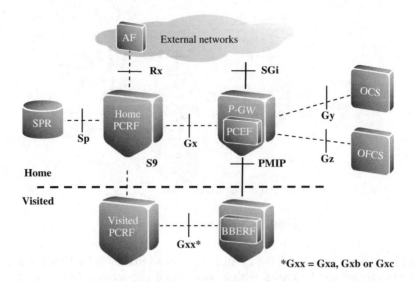

Figure 3.21   PCC functions in roaming with PMIP, home routed model

is defined in [9], and it re-uses the applicable parts from Rx, Gx and Gxx interfaces to convey the information between the PCRFs.

There are two different cases when the S9 interface is used. The first case, which is shown in Figure 3.21, applies when the roaming interface is based on PMIP, and the PCEF and BBERF are in different networks. In this scenario traffic is routed to the home network. In the second case, shown in Figure 3.22, the Local Breakout model is applied, and the P-GW resides in the visited network. The AF and Rx interface will be used from the same network that provides the service in question. The OCS will reside in the home network. As described above, the separate BBERF and Gxx interfaces apply only if PMIP is used from the P-GW in the visited network.

Table 3.7 lists the PCC related interfaces and the protocols, and the standards where they are specified.

3.7.2   QoS

The development of the SAE bearer model and the QoS concept started with the assumption that improvements compared to the existing 3GPP systems with, e.g. UTRAN access, should be made, and the existing model should not be taken for granted. Some potential areas had already been identified. It had not been easy for the operators to use QoS in the legacy 3GPP systems. An extensive set of QoS attributes was available, but it was to some extent disconnected from the application layer, and thus it had not been easy to configure the attributes in the correct way. This problem was emphasized by the fact that the UE was responsible for setting the QoS attributes for a bearer. Also, the bearer model had many layers, each signalling just about the same information. It was therefore agreed that for SAE, only a reduced set of QoS parameters and standardized characteristics would be specified. Also it was decided to turn the bearer set-up logic so that the network resource management is solely network controlled, and the network decides how the parameters

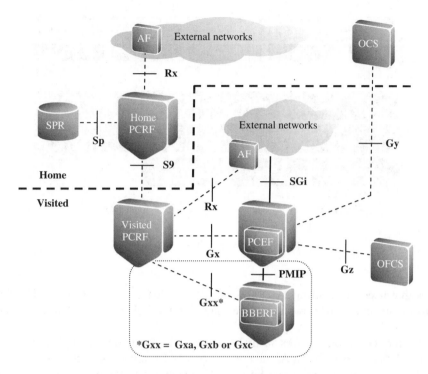

Figure 3.22 PCC functions in roaming, local breakout model

Table 3.7 Summary of PCC interfaces

Interface	Protocols	Specification
Gx	Diameter/SCTP/IP	29.212 [13]
Gxx (Gxa or Gxc)	Diameter/SCTP/IP	29.212 [13]
Rx	Diameter/SCTP/IP	29.214 [14]
S9	Diameter/SCTP/IP	29.215 [45]
Sp	*Not defined in Release 8*	N.A.
Gy =		32.240 [42]
Ro	Diameter/SCTP/IP	32.299 [46]
Gz =		32.251 [43]
Rf or	Diameter/SCTP/IP or	32.295 [47] or
Ga	GTP'/UDP or TCP/IP	32.299 [46]

are set, and the main bearer set-up logic consists of only one signalling transaction from the network to the UE and all interim network elements.

The resulting SAE bearer model is shown in Figure 3.23. The bearer model itself is very similar to the GPRS bearer model, but it has fewer layers. EPS supports the always-on concept. Each UE that is registered to the system has at least one bearer called the default bearer available, so that continuous IP connectivity is provided. The default bearer may have quite basic QoS capabilities, but additional bearers may be set up on demand for

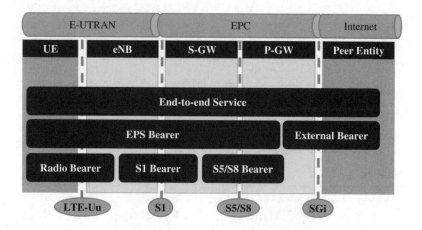

Figure 3.23   SAE Bearer model

services that need more stringent QoS. These are called dedicated bearers. The network may also map several IP flows that have matching QoS characteristics to the same EPS bearer.

The bearer set-up logic works so that the UE first signals on the application layer, on top of the default bearer, to an Application Server (AS) in the operator service cloud, e.g. with IMS, to set up the End-to-end Service. This signalling may include QoS parameters, or simply indication to a known service. The AS will then request the set-up of the corresponding EPS bearer through the PCC infrastructure. There is no separate signalling transaction for the EPS bearer layer, but the EPS bearer is set up together with the signalling for the lower layers, i.e. S5/S8 bearer, S1 Bearer and Radio Bearer. Furthermore, since the eNodeB is responsible for controlling the radio interface transmission in the uplink as well, the UE can operate based on very basic QoS information. The overall goal for network orientation in bearer set-up is to minimize the need for QoS knowledge and configuration in the UE.

Also the QoS parameters were optimized for SAE. Only a limited set of signalled QoS parameters are included in the specifications. They are:

- QoS Class Identifier (QCI): It is an index that identifies a set of locally configured values for three QoS attributes: Priority, Delay and Loss Rate. QCI is signalled instead of the values of these parameters. Ten pre-configured classes have been specified in two categories of bearers, Guaranteed Bit Rate (GBR) and Non-Guaranteed Bit-Rate (Non-GBR) bearers. In addition operators can create their own classes that apply within their network. The standard QCI classes and the values for the parameters within the class are shown in Table 3.8.
- Allocation and Retention Priority (ARP): Indicates the priority of the bearer compared to other bearers. This provides the basis for admission control in bearer set-up, and further in a congestion situation if bearers need to be dropped.
- Maximum Bit Rate (MBR): Identifies the maximum bit rate for the bearer. Note that a Release 8 network is not required to support differentiation between the MBR and GBR, and the MBR value is always set to equal to the GBR.

Table 3.8 QoS parameters for QCI

QCI	Resource type	Priority	Delay budget	Loss rate	Example application
1	GBR	2	100 ms	1e-2	VoIP
2	GBR	4	150 ms	1e-3	Video call
3	GBR	5	300 ms	1e-6	Streaming
4	GBR	3	50 ms	1e-3	Real time gaming
5	Non-GBR	1	100 ms	1e-6	IMS signalling
6	Non-GBR	7	100 ms	1e-3	Interactive gaming
7	Non-GBR	6	300 ms	1e-6	Application with TCP:
8	Non-GBR	8			browsing, email, file
9	Non-GBR	9			download, etc.

- Guaranteed Bit Rate (GBR): Identifies the bit rate that will be guaranteed to the bearer.
- Aggregate Maximum Bit Rate (AMBR): Many IP flows may be mapped to the same bearer, and this parameter indicates the total maximum bit rate a UE may have for all bearers in the same PDN connection.

Table 3.8 shows the QoS parameters that are part of the QCI class, and the nine standardized classes. The QoS parameters are:

- Resource Type: Indicates which classes will have GBR associated to them.
- Priority: Used to define the priority for the packet scheduling of the radio interface.
- Delay Budget: Helps the packet scheduler to maintain sufficient scheduling rate to meet the delay requirements for the bearer.
- Loss Rate: Helps to use appropriate RLC settings, e.g. number of re-transmissions.

References

[1] 3GPP TS 23.401, 'General Packet Radio Service (GPRS) enhancements for Evolved Universal Terrestrial Radio Access Network (E-UTRAN) access (Release 8)'.
[2] 3GPP TS 23.402, 'Architecture enhancements for non-3GPP accesses (Release 8)'.
[3] M. Poikselkä et al., ' The IMS: IP Multimedia Concepts and Services', 2nd edition, Wiley, 2006.
[4] 3GPP TS 33.401, 'Security Architecture (Release 8)'.
[5] 3GPP TS 23.203, 'Policy and charging control architecture (Release 8)'.
[6] 3GPP TS 36.413, 'Evolved Universal Terrestrial Radio Access (E-UTRA) and Evolved Universal Terrestrial Radio Access Network (E-UTRAN); Overall description (Release 8)'.
[7] 3GPP TS 36.423, 'Evolved Universal Terrestrial Radio Access Network (E-UTRAN); X2 Application Protocol (X2AP) (Release 8)'.
[8] 3GPP TS 29.274, 'Evolved GPRS Tunnelling Protocol (eGTP) for EPS (Release 8)'.
[9] 3GPP TS 36.413, 'Evolved Universal Terrestrial Radio Access (E-UTRA); S1 Application Protocol (S1AP) (Release 8)'.
[10] 3GPP TS 29.275, 'PMIP based Mobility and Tunnelling protocols (Release 8)'.
[11] 3GPP TS 29.061, 'Inter-working between the Public Land Mobile Network (PLMN) supporting packet based services and Packet Data Networks (PDN) (Release 8)'.
[12] 3GPP TS 29.272, 'MME Related Interfaces Based on Diameter Protocol (Release 8)'.
[13] 3GPP TS 29.212, 'Policy and charging control over Gx reference point (Release 8)'.
[14] 3GPP TS 29.214, 'Policy and charging control over Rx reference point (Release 8)'.
[15] 3GPP TS 24.301, 'Non-Access-Stratum (NAS) protocol for Evolved Packet System (EPS) (Release 8)'.
[16] H. Holma, A. Toskala, ' WCDMA for UMTS – HSPA Evolution and LTE', 5th edition, Wiley, 2010.

[17] T. Halonen, J. Romero, J. Melero, 'GSM, GPRS and EDGE Performance: Evolution Towards 3G/UMTS', 2nd edition, Wiley, 2003.
[18] 3GPP TS 23.060, 'General Packet Radio Service (GPRS); Service description; Stage 2 (Release 8)'.
[19] 3GPP TS 23.272, 'Circuit Switched (CS) fallback in Evolved Packet System (EPS); Stage 2 (Release 8)'.
[20] 3GPP TS 23.216, 'Single Radio Voice Call Continuity (SRVCC); Stage 2 (Release 8)'.
[21] 3GPP TS 29.118, 'Mobility Management Entity (MME) – Visitor Location Register (VLR) SGs interface specification (Release 8)'.
[22] 3GPP TS 29.280, '3GPP EPS Sv interface (MME to MSC) for SRVCC (Release 8)'.
[23] 3GPP TS 33.402, 'Security aspects of non-3GPP accesses (Release 8)'.
[24] IETF Internet-Draft, draft-arkko-eap-aka-kdf, 'Improved Extensible Authentication Protocol Method for 3rd Generation Authentication and Key Agreement (EAP-AKA)'. (J. Arkko, V. Lehtovirta, P. Eronen) 2008.
[25] 3GPP TS 23.234, '3GPP system to Wireless Local Area Network (WLAN) interworking; System description (Release 7)'.
[26] IETF RFC 4306, 'Internet Key Exchange (IKEv2) Protocol.' C. Kaufman, Editor, 2005.
[27] 3GPP TS 24.303, 'Mobility management based on Dual-Stack Mobile IPv6 (Release 8)'.
[28] 3GPP TS 29.273, 'Evolved Packet System (EPS); 3GPP EPS AAA interfaces (Release 8)'.
[29] 3GPP TS 24.302, 'Access to the Evolved Packet Core (EPC) via non-3GPP access networks (Release 8)'.
[30] 3GPP TS 24.304, 'Mobility management based on Mobile IPv4; User Equipment (UE) – foreign agent interface (Release 8)'.
[31] 3GPP2 Specification X.P0057, 'E-UTRAN – HRPD Connectivity and Interworking: Core Network Aspects (2008)'.
[32] 3GPP2 Specification X.P0022, 'E-UTRAN – HRPD Connectivity and Interworking: Access Network Aspects (E-UTRAN – HRPD IOS) (2008)'.
[33] 3GPP TS 29.276, 'Optimized Handover Procedures and Protocols between EUTRAN Access and cdma2000 HRPD Access (Release 8)'.
[34] 3GPP TS 29.277, 'Optimized Handover Procedures and Protocols between EUTRAN Access and 1xRTT Access (Release 8)'.
[35] 3GPP2 Specification A.S0008-C, 'Interoperability Specification (IOS) for High Rate Packet Data (HRPD) Radio Access Network Interfaces with Session Control in the Access Network (2007)'.
[36] 3GPP TS 23.002, 'Network architecture (Release 8)'.
[37] 3GPP TS 23.228, 'IP Multimedia Subsystem (IMS); Stage 2 (Release 8)'.
[38] 3GPP TS 24.229, 'IP multimedia call control protocol based on Session Initiation Protocol (SIP) and Session Description Protocol (SDP); Stage 3 (Release 8)'.
[39] 3GPP TS 33.203, '3G security; Access security for IP-based services (Release 8)'.
[40] IETF RFC 2617, 'HTTP Authentication: Basic and Digest Access Authentication', J. Franks, P. Hallam-Baker, J. Hostetler, S. Lawrence, P. Leach, A. Luotonen, L. Stewart (1999).
[41] 3GPP TS 29.213, 'Policy and Charging Control signalling flows and QoS parameter mapping (Release 8)'.
[42] 3GPP TS 32.240, 'Telecommunication management; Charging management; Charging architecture and principles (Release 8)'.
[43] 3GPP TS 32.251, 'Telecommunication management; Charging management; Packet Switched (PS) domain charging (Release 8)'.
[44] IETF RFC 4006, 'Diameter Credit-Control Application', H. Hakala, L. Mattila, J.-P. Koskinen, M. Stura, J. Loughney (2005).
[45] 3GPP TS 29.215, 'Policy and Charging Control (PCC) over S9 reference point (Release 8)'.
[46] 3GPP TS 32.299, 'Telecommunication management; Charging management; Diameter charging applications (Release 8)'.
[47] 3GPP TS 32.295, 'Telecommunication management; Charging management; Charging Data Record (CDR) transfer (Release 8)'.

4

Introduction to OFDMA and SC-FDMA and to MIMO in LTE

Antti Toskala and Timo Lunttila

4.1 Introduction

As discussed in Chapter 1, LTE multiple access is different to that of WCDMA. In LTE the downlink multiple access is based on the Orthogonal Frequency Division Multiple Access (OFDMA) and the uplink multiple access is based on the Single Carrier Frequency Division Multiple Access (SC-FDMA). This chapter will introduce the selection background and the basis for both SC-FDMA and OFDMA operation. The basic principles behind the multi-antenna transmission in LTE, using Multiple Input Multiple Output (MIMO) technology, is also introduced. The intention of this chapter is to illustrate the multiple access principles in a descriptive way without too much mathematics. For those interested in the detailed mathematical notation, two selected references are given that provide a mathematical treatment of the different multiple access technologies, covering both OFDMA and SC-FDMA.

4.2 LTE Multiple Access Background

A Single Carrier (SC) transmission means that information is modulated only to one carrier, adjusting the phase or amplitude of the carrier or both. Frequency could also be adjusted, but in LTE this is not affected. The higher the data rate, the higher the symbol rate in a digital system and thus the bandwidth is higher. With the use of simple Quadrature Amplitude Modulation (QAM), with the principles explained, for example in [1], the transmitter adjusts the signal to carry the desired number of bits per modulation symbol. The resulting spectrum waveform is a single carrier spectrum, as shown in Figure 4.1, with the spectrum mask influenced (after filtering) by the pulse shape used.

With the Frequency Division Multiple Access (FDMA) principle, different users would then be using different carriers or sub-carriers, as shown in Figure 4.2, to access the system simultaneously having their data modulation around a different center frequency. Care must be now taken to create the waveform in such a way that there is no excessive

LTE for UMTS: Evolution to LTE-Advanced, Second Edition. Edited by Harri Holma and Antti Toskala.
© 2011 John Wiley & Sons, Ltd. Published 2011 by John Wiley & Sons, Ltd.

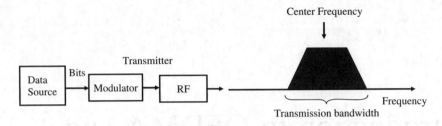

Figure 4.1 Single carrier transmitter

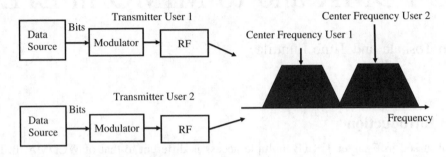

Figure 4.2 FDMA principle

interference between the carriers, nor should one be required to use extensive guard bands between users.

The use of the multi-carrier principle is shown in Figure 4.3, where data are divided on the different sub-carriers of one transmitter. The example in Figure 4.3 has a filter bank which for practical solutions (such as the ones presented later) is usually replaced with Inverse Fast Fourier Transform (IFFT) for applications where the number of sub-carriers is high. There is a constant spacing between neighboring sub-carriers. One of the approaches to multi-carrier is also the dual carrier WCDMA (dual cell HSDPA, as covered in Chapter 13), which sends two WCDMA carriers next to each other but does not use the principles explained later in this section for high spectrum utilization.

To address the resulting inefficiency from the possible guard band requirements, the approach is to choose the system parameters in such a way as to achieve orthogonality

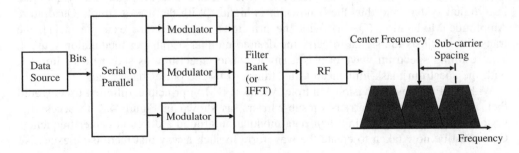

Figure 4.3 Multi-carrier principle

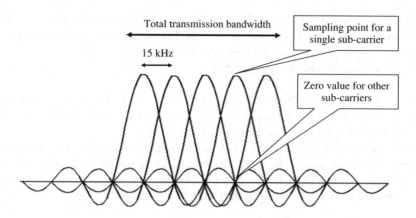

Figure 4.4 Maintaining the sub-carriers' orthogonality

between the different transmissions, and to create the sub-carriers so that they do not interfere with each other but their spectrums could still overlap in the frequency domain. This is what is achieved with the Orthogonal Frequency Division Multiplexing (OFDMA) principle, where each of the center frequencies for the sub-carriers is selected from the set that has such a difference in the frequency domain that the neighboring sub-carriers have zero value at the sampling instant of the desired sub-carrier, as shown in Figure 4.4. For LTE, the constant frequency difference between the sub-carriers has been chosen to be 15 kHz in Release 8 (an alternative of 7.5 kHz is planned to be supported in later releases in connection with broadcast applications such as mobile TV).

The basic principle of OFDMA was already known in the 1950s, at a time when systems were using analog technology, and making the sub-carriers stay orthogonal as a function of component variations and temperature ranges was not a trivial issue. Since the widespread use of digital technology for communications, OFDMA also became more feasible and affordable for consumer use. During recent years OFDMA technology has been widely adopted in many areas such as in digital TV (DVB-T and DVB-H) as well as in Wireless Local Area Network (WLAN) applications.

OFDMA principles have been used in the uplink part of LTE multiple access just as the SC-FDMA uses many of the OFDMA principles in the uplink direction to achieve high spectral efficiency, as described in the next section. The SC-FDMA in the current form, covered in a later section of this chapter, is more novel technology with publications from the late 1990s, such as those presented in [2] and the references therein.

The overall motivation for OFDMA in LTE and in other systems has been due to the following properties:

- good performance in frequency selective fading channels;
- low complexity of base-band receiver;
- good spectral properties and handling of multiple bandwidths;
- link adaptation and frequency domain scheduling;
- compatibility with advanced receiver and antenna technologies.

Many of these benefits (with more explanation provided in the following sections) could only be achieved following the recent developments in the radio access network architecture, meaning setting the radio related control in the base station (or NodeB in 3GPP terms for WCDMA), and as the system bandwidths are getting larger, beyond 5 MHz, receiver complexity also becomes more of an issue.

The OFDMA also has challenges, such as:

- Tolerance to frequency offset. This was tackled in LTE design by choosing a sub-carrier spacing of 15 kHz, which gives a large enough tolerance for Doppler shift due to velocity and implementation imperfections.
- The high Peak-to-Average Ratio (PAR) of the transmitted signal, which requires high linearity in the transmitter. The linear amplifiers have a low power conversion efficiency and therefore are not ideal for mobile uplinks. In LTE this was solved by using the SC-FDMA, which enables better power amplifier efficiency.

When looking back, the technology selections carried out for the 3rd generation system in the late 1990s, the lack of a sensible uplink solution, the lack of advanced antenna solutions (with more than a single antenna) and having radio resource control centralized in the Radio Network Controller (RNC) were the key factors not to justify the use of OFDMA technology earlier. There were studies to look at the OFDMA together with CDMA in connection with the 3rd generation radio access studies, such as are covered in [3]. The key enabling technologies that make OFDMA work better, such as base station based scheduling (Release 5 and 6) and Multiple Input Multiple Output (MIMO) (Release 7), have been introduced only in the later phase of WCDMA evolution. These enhancements, which were introduced in WCDMA between 2002 and 2007, allowed the OFDMA technology to be better used than would have been the case for the simple use of OFDMA only as a modulation method based on a traditional 2nd generation cellular network without advanced features.

4.3 OFDMA Basics

The practical implementation of an OFDMA system is based on digital technology and more specifically on the use of Discrete Fourier Transform (DFT) and the inverse operation (IDFT) to move between time and frequency domain representation. The resulting signal feeding a sinusoidal wave to the Fast Fourier Transform (FFT) block is illustrated in Figure 4.5. The practical implementations use the FFT. The FFT operation moves the signal from time domain representation to frequency domain representation. The Inverse Fast Fourier Transform (IFFT) does the operation in the opposite direction. For the sinusoidal wave, the FFT operation's output will have a peak at the corresponding frequency and zero output elsewhere. If the input is a square wave, then the frequency domain output contains peaks at multiple frequencies as such a wave contains several frequencies covered by the FFT operation. An impulse as an input to FFT would have a peak on all frequencies. As the square wave has a regular interval T, there is a bigger peak at the frequency $1/T$ representing the fundamental frequency of the waveform, and a smaller peak at odd harmonics of the fundamental frequency. The FFT operation can be carried out back and forth without losing any of the original information, assuming that the classical requirements for digital signal processing in terms of minimum sampling rates and word lengths (for the numerics) are fulfilled.

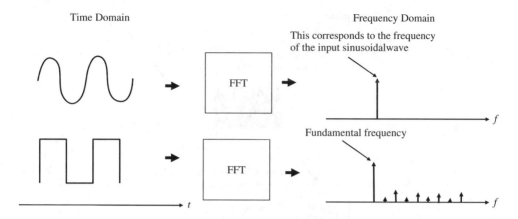

Figure 4.5 Results of the FFT operation with different inputs

The implementation of the FFT is well researched and optimized (low amount of multiplications) when one can stay with power of two lengths. Thus for LTE the necessary FFT lengths also tend to be powers of two, such as 512, 1024, etc. From the implementation point of view it is better to have, for example, a FFT size of 1024 even if only 600 outputs are used (see later the discussion on sub-carriers), than try to have another length for FFT between 600 and 1024.

The transmitter principle in any OFDMA system is to use narrow, mutually orthogonal sub-carriers. In LTE the sub-carrier spacing is 15 kHz regardless of the total transmission bandwidth. Different sub-carriers are orthogonal to each other, as at the sampling instant of a single sub-carrier the other sub-carriers have a zero value, as was shown in Figure 4.4. The transmitter of an OFDMA system uses IFFT block to create the signal. The data source feeds to the serial-to-parallel conversion and further to the IFFT block. Each input for the IFFT block corresponds to the input representing a particular sub-carrier (or particular frequency component of the time domain signal) and can be modulated independently of the other sub-carriers. The IFFT block is followed by adding the cyclic extension (cyclix prefix), as shown in Figure 4.6.

The motivation for adding the cyclic extension is to avoid inter-symbol interference. When the transmitter adds a cyclic extension longer than the channel impulse response, the effect of the previous symbol can be avoided by ignoring (removing) the cyclic extension at the receiver. The cyclic prefix is added by copying part of the symbol at the end and attaching it to the beginning of the symbol, as shown in Figure 4.7. The use of cyclic extension is preferable to simply a break in the transmission (guard interval) as the OFDM symbol then seems to be periodic. When the OFDMA symbol now appears as periodic due to cyclic extension, the impact of the channel ends up corresponding to a multiplication by a scalar, assuming that the cyclic extension is sufficiently long. The periodic nature of the signals also allows for a discrete Fourier spectrum enabling the use of DFT and IDFT in the receiver and transmitter respectively.

Typically the guard interval is designed to be such that it exceeds the delay spread in the environment where the system is intended to be operated. In addition to the channel delay spread, the impact of transmitter and receiver filtering needs to be accounted for

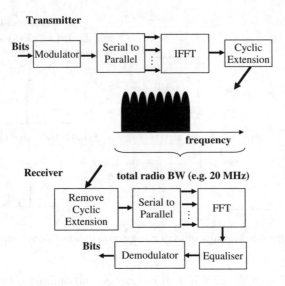

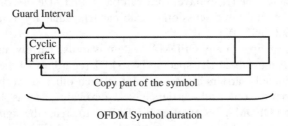

Figure 4.6   OFDMA transmitter and receiver

Figure 4.7   Creation of the guard interval for the OFDM symbol

in the guard interval design. The OFDMA receiver sees the OFDMA symbol coming as through a FIR filter, without separating individual frequency components like the RAKE receiver as described in [4]. Thus, similar to the channel delay spread, the length of the filter applied to the signal in the receiver and transmitter side will also make this overall 'filtering' effect longer than just the delay spread.

While the receiver does not deal with the inter-symbol interference, it still has to deal with the channel impact for the individual sub-carriers that have experienced frequency dependent phase and amplitude changes. This channel estimation is facilitated by having part of the symbols as known reference or pilot symbols. With the proper placement of these symbols in both the time and frequency domains, the receiver can interpolate the effect of the channel to the different sub-carriers from this time and frequency domain reference symbol 'grid'. An example is shown in Figure 4.8.

A typical type of receiver solution is the frequency domain equalizer, which basically reverts the channel impact for each sub-carrier. The frequency domain equalizer in OFDMA simply multiplies each sub-carrier (with the complex-valued multiplication) based on the estimated channel frequency response (the phase and amplitude adjustment

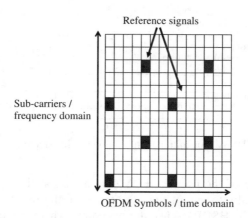

Figure 4.8 Reference symbols spread over OFDMA sub-carriers and symbols

each sub-carrier has experienced) of the channel. This is clearly a simpler operation compared with WCDMA and is not dependent on channel length (length of multipath in chips) as is the WCDMA equalizer. For WCDMA the challenge would be also to increase the chip rate from the current value of 3.84 Mcps, as then the amount of multi-path components separated would increase (depending on the environment) resulting in the need for more RAKE fingers and contributing heavily to equalizer complexity.

In WCDMA the channel estimation in the downlink is based on the Common Pilot Channel (CPICH) and then on pilot symbols on the Dedicated Channel (DCH), which are transmitted with the spread over the whole transmission bandwidth, and different cells separated by different spreading codes. As in the OFDMA system there is no spreading available, other means must be used to separate the reference symbols between cells or between different antennas. In the multi-antenna transmission, as discussed in further detail later in this chapter, the pilot symbols have different positions. A particular position used for a pilot symbol for one antenna is left unused for other antenna in the same cell. Between different cells this blanking is not used, but different pilot symbol patterns and symbol locations can be used.

The additional tasks that the OFDMA receiver needs to cover are time and frequency synchronization. Synchronization allows the correct frame and OFDMA symbol timing to be obtained so that the correct part of the received signal is dropped (cyclic prefix removal). Time synchronization is typically obtained by correlation with known data samples – based on, for example, the reference symbols – and the actual received data. The frequency synchronization estimates the frequency offset between the transmitter and the receiver and with a good estimate of the frequency offset between the device and base station, the impact can be then compensated both for receiver and transmitter parts. The device locks to the frequency obtained from the base station, as the device oscillator is not as accurate (and expensive) as the one in the base station. The related 3GPP requirements for frequency accuracy are covered in Chapter 11.

Even if in theory the OFDMA transmission has rather good spectral properties, the real transmitter will cause some spreading of the spectrum due to imperfections such as the clipping in the transmitter. Thus the actual OFDMA transmitter needs to have filtering similar to the pulse shape filtering in WCDMA. In the literature this filtering is often referred as windowing, as in the example transmitter shown in Figure 4.9.

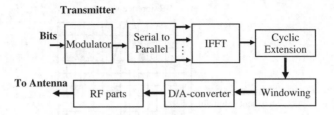

Figure 4.9 OFDMA transmitter with windowing for shaping the spectral mask

An important aspect of the use of OFDMA in a base station transmitter is that users can be allocated basically to any of the sub-carriers in the frequency domain. This is an additional element to the HSDPA scheduler operation, where the allocations were only in the time domain and code domain but always occupied the full bandwidth. The possibility of having different sub-carriers to allocated users enables the scheduler to benefit from the diversity in the frequency domain, this diversity being due to the momentary interference and fading differences in different parts of the system bandwidth. The practical limitation is that the signaling resolution due to the resulting overhead has meant that allocation is not done on an individual sub-carrier basis but is based on resource blocks, each consisting of 12 sub-carriers, thus resulting in the minimum bandwidth allocation being 180 kHz. When the respective allocation resolution in the time domain is 1 ms, the downlink transmission resource allocation thus means filling the resource pool with 180 kHz blocks at 1 ms resolution, as shown in Figure 4.10. Note that the resource block in the specifications refers to the 0.5 ms slot, but the resource allocation is done anyway with the 1 ms resolution in the time domain. This element of allocating resources dynamically in the frequency domain is often referred to as frequency domain scheduling or frequency domain diversity. Different sub-carriers could ideally have different modulations if one could adapt the channel without restrictions. For practical reasons it would be far too inefficient to try either to obtain feedback with 15 kHz sub-carrier resolution or to signal the modulation applied on a individual sub-carrier basis. Thus parameters such as modulation are fixed on the resource block basis.

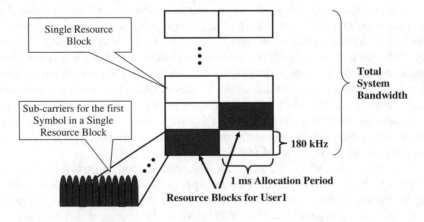

Figure 4.10   OFDMA resource allocation in LTE

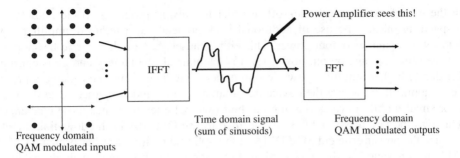

Figure 4.11 OFDMA signal envelope characteristics

The OFDMA transmission in the frequency domain thus consists of several parallel subcarriers, which in the time domain correspond to multiple sinusoidal waves with different frequencies filling the system bandwidth with steps of 15 kHz. This causes the signal envelope to vary strongly, as shown in Figure 4.11, compared to a normal QAM modulator, which is only sending one symbol at a time (in the time domain). The momentary sum of sinusoids leads to the Gaussian distribution of different peak amplitude values.

This causes some challenges to the amplifier design as, in a cellular system, one should aim for maximum power amplifier efficiency to achieve minimum power consumption. Figure 4.12 illustrates how a signal with a higher envelope variation (such as the OFDMA signal in the time domain in Figure 4.11) requires the amplifier to use additional back-off compared to a regular single carrier signal. The amplifier must stay in the linear area

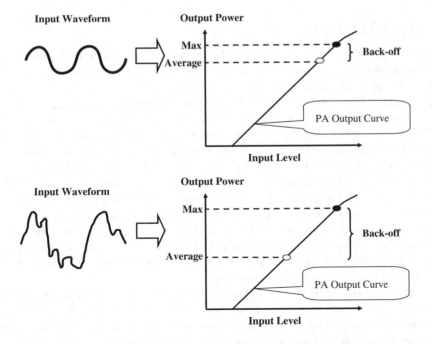

Figure 4.12 Power amplifier back-off requirements for different input waveforms

with the use of extra power back-off in order to prevent problems to the output signal and spectrum mask. The use of additional back-off leads to a reduced amplifier power efficiency or a smaller output power. This either causes the uplink range to be shorter or, when the same average output power level is maintained, the battery energy is consumed faster due to higher amplifier power consumption. The latter is not considered a problem in fixed applications where the device has a large volume and is connected to the mains, but for small mobile devices running on their own batteries it creates more challenges.

This was the key reason why 3GPP decided to use OFDMA in the downlink direction but to use the power efficient SC-FDMA in the uplink direction. Further principles of SC-FDMA are presented in the next section. From the research several methods are known to reduce the PAR, but of more significance – particularly for the amplifier – is the Cubic Metric (CM), which was introduced in 3GPP to better describe the impact to the amplifier. The exact definition of CM can be found from [5].

An OFDMA system is also sensitive to frequency errors as previously mentioned in section 4.2. The basic LTE sub-carrier spacing of 15 kHz facilitates enough tolerance for the effects of implementation errors and Doppler effect without too much degradation in the sub-carrier orthogonality. 3GPP has agreed that for broadcast only (on a dedicated carrier) an optional 7.5 kHz sub-carrier spacing can also be used, but full support of the broadcast only carrier has not been included in Release 8, 9 or 10. In Release 9 there is support for the shared carrier broadcast operation using the existing 15 kHz subcarrier spacing. The physical layer details for the 7.5 kHz case principles can already be found in the 36.2 series specifications starting from Release 8, but details – especially for the higher layer operation and performance requirements are to be considered for later releases after Release 10 finalization.

4.4 SC-FDMA Basics

In the uplink direction 3GPP uses SC-FDMA for multiple access, valid for both FDD and TDD modes of operation. The basic form of SC-FDMA could be seen as equal to the QAM modulation, where each symbol is sent one at a time similarly to Time Division Multiple Access (TDMA) systems such as GSM. Frequency domain generation of the signal, as shown in Figure 4.13, adds the OFDMA property of good spectral waveform in contrast to time domain signal generation with a regular QAM modulator. Thus the need for guard bands between different users can be avoided, similar to the downlink OFDMA principle. As in an OFDMA system, a cyclic prefix is also added periodically – but not after each symbol as the symbol rate is faster in the time domain than in OFDMA – to the transmission to prevent inter-symbol interference and to simplify the receiver design. The receiver still needs to deal with inter-symbol interference as the cyclic prefix now prevents inter-symbol interference between a block of symbols, and thus there will still be inter-symbol interference between the cyclic prefixes. The receiver will thus run the equalizer for a block of symbols until reaching the cyclic prefix that prevents further propagation of the inter-symbol interference.

The transmission occupies the continuous part of the spectrum allocated to the user, and for LTE the system facilitates a 1 ms resolution allocation rate. When the resource allocation in the frequency domain is doubled, so is the data rate, assuming the same level of overhead. The individual transmission (with modulation) is now shorter in time but wider in the frequency domain, as shown in Figure 4.14. The example in Figure 4.14

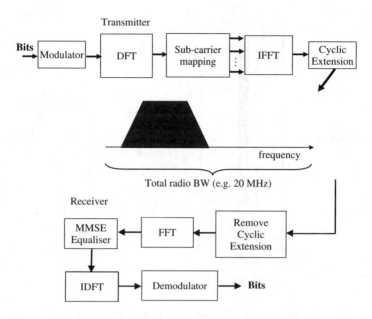

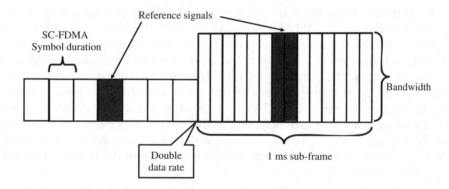

Figure 4.13 SC-FDMA transmitter and receiver with frequency domain signal generation

Figure 4.14 Adjusting data rate in a SC-FDMA system

assumes that in the new resource allocation the existing frequency resource is retained and the same amount of additional transmission spectrum is allocated, thus doubling the transmission capacity. In reality the allocations do not need to have frequency domain continuity, but can take any set of continuous allocation of frequency domain resources. The practical signaling constraints define the allowed amount of 180 kHz resource blocks that can be allocated. The maximum allocated bandwidth depends on the system bandwidth used, which can be up to 20 MHz. The resulting maximum allocation bandwidth is somewhat smaller as the system bandwidth definition includes a guard towards the neighboring operator. For example, with a 10 MHz system channel bandwidth the maximum resource allocation is equal to 50 resource blocks thus having a transmission bandwidth of 9 MHz. The relationship between the *Channel bandwidth* (*BW Channel*) and *Transmission bandwidth configuration* (N_{RB}) is covered in more detail in Chapter 14.

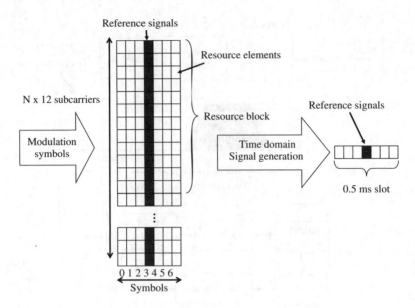

Figure 4.15 Resource mapping in SC-FDMA

The SC-FDMA resource block for frequency domain signal generation is defined using the same values used in the OFDMA downlink, based on the 15 kHz sub-carrier spacing. Thus even if the actual transmission by name is a single carrier, the signal generation phase uses a sub-carrier term. In the simplest form the minimum resource allocated uses 12 sub-carriers, and is thus equal to 180 kHz. The complex valued modulation symbols with data are allocated to the resource elements not needed for reference symbols (or control information) in the resource block, as shown in Figure 4.15. After the resource mapping has been done the signal is fed to the time domain signal generation that creates the SC-FDMA signal, including the selected length of the cyclic prefix. The example in Figure 4.15 assumes a particular length of cyclic prefix with the two different options introduced in Chapter 5.

As shown in Figure 4.15, reference symbols are located in the middle of the slot. These are used by the receiver to perform the channel estimation. There are different options for the reference symbols to be used; sometimes a reference symbol hopping pattern is also used, as covered in more detail in Chapter 5. Also specifically covered further in Chapter 5 are the sounding reference signals, which are momentarily sent over a larger bandwidth than needed, for the data to give the base station receiver information of a larger portion of the frequency spectrum to facilitate frequency domain scheduling in the uplink direction.

Different users are thus sharing the resources in the time as well as in the frequency domain. In the time domain the allocation granularity is 1 ms and in the frequency domain it is 180 kHz. The base station needs to control each transmission so that they do not overlap in the resources. Also to avoid lengthy guard times, timing advance needs to be used, as presented in Chapter 5. By modifying the IFFT inputs, the transmitter can place the transmission in the desired part of the frequency, as shown in Figure 4.16. The base station receiver can detect the transmission from the correct frequency/time resource. As all the

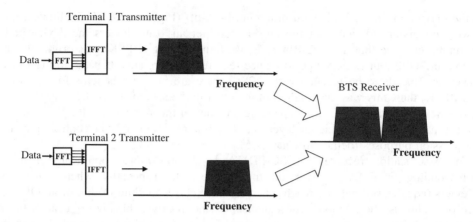

Figure 4.16 Multiple access with resource sharing in the frequency domain with SC-FDMA and frequency domain signal generation

uplink utilization is based on the base station scheduling, with the exception of the random access channel, the base station always knows which user to expect in which resource.

Since we are now transmitting in the time domain only a single modulation symbol at a time, the system retains its good envelope properties and the waveform characteristics are now dominated by the modulation method applied. This allows the SC-FDMA to reach a very low signal PAR or, even more importantly, CM facilitating efficient power amplifiers in the devices. The value of CM as a function modulation applied is shown in Figure 4.17. The use of a low CM modulation method such as Quadrature Phase Shift Keying (QPSK) allows a low CM value and thus the amplifier can operate close

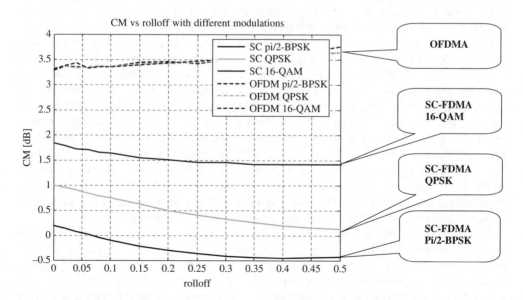

Figure 4.17 CM with OFDMA and SC-FDMA [6]. © 2006 IEEE

to the maximum power level with minimum back-off (Figure 4.17). This allows a good power conversion efficiency of the power amplifier and thus lowers the device power consumption. Note that the pi/2-Binary Phase Shift Keying (BPSK) was originally considered in 3GPP, but as 3GPP performance requirements are such that the full (23 dBm) power level needs to be reached with QPSK, there are no extra benefits for the use of pi/2-BPSK; thus, this was eventually not included in the specifications for user data. The modulation methods in LTE vary depending on whether the symbols are for physical layer control information or for higher layer data (user data or higher layer control signaling) transmission purposes (details in Chapter 5).

The base station receiver for SC-FDMA is slightly more complicated than the corresponding ODFMA receiver on the device side, especially when considering receivers (equalizers) that can reach a performance corresponding to that of an OFDMA receiver. This is the obvious consequence of the receiver having to deal with the inter-symbol interference that is terminated only after a block of symbols and not after every (long) symbol as in OFDMA. This increased need for processing power is, however, not foreseen to be an issue in the base station when compared to the device design constraints and was clearly considered to be outweighed by the benefits of the uplink range and device battery life with SC-FDMA. The benefits of a dynamic resource usage with a 1 ms resolution is also that there is no base-band receiver per UE on standby but the base station receiver is dynamically used for those users that have data to transmit. In any case the most resource consuming part both in uplink and downlink receiver chains is the channel decoding (turbo decoding) with the increased data rates.

4.5 MIMO Basics

One of the fundamental technologies introduced together with the first LTE Release is the Multiple Input Multiple Output (MIMO) operation including spatial multiplexing as well as pre-coding and transmit diversity. The basic principle in spatial multiplexing is sending signals from two or more different antennas with different data streams and by signal processing means in the receiver separating the data streams, hence increasing the peak data rates by a factor of 2 (or 4 with 4-by-4 antenna configuration). In pre-coding the signals transmitted from the different antennas are weighted in order to maximize the received Signal to Noise Ratio (SNR). Transmit diversity relies on sending the same signal from multiple antennas with some coding in order to exploit the gains from independent fading between the antennas. The use of MIMO has been included earlier in WCDMA specifications as covered in [4], but operating slightly differently than in LTE as a spreading operation is involved. The OFDMA nature is well suited for MIMO operation. As the successful MIMO operation requires reasonably high SNR, with an OFDMA system it can benefit from the locally (in the frequency/time domain) high SNR that is achievable. The basic principle of MIMO is presented in Figure 4.18, where the different data streams are fed to the pre-coding operation and then onwards to signal mapping and OFDMA signal generation.

The reference symbols enable the receiver to separate different antennas from each other. To avoid transmission from another antenna corrupting the channel estimation needed for separating the MIMO streams, one needs to have each reference symbol resource used by a single transmit antenna only. This principle is illustrated in Figure 4.19, where the reference symbols and empty resource elements are mapped to alternate between antennas. This principle can also be extended to cover more than two antennas, with the

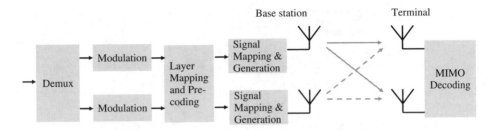

Figure 4.18 MIMO principle with two-by-two antenna configuration

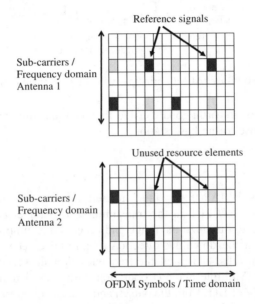

Figure 4.19 OFDMA reference symbols to support two eNodeB transmit antennas

first LTE Release covering up to four antennas. As the number of antennas increases, the required SNR also increases the resulting transmitter/receiver complexity and the reference symbol overhead.

Even LTE uplink supports the use of MIMO technology. While the device is using only one transmit antenna, the single user data rate cannot be increased with MIMO. The cell level maximum data rate can be doubled, however, by the allocation of two devices with orthogonal reference signals. Thus the transmission in the base station is treated like a MIMO transmission, as shown in Figure 4.20, and the data stream separated with MIMO receiver processing. This kind of 'virtual' or 'Multi-user' MIMO is supported in LTE Release 8 and does not represent any major implementation complexity from the device perspective as only the reference signal sequence is modified. From the network side, additional processing is needed to separate the users from each other. The use of 'classical' two antenna MIMO transmission is not particularly attractive due to the resulting device impacts, thus the multi-antenna device transmission was included later in Release 10 LTE-Advanced, as covered in Chapter 16. The SC-FDMA is well-suited

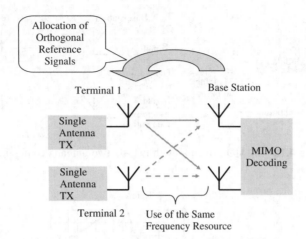

Figure 4.20   Multi-user MIMO principle with single transmit antenna devices

for MIMO use as users are orthogonal (inside the cell) and thus the local SNR may be very high for users close to the base station.

4.6   Summary

Both OFDMA and SC-FDMA are very much related in terms of technical implementation and rely on the use of FFT/IFFT in the transmitter and receiver chain implementation. The SC-FDMA is used to optimize the range and power consumption in the uplink while the OFDMA is used in the downlink direction to minimize receiver complexity, especially with large bandwidths, and to enable frequency domain scheduling with flexibility in resource allocation. Multiple antenna operation with spatial multiplexing has been a fundamental technology of LTE from the outset, and is well suited for LTE multiple access solutions. The mathematical principles of OFDMA and SC-FDMA were not included in this chapter, but can be found from different text books, some of which are included in the references, e.g. [7] for OFDMA and [8] for SC-FDMA.

References

[1] Proakis, J.G., 'Digital Communications', 3rd edition, McGraw-Hill Book Co., 1995.
[2] Czylwik, A., 'Comparison between adaptive OFDM and single carrier modulation with frequency domain equalisation', IEEE Vehicular Technology Conference 1997, VTC-97, Phoenix, USA, pp. 863–869.
[3] Toskala, A., Castro, J., Chalard, L., Hämäläinen, S., Kalliojärvi, K., 'Cellular OFDM/CDMA Downlink Performance in the link and system level', IEE Vehicular Technology Conference 1997, VTC-97, Phoenix, USA, pp. 855–859.
[4] Holma, H., Toskala, A., 'WCDMA for UMTS', 4th edition, Wiley, 2007.
[5] Holma, H., Toskala, A., 'HSDPA/HSUPA for UMTS', Wiley, 2006.
[6] Toskala, A, Holma, H., Pajukoski, K., Tiirola, E., 'UTRAN Long Term Evolution in 3GPP'. The 17th Annual IEEE International Symposium on Personal, Indoor and Mobile Radio Communications (PIMRC '06), in Proceedings, 2006, Helsinki, Finland.
[7] Schulze, H., Luders, C., 'Theory and Applications of OFDMA and CDMA', Wiley, 2005.
[8] Myung, H.G., Goodman, D.J., 'Single Carrier FDMA: A New Air Interface for Long Term Evolution', Wiley, 2008.

5

Physical Layer

Antti Toskala, Timo Lunttila, Esa Tiirola, Kari Hooli, Mieszko Chmiel
and Juha Korhonen

5.1 Introduction

This chapter describes the physical layer of LTE, based on the use of OFDMA and
SC-FDMA principles as covered in Chapter 4. The LTE physical layer is characterized by
the design principle of not reserving dedicated resources for a single user; resource usage
is based solely on dynamically allocated shared resources. This is analogous to resource
usage in the internet, which is packet based without user-specific resource allocation.
The physical layer of a radio access system has a key role of defining the resulting
capacity and ends up being a focal point when comparing different systems in terms
of expected performance. However, a competitive system requires an efficient protocol
layer to ensure good performance all the way to the application layer and to the end
user. The flat architecture adopted, covered in Chapter 3, also enables the dynamic nature
of the radio interface as all radio resource control is located close to the radio in the
base-station site. The 3GPP term for the base station used in rest of this chapter will be
'eNodeB' (similar to the WCDMA BTS term, which is 'Node B', where 'e' stands for
'evolved'). This chapter first covers the physical channel structures and then introduces the
channel coding and physical layer procedures. The chapter concludes with a description
of physical layer measurements and device capabilities as well as with a brief look at
physical layer parameter configuration aspects. In 3GPP specifications the physical layer
was covered in the 36.2 series, with the four key physical layer specifications being [1–4].
Many of the issues in this chapter apply to both FDD and TDD, but in some areas TDD
receives special solutions due to the frame being divided between uplink and downlink.
The resulting differences needed for a TDD implementation are covered in Chapter 15.

5.2 Transport Channels and their Mapping to the Physical Channels

By the nature of the design already discussed, the LTE contains only common transport
channels; a dedicated transport channel (Dedicated Channel, DCH, as in WCDMA) does

LTE for UMTS: Evolution to LTE-Advanced, Second Edition. Edited by Harri Holma and Antti Toskala.
© 2011 John Wiley & Sons, Ltd. Published 2011 by John Wiley & Sons, Ltd.

not exist. The transport channels are the 'interface' between the MAC layer and the physical layer. In each transport channel, the related physical layer processing is applied to the corresponding physical channels used to carry the transport channel in question. The physical layer is required to have the ability to provide dynamic resource assignment both in terms of data-rate variance and in terms of resource division between different users. This section presents the transport channels and their mapping to the physical channels.

- The Broadcast Channel (BCH) is a downlink broadcast channel that is used to broadcast the necessary system parameters to enable devices accessing the system. Such parameters include, for example, the cell's bandwidth, the number of transmit antenna ports, the System Frame Number and PHICH-related configuration.
- The Downlink Shared Channel (DL-SCH) carries the user data for point-to-point connections in the downlink direction. All the information (either user data or higher layer control information) intended for only one user or UE is transmitted on the DL-SCH, assuming the UE is already in the RRC_CONNECTED state. However, as in LTE, the role of BCH is mainly to inform the device of the scheduling of the system information. Control information intended for multiple devices is also carried on DL-SCH. In case data on DL-SCH are only intended for a single UE, then dynamic link adaptation and physical layer retransmissions can be used.
- The Paging Channel (PCH) is used to carry paging information for the device in the downlink direction in order to move the device from the RRC_IDLE state to the RRC_CONNECTED state.
- The Multicast Channel (MCH) is used to transfer multicast service content to the UEs in the downlink direction. 3GPP decided to provide full support in Release 9 (for shared carrier case).
- The Uplink Shared Channel (UL-SCH) carries the user data as well as device-originated control information in the uplink direction in the RRC_CONNECTED state. As with the DL-SCH, dynamic link adaptation and retransmissions are available.
- The Random Access Channel (RACH) is used in the uplink to respond to the paging message or to initiate the move from the RRC_CONNECTED state due to UE data transmission needs. There is no higher layer data or user data transmitted on RACH (as can be done with WCDMA) but it is used to enable UL-SCH transmission where, for example, actual connection set up with authentication and so forth will take place.

In the uplink direction the UL-SCH is carried by the Physical Uplink Shared Channel (PUSCH). The RACH is carried by the Physical Random Access Channel (PRACH). Additional physical channels exist but these are used only for physical layer control information transfer as covered in section 5.6 on control information. Transport channel mapping is illustrated in Figure 5.1.

In the downlink direction, the PCH is mapped to the Physical Downlink Shared Channel (PDSCH). The BCH is mapped to Physical Broadcast Channel (PBCH) but, as is shown

Figure 5.1 Mapping of the uplink transport channels to the physical channels

Figure 5.2   Mapping of the downlink transport channels to the physical channels

in Chapter 6 for the mapping of logical channels to transport channels, only part of the broadcast parameters is on BCH while the actual System Information Blocks (SIBs) are then on DL-SCH. The DL-SCH is mapped to the PDSCH and MCH is mapped to Physical Multicast Channel, as shown in Figure 5.2.

5.3   Modulation

In the uplink direction, modulation is carried out through a more traditional QAM modulator, as was explained in Chapter 4. The modulation methods available (for user data) are QPSK, 16QAM and 64QAM. The first two are available in all devices while support for 64QAM in the uplink direction is a UE capability, as covered in section 5.10. The different constellations are shown in Figure 5.3

PRACH modulation is phase modulation as the sequences used are generated from Zadoff–Chu sequences with phase differences between different symbols of the sequences – see section 5.7 for further details. Depending on the sequence chosen, the resulting Peak-to-Average Ratio (PAR) or the more practical Cubic Metric (CM) value is somewhat above or below the QPSK value. In the uplink direction the signal cubic metric (CM) was discussed in Chapter 4 with SC-FDMA.

The use of QPSK modulation allows good transmitter power efficiency when operating at full transmission power, as modulation determines the resulting CM (in case of SC-FDMA) and thus also the required device amplifier back-off. Devices will use lower maximum transmitter power when operating with 16QAM or 64QAM modulation. The actual Maximum Power Reduction (MPR) also depends on the bandwidth (number of physical resource blocks used) and in some cases frequency-band-specific Additional MPR (A-MPR) should be applied as instructed by the network to deal with specific emission limits in some regions and countries.

In the downlink direction the modulation methods for user data are the same as in the uplink direction. In theory an OFDM system could use different modulation for

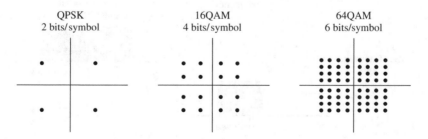

Figure 5.3   LTE modulation constellations

each sub-carrier. But to have channel quality information (and signaling) with such a granularity would not be feasible due to the resulting excessive overhead. If modulation was sub-carrier specific, one would have too many bits both in the downlink for informing the receiver of parameters for each sub-carrier and in the uplink the CQI feedback that would be needed in order to achieve sub-carrier level granularity in the adaptation would be too detailed.

Moreover, BPSK has been specified for control channels, which use either BPSK or QPSK for control information transmission. In a control channel, the modulation cannot be freely adapted because one needs to be able to receive it and a single signaling error must not prevent the detection of later control channel messages. This is similar to HSDPA/HSUPA where control channels have fixed parameterization to prevent error propagation due to frame-loss events. The exception is the uplink control data when multiplexed together with the user data. Here modulation for data and control is the same, even if 16QAM or 64QAM are utilized. This allows the multiplexing rules to be kept simpler.

5.4 Uplink User Data Transmission

The user data in the uplink direction is carried on the PUSCH. The PUSCH has the 10 ms frame structure and is based on the allocation of time and frequency domain resources with 1 ms and 180 kHz resolution. The resource allocation is coming from a scheduler located in the eNodeB, as illustrated in Figure 5.4. Thus there are no fixed resources for the devices; without prior signaling from the eNodeB only random access resources may be utilized. For this purpose the device needs to provide information for the uplink scheduler about its transmission needs (buffer status) as well as the available transmission power resources. This signaling is MAC layer signaling and is covered in detail in Chapter 6.

The frame structure adopts the 0.5 ms slot structure and uses the two-slot (one sub-frame) allocation period. The shorter 0.5 ms allocation period (as initially planned in 3GPP to minimize the round trip time) would have been too signaling intensive, especially with

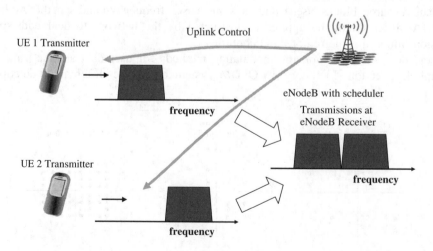

Figure 5.4 Uplink resource allocation controlled by eNodeB scheduler

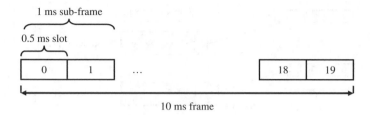

Figure 5.5   LTE FDD frame structure

a large number of users. The 10 ms frame structure is illustrated in Figure 5.5. The frame structure is basically valid for both for FDD and TDD, but TDD mode has additional fields for the uplink/downlink transition point(s) in the frame as covered in Chapter 15.

Within the 0.5 ms slot there are both reference symbols and user data symbols, in addition to the signaling, which is covered later. The momentary user data rate thus varies as a function of uplink resource allocation depending on the allocated momentary bandwidth. The allocation bandwidth may be between 0 and 20 MHz in steps of 180 kHz. The allocation is continuous, as uplink transmission is FDMA modulated with only one symbol being transmitted at the time. The slot bandwidth adjustment between consecutive TTIs is illustrated in Figure 5.6, where doubling the data rate results in double bandwidth being used. The reference symbols always occupy the same space in the time domain and thus a higher data rate results in a corresponding increase for the reference symbol data rate.

The cyclic prefix used in the uplink has two possible values depending on whether a short or extended cyclic prefix is applied. Other parameters stay unchanged and thus the 0.5 ms slot can accommodate either six or seven symbols as indicated in Figure 5.7. The data payload is reduced if an extended cyclic prefix is used, but in reality it is not going to be used too frequently as, in the majority of cases, the performance benefit in having seven symbols is far greater than possible degradation from the inter-symbol interference due to channel delay that is longer than the cyclic prefix.

The resulting instantaneous uplink data rate over a 1 ms sub-frame is a function of the modulation, the number of resource blocks allocated and the amount of control information overhead as well as the rate of channel coding applied. The instantaneous

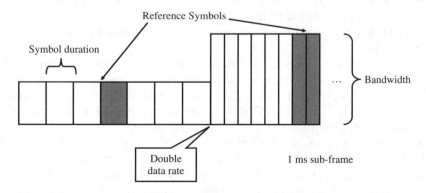

Figure 5.6   Data rate between TTIs in the uplink direction

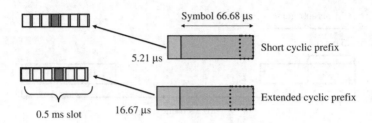

Figure 5.7   Uplink slot structure with short and extended cyclic prefix

uplink peak data rate range, when calculated from the physical layer resources, is between 700 kbps and 86 Mbps There is no multi-antenna uplink transmission, specified in Release 8, because using more than one transmitter branch in a UE is not seen as particularly attractive from the cost and complexity perspective. The instantaneous data rate for one UE in the LTE depends on:

- Modulation method applied, with 2, 4 or 6 bits per modulation symbol depending on the modulation order for QPSK, 16QAM and 64QAM respectively.
- Bandwidth applied. The momentary bandwidth may, of course, vary between the minimum allocation of 12 sub-carriers (one resource block of 180 kHz) and the system bandwidth, up to 1200 sub-carriers with 20 MHz bandwidth.
- Channel coding rate applied.
- The average data rate, then, also depends on the time domain resource allocation.

The cell or sector-specific maximum total data throughput can be increased with the Virtual Multiple Input Multiple Output (V-MIMO). In V-MIMO the eNodeB will treat transmission from two different UEs (with single transmit antenna each) as one MIMO transmission and separate the data streams from each other based on the UE specific uplink reference symbol sequences. Thus V-MIMO does not contribute to the single user maximum data rate. The maximum data rates, taking into account the UE categories, are presented in section 5.10, while the maximum data rates for each bandwidth are covered in Chapter 13.

The channel coding chosen for LTE user data was turbo coding. The encoder is parallel concatenated convolution coding (PCCC)-type turbo encoder, exactly the same as is being used in WCDMA/HSPA, as explained in [5]. The turbo interleaver was modified compared to that of WCDMA, to fit better LTE properties and slot structures and also to allow more flexibility for implementation in terms of parallel signal processing with increased data rates.

The LTE also uses physical layer retransmission combination, often referred to as Hybrid Automatic Repeat Request (HARQ). In physical layer HARQ operations the receiver stores the packets with failed CRC checks and combines the received packet when a retransmission is received. Both soft combining, with identical retransmissions, and combining with incremental redundancy are facilitated.

The channel coding chain for the uplink is shown in Figure 5.8, where the data and control information are separately coded and then mapped to separate symbols for transmission. As the control information has specific locations around the reference symbols, the physical layer control information is separately coded and placed in a predefined set of modulation symbols (but with the same modulation as data transmitted together). Thus

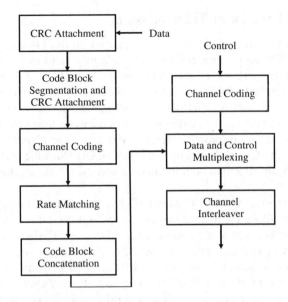

Figure 5.8 PUSCH channel coding chain

the channel interleaver in Figure 5.8 does not refer to true joint interleaving between control and data.

The data and control information are time multiplexed at the resource element level. Control is not evenly distributed but is intended to be either closest for the reference symbols in time domain or then filled in the top rows of Figure 5.9, depending on the type of control information, as covered in section 5.6. Data are modulated independently of control information, but modulation during the 1 ms TTI is the same.

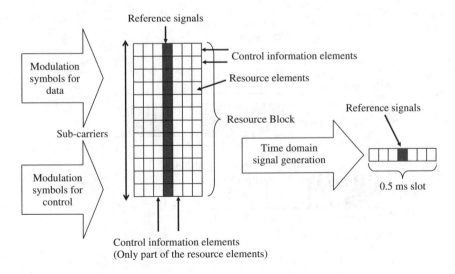

Figure 5.9 Multiplexing of uplink control and data

5.5   Downlink User Data Transmission

The user data rate in the downlink direction is carried on the Physical Downlink Shared Channel (PDSCH). The same 1 ms resource allocation is valid in the downlink direction as well and the sub-carriers are allocated in resource units of 12 sub-carriers, resulting in 180 kHz allocation units (physical resource blocks, PRBs). In the case of PDSCH, however, the multiple access is OFDMA, so each sub-carrier is transmitted as a parallel 15 kHz sub-carrier and thus the user data rate is dependent on the number of allocated sub-carriers (or resource blocks in practice) for a given user. The eNodeB does the resource allocation based on the Channel Quality Indicator (CQI) from the terminal. As with the uplink, the resources are allocated both in the time and the frequency domain, as illustrated in Figure 5.10.

The Physical Downlink Control Channel (PDCCH) informs the device about which resource blocks are allocated to it dynamically with 1 ms allocation granularity. The PDSCH data occupy between three and seven symbols per 0.5 ms slot depending on the allocation for PDCCH and depending on whether a short or extended cyclic prefix is used. Within the 1 ms sub-frame, only the first 0.5 ms slot contains PDCCH; the second 0.5 ms slot is purely for data (for PDSCH). For an extended cyclic prefix, six symbols are accommodated in the 0.5 ms slot, while a short cyclic prefix can fit seven symbols, as shown in Figure 5.11. The example in Figure 5.11 assumes three symbols for PDCCH but that can vary between one and three. With the smallest bandwidth of 1.4 MHz, the number of PDCCH symbols varies between two and four to enable sufficient signaling capacity and enough bits to allow good enough channel coding for range-critical cases.

In addition to the control symbols for PDCCH, space from the user data is reduced due to the reference signals, synchronization signals and broadcast data. As discussed in Chapter 4, due to the channel estimation it is beneficial when the reference symbols are distributed evenly in the time and frequency domains. This reduces the overhead needed, but requires some rules to be defined so that receiver and transmitter understand the resource mapping in similar manner. From the total resource allocation space over

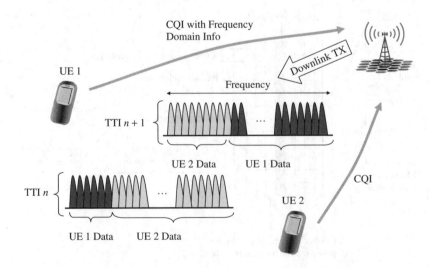

Figure 5.10   Downlink resource allocation at eNodeB

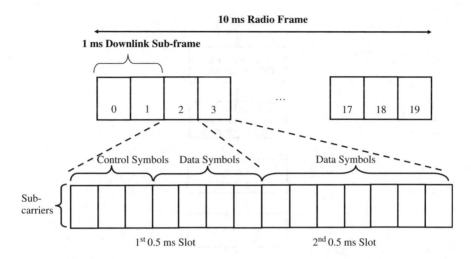

Figure 5.11 Downlink slot structure for bandwidths above 1.4 MHz

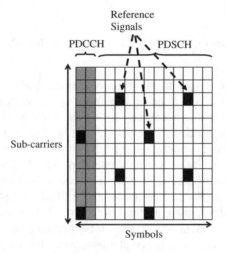

Figure 5.12 Example of downlink resource sharing between PDCCH and PDSCH

the whole carrier one needs to account for common channels, such as the Physical Broadcast Channel, which consume their own resource space. Figure 5.12 presents an example of PDCCH and PDSCH resource allocation.

The channel coding for user data in the downlink direction was also 1/3 mother rate turbo coding, as in the uplink direction. The maximum code block size for turbo coding is limited to 6144 bits to reduce the processing burden; higher allocations are then segmented to multiple code blocks. Higher block sizes would not add anything to performance as the turbo encoder performance improvement effect for big block sizes is saturated much earlier. In addition to turbo coding, the downlink uses physical layer HARQ with the same combining methods as in the uplink direction. The device categories reflect the amount

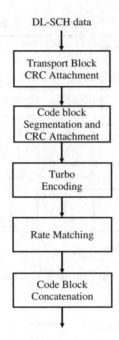

Figure 5.13   DL-SCH channel encoding chain

of soft memory available for retransmission combining. The downlink encoding chain is illustrated in Figure 5.13. There is no multiplexing to same physical layer resources with PDCCH as they have their own resources during the 1 ms sub-frame.

Once the data have been encoded, the code words are provided onwards for scrambling and modulation functionality. Scrambling in the physical layer should not be confused with the ciphering functionality; it is simply intended to avoid a wrong device successfully decoding the data if the resource allocation happens to be identical between cells. The modulation mapper applies the desired modulation (QPSK, 16QAM or 64QAM) and then symbols are fed for the layer mapping and precoding. Where there are multiple transmit antennas (two or four) the data are divided into the same number of different streams and then mapped to the correct resource elements available for PDSCH. Then the actual OFDMA signal is generated, as shown in Figure 5.14 with an example of two antenna transmissions. If there is only a single transmit antenna available then obviously the layer mapping and precoding functionalities do not have a role in signal transmission.

The downlink resulting instantaneous data rate depends on:

- Modulation – the same methods being possible as in the uplink direction.
- The allocated number of sub-carriers. In the 1.4 MHz case the overhead is the largest due to the common channels and synchronization signals. Note that in the downlink the resource blocks are not necessary having continuous allocation in frequency domain. The range of allocation is the same as in the uplink direction from 12 sub-carriers (180 kHz) up to the system bandwidth with 1200 sub-carriers.
- Channel encoding rate.
- Number of transmit antennas (independent streams) with MIMO operation.

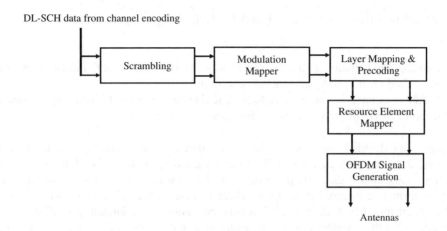

DL-SCH data from channel encoding

Figure 5.14 Downlink signal generation

The instantaneous downlink peak data rate (assuming all resources to a single user and counting only the physical layer resources available) ranges between 0.7 Mbps and 170 Mbps; 300 Mbps or higher could be expected using 4×4 antenna MIMO operation. There is no limit on the smallest data rate, and should the smallest allocation unit (one resource block) be too high then padding could be applied (in higher layers). Section 5.10 presents the maximum data rates taking the UE categories into account. The possible data rates for different bandwidth/coding/modulation combinations are presented in Chapter 10.

5.6 Uplink Physical Layer Signaling Transmission

Uplink L1/L2 control signaling is divided into two classes in the LTE system:

- Control signaling in the absence of UL data, which takes place on the PUCCH (Physical Uplink Control Channel).
- Control signaling in the presence of UL data, which takes place on PUSCH (Physical Uplink Shared Channel).

The simultaneous transmission of PUCCH and PUSCH is not allowed due to single carrier limitations. This means that separate control resources are defined for cases with and without UL data. Parallel transmission in the frequency domain (bad for the transmitter envelope) or pure time division (bad for control channel coverage) was considered as a possible alternative. The selected approach maximizes the link budget for PUCCH and maintains single carrier properties on the transmitted signal.

PUCCH is a shared frequency/time resource reserved exclusively for UEs transmitting only L1/L2 control signals. It has been optimized for a large number of simultaneous UEs with a relatively small number of control signaling bits per UE.

PUSCH carries the UL L1/L2 control signals in cases when the UE has been scheduled for data transmission. PUSCH is capable of transmitting control signals and supporting a large range of signaling sizes. Data and different control fields such as ACK/NACK and CQI are separated by means of TDM by mapping them into separate modulation symbols prior to the DFT. Different coding rates for control are achieved by occupying different number of symbols for each control field.

There are two types of uplink L1 and L2 control-signaling information as discussed in [6]:

- data-associated signaling (for example, transport format and HARQ information), which is associated with uplink data transmission; and
- non-data-associated signaling (ACK/NACK due to downlink transmissions, downlink CQI, and scheduling requests for uplink transmission).

It has been decided there is no data-associated control signaling in LTE UL. Furthermore, it is assumed that eNodeB is not required to perform blind transport format detection. This means that UE just obeys the UL scheduling grant with no freedom in transport format selection. Furthermore, there is a new data indicator (1 bit) included in the UL grant together with implicit information about the redundancy version [7]. This guarantees, that the eNodeB always has exact information about the UL transport format.

5.6.1 *Physical Uplink Control Channel, PUCCH*

From a single UE perspective, PUCCH consists of a frequency resource of one resource block (12 sub-carriers) and a time resource of one sub-frame. To handle coverage-limited situations, transmission of ACK/NACK spans the full 1 ms sub-frame. To support extreme coverage-limited cases it has been agreed that ACK/NACK repetition is supported in LTE UL. Slot-based frequency hopping on the band edges symmetrically over the center frequency is always used on PUCCH, as shown in Figure 5.15. Frequency hopping provides the necessary frequency diversity needed for delay-critical control signaling.

Different UEs are separated on PUCCH by means of Frequency Division Multiplexing (FDM) and Code Division Multiplexing (CDM). FDM is used only between the resource blocks whereas code division multiplexing is used inside the PUCCH resource block.

There are two ways to realize CDM inside the PUCCH resource block:

- by means of cyclic shifts of a CAZAC[1] sequence;
- by means of blockwise spreading with orthogonal cover sequences.

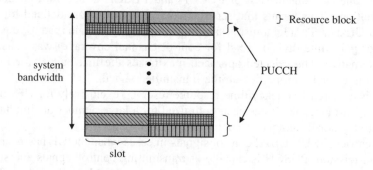

Figure 5.15   PUCCH resource

---

[1] The applied sequences are not true CAZAC but computer searched Zero-Autocorrelation (ZAC) sequences. The same sequences are applied as reference signals with bandwidth allocation of one resource block.

The main issue with CDM is the well-known near-far problem. Orthogonality properties of the considered CDM techniques were carefully studied during the Work Item phase of LTE standardization. We note that:

- channel delay spread limits the orthogonality between cyclically shifted CAZAC sequences;
- channel Doppler spread limits the orthogonality between blockwise spread sequences.

Orthogonality properties are optimized by means of staggered and configurable channelization arrangement (see more details in section 5.6.2.1), proper configuration of block spreading and a versatile randomization arrangement including optimized hopping patterns used for the control channel resources and the applied CAZAC sequences.

5.6.1.1 Sequence Modulation

Control signaling on PUCCH is based on sequence modulation. Cyclically shifted CAZAC sequences take care of CDM and convey control information. Figure 5.16 shows a block diagram of the sequence modulator configured to transmit periodic CQI on PUCCH. On the PUCCH application CAZAC sequences of length 12 symbols (1 RB) are BPSK or QPSK modulated thus carrying one or two information bits per sequence. Different UEs can be multiplexed into the given frequency/time resource by allocating different cyclic shifts of the CAZAC sequence for them. There are six parallel channels available per RB, assuming that every second cyclic shift is in use.

5.6.1.2 Block-wise Spreading

Block-wise spreading increases the multiplexing capacity of PUCCH by a factor of the spreading factor (SF) used. The principle of block-wise spreading is shown in Figure 5.17,

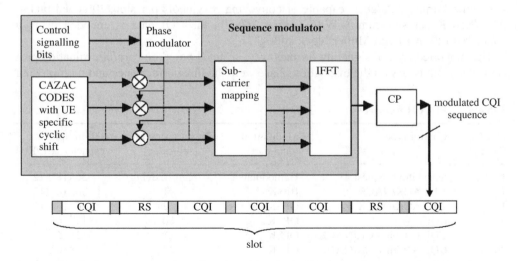

Figure 5.16 Block diagram of CAZAC sequence modulation applied for CQI

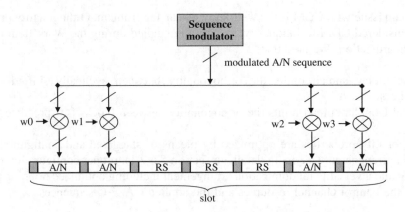

Figure 5.17 Principle of block spreading applied for ACK/NACK, spreading SF = 4

which illustrates the block spreading operation for the ACK/NACK data sequence transmitted on PUCCH. A separate block-spreading operation is made for the reference signal and ACK/NACK data parts but, for simplicity, block processing related to reference symbol (RS) part is omitted in Figure 5.17. In the example of Figure 5.17, the spreading factors applied to the ACK/NACK data and RS parts equal four and three, respectively. Walsh-Hadamard codes are used as block spreading codes with SF = 4 and SF = 2, whereas DFT codes are used when SF equals three.

5.6.1.3 PUCCH Formats

The available PUCCH formats have been summarized in Table 5.1. PUCCH Format 1/1a/1b is based on the combination of CAZAC sequence modulation and block-wise spreading whereas PUCCH Format 2/2a/2b utilizes only CAZAC sequence modulation. As a result, Format 1/1a/1b can only carry one information symbol (1 to 2 bits) per slot while Format 2/2a/2b is capable of conveying 5 symbols per slot (20 coded bits + ACK/NACK per sub-frame). With Format 2/2a/2b, the CQI data is encoded using a punctured (20, N) Reed-Muller block codes.

The supported control signaling formats were selected based on careful evaluation process. The main issues on the evaluation phase were the link performance and multiplexing

Table 5.1 PUCCH formats

PUCCH Formats	Control type	Modulation (data part)	Bit rate (raw bits/ subframe)	Multiplexing capacity (UE/RB)
1	Scheduling request	Unmodulated	−(on/off keying)	36, 18*, 12
1a	1-bit ACK/NACK	BPSK	1	36, 18*, 12
1b	2-bit ACK/NACK	QPSK	2	36, 18*, 12
2	CQI	QPSK	20	12, 6*, 4
2a	CQI + 1-bit ACK/NACK	QPSK	21	12, 6*, 4
2b	CQI + 2-bit ACK/NACK	QPSK	22	12, 6*, 4

*Typical value

capacity as well as the compatibility with other formats. It is also noted that the number of reference signal blocks were optimized separately for different formats.

Two different approaches were selected for signaling the ACK/NACK and CQI on PUCCH (Format 2a/2b):

- Normal cyclic prefix: ACK/NACK information is modulated in the second CQI reference signals of the slot. The RS modulation follows the CAZAC sequence modulation principle.
- Extended cyclic prefix: ACK/NACK bits and the CQI bits are jointly coded. No information is embedded in any of the CQI reference signals.

The main reason for having different solution with normal and extended cyclic prefix lengths was that with extended cyclic prefix there is only one reference signal per slot and hence the method used with the normal cyclic prefix cannot be utilized.

Support for Format 2a/2b is configurable in the LTE UL system. To guarantee ACK/NACK coverage, the eNodeB can configure a UE to drop the CQI in cases when ACK/NACK and CQI would appear in the same sub-frame on PUCCH. In this configuration, Format 1a/1b is used instead of Format 2a/2b.

5.6.1.4 Scheduling Request

One of the new features of the LTE uplink system is the fast uplink scheduling request mechanism's support for the active mode UEs (RRC_CONNECTED state) being synchronized by the eNodeB but having no valid uplink grant on PUSCH available. The supported scheduling request procedure is presented in Figure 5.18 [8].

The UE indicates the need for an uplink resource using the scheduling request indicator. During the Release 8 LTE standardization process, the contention-based synchronized RACH and non-contention-based scheduling request indicator mechanisms were compared. It was pointed out that the non-contention-based approach is more suitable for LTE UL due to the fact that it provides better coverage, lower system overhead and better delay performance than the contention-based approach [9].

The Scheduling Request Indicator (SRI) is transmitted using PUCCH Format 1. On-off keying-based signaling is applied with SRI; that is, only the positive SRI is transmitted. This is transmitted using the ACK/NACK structure [1], the only difference between the

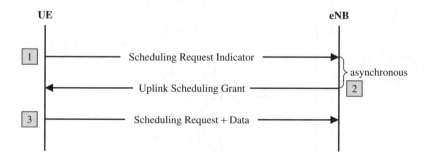

Figure 5.18   Scheduling request procedure

SRI and the ACK/NACK formats is that data part is not modulated with SRI. The benefit of this arrangement is that SRI and ACK/NACK can share the same physical resources.

5.6.2 PUCCH Configuration

Figure 5.19 shows the logical split between different PUCCH formats and the way in which the PUCCH is configured in the LTE specifications [1]. The number of resource blocks in a slot reserved for PUCCH transmission is configured by the N_{RB}^{HO} -parameter. This broadcast-system parameter can be seen as the maximum number of resource blocks reserved for PUCCH while the actual PUCCH size changes dynamically based on PCFICH transmitted on downlink control channel. The parameter is used to define the frequency-hopping PUSCH region. The number of resource blocks reserved for periodic CQI (i.e., PUCCH Format 2/2a/2b) is configured by another system parameter, $N_{RB}^{(2)}$.

In general it makes sense to allocate separate PUCCH resource blocks for PUCCH Format 1/1a/1b and Format 2/2a/2b. However, with narrow system bandwidth options such as a 1.4 MHz case, this would lead to unacceptably high PUCCH overhead [10]. Therefore, sharing the PUCCH resources block between Format 1/1a/1b and Format 2/2a/2b users is supported in the LTE specifications. The mixed resource block is configured by the

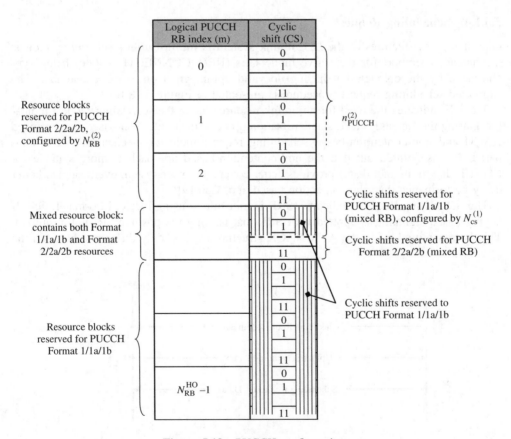

Figure 5.19   PUCCH configuration

broadcasted system parameter $N_{cs}^{(1)}$, which is the number of cyclic shifts reserved for PUCCH Format 1/1a/1b on the mixed PUCCH resource block.

Resources used for transmission of PUCCH formats 2/2a/2b are identified by a resource index $n_{PUCCH}^{(2)}$, which is mapped directly into a single CS resource. This parameter is explicitly signaled via UE-specific higher layer signaling.

5.6.2.1   Channelization and Resource Allocation for PUCCH Format 1/1a/1b

PUCCH Format 1/1a/1b resources are identified by a resource index $n_{PUCCH}^{(1)}$. Direct mapping between the PUCCH cyclic shifts and the resource indexes cannot be used with PUCCH Format 1/1a/1b due to the block-spreading operation. Instead, PUCCH channelization is used to configure the resource blocks for PUCCH Format 1/1a/1b. The purpose of channelization is to provide a number of parallel channels per resource block with optimized and adjustable orthogonality properties. The format 1/1a/1b channelization structure is configured by means of the broadcasted system parameter, *Delta_shift*.

The number of PUCCH format 1/1a/1b resources per resource block, denoted as $N_{PUCCH\ Format\ 1}^{RB}$, can be calculated as follows:

$$N_{PUCCH\ Format\ 1}^{RB} = \frac{N_{RS}^{PUCCH} * 12}{Delta\_shift}, \tag{5.1}$$

where *Delta_shift* -parameter is the cyclic shift difference between two adjacent ACK/NACK resources using the same orthogonal cover sequence [11] and the parameter N_{RS}^{PUCCH}, is the number of reference signals on PUCCH Format 1/1a/1b (N_{RS}^{PUCCH} equals to tree with normal CP and two with extended CP). Three values are allowed for the *Delta_shift* parameter, namely 1, 2, or 3. This means that the number of PUCCH Format 1/1a/1b resources per RB equals to 36, 18 or 12, with normal CP length.

An example of the PUCCH channelization within the resource block following the staggered resource structure is shown in Figure 5.20. In this example, *Delta_shift* is set to two and normal CP length is assumed.

Cyclic shift	Orthogonal cover code		
	0	1	2
0	0		12
1		6	
2	1		13
3		7	
4	2		14
5		8	
6	3		15
7		9	
8	4		16
9		10	
10	5		17
11		11	

Figure 5.20   The principle of Format 1/1a/1b channelization within one resource block, *Delta_ shift* = 2, normal CP

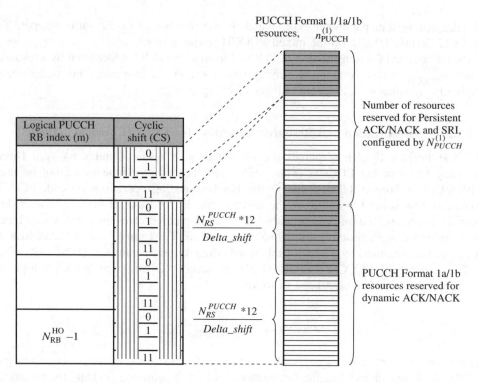

Figure 5.21   Configuration of PUCCH Format 1/1a/1b resource

The configuration of the PUCCH Format 1/1a/1b resource is shown in Figure 5.21. PUCCH Format 1/1a/1b resources are divided into available PUCCH resource blocks and are subject to channelization within the RB as described earlier. Before that, the Format 1/1a/1b resource is split into the persistent and dynamic parts. This is done using the broadcasted system parameter $N_{PUCCH}^{(1)}$, which is the number of resources reserved for persistent Format 1/1a/1b resources. These resources are used by the SRI and ACK/NACK related to persistently scheduled PDSCH. Both resources are allocated explicitly by means of resource index $n_{PUCCH}^{(1)}$. Dynamic Format 1/1a/1b resources are placed at the end of logical PUCCH resources. Allocation of these ACK/NACK resources, which relate to dynamically scheduled PDSCH, is made implicitly based on the PDCCH allocation.

The idea in implicit allocation of dynamic ACK/NACK resources is to have one-to-one mapping to the lowest PDCCH control channel element (CCE) index. The total number of CCEs depends on the system bandwidth and the number of OFDM symbols allocated for control signaling in a DL sub-frame, which is signaled in each sub-frame using PCFICH (1, 2, or 3 OFDM symbols/sub-frame for bandwidths above 1.4 MHz, for 1.4 MHz case 2, 3 or 4 OFDMA symbols may be occupied). There has to be a dedicated ACK/NACK resource for each CCE. This means, for example, that with 20 MHz system bandwidth the number of CCE can be up to 80 if three OFDM symbols are allocated for control signaling in a sub-frame.

5.6.2.2 Mapping of Logical PUCCH Resource Blocks onto Physical PUCCH Resource Blocks

Mapping of logical resource blocks, denoted by m, into physical PUCCH resource blocks is shown in Figure 5.22. Taking into account the logical split between different PUCCH Formats, we note that PUCCH Format 2/2a/2b is located at the outermost resource blocks of the system bandwidth. ACK/NACK is reserved for persistently scheduled PDSCH and SRI is located on the PUCCH resource blocks next to periodic CQI while the ACK/NACK resources reserved to dynamically scheduled PDSCH are located at the innermost resource blocks reserved for PUCCH.

An interesting issue from the system perspective is the fact that PUCCH defines the uplink system bandwidth. This is due to the fact that PUCCH always exists and it is located at both edges of the frequency spectrum. We note that proper PUCCH configuration allows the active uplink system bandwidth to be narrowed down by the resolution of two resource blocks. This can be done in such a way that PUCCH Format 2/2a/2b resource is over-dimensioned and at the same time the pre-defined PUCCH Format 2/2a/2b resources placed at the outermost RBs are left unused. Figure 5.23 shows the principle of this configuration.

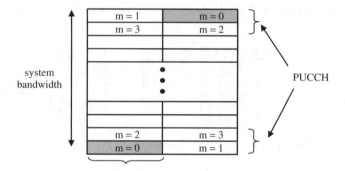

Figure 5.22   Mapping of logical PUCCH RBs onto physical RBs [1]

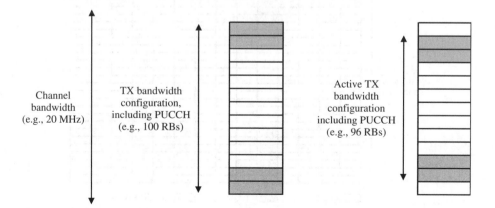

Figure 5.23   Changing the UL system bandwidth via the PUCCH configuration

5.6.3 Control Signaling on PUSCH

PUSCH carries the uplink L1/L2 control signals in the presence of uplink data. Control signaling is realized by a dedicated control resource, which is valid only during the uplink sub-frame when UE has been scheduled for data transmission on PUSCH. The main issues related to control signal design on PUSCH are

- how to arrange multiplexing between uplink data and different control fields; and
- how to adjust the quality with L1/L2 signals transmitted on PUSCH.

Figure 5.24 shows the principle of control and data multiplexing within the SC-FDMA symbol (block). In order to maintain the single carrier properties of transmitted signal data and different control symbols are multiplexed before the DFT. Data and different control fields (ACK/NAK, CQI/PMI, Rank Indicator) are coded and modulated separately before they are multiplexed into the same SC-FDMA symbol block. Block-level multiplexing was also considered but would have resulted in large control overheads [12]. Using the selected symbol level multiplexing scheme the ratio between the data symbols and control symbols can be accurately adjusted within each SC-FDMA block.

Figure 5.25 shows the principle behind the way in which uplink data and different control fields are multiplexed on PUSCH. The actual mix of different L1/L2 control

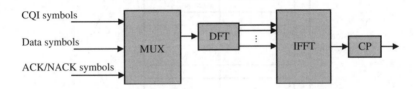

Figure 5.24   Principle of data and control modulation

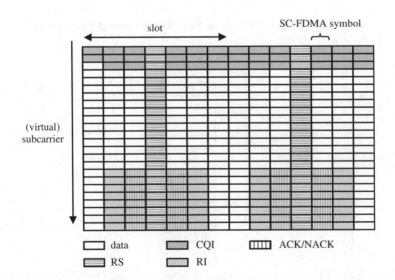

Figure 5.25   Allocation data and different control fields on PUSCH

signals and their size varies from sub-frame to sub-frame. Both the UE and the eNodeB have the knowledge about the number of symbols reserved by the control part. The data part of PUSCH is punctured/rate matched according to the number of control symbols allocated in the given sub-frame.

Control and data multiplexing are carried out in such a way that control is present at both slots of the sub-frame. This guarantees that control channels can benefit from the frequency hopping when it is applied. ACK/NACK is placed at end of SC-FDMA symbols next to the reference signals. There can be at most two SC-FDMA symbols per slot allocated to ACK/NACK signaling. The same applies to Rank Indicator, which is placed on the SC-FDMA symbols next to ACK/NACK. CQI/PMI symbols are placed at the beginning of the SC-FDMA symbols and they are spread over all the available SC-FDMA symbols.

CQI/PMI transmitted on PUSCH uses the same modulation scheme as the data part. ACK/NACK and Rank Indicator are transmitted in such a way that the coding, scrambling and modulation maximize the Euclidean distance at the symbol level. This means that a modulation symbol used for ACK/NACK carries at most two bits of coded control information regardless of the PUSCH modulation scheme. The outermost constellation points with the highest transmission power are used to signal the ACK/NACK and RI in the case of 16QAM and 64QAM. This selection provides a small power gain for ACK/NACK and RI symbols, compared to PUSCH data using higher order modulation.

Four different channel coding approaches are applied with control signals transmitted on PUSCH:

- repetition coding only: 1-bit ACK/NACK;
- simplex coding: 2-bit ACK/NACK/RI;
- (32, N) Reed-Muller block codes: CQI/PMI < 11 bits;
- tail-biting convolutional coding (1/3): CQI/PMI ≥ 11 bits.

An important issue related to control signaling on PUSCH is how to keep the performance of control signaling at the target level. Power control will set the SINR target of PUSCH according to the data channel, so the control channel has to adapt to the SINR operation point set for data.

One way to adjust the available resources would be to apply different power offset values for data and different control parts. The problem of the power offset scheme is that single carrier properties are partially destroyed [13], so this scheme is not used in the LTE UL system. Instead, a scheme based on a variable coding rate for the control information is used. This is achieved by varying the number of coded symbols for control channel transmission. In order to minimize the overall overhead from the control signaling the size of physical resources allocated to control transmission is scaling according to PUSCH quality. This is realized in such a way that the coding rate to use for the control signaling is given implicitly by the Modulation and Coding Scheme (MCS) of PUSCH data. The linkage between data MCS and the size of the control field is depending on the number of controlling signaling bits and on the semi-statically configured offset parameter related to coding rate adjustment of control channel and is used to achieve desired B(L)ER operation point for a given control signalling type. Equations are given in [14] to determine the amount of coded symbols needed, covering both the ACK/NACK feedback and rank indication case as well as for the channel quality feedback transmission.

Offset-parameter is used to adjust the quality of control signals with respect to the PUSCH data channel. It is a UE-specific parameter configured by higher layer signaling. Different control channels need their own offset-parameter setting. There are some issues that need to be taken into account when configuring the offset-parameter:

- the BLER operation point for the PUSCH data channel;
- the B(L)ER operation point for the L1/L2 control channel;
- the difference in coding gain between control and data parts, due to different coding schemes and different coding block sizes (no coding gain with 1-bit ACK/NACK);
- the DTX performance.

Different BLER operation points for data and control parts occur due to the fact that HARQ is used for the data channel whereas control channels do not benefit from HARQ. The greater the difference in BLER operation point between data and control channels the larger is the offset parameter (and vice versa). Similar behavior also relates to packet size. The highest offset values are needed with ACK/NACK signals due to the lack of coding gain.

5.6.4 Uplink Reference Signals

In addition to the control and data signaling, there are reference signals as discussed in Section 5.4. The eNodeB needs to have some source of know data symbols to facilitate coherent detection, like in the WCDMA where uplink Physical Dedicated Control Channel (PDCCH) was carrying pilot symbols in the uplink direction. In LTE uplink, reference signals (RS) are used as demodulation reference signals (DM RS) on PUCCH and PUSCH. The new purpose of the reference signals, not part of WCDMA operation, is to use them as sounding reference signals (SRS). Additionally, reference signals are used for sequence modulation on PUCCH as discussed in section 5.6.1.1. The sequences used as reference signals are discussed in section 5.6.4.1, while demodulation reference signals and sounding reference signals are considered in sections 5.6.4.2 and 5.6.4.3 respectively.

5.6.4.1 Reference Signal Sequences

The most important properties for the RS sequences in LTE uplink are:

- favourable auto- and cross-correlation properties;
- sufficient number of sequences;
- flat frequency domain representation facilitating efficient channel estimation;
- low cubic metric values comparable to the cubic metric of QPSK modulation.

The sequences need also to be suitable for supporting the numerous bandwidth options in uplink allocations. This means that sequences of various lengths, multiples of 12, are needed.

Constant Amplitude Zero Autocorrelation Codes (CAZAC) such as Zadoff–Chu [15] and Generalized Chirp-Like [16] polyphase sequences have most of the required properties. There exist also a reasonable number of Zadoff–Chu sequences when the sequence length is a prime number. However, the sequence lengths needed in LTE UL are multiples of 12, for which only a modest number of Zadoff–Chu sequences exist. To obtain a sufficient number of RS sequences, computer-generated sequences are used for sequence lengths of 12 and 24. They are constructed from the QPSK alphabet in the frequency domain. Longer sequences are derived from Zadoff–Chu sequences where the length is a prime number. They are circularly extended in frequency domain to the desired length. These sequences are frequently referred to as extended Zadoff–Chu sequences.

As a result, there are 30 reference signal sequences available for sequence lengths of 12, 24, and 36 and a larger number for longer sequence lengths. The RS sequences do not have a constant amplitude in time and, thus, they are not actually CAZAC sequences. However, they have acceptable cubic metric values and zero autocorrelation and, thus, may be referred to as Zero Autocorrelation (ZAC) sequences.

The RS sequences have periodic autocorrelation function that is zero except for the zero shift value. In other words, the cyclic, or circular, shifts of a sequence, illustrated in Figure 5.26, are orthogonal to each other. This provides a convenient way to derive multiple orthogonal sequences from a single RS sequence, which is used in LTE to multiplex UEs. However, to maintain the orthogonality, the time difference between the signals arriving at the base station should not exceed the time interval corresponding to the cyclic shift separation. To accommodate the multipath delay spread, the minimum time separation between the cyclic shifts available in LTE is 5.56 μs for demodulation RS and 4.17 μs for sounding RS. Correspondingly, 12 and eight cyclic shifts are specified for DM RS and SRS, respectively, with constant time separation between cyclic shifts irrespective of the reference signal bandwidth.

5.6.4.2 Demodulation Reference Signals

Demodulation reference signals are primarily used for channel estimation. They are needed for coherent detection and demodulation and have the same bandwidth as the uplink data transmission. There is one demodulation reference signal in every 0.5 ms slot on PUSCH, whereas on PUCCH, there are two or three reference signals per slot depending on the PUCCH format used. For PUSCH, DM RS occupies the fourth SC-FDMA symbol in the slot, and the RS sequence length equals the number of allocated sub-carriers. The reference signal locations for PUCCH are discussed in section 5.6.1.

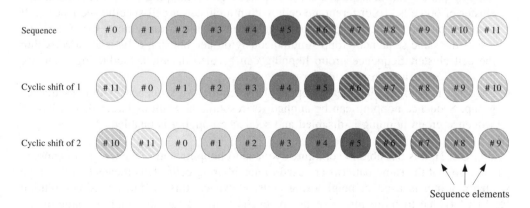

Figure 5.26   Cyclic shifts of a sequence

There are RS sequences of numerous lengths with 30 or more sequences for each sequence length. For simplicity, sequences of various lengths are arranged into groups of 30 sequences. Each sequence group contains RS sequences for all supported sequence lengths, with one RS sequence for allocations from one to five PRBs and two RS sequences for larger allocations. As a result, the used demodulation reference signal is defined with four parameters:

- Sequence group, with 30 options. This is a cell-specific parameter.
- Sequence, with two options for sequence lengths of 6 PRBs or longer. This is a cell-specific parameter.
- Cyclic shift, with 12 options. This has both terminal- and cell-specific components, and eight different values can be configured with the uplink allocation.
- Sequence length, given by the uplink allocation.

Cyclic shifts are used to multiplex reference signals from different terminals within a cell, whereas different sequence groups are typically used in neighboring cells. Cyclic shifts provide good separation within a cell but a more complicated interference scenario occurs between cells. Simultaneous uplink allocations on neighboring cells can have different bandwidths and can be only partially overlapping in frequency. This alone prevents effective optimization of RS cross-correlations between cells. Thus, multiple hopping methods are included to LTE to randomize inter-cell interference for reference signals. The pseudo-random hopping patterns are cell specific and derived from the physical layer cell identity. LTE supports for PUSCH and PUCCH:

- Cyclic shift hopping, which is always used. A cell-specific cyclic shift is added on top of UE-specific cyclic shifts. Cyclic shift hops for every slot on PUSCH. Inter-cell interference is expected to be more significant on PUCCH than on PUSCH due to the CDM applied on PUCCH. To enhance inter-cell interference randomization, cyclic shift hops for every SC-FDMA symbol on PUCCH. Cyclic shift hopping is applied also for SC-FDMA symbols carrying control data due to the sequence modulation used on PUCCH.
- Sequence group hopping. A sequence group hopping pattern is composed of a group hopping pattern and a sequence shift. The same group hopping pattern is applied to a cluster of 30 cells. To differentiate cells within a cluster, a cell-specific sequence shift is added on top of the group hopping pattern. With this arrangement, the occasional use of the same sequence group simultaneously on neighboring cells is avoided within the cell cluster. Sequence group hopping can be also disabled, facilitating sequence planning. Sequence group hops for every slot.
- Sequence hopping, which means hopping between the two sequences within a sequence group. Sequence hopping can be applied for resource allocations larger than 5 RBs if sequence group hopping is disabled and sequence hopping is enabled.

On PUSCH, it is possible to configure cyclic shift hopping and sequence group hopping patterns so that the same patterns are used in neighboring cells. This means that the same sequence group is used in neighboring cells. However, this is not a feasible solution for PUCCH due to more intensive use of cyclic shifts. Thus, the hopping patterns are cell specific and are derived from the cell identity on PUCCH. Therefore, the sequence group is configured separately for PUCCH and PUSCH, with an additional configuration parameter for PUSCH.

5.6.4.3 Sounding Reference Signals

Sounding Reference Signal (SRS) is used to provide information on uplink channel quality on a wider bandwidth than the current PUSCH transmission or when terminal has no transmissions on PUSCH. The channel is estimated on eNodeB and the channel information obtained can be used in the optimization of uplink scheduling as part of the uplink frequency domain scheduling operation. Thus, SRS is in a sense an uplink counterpart for the CQI reporting of downlink channel. SRS can be used also for other purposes, for example, to facilitate uplink timing estimation for terminals with narrow or infrequent uplink transmissions. Sounding reference signal is transmitted on the last SC-FDMA symbol of the sub-frame, as shown in Figure 5.27. Note that the SRS transmission does not need to be in the frequency area used by the PUSCH for actual data transmission but it may locate elsewhere as well.

On sounding reference signal, distributed SC-FDMA transmission is used. In other words, UE uses every second sub-carrier for transmitting reference signal as illustrated in Figure 5.28. The related sub-carrier offset defines a transmission comb for the distributed transmission. The transmission comb provides another mean to multiplex UE reference signals in addition to the cyclic shifts. Sounding reference signal uses the same sequences as demodulation reference signals. SRS sequence lengths are multiples of 24, or, correspondingly, SRS bandwidths are multiples of 4 RBs. This follows from the available RS sequence lengths combined with definition of eight cyclic shifts for SRS.

SRS transmissions can be flexibly configured. An SRS transmission can be a single transmission or transmissions can be periodic with the period ranging from 2 ms to 320 ms. There can be up to four different SRS bandwidth options available, depending on the system bandwidth and cell configuration. SRS transmission can also hop in frequency. This is particularly beneficial for the terminals on the cell edge, which cannot support wideband SRS transmissions. Frequency hopping can also be limited to a certain portion of the system bandwidth, which is beneficial for inter-cell interference coordination [17]. SRS configuration is explicitly signaled via terminal specific higher layer signaling.

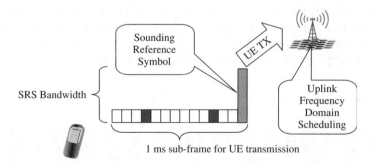

Figure 5.27   Sounding reference signal transmission in the frame

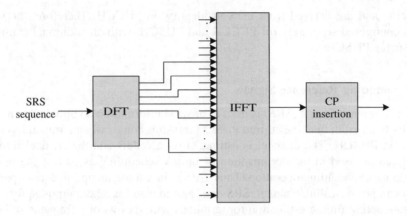

Figure 5.28   Sub-carrier mapping for sounding reference signal

Sounding reference signal transmissions from different terminals can be multiplexed in multiple dimensions:

- In time; periodic SRS transmissions can be interleaved into different sub-frames with sub-frame offsets.
- In frequency. To facilitate frequency division multiplexing, the available SRS bandwidths follow a tree structure. This is illustrated in Figure 5.29, where a set of available SRS bandwidths is shown for a certain cell configuration. SRS frequency hopping pattern follows also the tree structure, as shown in Figure 5.30 with an illustrative example based on the SRS bandwidth set of Figure 5.29.
- With cyclic shifts; up to eight cyclic shifts can be configured. However, the cyclic shift multiplexed signals need to have the same bandwidth to maintain orthogonality. Due to the intensive use of cyclic shifts, the sequence group configured for PUCCH is used also for SRS.
- Transmission comb in the distributed transmission; two combs are available. Contrary to cyclic shifts, the transmission comb does not require the multiplexed signals to occupy the same bandwidth.

In addition to the terminal-specific SRS configuration, the cell-specific SRS configuration defines the sub-frames, which can contain SRS transmissions as well as the set of SRS bandwidths available in the cell. Typically SRS transmissions should not extend into the frequency band reserved for PUCCH. Therefore, multiple SRS bandwidth sets are needed for supporting flexible cell specific PUCCH configuration.

10 MHz bandwidth

PRB index

0 1 2 3 4 5 6 7 8 9 10 11 12 13 14 15 16 17 18 19 20 21 22 23 24 25 26 27 28 29 30 31 32 33 34 35 36 37 38 39 40 41 42 43 44 45 46 47 48 49

SRS BW options

40 RBs									
20 RBs					20 RBs				
4 RBs	4 RBs	4 RBs	4 RBs	4 RBs	4 RBs	4 RBs	4 RBs	4 RBs	4 RBs

Figure 5.29   A SRS bandwidth set with a tree structure

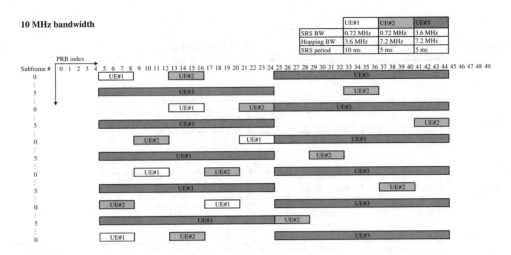

Figure 5.30   An example on SRS frequency-hopping patterns

5.7   PRACH Structure

5.7.1   Physical Random Access Channel

Random access transmission is the only non-synchronized transmission in the LTE uplink. Although terminal synchronizes to the received downlink signal before transmitting on RACH, it cannot determine its distance from base station. Thus, timing uncertainty caused by two-way propagation delay remains on RACH transmissions.

An appropriately designed Physical Random Access Channel (PRACH) occurs reasonably frequently, provides a sufficient number of random access opportunities, supports the desired cell ranges in terms of path loss and uplink timing uncertainty, and allows for sufficiently accurate timing estimation. Additionally, PRACH should be configurable to a wide range of scenarios, both in terms of RACH load and physical environment. For example, LTE is required to support cell ranges up to 100 km, which translates to 667 μs two-way propagation delay, as facilitated in the timing advance signaling range in MAC layer.

In LTE frame structure type 1 (FDD), only one PRACH resource can be configured into a sub-frame. The periodicity of PRACH resources can be scaled according to the expected RACH load, and PRACH resources can occur from every sub-frame to once in 20 ms. PRACH transmission consists of a preamble sequence and a preceding cyclic prefix with four different formats, shown in Figure 5.31. Multiple preamble formats are needed due to the wide range of environments. For example, the long CP in preamble formats 1 and 3 facilitate increased timing uncertainty tolerance for large cell ranges whereas the repeated preamble sequence in formats 2 and 3 compensates for increased path loss. The guard period, which is necessary after the unsynchronized preamble, is not explicitly specified but the PRACH location in the sub-frame structure provides a sufficient guard period. Particular considerations are needed only in very special cases. For each cell, 64 preamble sequences are configured and so there are 64 random access opportunities per PRACH resource. PRACH occupies the 1.08 MHz bandwidth, which provides reasonable resolution for timing estimation.

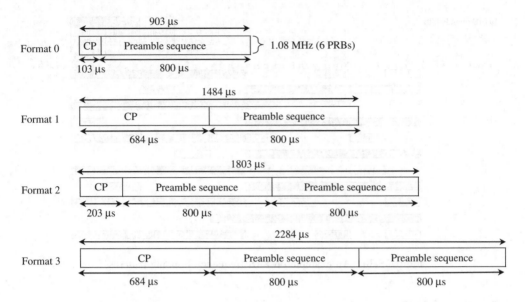

Figure 5.31 LTE RACH preamble formats for FDD

5.7.2 Preamble Sequence

Zadoff–Chu sequences [15] belonging to constant amplitude zero autocorrelation codes (CAZAC) are used as RACH preamble sequences due to several desirable sequence properties:

- They have a periodic autocorrelation function that is zero except for the zero shift value. This is desirable for preamble detection and timing estimation. The cyclic, or circular, shifts of a sequence are also orthogonal to each other. As a result, multiple preamble sequences are obtained from a single Zadoff–Chu sequence with cyclic shifts.
- As the preamble sequence length is set to a prime number of 839, there are 838 sequences with optimal cross-correlation properties.
- Sequences also have reasonable cubic metric properties.

The cyclic shifts used as different preambles need to have a sufficient separation. The cyclic shift separation need to be sufficient wide to accommodate for the uplink timing uncertainty as illustrated in Figure 5.32. The propagation delay and, thus, the cyclic separation are directly related to the cell range. To facilitate for the wide range of cell ranges supported by LTE, 16 different cyclic shift separations can be configured for a cell, providing 1 to 64 preambles from a single Zadoff–Chu sequence.

A particular high speed mode, or a restricted set, is defined for RACH due to the peculiar properties of Zadoff–Chu sequences in case of Doppler spread. Essentially, high-speed mode is a set of additional restrictions on the cyclic shifts that can be used as preambles.

The Discrete Fourier transform of the Zadoff–Chu sequence is also a Zadoff–Chu sequence in the frequency domain. Due to the Doppler, the transmitted preamble sequence spreads on cyclic shifts adjacent in frequency to the transmitted cyclic shift and particularly on the cyclic shift neighboring the transmitted cyclic shift in frequency. In particular, a Doppler frequency corresponding to the inverse of sequence length −1.25 kHz for

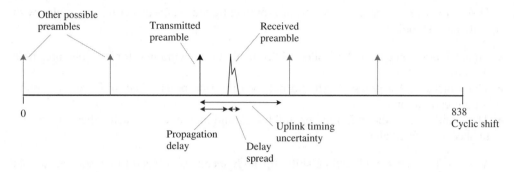

Figure 5.32   Illustration of cyclic shift separation N_{CS} between preamble sequences

LTE – transforms the transmitted cyclic shift (of sequence) completely to the cyclic shift neighboring the transmitted one in frequency. As a result, the received sequence is orthogonal with the transmitted sequence for Doppler frequency of 1.25 kHz.

There is a tractable one-to-one relation between the cyclic shifts neighboring each other in frequency and the cyclic shifts in time. Thus, three uplink timing uncertainty windows can be defined for a preamble, with the windows separated by a shift distance as shown in Figure 5.33. The shift distance is a function of the Zadoff–Chu sequence index. The shifting of the timing uncertainty windows is circular, as illustrated in Figure 5.33.

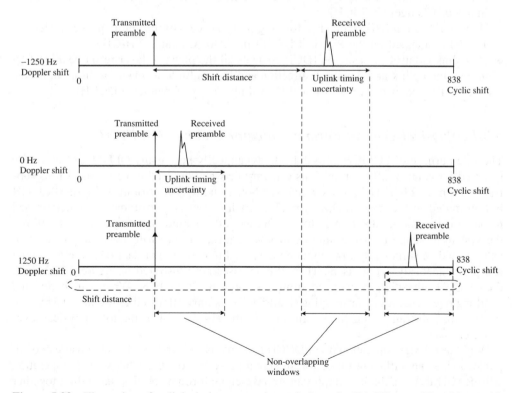

Figure 5.33   Illustration of uplink timing uncertainty windows for RACH preamble with considerable Doppler

There are several consequences from aforementioned when detecting preambles with considerable Doppler:

- Signal energy needs to be collected from all three windows for reliable preamble detection.
- The windows of a preamble should not overlap each other to allow for initial uplink timing estimation.
- The windows of the different preambles should not overlap each other to prevent unnecessary false alarms.

As a result, the preamble cyclic shifts for high speed mode need to be selected so that the timing uncertainty windows do not overlap each other for each preamble as well as between preambles. Although these requirements as well as the dependency of the shift distance on the sequence index complicate the calculation of cyclic shift for preambles, tractable equations were found and standardized [18].

5.8 Downlink Physical Layer Signaling Transmission

The control information in the downlink direction is carried using three different types of control messages:

- The Control Format Indicator (CFI), which is indicating the amount of resources devoted to control channel use. The CFI is mapped to the Physical Control Format Indicator Channel (PCFICH).
- HARQ Indication (HI), which is informing of the success of the uplink packets received. The HI is mapped on the Physical HARQ Indicator Channel (PHICH).
- Downlink Control Information (DCI) controls all the physical layer resource allocation in both the uplink and downlink direction and has multiple formats for different needs. The DCI is mapped on the Physical Downlink Control Channel (PDCCH).

5.8.1 Physical Control Format Indicator Channel (PCFICH)

The sole purpose of PCFICH is to indicate dynamically how many OFDMA symbols are reserved for control information. This can vary between one and three for each 1 ms sub-frame. From PCFICH UE knows which symbols to treat as control information. PCFICH is transmitted in the first OFDM symbol of each downlink sub-frame in predetermined resource elements, modulation is fixed. The use of dynamic signaling capability allows the system to support a large number of low data-rate users (VoIP for example) as well as to provide a sufficiently low signaling overhead in case higher data rates are used by a few simultaneously active users. The extreme cases are illustrated in Figure 5.34, where the PDCCH allocation is changed from 1 symbol to 3 symbols. When calculating the resulting overhead, it is worth noting that PDCCH is only allocated to the first 0.5 ms slot in the 1 ms sub-frame, thus the overhead is from 1/14 to 3/14 of the total physical layer resource space.

With the 1.4 MHz operation, the PDCCH resource is 2, 3 or 4 symbols to unsure enough payload size and sufficient range for all signaling scenarios. Especially for the operation with RACH in big cells it is important to have enough room for channel coding together with signaling.

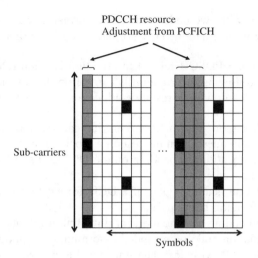

Figure 5.34 PDCCH resource allocation from PCFICH for bandwidths above 1.4 MHz

Table 5.2 PDCCH formats and their size

PDCCH format	Number of CCEs	Number of resource-element groups	Number of PDCCH bits
0	1	9	72
1	2	18	144
2	4	36	288
3	8	72	576

5.8.2 Physical Downlink Control Channel (PDCCH)

The UE will obtain from the Physical Downlink Control Channel (PDCCH) information for both uplink and downlink resource allocations the UE may use. The DCI mapped on the PDCCH has different formats and depending on the size and required coverage DCI is transmitted using one or more Control Channel Elements (CCEs). CCE is equal to nine resource elements groups. Each group in turn consists of four resource elements. The different PDCCH formats are shown in Table 5.2, where it can be seen that as PDCCH is using QPSK modulation then a single resource element carries two bits and there are eight bits in a resource element group.

The UE will listen to the set of PDCCHs and tries to decode them (checking all monitored formats) in all sub-frames except during those where DRX is configured. The set of PDCCHs to monitor is up to six channels. Depending on the network parameterization, some of the PDCCHs are so-called common PDCCHs and may then contain also power control information.

The DCI mapped to PDCCH has four different formats and further different variations for each format. It may provide the control information for the following cases:

- PUSCH allocation information (DCI Format 0);
- PDSCH information with one codeword(DCI Format 1 and its variants);

- PDSCH information with two codewords (DCI Format 2 and its variants);
- uplink power control information (DCI Format 3 and its variants).

The PDCCH containing PDSCH related information is often referred to as the downlink assignment. The following information is carried on the downlink assignment when providing PDSCH-related downlink resource-allocation information:

- Resource block allocation information. This indicates the position of the resources allocated for the user in question in the resource block domain. There are three types of downlink resource allocation. Depending on the DCI format either resource allocation type 0/1 or type 2 is used.
- The resource allocation header is part of a DCH format to provide information about the type of DCI allocation being used.
- The modulation and coding scheme used for downlink user data. The five-bit signaling indicates the modulation and coding index determining the modulation order and the Transport Block Size (TBS) index. The transport block size can be derived based on these parameters and the number of allocated resource blocks.
- The HARQ process number needs to be signaled as the HARQ retransmission, from the eNodeB's point of view, is asynchronous and the exact transmission instant depends on the eNodeB's scheduler functionality. Without the process number UE could confuse the different processes and combine wrong data. This also prevents error propagation from this part if the control signaling for a single TTI is lost. The number of HARQ processes was fixed at eight both in the uplink and the downlink.
- A new data indicator to tell whether the transmission for the particular process is a retransmission or not. This follows similar principles to those applied with HSDPA.
- Redundancy version is a HARQ parameter that can be used with incremental redundancy to tell which retransmission version is used.
- The power control commands for the PUCCH are included on the PDCCH as well. The power control command has two bits and it can thus use two steps up and downwards to adjust the power.
- Flag for the use of localized or distributed Virtual Resource Blocks (VRB) are used in some DCH formats. Localized VRB refers to a continuous allocation while with distributed VRB there may be one or two gaps (two only with larger bandwidths) with the mapping on the actual physical resource blocks.

When MIMO operation is involved, there are MIMO-specific signaling elements involved, as discussed in section 5.9.7.

The PDCCH containing PUSCH-related information is also know as the uplink grant. The following information is carried on the uplink grant:

- Flag for differentiation between an uplink grant and a compact downlink assignment.
- Hopping flag and resource block assignment and hopping resource allocation. The number of bits for this depends on the bandwidth to be used. Uplink resource allocation is always contiguous and it is signaled by indicating the starting resource block and the size of the allocation in terms of resource blocks.
- Modulation and coding scheme and the redundancy version.
- New data indicator, which is intended to be used for synchronizing the scheduling commands with the HARQ ACK/NACK message status.

- TPC command for the scheduled PUSCH, which can represent four different values.
- Cyclic shift for the Demodulation Reference Symbols – three bits.
- Aperiodic CQI report request.

Besides these purposes, the PDCCH can carry also power control information for several users. The options supported are both the single-bit and two-bit formats. The maximum number of users listening to a single PDCCH for power control purposes is N. Higher layers indicate to the device which bit or bits it needs to receive for the power control purposes.

As mentioned earlier, the PCFICH indicated the amount of resources reserved for the PDCCH on a 1 ms basis. The UEs in the cell check the PCFICH and then blindly decode the PDCCH to see which of the control channels (if any) is intended for them. The user identification is based on using UE-specific CRC, which is generated after normal CRC generation by masking the CRC with the UE ID. The PDCCH channel coding is based on the same 1/3-mother rate convolutional coding as other convolutionally encoded channels. The PDCCH encoding chain is illustrated in Figure 5.35.

5.8.3 Physical HARQ Indicator Channel (PHICH)

The task for the Physical HARQ Indicator Channel (PHICH) is simply to indicate in the downlink direction whether an uplink packet was correctly received or not. The device will decode the PHICH based on the uplink allocation information received on the PDCCH.

The PHICH transmission duration and the number of PHICH groups are configured via PBCH signaling. The PHICH transmission is BPSK modulated and Walsh sequences are used to separate different ACK/NACK transmissions within one PHICH group. One PHICH group uses spreading factor 4 or 2 for the normal and extended CP accordingly. This allows up to 8 (normal CP) or 4 (extended CP) ACK/NACKs to be multiplexed

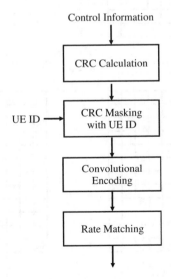

Figure 5.35   PDCCH channel coding chain

within one PHICH group of 12 resource elements. The reason to select different spreading factors for the normal and extended CP is that the extended CP is expected to be used in more frequency-selective radio channels where the lower spreading factor provides better protection against inter-code interference.

5.8.4 Cell-specific Reference Signal

The main purpose of the Cell-specific Reference Signal (CRS) is to facilitate the UE channel estimation for demodulation of data and control signaling. In addition, it is used for DL mobility and CSI measurements. There are also many other auxiliary UE functions that can use the CRS such as:

- time tracking;
- frequency tracking;
- physical cell identity and CP length verification during cell search.

The CRS has an incremental time-frequency structure dependent on the number of antenna ports as illustrated in Figure 5.36.

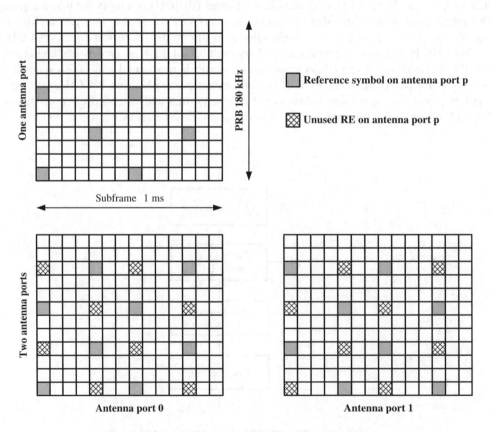

Figure 5.36 CRS structure for 1 and 2 antenna ports

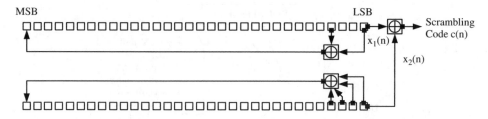

Figure 5.37   Gold sequence generator

Degree-31 Gold sequences, generated as shown in Figure 5.37, are used for scrambling the CRS. The same scrambling principle is used for other LTE channels and signals with an appropriate corresponding set of initialization values and a re-initialization period.

5.8.5 *Downlink Transmission Modes*

For robust and efficient system operation, it is important that the UE knows beforehand the type of transmission to expect. If the transmission mode could change dynamically from one sub-frame to another the UE would need to monitor all the possible DCI formats simultaneously, leading to a considerable increase in the number of blind decodings and receiver complexity (and likely for increased number of signaling errors). Furthermore, the UE would not be able to provide meaningful channel feedback because, for example, the CQI value depends on the transmission mode assumed.

Therefore each UE is configured semi-statically via RRC signaling to one transmission mode. The transmission mode defines what kind of downlink transmissions the UE should expect – for example, transmit diversity or closed loop spatial multiplexing – and restricts the channel feedback to modes corresponding to the desired operation. In LTE Release 8 seven transmission modes have been defined:

1   Single-antenna port; port 0. This is the simplest mode of operation with no precoding. It is used by eNodeBs equipped with only one transmit antenna per cell.
2   Transmit diversity. Transmit diversity with two or four antenna ports (APs) uses SFBC and SFBC-FSTD respectively as presented in Figure 5.38.
3   Open-loop spatial multiplexing. This is an open loop mode with the possibility of doing rank adaptation based on the RI feedback. Figure 5.39 presents the basic

$$
\begin{array}{c}
REs \;\; \rightarrow \\
\begin{array}{c} APs \\ \downarrow \end{array}
\begin{bmatrix} S_1 & S_2 \\ -S_2^* & S_1^* \end{bmatrix}
\end{array}
$$

$$
\begin{array}{c}
REs \;\; \rightarrow \\
\begin{array}{c} APs \\ \downarrow \end{array}
\begin{bmatrix} S_1 & S_2 & 0 & 0 \\ 0 & 0 & S_3 & S_4 \\ -S_2^* & S_1^* & 0 & 0 \\ 0 & 0 & -S_4^* & S_3^* \end{bmatrix}
\end{array}
$$

Figure 5.38   SFBC and SFBC-FSTD transmit diversity precoding

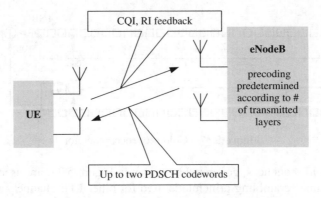

Figure 5.39 Illustration of open-loop spatial multiplexing operation

principle of open-loop spatial multiplexing. In the case of rank = 1, transmit diversity is applied in a similar way to transmission mode 2. With higher rank spatial multiplexing, with up to four layers with a large delay, CDD and predetermined precoder selection is used. In this mode, the eNode B informs the UE via DL control signaling about the number of transmitted codewords and the number of spatial layers. If two transport blocks are transmitted in a sub-frame to the UE, they use separate HARQ processes and can have different transport block sizes and redundancy versions.

4   Closed-loop spatial multiplexing. This is a spatial multiplexing mode with precoding feedback supporting dynamic rank adaptation. In Transmission Mode 5, the eNode B with the use of DCI 2 can either select wideband precoding with one of the specified precoding matrices or can apply frequency selective precoding by confirmation of the PUSCH reported PMIs. Closed-loop spatial multiplexing principles are shown in Figure 5.40.

5   Multi-user MIMO. Transmission mode for downlink MU-MIMO operation. The MU-MIMO transmission scheme uses rank = 1, DCI 1D informs the UE whether the PDSCH power is reduced by 3 dB to facilitate PDSCH detection in case of SDMA.

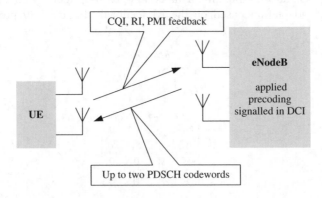

Figure 5.40 Illustration of closed-loop spatial multiplexing operation

6   Closed-loop Rank = 1 precoding. Closed-loop precoding similar to transmission mode 5 without the possibility of spatial multiplexing – that is, the rank is fixed to one. This mode, compared to TM 4, has reduced DCI overhead associated with closed-loop rank = 1 precoding.

7   Single-antenna port; port 5. This mode can be used in a beamforming operation when UE specific reference signals (URS) are in use. Precoding is not specified in this case and the UE assumes that both the URS and the PDSCH undergo the same precoding operation. The scrambling of URS and PDSCH is initialized, among others with the UE identity (C-RNTI). So TM 7 can support MU-MIMO in a standard transparent way.

Table 5.3 shows which DCI formats and which PDSCH transmission schemes are available in each Transmission Mode (TM).

Transmission mode 8 was added in Release 9. It is facilitating the combination of UE specific reference symbols (with beamforming) and multi-stream transmission with MIMO. At least initially this is foreseen as a TDD feature and is specified as a UE capability for the FDD mode of operation.

5.8.6   Physical Broadcast Channel (PBCH)

The Physical Broadcast Channel (PBCH) is carrying the system information needed to access the system, such as the cell's bandwidth, as covered in more details in Chapter 6. The channel is always provided with 1.08 MHz bandwidth as shown in Figure 5.41 so the PBCH structure is independent of the actual system bandwidth being used, like other channels/signals needed for initial system access. The PBCH is convolutionally encoded as

Table 5.3   DCI formats, transmission schemes and resource allocations per transmission mode

Transmission Mode	Monitored DCI format	PDSCH transmission scheme	PDSCH resource allocation
TM 1	DCI 1A	Single antenna port 0	Type 2
	DCI 1	Single antenna port 0	Type 0/1
TM 2	DCI 1A	Transmit Diversity	Type 2
	DCI 1	Transmit Diversity	Type 0/1
TM 3	DCI 1A	Transmit Diversity	Type 2
	DCI 2A	Open-loop spatial multiplexing or transmit diversity	Type 0/1
TM 4	DCI 1A	Transmit Diversity	Type 2
	DCI 2	Closed-loop spatial multiplexing or transmit diversity	Type 0/1
TM 5	DCI 1A	Transmit Diversity	Type 2
	DCI 1D	Multi-user MIMO	Type 2
TM 6	DCI 1A	Transmit Diversity	Type 2
	DCI 1B	Closed-loop rank = 1 precoding	Type 2
TM 7	DCI 1A	Single antenna port 0 or transmit diversity if more than one cell-specific AP configured	Type 0/1
	DCI 1	Single antenna port 5 including standard transparent multi-user MIMO	Type 2

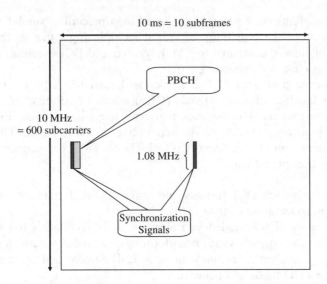

Figure 5.41   PBCH location at the center frequency

the data rate is not that high. As discussed in Chapter 6, the broadcast information is partly carried on the PBCH, where the Master Information Block (MIB) is transmitted while the actual System Information Blocks (SIBs) are then on the PDSCH. The 600 sub-carriers in Figure 5.41 only need 9 MHz (50 resource blocks) in the resource domain but the system bandwidth needed to cover for sufficient attenuation for adjacent operator makes the total bandwidth needed 10 MHz as discussed in Chapter 14. With a 1.4 MHz system bandwidth there are no resource blocks on either side of the PBCH in the frequency domain in use as effectively only six resource blocks may then be in use to meet the spectrum mask requirements.

5.8.7   Synchronization Signal

There are 504 Physical Cell Identity (PCIs) values in LTE system, which could be compared with the 512 primary scrambling codes in WCDMA. The Primary Synchronization Signal (PSS) and the Secondary Synchronization Signals (SSS) are transmitted, like PBCH, always with the 1.08 MHz bandwidth, located in the end of first and eleventh slots (slots 0 and 10) of the 10 ms frame, as shown in Figure 5.42.

The PSS and SSS jointly indicate the space of 504 unique PCIs. The PCIs form 168 PCI groups, each having three PCIs (thus giving a total of 504 PCIs). The location and structure of the PCIs allows a sample to be taken from the center frequency (with bandwidth of 1.08 MHz) and for a maximum duration of 5 ms contains the necessary information needed for cell identification.

5.9   Physical Layer Procedures

The key physical layer procedures in LTE are power control, HARQ, timing advance and random access. Timing advance, as such, is based on the signalling in the Medium

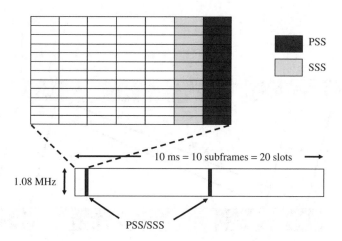

Figure 5.42   Synchronization signals in the frame

Access Control (MAC) layer, as shown in the MAC section in Chapter 6, but as it is directly related to the physical layer the timing advance details are covered in this chapter. The big contrast to WCDMA is that there are no physical layer issues related to macro-diversity since the UE is only connected to one base station at a time and hard handover is applied. Also specific means for dealing with inter-system and inter-frequency measurements like compressed mode is not needed in LTE, as LTE is by nature having discontinuous operation that will facilitate the measurements by the help of scheduling.

5.9.1   HARQ Procedure

The HARQ in LTE is based on the use of stop-and-wait HARQ procedure. Once the packet is being transmitted from the eNodeB, the UE will decode it and provide feedback in the PUCCH, as described in section 5.6. In case of negative acknowledgement (NACK) the eNodeB will send a retransmission. The UE will combine the retransmission with the original transmission and will run again the turbo decoding. Upon successful decoding (based on CRC check) the UE will send positive acknowledgement (ACK) for the eNode. After that eNodeB will send a new packet for that HARQ process. Due to the stop-and-wait way of operating, one needs to have multiple HARQ processes to enable continuous data flow. In LTE the number of processes is fixed to 8 processes both for uplink and downlink direction. The example case of single user continuous transmission is illustrated in Figure 5.43. In case of multiple users, it is dependent on the eNodeB scheduler when a retransmission is sent in the uplink or downlink direction, as a retransmission also requires resources to be allocated.

The HARQ operation in LTE supports both Chase combining and the use of incremental redundancy. The use of Chase combining means that retransmission has exactly the same rate matching parameters as the original transmission and thus exactly the same symbols are transmitted. In case of incremental redundancy the retransmission may have different rate matching parameters like the original transmission. The minimum delay between the end of a packet and start of a retransmission is 7 ms. The UE will send the ACK/NACK for a packet in frame n, in the uplink frame $n + 4$. This leaves around 3 ms processing time for

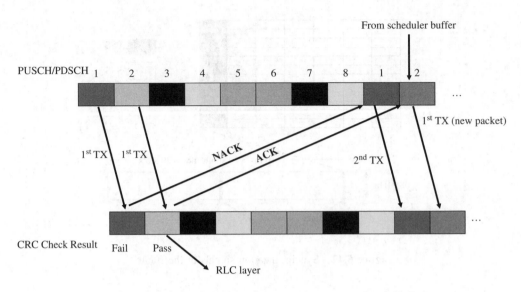

Figure 5.43 LTE HARQ operation with eight processes

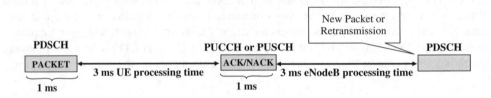

Figure 5.44 LTE HARQ timing for a single downlink packet

the UE, depending on the uplink/downlink timing offset controlled by the timing advance procedure. The downlink timing for a single downlink packet transmitted is shown in Figure 5.44. The retransmission instant in the downlink is subject to the scheduler in eNodeB and thus the timing shown in Figure 5.44 indicates the earliest moment for a retransmission to occur.

5.9.2 Timing Advance

The timing control procedure is needed so that the uplink transmissions from different users arrive at the eNodeB essentially within the cyclic prefix. Such uplink synchronization is needed to avoid interference between the users with uplink transmissions scheduled on the same sub-frame. The eNodeB continuously measures the timing of the UE uplink signal and adjusts the uplink transmission timing as shown in Figure 5.45.

Timing advance commands are sent only when a timing adjustment is actually needed. The resolution of a timing advance command is $0.52\,\mu s$, and timing advance is defined relative to the timing of downlink radio frame received on UE.

The timing advance value is measured from RACH transmission when UE does not have a valid timing advance – that is, the uplink for the UE is not synchronized. Such situations

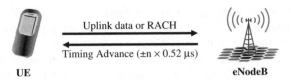

Figure 5.45   Uplink timing control

include system access, when UE is in RRC_IDLE state or when UE has had an inactivity period exceeding the related timer, non-synchronized handover, and after radio link failure. eNodeB can also assign to UE a dedicated – that is, contention-free – preamble on RACH for uplink timing measurement when eNodeB wants to establish uplink synchronization. Such situations are faced at handover or when downlink data arrive for a non-synchronized UE. From the range defined for timing advance, cell sizes beyond 100 km would be facilitated by leaving some resources unused.

5.9.3   Power Control

The LTE power control is slow power control for the uplink direction. In the downlink direction there is no specified power control. As the bandwidth varies due to data rate changes, the absolute transmission power of the UE will change accordingly. The power control is actually now not controlling the absolute power but the power spectral density (PSD), thus rather power per Hz, for a particular device. What facilitates the use of slower rate for power control is the use of orthogonal resources in the LTE uplink, which avoids the near-far problem that required fast power control in WCDMA. The key motivation for the power control is to reduce terminal power consumption and also avoid too large dynamic range in the eNodeB receiver rather than to mitigate interference. In the receiver, the PSDs of different users have to be reasonably close to each other so the receiver A/D converter has reasonable requirements and also the interference due to the less than ideal spectrum shape of the UE transmitter is kept under control. The LTE uplink power control principle is illustrated in Figure 5.46 where at the change of data rate the PSD stays constant but the resulting total transmission power is adjusted relative to the data rate change.

The actual power control is based on estimating the path loss, taking into account cell-specific parameters and then applying the (accumulated) value of the correction factor received from eNodeB. Depending on the higher layer parameter settings, the power

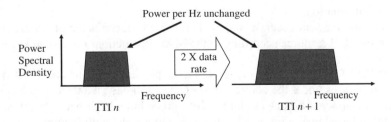

Figure 5.46   LTE uplink power with data rate change

control command is either 1 dB up or down or then the set of [−1 dB, 0, +1 dB, +3 dB] is used. The specifications also include absolute value-based power control but that is not expected to be taken into use in the first phase networks based on text case prioritization. The total power control dynamic range is slightly smaller than in WCDMA and now the devices have a minimum power level of −41 dBm compared to −50 dBm with WCDMA.

5.9.4 Paging

To enable paging, the UE is allocated a paging interval and a specific sub-frame within that interval where the paging message could be sent. The paging is provided in the PDSCH (with allocation information on the PDCCH). The key design objective in paging is to ensure a sufficient DRX cycle for devices to save power and on the other hand ensure fast enough response time for the incoming call. The E-UTRAN may parameterize the duration of the paging cycle to ensure sufficient paging capacity, as covered in more detail in Chapter 6.

5.9.5 Random Access Procedure

The LTE Random Access (RACH) operation resembles WCDMA's because both use preambles and a similar ramping of preamble power. The initial power is based on the measured path loss in DL, and power ramping is necessary because of UE's relatively coarse accuracy in path-loss measurement and absolute power setting and in order to compensate for UL fading. Although LTE PRACH resources are separate from PUSCH and PUCCH, power ramping is useful for simultaneous detection of different preamble sequences and for minimizing the interference due to asynchronous PRACH transmission at the adjacent PUCCH and PUSCH resources. The steps of the physical layer procedure are as follows:

- Transmit a preamble using the PRACH resource, preamble sequence, and power selected by MAC.
- Wait for the RACH response with matching preamble information (PRACH resource and preamble sequence). In addition to the preamble information, the response contains also the information on the uplink resource to use for further information exchange as well as the timing advance to be used. This is the fundamental difference from the WCDMA RACH procedure where, after acknowledging a preamble, the UE continues with a frame of data of 10 or 20 ms duration, or even longer, as described for the Release 8 HSPA operation in Chapter 17. In LTE the device will instead move directly to the use of UL-SCH upon reception of the random access response, which has the necessary information.
- If no matching random access response is received, transmit the preamble in the next available PRACH resource according to the MAC instructions, as shown in Figure 5.47.

Although the LTE specification suggests that the physical layer just transmits preambles and detects responses under the control of MAC, we will describe the complete procedure.

Two fundamentally different random access procedures have been specified for LTE. The contention-based procedure is what we normally expect with random access: UEs are transmitting randomly selected preambles on a common resource in order to establish a

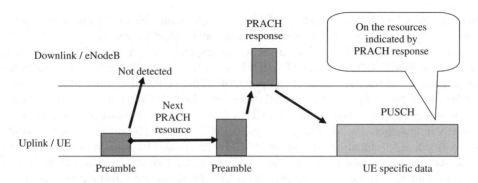

Figure 5.47   Power ramping in random access procedure

network connection or requesting resources for uplink transmission. Non-contention-based random access is initiated by the network for synchronizing the UE's uplink transmission, and the network can identify the UE from the very first uplink transmission. This procedure is nevertheless included under LTE random access because it uses PRACH resources. Both procedures are common for TDD and FDD systems.

5.9.5.1   Contention- and Non-contention-based Random Access

The contention-based procedure follows the signaling diagram on the left side of Figure 5.48.

In the first step, UE transmits a preamble sequence on PRACH. The details of PRACH and preamble sequences are explained in section 5.7. In total, 64 sequences are reserved for each cell, and these are grouped for the non-contention-based and contention-based procedures. The group reserved for the contention-based procedures is further divided

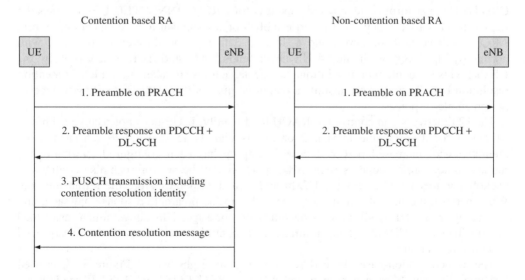

Figure 5.48   The contention and non-contention-based random access procedures

into two: by selecting the proper group, the UE sends one bit of information about the transport block size that the UE desires to send on PUSCH in step 3.

In the second step, UE receives a preamble response on a DL-SCH resource that is assigned on PDCCH. The identity RA-RNTI used for this assignment is associated with the frequency and time resource of the preamble. This permits bundling of the responses that are meant for preambles transmitted in the same PRACH frequency and time resource, which is important for saving the PDCCH resources. eNodeB transmits the response in a time window that can be configured up to the length of 10 ms. The flexible window allows freedom for dimensioning of the RACH receiver and for scheduling the responses.

The response (signaling part of the MAC layer as covered in Chapter 6) lists the sequence numbers of the observed preambles and the following information is also given for each acknowledged preamble:

- A grant for the first transmission on PUSCH, including information on the need for frequency hopping, power control command for uplink transmission and information on the need for CQI transmission and whether the PUSCH transmission needs to be delayed by one sub-frame from the nominal value.
- A timing alignment command.
- A temporary allocation for identity called temporary CRNTI, which is used for addressing PUSCH grants and DL-SCH assignments in steps 3 and 4 of the procedure.

It is expected that the typical probability of preamble collisions, meaning that two or more UEs are transmitting the same preamble sequence in the same frequency and time resource, could be around 1%. These collisions are resolved in steps 3 and 4: UE includes its identity in the first message that it transmits on PUSCH in step 3 and it expects in step 4 an acknowledgement that eNodeB has received the identity. There are two forms of acknowledgement: it can be either (1) a PUSCH grant or DL-SCH assignment addressed with CRNTI if UE had included CRNTI to the message of step 3, or (2) UE's identity can be acknowledged with a message sent on a DL-SCH resource assigned with the temporary CRNTI. The first form of acknowledgement is for RRC_CONNECTED UEs whereas the second form is used when UE tries to establish or re-establish RRC connection. Hybrid ARQ is utilized both in step 3 and 4. In step 3 there is no difference compared with normal hybrid ARQ, but in step 4 UE never sends NAK, and ACK is sent only by the UE that wins the contention resolution. No special actions are taken after a lost contention resolution but UE simply retransmits a preamble, just as it would after failing to receive the preamble response.

The LTE network can control the RACH load rapidly. If UE does not receive acknowledgement for its preamble in step 2 or for its identity in step 4, UE retransmits the preamble with increased power if power ramp-up has been configured. Normally the retransmission can be done as soon as UE is ready but the network can also configure a backoff parameter that forces the UE to add a random delay before the retransmission. When needed, the backoff parameter is included in the preamble response message, and the setting is obeyed by all the UEs decoding the message. This allows much faster load control than in WCDMA where a similar load control parameter is in the broadcasted system information.

The non-contention-based procedure, shown in the right half of Figure 5.43, is used for time alignment during handover and when an RRC_CONNECTED UE needs to be synchronized for downlink data arrival. UE receives in the handover command or through

PDCCH signaling an index of its dedicated preamble sequence that it is allowed to transmit on PRACH. Besides the sequence index, some restrictions for the frequency and time resource can be signaled so that the same sequence can be simultaneously allocated for UEs that transmit on different PRACH sub-frames or, in case of TDD, at different PRACH frequencies. The preamble responses in the contention-based and non-contention-based procedures are identical and they can thus be bundled to the same response message. As eNodeB knows the identity of the UE that has sent the dedicated preamble, the contention resolution with steps 3 and 4 is not needed.

The non-contention-based procedure provides delay and capacity enhancements compared with the contention-based procedure. As the preamble collisions are absent and the contention resolution is not needed, a shorter delay can be guaranteed, which is especially important for handover. The sequence resource is in effective use because it is assigned to the UE only when needed and can be released as soon as eNodeB detects that the UE has received the preamble response.

An unsuccessful Random Access procedure ends based on the preamble count or RRC timers. The preamble count is decisive only with two causes of random access: (a) an RRC_CONNECTED UE, lacking scheduling request resources, requests resources because of UL data arrival, or (b) an RRC_CONNECTED UE needs to be synchronized because of DL data arrival. If Random Access has been started because of RRC connection establishment or re-establishment or because of handover, the procedure continues until success or MAC reset in the expiry of the RRC timer corresponding to the cause of random access.

5.9.6 Channel Feedback Reporting Procedure

The purpose of the channel state feedback reporting is to provide the eNodeB with information about the downlink channel state in order to help optimizing the packet scheduling decision. The principle of the channel state feedback reporting procedure is presented in Figure 5.49. The channel state is estimated by the UE, based on the downlink transmissions (reference symbols and so forth), and is reported to the eNodeB using PUCCH or PUSCH. The channel state feedback reports contain information about the scheduling and link adaptation-related parameters (MCS/Transport Block Size (TBS), MIMO

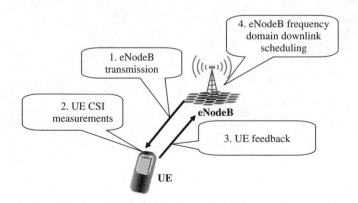

Figure 5.49 Channel State Information (CSI) reporting procedure

related parameters) that the UE can support in the data reception. The eNodeB can then take advantage of the feedback information in the scheduling decision in order to optimize the usage of the frequency resources.

In general the channel feedback reported by the UE is just a recommendation and the eNodeB does not need to follow it in the downlink scheduling. In LTE the channel feedback reporting is always fully controlled by the eNodeB and the UE cannot send any channel state feedback reports without eNodeB knowing it beforehand. The corresponding procedure for providing information about the uplink channel state is called channel sounding and it is carried out using the Sounding Reference Signals (SRS) as presented in section 5.6.4

The main difference of the LTE channel state information feedback, when compared to the WCDMA/HSDPA, is the frequency selectivity of the reports – information regarding the distribution of channel state over the frequency domain can also be provided. This is an enabler for Frequency Domain Packet Scheduling (FDPS), a method that aims to divide radio resources in the frequency domain for different users so that system performance is optimized. The gain from the FDPS is illustrated in Figure 5.50. As the UE speed increases, the CSI reports become more inaccurate and become outdated faster leading to reduced gains in high mobility.

5.9.6.1   Channel Feedback Report Types in LTE

In LTE the UE can send three types of channel feedback information:

- CQI – Channel Quality Indicator.
- RI – Rank Indicator.
- PMI – Precoding Matrix Indicator.

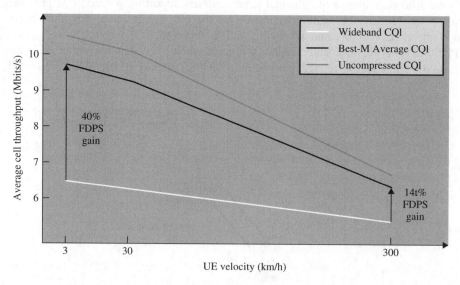

Figure 5.50   Comparison of the average cell throughputs for different CQI schemes and velocities

The most important part of channel information feedback is the Channel Quality Indicator (CQI). The CQI provides the eNodeB with information about the link adaptation parameters that the UE expects it can support at the time of generating the CQI feedback (taking into account the transmission mode, the receiver type, number of antennas and the interference situation experienced by the UE). The CQI is defined as a table containing 16 entries with modulation and coding schemes (MCSs). The UE reports back to the eNodeB the highest CQI index corresponding to the MCS and TBS for which the estimated received DL transport block BLER does not exceed 10%. The CQI operation is very similar to HSDPA CQI use as covered in [5]. Note that there are many more possibilities for MCS and TBS size values than the 15 indicated by the CQI feedback in Table 5.4.

Rank Indicator (RI) is the UE's recommendation for the number of layers (streams) to be used in spatial multiplexing. Rank Indicator is only reported when the UE is operating in MIMO modes with spatial multiplexing (transmission modes 3 and 4). In single antenna operation or TX diversity it is not reported. The RI can have values 1 or 2 with two-by-two antenna configuration and from one up to four with four-by-four antenna configuration. The RI is always associated with one or more CQI reports meaning that the reported CQI is calculated assuming that particular RI value. Since the rank typically varies more slowly than the CQI it is normally reported less often. Rank indicator always describes the rank on the whole system band – frequency-selective RI reports are not possible.

The PMI provides information about the preferred precoding matrix in codebook-based precoding. Like RI, PMI is only relevant to the MIMO operation. The MIMO operation with PMI feedback is called Closed Loop MIMO. The PMI feedback is limited to transmission modes 4, 5, and 6. The number of precoding matrices in the codebook depends on the number of eNodeB antenna ports: in the case of two antenna ports there are seven matrices to choose from altogether, whereas with four antenna ports the total number can

Table 5.4   CQI table

CQI index	Modulation	Coding rate $\times 1024$	Bits per resource element
0		out of range	
1	QPSK	78	0.1523
2	QPSK	120	0.2344
3	QPSK	193	0.3770
4	QPSK	308	0.6016
5	QPSK	449	0.8770
6	QPSK	602	1.1758
7	16QAM	378	1.4766
8	16QAM	490	1.9141
9	16QAM	616	2.4063
10	64QAM	466	2.7305
11	64QAM	567	3.3223
12	64QAM	666	3.9023
13	64QAM	772	4.5234
14	64QAM	873	5.1152
15	64QAM	948	5.5547

be up to 64 depending on the RI and the UE capability. The PMI reporting can be either wideband or frequency selective depending on the CSI feedback mode.

5.9.6.2 Periodic and Aperiodic Channel State Feedback Reporting

Although in principle the UE has up-to-date information about the changes in channel state, channel state feedback report initiated by the UE would raise several issues. First of all, in order to detect the reports there would be a need to perform blind decoding at the eNodeB, which is not desirable from the receiver implementation point of view. Secondly, as the eNodeB is fully in charge of the scheduling decisions, UE-initiated reports would quite likely often be unnecessary. Furthermore, the reports initiated by UEs would complicate the UL resource allocation considerably, leading to increased signaling overhead. Hence in was agreed in the LTE standardization that the channel-state feedback reporting is always fully controlled by the eNodeB, i.e the UE cannot send any channel-state feedback reports without eNodeB knowing beforehand.

In order fully to exploit gains from frequency-selective packet scheduling, detailed CSI reporting is required. However, as the number of UEs reporting the channel-state feedback increases, the uplink signaling overhead becomes significant. Furthermore the PUCCH, which is supposed primarily to carry the control information, is rather limited in capacity: payload sizes of only up to 11 bits/sub-frame can be supported. On PUSCH there are no similar restrictions on the payload size, but since PUSCH is a dedicated resource only one user can be scheduled on a single part of the spectrum.

To optimize the usage of the uplink resources while allowing for also detailed frequency selective CSI reports, a two-way channel state feedback reporting scheme has been adopted in LTE. Two main types of reports are supported: Periodic and Aperiodic. A comparison of the main features of the two reporting options is presented in Table 5.5

The baseline mode for channel information feedback reporting is Periodic reporting using PUCCH. The eNodeB configures the periodicity parameters and the PUCCH

Table 5.5 Comparison of periodic and aperiodic channel information feedback reporting

	Periodic reporting	Aperiodic reporting
When to send	Periodically every 2–160 ms	When requested by the eNodeB
Where to send	Normally on PUCCH, PUSCH used when multiplexed with data	Always on PUSCH
Payload size of the reports	4–11 bits	Up to 64 bits
Channel Coding	Linear block codes	Tail-biting convolutional codes
CRC protection	No	Yes, 8 bit CRC
Rank Indicator	Sent in a separate sub-frame at lower periodicity	Sent separately encoded in the same sub-frame
Frequency selectivity of the CQI	Only very limited amount of frequency information	Detailed frequency-selective reports are possible
Frequency selectivity of the PMI	Only wideband PMI	Frequency selective PMI reports are possible

resources via higher layer signaling. The size of a single report is limited up to about 11 bits depending on the reporting mode and the reports contain little or no information about the frequency domain behavior of the propagation channel. Periodic reports are normally transmitted on PUCCH. However, if the UE is scheduled in the UL the Periodic report moves to PUSCH. The reporting period of RI is a multiple of CQI/PMI reporting periodicity. RI reports use the same PUCCH resource (PRB, Cyclic shift) with PUCCH formats 2/2a/2b.

When the eNodeB sees the need for more precise channel-state feedback information it can at any time request the UE to send an Aperiodic channel-state feedback report on PUSCH. An aperiodic report can be either piggybacked with data or sent alone on PUSCH. Using the PUSCH gives the possibility to transmit large and detailed reports. In case the transmission of Periodic and Aperiodic reports from the same UE would collide, only the Aperiodic report is sent.

The two modes can also be used to complement each other. The UE can, for example, be configured to send Aperiodic reports only when it is scheduled, while periodic reports can provide coarse channel information on a regular basis.

5.9.6.3 CQI Compression Schemes

Compared to the WCDMA/HSPA, the main new feature in the channel feedback is the frequency selectivity of the report. This is an enabler for the Frequency Domain packet Scheduling (FDPS). Providing a full 4-bit CQI for all the PRBs would mean excessive UL signaling overhead of hundreds of bits per sub-frame, so some feedback compression schemes are used.

In order to reduce feedback, the CQI is reported on a *sub-band* basis. The size of the sub-bands varies depending on the reporting mode and system bandwidth from two consecutive PRBs up to whole-system bandwidth.

The main CQI compression methods are:

- wideband feedback;
- best-M average (UE-selected sub-band feedback);
- higher layer-configured sub-band feedback.

Delta compression can also be used in combination with the above options. For example, in the case of closed-loop MIMO, CQI for the second codeword can be signaled as a 3-bit delta relative to the CQI of the first codeword. When the number of sub-bands is large this leads to considerable savings in signaling overhead.

5.9.6.4 Wideband Feedback

The simplest way to reduce the number of CQI bits is to use only wideband feedback. In wideband feedback only a single CQI value is fed back for the whole system band. Since no information about the frequency domain behavior of the channel is included, wideband feedback cannot be utilized in FDPS. However, on many occasions this is still sufficient: for example, PDCCH link adaptation or TDM-like scheduling in lowly loaded cells do not benefit from frequency-selective CQI reporting. Moreover, when the number of scheduled UEs is high, the total UL signaling overhead due to detailed CQI reports may become excessive and the wideband CQI reports are the only alternative.

5.9.6.5 Best-M Average

Best-M average is an effective compromise between the system performance and the UL feedback signaling overhead. The principle of the Best-M average compression is shown in Figure 5.51. In Best-M average reporting the UE first estimates the channel quality for each sub-band. Then it selects the M best ones and reports back to the eNodeB a single average CQI corresponding to the MCS/TBS the UE could receive correctly assuming the eNodeB schedules the UE on those M sub-bands. The parameter M depends on the system bandwidth and corresponds to roughly 20% of the whole system bandwidth.

5.9.6.6 Higher Layer-configured Sub-band Feedback

In Higher Layer-Configured sub-band feedback a separate CQI is reported for each sub-band using delta compression. This will result in the best performance at the cost of feedback overhead: the payload size of the reports can be as large 64 bits. To keep the signaling on a manageable level the sub-band sizes with Full Feedback reporting are twice as large as with Best-M average. This will limit the accuracy and performance in very frequency selective channels, where Best-M average may be a better alternative.

5.9.7 Multiple Input Multiple Output (MIMO) Antenna Technology

In the Release 8 LTE specifications there is the support for multi-antenna operation both in terms of transmit diversity as well as spatial multiplexing with up to four layers. The use of MIMO with OFDMA has some favorable properties compared to WCDMA due to the ability to effectively cope with multi-path interference. In WCDMA, MIMO had been introduced in Release 7 but not deployed in reality. In LTE, MIMO is part of the first release and also part of the device categories with the exception of the simplest

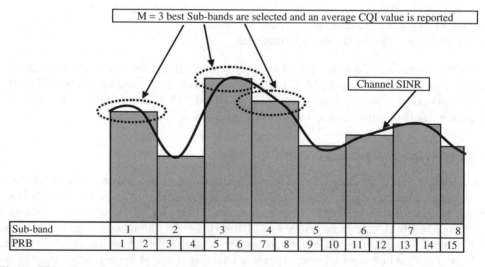

Figure 5.51 The principle of the Best-M average compression. An average CQI value is reported for the M best sub-bands

device type, as shown in section 5.9. The use of more than a single antenna in a eNodeB will not result in a problem with non-MIMO LTE devices as all devices can cope with transmit diversity of up to four antennas. Thus full transmit power can be always utilized regardless of the number of transmit antennas included. As shown in Chapter 4, the reference symbols are transmitted so that they have separate time and frequency domain resources for each antenna port to allow for good separation of different antennas and robust channel estimation, while the actual user data are then overlapping in time and frequency domain.

Perhaps the most attractive mode of MIMO operation is spatial multiplexing, which allows for the peak rates to be increased by a factor of two or four compared to the non-MIMO case depending on the eNodeB and the UE antenna configuration. In LTE the MIMO modes supporting spatial multiplexing are:

- Transmission mode 3 – open-loop spatial multiplexing. The UE reports the RI but no precoding feedback. Based on the RI the eNodeB scheduler can select the number of layers used in spatial multiplexing. In the case of rank = 1 TX diversity is used. With higher rank large delay CDD is applied with deterministic precoding.
- Transmission mode 4 – closed-loop spatial multiplexing. Here the UE reports both the RI and index of the preferred precoding matrix. Dynamic rank adaptation is supported on the eNodeB signals the applied precoding vector to the UE in the Downlink grant.

The UE provides feedback for the MIMO operation (as discussed in section 5.6.9.1) and the eNodeB selects the precoding vectors and the number of layers to use accordingly, as illustrated in Figure 5.52. This has some similarity with the MIMO solution in WCDMA, which is based on the closed-loop feedback from the UE. In LTE some of the parameters in the downlink assignments are intended for MIMO control either for informing the device of the precoding applied or whether or not to swap the transport blocks between the codewords in the retransmissions.

One specific mode of MIMO operation is the Multi-User MIMO (MU-MIMO) where the eNodeB is sending different data streams intended for different devices from two antenna ports (from the specifications, four antenna ports would also be possible). The principle of MU-MIMO is illustrated in Figure 5.53.

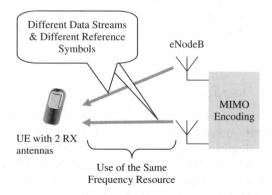

Figure 5.52   Single user MIMO principle

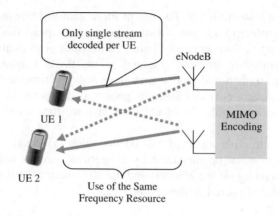

Figure 5.53 Multiuser MIMO transmission principle

5.9.8   Cell Search Procedure

The first operation performed by a LTE device when the power is turned on is the cell search. In LTE the cell-search procedure is based on the use of synchronization signals. As with WCDMA, there are primary and secondary synchronization signals (in WCDMA specification called synchronization channels). The cell search steps are as follows:

1   The device will look for the primary synchronization signal at the center frequencies possible at the frequency band in question. Three different possibilities for the primary synchronization signal exist as described in section 5.8.6.
2   Once the primary synchronization signal has been detected, the UE will look for the secondary synchronization signal.
3   Once one alternative is detected from the 168 possible secondary synchronization signals, the UE has figured out the Physical Cell ID (PCI) value from the address space of 504 IDs.

The UE has information from the PCI about the parameters used for downlink reference signals and thus UE can decode the PBCH. All this is independent of the actual system bandwidth in question. The LTE network specifications support also as part of the Self-Organization Networks (SON) functionality automated PCI planning, as described in section 5.12.

5.9.9   Half-duplex Operation

The LTE specifications also enable the half-duplex operation which is basically the FDD mode operation (separate transmission and reception frequency) but transmission and reception do not occur simultaneously, as happens in TDD mode. The intention in 3GPP has been to have an option for the cases where, due to the frequency arrangements, the resulting requirements for the duplex filter would be unreasonable, resulting in excessively high cost and high power consumption, or the case when a terminal supports a switch for legacy paired bands being refarmed to LTE anyway. Figure 5.54 illustrates the

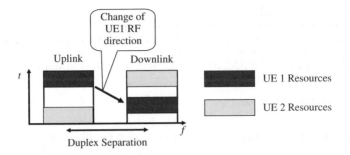

Figure 5.54   LTE FDD half-duplex operation

Duplex operation principle. It remains to be seen whether the performance requirements for some bands will be specified with the half-duplex operation in mind. The eNodeB would obviously need to be aware if some devices would be based on half-duplex operation. The impact for the data rates would be that uplink and downlink would no longer be independent but the available data rate in the other transmission direction would depend on the resources allocated for the other direction. As with TDD, one would need to schedule the data on a device basis in such a manner that there is no conflict between uplink and downlink allocations for one UE. Time would also be needed for the UE to change between transmit and reception. One still avoids aligning uplink/downlink allocation for all the users in the cell, like in TDD operation.

5.10   UE Capability Classes and Supported Features

In LTE five device-capability classes are defined. The supported data ranges run from 5 Mbps to 75 Mbps in the uplink direction and from 10 to 300 Mbps in the downlink direction. All devices support the 20 MHz bandwidth for transmission and reception, assuming that this has been specified for the given frequency band. It is foreseen that, in most cases for frequency bands below 1 GHz, the interest is with the smallest bandwidths and support for up to 20 MHz will not be specified. For bands above 1 GHz, in many cases, bandwidths below 5 MHz are not needed. Only a category 5 device will do 64QAM in the uplink; others use QPSK and 16QAM. The receiver diversity and MIMO are in all categories, except in category 1, which does not support MIMO. The UE categories are shown in Table 5.6. The step in the data rates up to 300 Mbps with category 5 is achieved with the four-antenna MIMO transmission, which is not supported by the other categories.

The actual device capabilities are behind other signaling as well than just these categories. While these categories mainly define the boundaries for the data rates, the test case prioritization in 3GPP also defines what the first-phase devices will support and what they will not. 3GPP has put three different levels into use for prioritization of test cases based on the input created by the operators planning to put the system into use. The operator priorization for Release 8 can be found from [19]. The levels are:

- High priority, which refers to the features that are planned to be in the first-phase deployments. This is the highest test-case category in terms of number of features.

Table 5.6 LTE device categories

	Category 1	Category 2	Category 3	Category 4	Category 5
Peak rate DL/UL	10/5 Mbps	50/25 Mbps	100/50 Mbps	150/50 Mbps	300/75 Mbps
Modulation DL	QPSK/16QAM/64QAM				
Modulation UL	QPSK/16QAM	QPSK/16QAM	QPSK/16QAM	QPSK/16QAM	QPSK/16QAM + 64QAM
MIMO DL	Optional	2×2	2×2	2×2	4×4

- Medium priority, which refers to features planned to be put into use a few years after initial deployments. Originally this priority level contained features like the large delay CDD but this turned out to be difficult or impossible to deploy later if devices were not able to reach, for example, PBCH based on this. Similarly the extended cyclic prefix as originally in this class but deploying it later was found problematic if there was an existing population of devices not able to support it (due to lack of test cases).
- Low priority features, which refers to features where no immediate plans exists for deployment from the 3GPP operator perspective. The UE-specific reference symbols for FDD mode operation are an example, reflecting the lack of interest for adaptive antenna arrays in the industry for Release 8 FDD. This is similar to WCDMA where support for dedicated pilots was removed from a later version of the specifications due to lack of practical interest for introducing adaptive antennas. (WCDMA system contains other means for dealing with adaptive antenna arrays as covered in [5].) In TDD operation mode, this feature is a high-priority test case, reflecting the fact that many sites using the 1.28 Mcps TDD (TD-SCDMA) in China are based on adaptive antenna arrays. To enable smooth evolution toward LTE, LTE devices operating in TDD mode should permit the same antenna structures to be used along with other aspects of LTE TDD mode and TD-SCDMA co-existence, as covered in Chapter 15.

5.11 Physical Layer Measurements

5.11.1 eNodeB Measurements

As all the radio functionalities are located in eNodeB, there are not too many eNodeB measurements that would need to be reported over any interface as there is no separate centralized RRM functionality like the radio network controller in WCDMA. The eNodeB measurements specified in the physical layer specifications in Release 8 in the downlink and uplink are as follows [4]:

- The power used (power contribution) for the resource elements that transmit cell-specific reference signals from the eNodeB (in the system bandwidth).
- Received interference power per physical resource block.
- Thermal noise power over the system bandwidth.

The purpose of this measurement is to enable the handover to take into account the relative base-station strengths while facilitating inter-cell interference coordination. eNodeB may use this information for different purposes in addition to other information available to it.

In 3GPP there are additional indicators as part of the Operation and Maintenance (O&M) specifications to support monitoring the system performance.

5.11.2 UE Measurements and Measurement Procedure

The following measurements should be performed inside the LTE system for the UE:

- Reference signal received power (RSRP), which, for a particular cell, is the average of the power measured (and average between receiver branches) of the resource elements that contain cell-specific reference signals.
- Reference Signal Received Quality (RSRQ) is the ratio of the RSRP and E-UTRA carrier received signal strength indicator to the reference signals.
- E-UTRA Carrier Received Signal Strength Indicator (RSSI), which is the total received wideband power on a given frequency. Thus it includes the noise 'from the whole universe' on the particular frequency, which may include noise from interfering cells or any other source of noise. E-UTRA RSSI is not reported by the UE as individual measurement (as indicated in the early versions of [4] until June 2008) but it is only used in calculating the RSRQ value inside the UE.

Scheduling will create sufficient DTX/DRX gaps for the device to perform the measurement. In WCDMA specific frames were in use with compressed mode when one needed to turn of transmitter and/or receiver to measure other frequencies, as covered in [5]. Most of the measurements were related to the Common Pilot Channel (CPICH).

For the inter-system case, measurements are as follows:

- UTRA FDD CPICH Common Pilot Channel (CPICH) Received Signal Code Power (RSCP), which represents the power measured on the code channel used to spread the primary pilot channel on WCDMA.
- The UTRA FDD (and TDD) carrier Received Signal Strength Indicator (RSSI) is the corresponding wideband power measurement as defined for LTE as well.
- UTRA FDD CPICH Ec/No is the quality measurement, like RSRQ in LTE, and provides the received energy per chip energy over the noise.
- The GSM carrier RSSI represents the wideband power level measured on a particular GSM carrier.
- UTRA TDD Primary Common Control Physical Channel (P-CCPCH) RSCP is the code power of the UTRA TDD broadcast channel.
- CDMA2000 1xRTT and HRPD Pilot Strengths are the power on the respective pilot channel (code) when considering the handover to cdma2000.

5.12 Physical Layer Parameter Configuration

The configuration of the physical layer parameters in a particular cell, for a connection, is the responsibility of a particular eNodeB. Several issues will arise from operations and

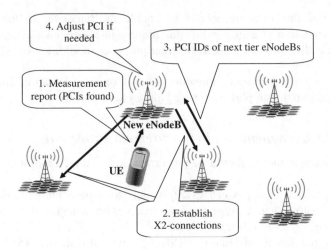

Figure 5.55   Self-configuration for PCI

maintenances (O&M) settings, such as the cyclic prefix lengths to be used. For some parameters 3GPP has been developing Self Organizing Network (SON) solutions. In the physical layer this covers the Physical Cell ID (PCI), as shown in Figure 5.55. When installing a new cell, the principle is that the cell could select the PCI randomly and, when obtaining the first measurement report from any UE, it learns the PCIs in use nearby. After that, eNodeB knows the neighbors and it can establish the X2 connections (UE needs to be then to be instructed to decode the BCH as well to get the global cell ID after which O&M system can provide the connection information for X2-creation). Once the X2 connections provide information of the PCI values used in nearby cells, the cell can confirm whether the selected PCI needs to be adjusted or not. Alternatively the PCI could be obtained directly from O&M, avoiding initial conflicts for PCIs between nearby cells.

5.13   Summary

The LTE physical layer has been built on top of OFDMA (for downlink) and SC-FDMA (for uplink) technologies. Resource allocation is dynamic with 1 ms resolution in time and 180 kHz in the frequency domain, making radio resource utilization matching fit the packet nature of the communication. The physical layer facilitates advanced features known from HSDPA/HSUPA, such as physical layer retransmission, link adaptation, multi-antenna transmission and physical layer eNodeB-based scheduling. There are also new properties, taking into account possibilities with new radio access technology, such as frequency domain scheduling, which is facilitated both in terms of physical feedback methods and in terms of signaling to make the uplink and downlink resource allocations to the advantage of the frequency domain element. The maximum data rates within the Release 8 UE categories run up to 300 Mbps in the downlink and 75 Mbps in the uplink direction. The physical layer design adapts well to the support of different system bandwidths because common channels are not dependent on the actual system bandwidth used but are always based on the use of resources within the 1.08 MHz block at the center frequency.

References

[1] 3GPP Technical Specification, TS 36.211, 'Evolved Universal Terrestrial Radio Access (E-UTRA); Physical channels and modulation', 3GPP, v 8.4.0 September 2008.

[2] 3GPP Technical Specification, TS 36.212, 'Evolved Universal Terrestrial Radio Access (E-UTRA); Multiplexing and channel coding', 3GPP, v 8.4.0 September 2008.

[3] 3GPP Technical Specification, TS 36.213, 'Evolved Universal Terrestrial Radio Access (E-UTRA); Physical layer procedures', 3GPP, v 8.4.0 September 2008.

[4] 3GPP Technical Specification, TS 36.214, 'Evolved Universal Terrestrial Radio Access (E-UTRA); Physical layer measurements', 3GPP, v 8.4.0 September 2008.

[5] H. Holma, A. Toskala 'WCDMA for UMTS', 5th edition, Wiley 2010.

[6] 3GPP technical report, TR 25.814, 'Physical Layer Aspect for Evolved Universal Terrestrial Radio Access (UTRA)', 3GPP.

[7] 3GPP Tdoc R1-081126, 'Way Forward on RV and NDI signaling', Panasonic et al.

[8] 3GPP Tdoc R1-070041, 'Dynamic Contention Free Scheduling Request', Motorola.

[9] 3GPP Tdoc R1-070378, 'UL Resource Request for LTE System', Nokia.

[10] 3GPP Tdoc R1-080931, 'ACK/NACK Channelization for PRBs containing both ACK/NACK and CQI', Nokia Siemens Networks, Nokia, Texas Instruments.

[11] 3GPP Tdoc R1-080035, 'Joint Proposal on Uplink ACK/NACK Channelization', Samsung, Nokia, Nokia Siemens Networks, Panasonic, TI.

[12] 3GPP Tdoc R1-062840, 'TDM based Multiplexing Schemes between L1/L2 Control and UL Data', Nokia.

[13] 3GPP Tdoc R1-072224, 'Uplink Control Signal Transmission in Presence of Data', Samsung.

[14] 3GPP Tdoc R1-081852, 'Linkage between PUSCH MCS and Amount of Control Resources on PUSCH', Nokia Siemens Networks, Nokia.

[15] D. C. Chu, 'Polyphase codes with good periodic correlation properties', IEEE Transactions Information Theory, vol. 18, pp. 531–532, July 1972.

[16] B. M. Popovic, 'Generalized chirp-like polyphase sequences with optimum correlation properties', IEEE Transactions Information Theory, vol. 38, pp. 1406–1490, Jul. 1992.

[17] 3GPP Tdoc R1-082570, 'Signalling for SRS Hopping Bandwidth', NTT DoCoMo, Panasonic.

[18] 3GPP Tdoc R1-074977, 'Specification of Formula for Restricted Cyclic Shift Set', LG Electronics, Nokia, Nokia Siemens Networks, Panasonic, Huawei, Texas Instruments.

[19] 3GPP Tdoc R5-083674, 'LTE Features and Test Prioritisation', NTT DoCoMo.

6

LTE Radio Protocols

Antti Toskala, Woonhee Hwang and Colin Willcock

6.1 Introduction

The role of LTE radio interface protocols is to set up, reconfigure and release the Radio Bearer, which provides the means for transferring the EPS bearer (as presented in Chapter 3). The LTE radio interface protocol layers above the physical layer include Layer 2 protocols, Medium Access Control (MAC), Radio Link Control (RLC) and Packet Data Convergence Protocol (PDCP). Layer 3 consists of the Radio Resource Control (RRC) protocol, which is part of the control plane. The protocol layer above (for the control plane) is the Non-Access Stratum (NAS) protocol, which terminates in the core network side and was addressed in Chapter 3. This chapter describes the general radio interface protocol architecture. The main functions of MAC, RLC, PDCP and RRC layers will be introduced, including the principles of ASN.1 used in the RRC protocol. The radio protocol aspects of the control plane of the X2 interface, the interface between eNodeBs, is also covered. From the Release 9 enhancements, the options added for the CS fallback are covered in more detail. This chapter concludes with the introduction of the LTE Inter Operability Testing (IOT) bits intended for early UE handling in LTE.

6.2 Protocol Architecture

The overall LTE radio interface protocol architecture is shown in Figure 6.1. This covers only the protocol part of radio access in LTE. There are also protocols in the core network between UE and the core network but they are transparent to the radio layers. Those protocols are generally referred to as Non-Access Stratum (NAS) signaling.

The physical layer carries the transport channels provide by the MAC layer. The transport channels, listed in Chapter 5, describe how and with what characteristics data are carried from the radio interface on the physical channels. The MAC layer offers the logical channels to the RLC layer. The logical channels characterize the type of data to be transmitted and are covered in detail in section 6.4. Above the RLC layer there is the PDCP layer, now both for the control and user plane in contrast to WCDMA where it was only for the user plane data. The Layer 2 provides radio bearers to the higher layers.

LTE for UMTS: Evolution to LTE-Advanced, Second Edition. Edited by Harri Holma and Antti Toskala.
© 2011 John Wiley & Sons, Ltd. Published 2011 by John Wiley & Sons, Ltd.

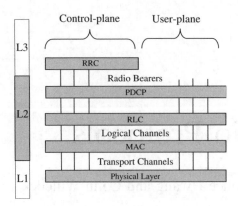

Figure 6.1   LTE Radio Protocol Stacks

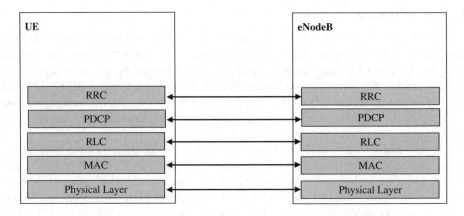

Figure 6.2   LTE control plane radio protocols in LTE architecture

Signaling Radio Bearers (SRBs) carry the RRC signaling messages. The user plane Radio
Bearers (RBs) carry the user data. As described in Figure 6.2 on radio protocol archi-
tecture for control plane, MAC, RLC, RRC and PDCP are all located in eNodeB. With
WCDMA RRC and PDCP were both in RNC and the MAC layer was either in a NodeB
(especially for HSPA) or in a Radio Network Controller (RNC) in the case of Release'99
WCDMA.

The control plane signaling between different network elements is carried on the X2
interface for the inter-eNodeB communications. For the traffic between Mobility Management Entity (MME) and eNodeB, the S1_MME is used for control plane mobility
management-related signaling. The control plane interfaces inside E-UTRAN and between
E-UTRAN and MME are shown in Figure 6.3 and were introduced in Chapter 3. All the
radio mobility-related decisions are still taken in eNodeB but the MME acts as the mobility anchor when the UE moves between different eNodeBs. The X2 protocol stack itself is
not shown but the architectural aspects are shown instead. The full protocol stack for X2,

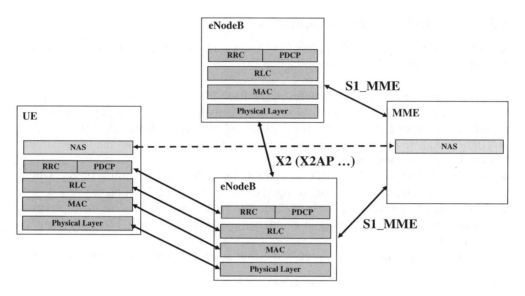

Figure 6.3   LTE control plane radio protocols in LTE architecture

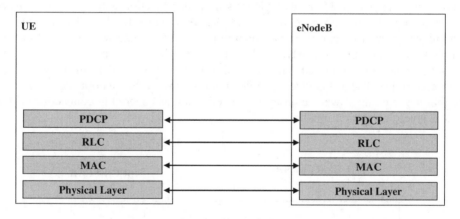

Figure 6.4   LTE user plane radio protocols in LTE architecture

including the underlying transport layers, is described in Chapter 3, while more details are discussed later in this chapter about radio-related signaling in the X2 Application Protocol (X2AP). Note also that most of the RRC signaling traffic goes through the PDCP layer, thus PDCP is part of the control plane as well.

For the user plane, all the network-side user-plane radio protocols are located in eNodeB as shown in Figure 6.4. This was possible from the system design point of view because the radio technology chosen does not need macro-diversity, so a flat architecture was easily facilitated without the need to provide user data continuously over the X2 interface.

6.3   The Medium Access Control

The MAC layer maps the logical channels to transport channels, as shown in Figure 6.5. The other tasks of the MAC layer in LTE are:

- MAC layer multiplexing/demultiplexing of RLC PDUs belonging to one or different radio bearers into/from transport blocks (TB) delivered to/from the physical layer on transport channels; also padding if a PDU is not completely filled with data.
- Traffic volume measurement reporting, to provide information about the traffic volume experience to the RRC layer.
- Error correction through HARQ, to control the uplink and downlink physical layer retransmission handling in eNodeB together with scheduling functionality.
- Priority handling between logical channels of one UE and UEs by means of dynamic scheduling, thus the scheduling in eNodeB is considered as MAC layer functionality, as with HSPA.
- Transport format selection (as part of the link adaptation functionality in the eNodeB scheduler).

Compared to WCDMA there is no ciphering functionality in the MAC layer, nor is there transport-channel-type switching as the user data are only transmitted over a single type of transport channel (UL-SCH or DL-SCH). As user identification is based on the physical layer signaling, there is no need for use of the MAC layer for UE identification. The downlink MAC layer functionality is shown in Figure 6.5. is otherwise identical to that in the uplink direction but, obviously, the BCCH and PCCH are not present there and only one user is considered in the UE uplink MAC structure. The MAC layer details in 3GPP are covered in [1]. The MAC layer specification also covers the random access procedure description, including power-ramping parameterization as covered in connection with the physical layer description in Chapter 5.

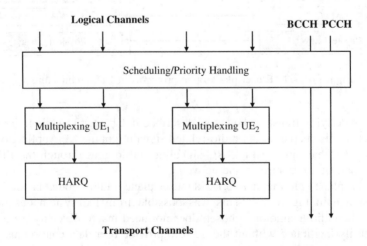

Figure 6.5   MAC Layer (downlink)

6.3.1 Logical Channels

The service that the MAC layer provides to the RLC layer takes place by means of the logical channels. Different logical channels are defined for different data transfer services in the uplink direction and the downlink direction.

The uplink logical channels mapping to transport channels is shown in Figure 6.6. The following logical channels are defined in LTE uplink:

● The Common Control Channel (CCCH) transfer controls information between the UE and the network. It is used when no RRC connection exists between the UE and the network.
● The Dedicated Control Channel (DCCH) is a point-to-point channel for dedicated control information between the UE and the network.
● The Dedicated Traffic Channel (DTCH) carriers all the user data for point-to-point connection.

In the uplink direction all the logical channels are mapped to the Uplink Shared Channel (UL-SCH). There is no logical channel mapped on the Random Access Channel (RACH) as that is not carrying any information above the MAC layer.

The following logical channels are defined in the LTE downlink, with the mapping to the downlink transport channels as illustrated in Figure 6.7:

● The CCCH, DCCH and DTCH have the same functionality as their uplink counterparts, now just delivering the control or user data information in the downlink direction. They are mapped to the Downlink Shared Channel (DL-SCH) in the transport channels.
● The Multicast Control Channel and the Multicast Traffic Channel are not included in Release 8 version of the specifications but are part of Release 9 LTE specification version. They carry multicast data (and related control information) in a point-to-multipoint fashion, which is similar to the Multimedia Broadcast Multicast Service (MBMS) part of WCDMA in Release 6.
● The Broadcast Control Channel (BCCH) carries the broadcast information, such as the necessary parameters for system access. It uses the Broadcast Channel (BCH) as the transport channel for the Master Information Block (MIB) while the actual System Information Blocks (SIBs) are mapped on DL-SCH.
● The Paging Control Channel carries paging information, to enable the network to page a device not in connected mode. It is mapped to the Paging Channel (PCH).

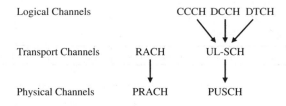

Figure 6.6 Mapping of the uplink logical and transport channels

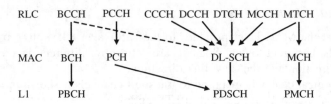

Figure 6.7   Mapping of the downlink logical and transport channels

6.3.2   Data Flow in MAC Layer

The MAC layer receives the data from the RLC layer, in the form of MAC Service Data Units (SDUs). The MAC Payload Data Unit (PDU) then consists of the MAC header, MAC SDUs and MAC control elements. The MAC control elements carry important control information, which is used to take care of several control functionalities, as shown in Figure 6.8. In the uplink direction (in connection with UL-SCH) the MAC payload can contain several control information elements besides data (in MAC SDU). These include:

- Buffer status report, to tell how much data there is in the UE waiting for transmission and information of the priority of the data in the buffer.
- Power headroom report to indicate the available uplink transmission power resources.
- Contention resolution procedure information (CCCH and C-RNTI).

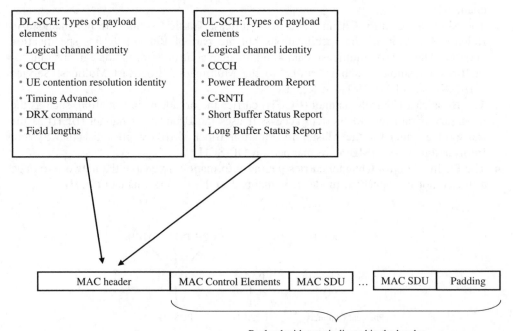

Figure 6.8   MAC PDU structure and payload types for DL-SCH and UL-SCH

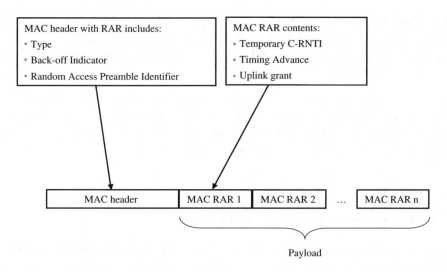

Figure 6.9   MAC PDU structure with random access response

In the downlink direction (in connection with DL-SCH), the following control information can be carried in the MAC control elements.

- Control of the DRX operation (when to start, when to stop and so forth). Timing advance commands to adjust uplink timing.
- Contention resolution information. The MAC header and payload in case of RACH are different as the key is to enable the access procedure to be completed and deal with the potential collisions or overload situations in the cell that might have occurred. The MAC layer random access signaling is indicated in Figure 6.9 and it consists of the necessary control information needed to enable data transmission after a successful RACH procedure (detection of preamble after the ramping as covered in Chapter 5). The information transmitted in the MAC Random Access Response (RAR) covers:
  - Timing advance indicating the necessary adjustment needed for the uplink transmission to ensure different users fit without overlap or extra guard periods for the 1 ms resource allocations.
  - Uplink grant indicates the detailed resources (time and frequency domain) to be used in the uplink direction.
  - Temporary C-RNTI provides the UE with a temporary identity to be used in completing random access operations.
- The header indicates possible back-off and the identification of the preamble received, which UE had chosen (randomly) in connection with the physical layer random access procedure as described in Chapter 5.

6.4   The Radio Link Control Layer

The RLC layer has the following basic functionalities:

- Transferring the PDUs received from higher layers – from the RRC (Common Control Channel) or the PDCP (other cases, including user plane).

- Then, depending on the RLC mode used, error correction with ARQ, concatenation/segmentation, in-sequence delivery and duplicate detection may occur.
- Protocol error handling to detect and recover from protocol error states caused by, for example, signaling errors.

Fundamentally, the key difference of the LTE RLC specifications [2] when compared with WCDMA is the lack of ciphering functionality in RLC. The re-segmentation before RLC retransmission is also enabled. Thus, compared to the order of sequences in [3], the buffer occurs before segmentation/concatenation functionality.

6.4.1   RLC Modes of Operation

The RLC can be operated in three different modes:

- Transparent Mode (TM). In the TM mode the RLC only delivers and receives the PDUs on a logical channel but does not add any headers to it and thus it does not track received PDUs between receiving and transmitting entities. The TM mode of operation is only suited for services that do not use physical layer retransmissions or that are not sensitive to delivery order. Thus from the logical channel only BCCH, CCCH and PCCH can be operated in TM mode. In WCDMA the AMR speech call (and CS video) was used in the transparent mode as well as there were no retransmissions. In LTE, however, all point-to-point user data basically has physical layer retransmissions. Thus a minimum in-sequence delivery of the packet to PDCP layer is expected and TM mode cannot be used.
- The Unacknowledged Mode (UM) of operation provides more functionality, including in-sequence delivery of data, which might be received out-of-sequence due to HARQ operation in lower layers. The UM Data (UMD) is segmented or concatenated to suitable size RLC SDUs and the UMD header is then added. The RLC UM header includes the sequence number for facilitating in-sequence deliver (as well as duplicate detection). As shown in Figure 6.10, the receiving side will then based on the header information perform the re-ordering and then necessary reassembly in response to the segmentation or concatenation applied. Besides the DCCH and DTCH, the UM RLC has been planned to be used for multicast channels (MCCH/MTCH) expected to be completed beyond Release 8.

The Acknowledged Mode (AM) of RLC operation provides retransmission, in addition to the UM mode functionalities, if PDUs are lost due to the operation of lower layers. The AM Data (AMD) can also be resegmented to fit the physical layer resources available for retransmissions, as indicated in Figure 6.11. The header contains now information about the last correctly received packet on the receiving side along with the sequence numbers, as with the UM operation.

6.4.2   Data Flow in the RLC Layer

When carrying out the AM RLC operation, the RLC layer receives data from the PDCP layer. The data is stored in the transmission buffer and then, depending on the resources available, segmentation or concatenation is used. In the downlink direction

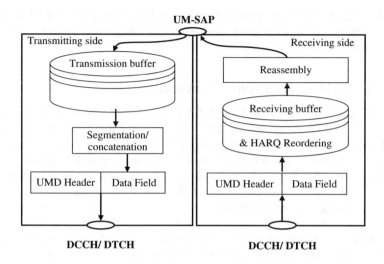

Figure 6.10   RLC UM operation

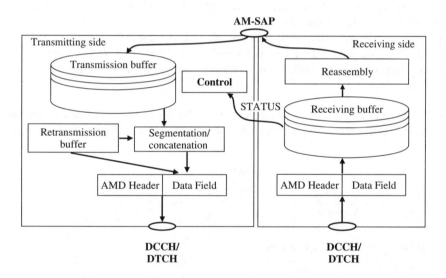

Figure 6.11   RLC AM mode operation

(DTCH/DCCH) the following methods of control are available for AM RLC purposes, either in the AMD PDU or in the status PDU:

- a ten-bit Sequence Number (SN), to enable a sufficiently long in-sequence delivery window;
- last correctly received SN and not correctly received SN(s) (detected as lost).

When the receiving entity receives a RLC PDU, it checks for possible duplicated parts and will then forward further for reassembly (assuming no SNs are missing in between) and provides data to the next protocol layer (in this case for the PDCP) for further processing. If there was a SN missing, the receiving entity will request a retransmission from the transmitting side.

6.5   Packet Data Convergence Protocol

The Packet Data Convergence Protocol (PDCP) is located above the RLC layer of the user plane and the PDCP is used for most of the RRC messages. The key difference to WCDMA is that now all user data is going via the PDCP layer, due to the fact that ciphering is now in PDCP and the PDCP itself is located in eNodeB. In some early plans of LTE architecture, PDCP was planned to be in the other side of the S1 interface (on the core network) but was later placed in eNodeB along with all the other radio protocols. The key functionalities of the PDCP are:

- Header compression and respectively decompression of the IP packets. This is based on the Robust Header Compression (ROHC) protocol, which is specified in the Internet Engineering Task Force (IETF) [4–9] and is also part of the WCDMA PDCP layer. Header compression is more important for smaller IP packets in question, especially in connection with the VoIP service, as the large IP header could be a significant source of overhead for small data rates.
- Ciphering and deciphering both the user plane and most of the control plane data, a functionality that, in WCDMA, was located in the MAC and RLC layers.
- Integrity protection and verification, to ensure that control information is coming from the correct source.

The PDCP layer receives PDCP SDUs from the NAS and RRC and, after ciphering and other actions, as shown in Figure 6.12 for operation for the packets that are associated

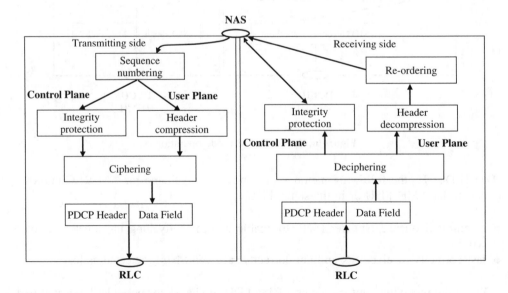

Figure 6.12   PDCP Layer operation for the packets associated with PDCP SDU

with a PDCP SDU, the data is forwarded to the RLC layer. On the receiving side, the data is received from the RLC layer. Besides the functionality listed above, the PDCP layer has specific functions in connection with handover events (intra-LTE). The PDCP does the in-order delivery function in the downlink direction and detects duplicates. In the uplink direction the PDCP retransmits all the packets that have not been indicated by lower layers to be completed as lower layers will flush al the HARQ buffers with handover. In the downlink direction the PDCP layer will forward the non-delivered packets to the new eNodeB as described in Chapter 7. This is to ensure that no data are lost in the connection of a handover event between LTE eNodeBs. The LTE PDCP specification in 3GPP is covered in [10].

6.6   Radio Resource Control (RRC)

Radio Resource Control messages are a major part of control information exchanged between UE and EUTRAN. Comparing with the RRC in UTRAN, the RRC in EUTRAN was simplified significantly by reducing the number of messages and by removing the redundancies in the messages. The same protocol specification language is being used with RRC as in WCDMA – Abstract Syntax Notation One (ASN.1) – as it has been proven to facilitate evolution efficiently between different release versions with the extension capability inside ASN.1. The encoding of the X2 and S1 interface messages also uses ASN.1. The LTE RRC specification in [11] has the ASN.1 message descriptions at the end. The introduction to the ASN.1 basics is given in section 6.8.

6.6.1   *UE States and State Transitions Including Inter-RAT*

Unlike the UTRAN (WCDMA), UE states in EUTRAN are simplified significantly and there are only two states: RRC_CONNECTED and RRC_IDLE, depending on whether the RRC connection has been established or not.

In the RRC_IDLE state, the UE monitors a paging channel to detect incoming calls, acquires system information and performs neighboring cell measurement and cell (re-)selection. In this state, a UE specific DRX may be configured by the upper layer and the mobility is controlled by the UE.

In the RRC_CONNECTED state, the UE transfers/receives data to/from the network. For this, the UE monitors control channels, which are associated with the shared data channel to determine if data are scheduled for it and it provides channel quality and feedback information to the eNodeB. Also in this state, the UE performs neighboring cell measurement and measurement reporting based on the configuration provided by the eNodeB. Unlike the UTRAN system, the UE can acquire system information from BCCH during the RRC_CONNECTED state. At lower layers the UE may be configured with a UE-specific DRX and mobility is controlled by the network, i.e. handover.

6.6.1.1   E-UTRAN States and Inter-RAT State Mobility

Figure 6.13 shows the mobility support between EUTRAN, UTRAN and GERAN. (Typically, the UE is only in Cell_FACH for a very short time). Therefore, 3GPP has not defined the direction transition from 3G Cell_FACH to LTE.

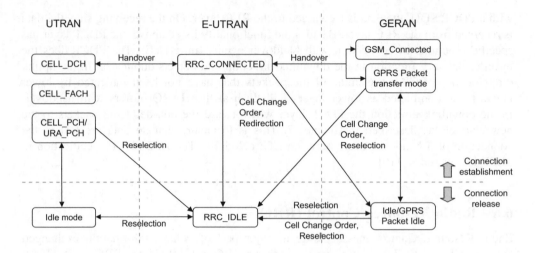

Figure 6.13 E-UTRAN RRC states and state transitions among 3GPP systems

6.6.1.2 Signaling Radio Bearers (SRB)

Signaling Radio Bearers (SRBs) are special radio bearers that convey only RRC messages and NAS messages. Three SRBs are defined. SRB0 is used for RRC messages using CCCH, for example during RRC Connection setup or during radio link failure. Thus, for instance, the following messages are transferred via SRB0: RRC Connection Request messages, RRC Connection Setup messages, RRC Connection Reject messages, RRC Connection Reestablishment Request messages, RRC Connection Reestablishment messages, and RRC Connection Reestablishment Reject messages. Once a RRC connection is established, SRB1 is used to transfer both RRC messages using DCCH and NAS messages until the security is activated. Once the security is successfully activated, SRB2 is setup and NAS messages are transferred via SRB2 while RRC messages are still transferred via SRB1. SRB2 has a lower priority than SRB1.

6.6.2 RRC Functions and Signaling Procedures

The following functions are provided by RRC protocol layer:

- broadcast of system information;
- paging;
- establishment, maintenance and release of an RRC connection between the UE and e-UTRAN;
- security functions including key management;
- establishment, configuration, maintenance and release of point to point Radio Bearers;
- UE measurement reporting and control of the reporting;
- handover;
- UE cell selection and reselection and control of cell selection and reselection;
- context transfer between eNodeBs;
- NAS direct message transfer between network and UE;

- UE capability transfer;
- generic protocol error handling;
- support of self-configuration and self-optimization.

6.6.2.1 Broadcast of System Information

System information contains both non-access stratum-related (NAS) and access stratum-related (AS) information. Based on the characteristics and uses of the information, the system information elements are grouped together into a master information block (MIB) and different system information blocks (SIBs).

As the MIB is the most important information block, the MIB is transferred on the BCH every 40 ms and is repeated within 40 ms. The first transmission of the MIB is scheduled at SFN mod 4 = 0 in the sub-frame #0. The UE acquires the MIB to decode SCH. The MIB contains DL system bandwidth, and PHICH configuration and system frame number (SFN).

SIB1 is scheduled in a fixed manner with a periodicity of 80 ms and is repeated within 80 ms. The first transmission of the SIB1 is scheduled at SFN mod 8 = 0 in the sub-frame #5. SIB 1 contains cell access-related information (for example, PLMN identity list, tracking area code, and cell identity), information for cell selection (for example, minimum required Rx level in the cell and offset), p-Max, frequency band indicator, scheduling information, TDD configuration, SI-window length and system information value tag.

All other SIBs except SIB1 are contained in SI message(s). Each SI message is transmitted periodically in time domain windows (SI-Window) and SI-windows for different SI messages do not overlap. The length of SI-window is defined in SIB1 and common for all SI messages.

Figure 6.14 shows how UE can find each SI message to read actual SIBs in it. SIB2 is not listed in the scheduling information in SIB1 but the first SI message contains SIB2 as the first entry.

SIB2 contains radio resource configuration information common to all UEs. More specifically, SIB2 includes access barring information, radio resource configuration of common channels (RACH configuration, BCCH configuration, PCCH configuration, PRACH configuration, PDSCH configuration, PUSCH configuration and PUCCH configuration, sounding reference signal configuration, UL power control information), timers and constants, which are used by UEs, MBSFN sub-frame configuration, time alignment timer and frequency information (UL EARFCN, UL bandwidth and additional spectrum emission).

SIB3 contains cell-reselection information, which is common for intra-frequency, inter-frequency and/or inter-RAT cell re-selection. Speed-dependent scaling parameters are also included in SIB3 to provide different set of re-selection parameters depending on UE speed.

SIB4 contains neighbor cell-related information only for intra-frequency cell re-selection. Thus the intra-frequency blacklisted cell list is also included in SIB4 in addition to the intra-frequency neighboring cell list with Q_{offset}.

SIB5 contains information relevant only for inter-frequency cell re-selection, like EUTRAN inter-frequency neighboring cell-related information and the EUTRAN inter-frequency blacklisted cell list.

SIB6 contains information relevant only for cell re-selection to the UTRAN. Thus UTRA FDD and TDD frequency information is included in SIB6.

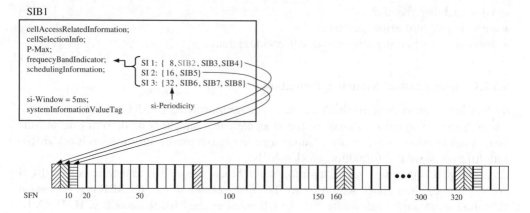

determine integer value $x = (n-1)*w$, where n is the order of entry in the list of SI message and w is the *si-WindowLength*

si-Window starts at subframe #a, where a = x mod 10, in the next radio frame for which SFN mod T = FLOOR $(x/10)$ where T is the si-Periodicity of the concerned SI message

Figure 6.14 Acquisition of SI message

SIB7 contains information relevant only for cell re-selection to the GERAN like GERAN neighboring frequency list.

SIB8 contains information relevant only for cell re-selection to the CDMA2000 system. Thus SIB8 contains CDMA2000 system time information, HRPD related parameters and 1xRTT related parameters.

SIB9 contains a home eNodeB identifier, which is maximum 48 octets, so that UE can display home eNodeB identifier to help manual home eNodeB selection.

SIB10 contains Earthquake and Tsunami Warning System (ETWS) primary notification and SIB 11 contains ETWS secondary notification.

In case the system information is modified, the SI messages may be repeated during the *modification period*. The *modification period* starts at SFN mod *modification period* = 0. During the first modification period, the system information modification indicator is sent to UE by the paging message and in the next modification period, the network transmits the updated system information.

The *modification period* is calculated as follows:

Modification period (in number of radio frames)

= modificationPeriodCoeff* defaultPagingCycle DIV 10 ms

where modificationPeriodCoeff: signaled in BCCH-Configuration in SIB2, value = 1, 2, 4, 8 and the defaultPagingCycle is 320 ms, 640 ms, 1280 ms, 2560 ms (signaled in PCCH-Configuration in SIB2).

The value tag in SIB1 indicates a change in SI messages. The SI message is considered to be valid for a maximum of 3 hours from the moment it was acquired.

6.6.2.2   Paging

The main purpose of the paging message is to page a UE in RRC_IDLE mode for a mobile terminated call. The paging message can also be used to inform UEs, which are in RRC_IDLE as well as in RRC_CONNECTED, that system information will be changed or that the ETWS notification is posted in SIB10 or SIB11.

As a UE may use Discontinuous Reception (DRX) in idle mode in order to reduce the power consumption, the UE should be able to calculate when to wake up and to check the correct sub-frame. For this, the UE stores the default paging cycle and the number for the paging group when it receives the necessary information from the SIB2 and applies the calculation as described in section 7 in TS36.304.

The UE can be paged either by S-TMSI (temporary UE ID allocated by MME) or IMSI. In Release 8, IMEI paging is not considered.

For CS fallback, if the UE has registered to the CS core network and received a temporary identity from the CS core network, CS paging can be delivered to the UE with the corresponding core network indicator. (CS or PS). In this case, the UE should deduce the correct CS UE identity from the PS UE identity included in the paging message and include the CS UE identity in the paging response once it is moved to another RAT. The details for the CS fallback procedure can be found in the 6.6.2.8 Handover section, including the latest additions in Release 9.

6.6.2.3   UE Cell Selection and Reselection and Control of Cell Selection and Reselection

Based on idle mode measurements and cell selection parameters provided in SIBs, the UE selects the suitable cell and camps on it. In LTE systems, the priority parameter has been newly introduced. The eNodeB can provide a priority per LTE frequency and per RAT in SIBs or the RRC connection Release message. In case a UE is camping on a cell on the highest priority frequency, the UE does not require measuring cells in other frequencies or RATs as long as the signaling strength of the serving cell is above a certain level. On the other hand, if the UE is camped on a cell in the low priority layer, the UE should measure other cells with higher priority frequency or RAT regularly.

This priority concept is also adapted to Release 8 UTRAN and GERAN (or GSM) systems for inter-RAT cell reselection.

The LTE idle mode mobility is explained in Chapter 7 in more detail.

6.6.2.4   Establishment, Maintenance and Release of an RRC Connection between the UE and e-UTRAN

The RRC Connection Setup procedure, as shown in Figure 6.15 is triggered by a request from the UE NAS layer due to various reasons like mobile originated call, NAS signaling transfer or paging response. Consequently the RRC connection is established between UE and eNodeB and SRB1 is setup. In case the network is overloaded, the eNodeB can set access class bearer parameters in SIB2 appropriately and/or reject RRC Connection Request with a wait time.

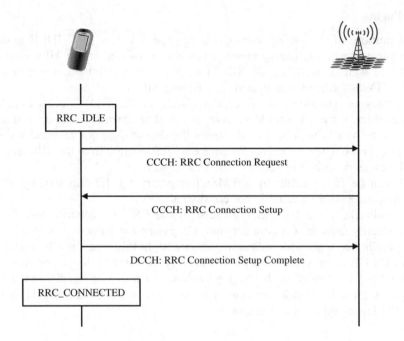

Figure 6.15 RRC Connection Setup procedure

If a UE has a valid S-TMSI, the UE includes it in the RRC connection request message. Otherwise the UE includes a 40 bits random value. Due to the limited size of messages, only five establishment causes are defined in the RRC Connection Request message: emergency, high-priority access, mobility-terminated access, mobile-originated signaling and mobile-originated data. The NAS service-request message is conveyed in the RRC Connection Setup Complete message.

SIB1 broadcasts, at most, six PLMN identities (which can facilitate also the network sharing case). The UE selects one and reports it in the RRC Connection Setup Complete message. In case the UE is registered to an MME, the UE also includes the identity of the registered MME in the RRC Connection Setup Complete message. Then the eNodeB finds/selects the correct MME (which stores UE idle mode context) in case that S1-flex is deployed and starts the S1 connection setup.

After a successful RRC Connection setup procedure, the UE moves to the RRC_CONNECTED state.

6.6.2.5 Security Functions Including Key Management

The security keys for Access Stratum (AS), covering user data and RRC control signaling are different from those used on the Evolved Packet Core (EPC) side. From the eNode B perspective, the following keys are necessary (ignoring the keys used for NAS signaling):

- K_{eNodeB} is derived by the UE and MME from the 'master key' (K_{ASME}) and then provided by the MME to eNodeB. K_{eNodeB} is used to derive the necessary keys for AS traffic and also for the derivation of K_{eNodeB^*} during handover.

- **K**$_{eNodeB*}$ is derived by the UE and the source eNodeB from either K$_{eNodeB*}$ or from a valid NH, from K$_{eNodeB*}$ the UE and the target eNodeB will derive a new K$_{eNodeB}$ for AS traffic at handover.
- **K**$_{UPenc}$ is used for the user plane traffic ciphering and derived from K$_{eNodeB}$.
- **K**$_{RRCint}$ is derived from K$_{eNodeB}$ to be used for RRC message integrity handling.
- **K**$_{RRCenc}$ is derived from K$_{eNodeB}$ to be used for RRC message ciphering.

Next Hop (**NH**) is an intermediate key that is used in the derivation of K$_{eNodeB*}$ for the provision of security for intra-LTE handover purposes. There is a related counter, Next Hop Chaining Counter (**NCC**), which allows it to be determined whether the next **K**$_{eNodeB*}$ needs to be based on the current K$_{eNodeB}$ or if a fresh NH is needed. If no fresh NH is available, the K$_{eNodeB*}$ is derived by the UE and source eNodeB from the target PCI and K$_{eNodeB}$ (horizontal key derivation) or if a fresh NH is available, then derivation is based on the target PCI and new NH provided (vertical key derivation).

The EPC and UE will share the same master key after the security procedure is completed, and this master key (K$_{ASME}$) is not provided outside the EPC but only the keys derived from it and necessary for the traffic between eNodeB and UE are delivered to eNodeB. The eNodeB retains a given key as long as a UE is connected to it but will delete the keys when the UE is moving to idle mode (or to another eNodeB). The integrity and ciphering algorithm can be changed only upon handover. The AS keys are changed upon every handover and connection re-establishment. In the LTE system, a handover can be performed only after security is activated.

RRC integrity and ciphering are always activated together and never de-activated, but it is possible to use NULL ciphering algorithm for specific cases, such as in connection with emergency calls (note that some specific aspects of emergency calls are completed in Release 9). The security algorithms also use a sequence number as inputs in addition to the keys – this sequence number consists of PDCP Sequence Number and Hyper Frame Number (HFN).

6.6.2.6   Establishment, Maintenance and Release of Point-to-Point Radio Bearers

The RRC connection reconfiguration procedure, as shown in Figure 6.16 is used to maintain and to modify radio bearers and to release data radio bearers. (i.e. SRBs cannot be released by the RRC connection reconfiguration procedure).

The parameters defined in the radio resource configuration information element group in the RRC Connection Reconfiguration message deal mainly with the configuration of radio bearer. In case RRC Connection Reconfiguration message is used to setup a new Data Radio Bearer (DRB), the corresponding NAS message is also included in the same message.

6.6.2.7   UE Measurement Reporting and Control of the Reporting

The Measurement Configuration parameter in the RRC Connection reconfiguration message is used to configure measurements in the UE (for measurements inside LTE or for measurements from other Radio Access Technologies (RATs) and the Measurement report message is used for reporting.

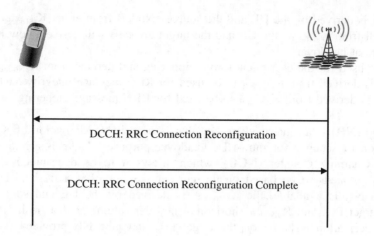

Figure 6.16   RRC connection reconfiguration procedure

The measurement configuration consists of the following parameters.

1   Measurement objects: The objects on which the UE performs the measure-
     ments. E-UTRAN configures only a single measurement object for a given
     frequency.
     ● For intra-frequency (measurements at the same downlink carrier frequency as the
       serving cell) and inter-frequency measurement (measurements at frequencies that
       differ from the downlink carrier frequency of the serving cell), a measurement
       object is a single EUTRA carrier frequency. For this carrier frequency, EUTRAN
       can configure a list of cell-specific offsets and a list of blacklisted cells.
     ● For inter-RAT UTRA measurements, a measurement object is a set of cells on a
       single UTRA carrier frequency.
     ● For inter-RAT GERAN measurements, a measurement object is a set of GERAN
       carrier frequencies.
     ● For inter-RAT CDMA2000 measurements, a measurement object is a set of cells
       on a single CDMA2000 carrier frequency.
2   Reporting configuration: The reporting configuration contains reporting criteria
     and reporting format. The reporting criteria are criteria for UE to trigger a measure-
     ment reporting and can be either event triggered or periodic. Periodic reporting, in
     particular, is used for the measurement for automatic neighbor cell search. Report-
     ing format addresses the quantities that should be included in the measurement
     report (for example, the number of cells to report).
     ● Reporting criteria: For event triggered report, the criteria for EUTRA measure-
       ments are A1, A2, A3, A4 and A5. In case of inter-RAT measurement, B1 and
       B2 are used.
     ● Event A1: Serving becomes better than absolute threshold.
     ● Event A2: Serving becomes worse than absolute threshold.
     ● Event A3: Neighbor becomes amount of offset better than serving.
     ● Event A4: Neighbor becomes better than absolute threshold.

- Event A5: Serving becomes worse than absolute threshold1 and Neighbor becomes better than another absolute threshold 2.
- Event B1: Neighbor becomes better than absolute threshold.
- Event B2: Serving becomes worse than absolute threshold 1 and Neighbor becomes better than another absolute threshold 2.

3   Measurement identities: Measurement identity binds one measurement object with one reporting configuration. It is used as a reference number in the measurement report.

4   Quantity configurations: One quantity configuration can be configured per RAT (that is, EUTRA, UTRAN, GERAN and CDMA2000) and contains the filter coefficient for the corresponding measurement type.

5   Measurement gaps: Periods where UE may use to perform measurement. Measurement gaps are used for inter-frequency and inter-RAT measurements.

6.6.2.8   Handover

Usually eNodeB triggers handover based on the measurement result received from the UE. A handover can be performed inside EUTRAN or to EUTRAN from an other RAT or from EUTRAN to an other RAT and is classified as intra-frequency intra-LTE handover, inter-frequency intra-LTE handover, inter-RAT towards LTE, inter-RAT towards UTRAN handover, inter-RAT towards GERAN handover, and inter-RAT towards cdma 2000 system handover.

Intra-LTE handover
For intra-LTE handover, the source and target are both in the LTE system and the RRC connection reconfiguration message with the mobility control information (including parameters necessary for handover) used as a handover command. If an X2 interface exists between the source eNodeB and the target eNodeB, the source eNodeB sends an X2: Handover Request message to target eNodeB to prepare the target cell. The target eNodeB builds a RRC Connection reconfiguration message and transfers the Handover Request Acknowledge message to source eNodeB in the X2, as shown in Figure 6.17.

Inter-radio access technology handover to other Radio Access Technology
Based on UE measurement reporting, the source eNodeB can decide the target Radio Access Technology (RAT) and currently In Release 8 the target RAT can be UTRAN, GERAN or CDMA2000 system. To perform the inter-RAT handover to another RAT, the actual handover command message is built by the target RAT and is sent to the source eNodeB transparently. Source eNodeB includes this actual handover command in the Mobility From EUTRA Command message as a bit string, and sends it to the UE.

   As the LTE system supports the packet service only, if a multi-mode UE wants to start a CS service, CS fallback can be performed. For a CS fallback to the UTRAN system, a handover procedure is mainly used. For a CS fallback to the GERAN system, either a handover procedure or a cell-change order procedure are used. For a CS fallback to the CDMA 2000 system, the RRC Connection release procedure is used. To align the behavior of cell-change order with inter-RAT handover, the cell-change order procedure uses the Mobility From EUTRA Command message. Also when no VoIP service is supported in the target RAT, SR-VCC (Single Radio Voice Call Continuity) can be used if supported.

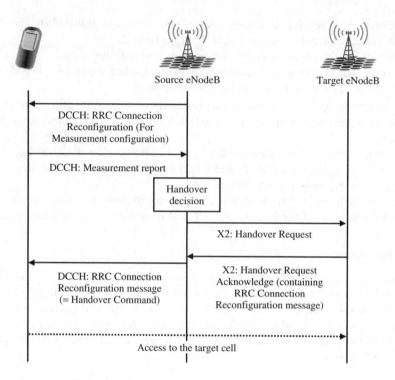

Source eNodeB   Target eNodeB

DCCH: RRC Connection
Reconfiguration (For
Measurement configuration)

DCCH: Measurement report

Handover
decision

X2: Handover Request

DCCH: RRC Connection
Reconfiguration message
(= Handover Command)

X2: Handover Request
Acknowledge (containing
RRC Connection
Reconfiguration message)

Access to the target cell

Figure 6.17   Inter-eNodeB Handover procedure

Inter-radio access technology handover from other Radio Access Technology
The handover is triggered based on the criteria defined in the source RAT system. When
the target eNodeB receives the S1: Handover Request message from MME, it allocates
necessary resources, builds a RRC Connection Reconfiguration message (= Handover
Command) and sends it in the S1: Handover Request Acknowledge message to MME.
Thus this RRC Connection Reconfiguration message will be transferred to the UE via the
source RAT.

6.6.2.9   CS Fallback

As initial LTE deployment may not support voice service as such, 3GPP agreed to provide
a means to fall back to the GERAN, UTRAN or cdma2000 system for a voice service.

As UE has to move to GERAN, UTRAN or cdma2000 systems if it uses CS-domain services, the CS fallback function is available only if the E-UTRAN coverage is overlapped
by either GERAN, UTRAN or cdma2000 coverage.

CS fallback is based on the following architecture and an SGs interface between MSC
server and MME is required, as shown in Figure 6.18. Via SGs interface, the UE performs
combined EPC and IMSI attachment and detachment procedure. Also for MT CS fallback
service, CS paging is delivered via SGs interface to MME and finally to the UE. The SGs
interface is based on the Gs interface.

The SGs interface is also used to deliver MO SMS and MT SMS and in this case
CS fallback does not occur. Support for SMS over SGs is mandatory for UE, MME and

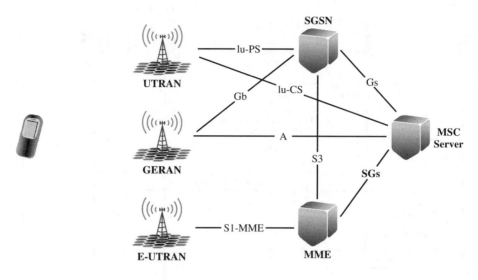

Figure 6.18 CSFB architecture

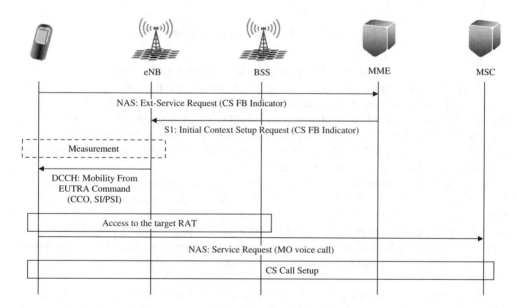

Figure 6.19 CS fallback with Cell Change Order

MSC, which support CS fallback, whereas there is no support for CS fallback for those that support SMS over SGs.

The use of CS fallback solution is intended to be a temporary solution until the IMS-based solution for voice is available but still the solution is important for enabling a smooth transition to LTE. The original intention was to use only PS handover (to UMTS) and cell change order (to GSM) as the mandatory solutions and those methods are illustrated in Figure 6.19 and Figure 6.20.

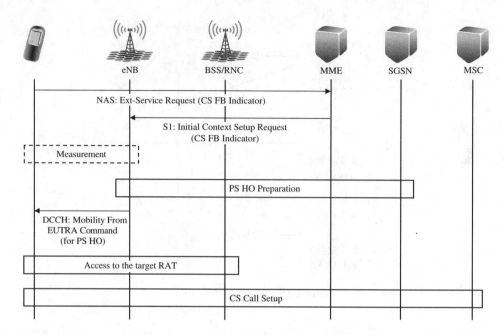

Figure 6.20   CS fallback with PS handover

For an incoming CS call, the UE is registered to the CS core network in addition to the MME. If a UE receives CS paging or a UE wants to initiate a mobile-originated CS call, the UE indicates that NAS: Ext-Service Request message is for the CS fallback to the MME. Then the MME indicates that S1:Initial Context Setup Request message is for the CS Fallback (CS FB). Depending on the target system and the capability of target cell, eNodeB performs PS handover or cell-change order procedure and the UE starts the CS call setup once it moves in to the target cell.

However further solutions were allowed later as it was assumed that, in all cases, inter-RAT mobility is not the first feature to be deployed in the commercial network. Thus RRC Connection Release with redirection information was added as a further option in Release 8. And to improve the performance of the basic RRC Connection Release with a redirection information solution, further enhancement was added as an additional option in Release 9 to make it possible to provide system information about the target system.

All options for CS fallback to GSM, WCDMA and cdma2000 are listed in Table 6.1. The challenge created by multiple options is that it will be difficult to ensure a 'universal' solution, available in all devices at the beginning, as not all of the solutions are likely to be available for testing with UEs in the first phase.

The fundamental challenge with the solutions based on RRC Connection Release with Redirection is the trade-off between reliability and needed time to setup the CS call. When forcing the UE to start from IDLE in the target system (for example, RRC Connection Release with redirection) it takes longer than the native call setup time in the target system because there will always be an additional delay because the UE has not been connected to the target system prior to the call set-up. The following sections looks in more detail at the different CS fallback options.

Table 6.1 CS fallback options in Release 8 and 9

Target system	Solutions	Release
UTRAN	RRC Connection Release with Redirection without Sys Info	Release 8
	RRC Connection Release with Redirection with Sys Info	Release 9
	PS handover with DRB(s)	Release 8
GSM	RRC Connection Release with Redirection without Sys Info	Release 8
	RRC Connection Release with Redirection with Sys Info	Release 9
	Cell change order without NACC	Release 8
	Cell change order with NACC	Release 8
	PS handover	Release 8
cdma2000 (1xRTT)	RRC Connection Release with Redirection	Release 8
	enhanced 1xCSFB	Release 9
	enhanced 1xCSFB with concurrent HRPD handover	Release 9
	dual receiver 1xCSFB (RRC Connection Release without Redirection)	Release 9

CS fallback to UMTS

To perform CS fallback to UMTS, RRC Connection Release with redirection information is the least-effort solution because all the network and UE will support RRC Connection Release procedure. In this solution, if the UE starts a CS call setup request via an extended service request, or receives a CS paging, the UE's RRC Connection in the LTE system is released and the UE is redirected to the UMTS system. In this redirection information, the eNodeB can indicate a target frequency. Then the UE moves to the target RAT and searches for a suitable cell by utilizing the frequency information in the RRC Connection Release. Once the UE finds a suitable cell, it starts from IDLE and acquires necessary system information blocks to access the cell. This is the Release 8 basic redirection solution as illustrated in Figure 6.21. Acquisition time for the system information block can be very long depending on the implementation of system information block scheduling and the sizes of the system information blocks.

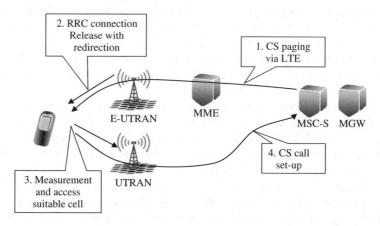

Figure 6.21 CS fall-back to UTRAN with RRC Connection Release with redirection information

If the system information is well scheduled and the system information blocks are compact, acquiring system information blocks in the target system will be relatively fast. However this may not always be the case and in some implementations, system information block scheduling may not be efficient and system information block acquisition may take even more than two seconds. Thus in Release 9 enhancement, the eNodeB is allowed to provide UTRAN system information blocks which IDLE UEs needs to acquire access to the cell (SIB1, SIB3, SIB7 and SIB11) for a maximum of 16 cells so that the UE can omit the system information block acquisition process after the cell selection (assuming the information was accurate). As the eNodeB can provide system information for the multiple cells, the eNodeB can skip the security activation and measurement procedure in the source side and consequently speed up the redirection preparation on the source side without endangering reliability too much. However as eNodeB has to have the up-to-date UTRAN system information, this may require new functionalities in the network nodes as well. (eNodeB, RNC and SGSN). The most efficient CS fallback solution to UMTS is PS handover. In LTE, after security is activated, as the default bearer is always set up, PS handover can always be used for CS fallback. As the default bearer is typically a non-GBR bearer, having the default bearer in the target system does not require any resources to be reserved. For reasons of reliability, the target cell for PS handover is usually selected according to the measurement result. However if deployment is assured, blind handover can also be performed to reduce the CS fallback delay. A delay in measurement can occur depending on how many frequencies the UE has to measure. When the PS handover procedure is completed, the UE is already connected to the target, which makes the procedure clearly faster than any type of redirection. In Release 9 the option of 'Signaling Radio Bearer only PS handover' was originally included in the specifications but was later removed to minimize the CS fallback options.

CS fallback to GSM

Like CS fallback to UMTS, RRC Connection Release with redirection information is the least-effort solution for CS fallback to GSM. In this solution, the UE RRC Connection in the LTE system is released and the UE is redirected to the GSM system to set up the CS service like CS fallback to UMTS. In this redirection information, the eNodeB can indicate a target frequency. Then the UE moves to the target RAT and searches for a suitable cell by using the frequency information in the RRC Connection Release. Once the UE finds a suitable cell, it starts from IDLE and acquires the necessary system information blocks to access the cell. With the Release 9 enhancement, the eNodeB can include a selected set of system information blocks for maximum 32 cells for GSM to avoid the measurement procedure in the LTE side for reduced delay, without endangering the reliability too much, as shown in Figure 6.22.

In addition to RRC Connection Release with redirection information, the cell change order and the PS handover can be used for CS fallback. To perform either the cell change order or the PS handover, the eNodeB sends the same message – the Mobility From EUTRA Command message. However the contents of the message will be different depending on the purpose. The UE behavior is also different. In case of the cell change order, the UE moves to the target GERAN cell and starts from IDLE. If NACC is used, the eNodeB provides GERAN system information blocks in the Mobility From EUTRA Command message and the UE can skip the system information block acquisition step. If the Mobility From EURA Command is used for PS handover, as in the UTRAN case,

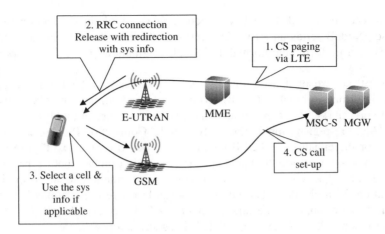

Figure 6.22 CS fall-back to GSM with RRC Connection Release with redirection information with sys info

the target GERAN cell is prepared to accept the UE during the handover preparation and the UE will still be CONNECTED in the target GERAN cell after the successful handover.

As in the UMTS case, all three solutions can be performed after the UE measurement and choosing the target cell or blindly without UE measurement procedure if the actual deployment allows this.

The most efficient CS fallback solution to GSM is PS handover. However, PS handover has not been widely deployed in GERAN network. Thus 3GPP decided to mandate the cell change order as the CS fallback to GSM solution. An alternative is the UE capability signaling informing network.

CS fallback to the cdma2000 system

The CS fallback mechanisms to cdma2000 system, 1xRTT system, was defined as well and this is called 1xCSFB. As in the CS fallback to the 3GPP systems, the UE initiates the NAS signaling to enable the delivery of CS-domain services and the MME indicates to the eNodeB that CS fallback is required. Consequently, the eNodeB executes one of the following CS fallback to 1xRTT procedures depending on network support and UE capabilities:

- The RRC Connection Release, with redirection to 1xRTT (Basic Release 8 mechanism), is the default procedure and the eNodeB can solicit 1xRTT measurements from the UE before performing RRC Connection Release with redirection.
- Enhanced 1xCSFB: 1xRTT handover signaling is tunneled between the UE and the 1xRTT network. The eNodeB sends a Handover From EUTRA Preparation Request message to trigger the UE to send UL Handover Preparation Transfer message containing 1xRTT dedicated information, which is transferred to the 1xRTT network, And the response from the 1xRTT network triggers the eNodeB to send a Mobility From EUTRA Command message containing 1xRTT channel assignment message. Based on this information, UE acquires a traffic channel in the 1xRTT network.

The dual receiver 1xCSFB: RRC connection release without redirection information. In this case eNodeB sends an RRC Connection Release without redirection information and the UE then performs the normal 1xCS call origination or termination procedure in the 1xRTT.

The eNodeB advertises that it supports each CS fallback to the 1xRTT mechanism in the system information block 8 (SIB8). To originate or terminate the CS services in the 1xRTT network, a 1xCSFB capable terminal may pre-register in the 1xRTT network via the E-UTRAN. Pre-registration applies only to RRC Connection Release with redirection information or the enhanced 1xCSFB mechanisms. Based on the 1xRTT pre-registration parameters in system information block 8, the UE determines whether pre-registration is needed or not. The UE acquires the necessary information to perform the 1xRTT pre-registration from eNodeB. The UE is responsible for re-registrations if needed. Pre-registration does not apply to the dual receiver 1xCSFB because the UE register directly in the 1xRTT network using the normal 1xCS registration procedure.

6.6.2.10   Context Transfer between eNodeBs

During the HO preparation, the source eNodeB should transfer the context of the UE to the target eNodeB so that the target eNodeB can take into account the configuration in the source cell and the handover command can contain only differences from the source configuration. This contains AS configuration, RRM configuration and AS Context. The AS configuration contains the UE AS configuration configured via the RRC message exchange. Examples of RRC message exchanges include measurement configuration, radio resource configuration, security configuration, system information about the source cell, the UE ID. The RRM configuration contains the UE specific RRM information. The AS Context contains the UE Radio Access Capability and information for re-establishment.

6.6.2.11   NAS Direct Message Transfer between Network and UE

The NAS messages are transferred mainly via dedicated RRC messages – DL information transfer and UL Information transfer. However, during a bearer set-up, necessary NAS messages should be transferred via RRC Connection Complete message and RRC Connection Reconfiguration message.

6.6.2.12   UE Capability Transfer

The UE transfers NAS UE capability (for example, security and UE capability) directly to MME via the NAS message and this is forwarded to the eNodeB during the initial UE context setup. As UE capability may contain UTRAN and GERAN capability in addition to LTE capability, the information can be rather large in size. To reduce the overhead on the air interface during the transition from RRC_IDLE mode to RRC_CONNECTED mode, the MME stores the UE AS capability as well and provides it to eNodeB during the initial UE context setup. However, if UE AS capability has been changed during RRC_IDLE or the MME doesn't have the valid UE AS capability, the MME doesn't provide the UE AS capability during the initial UE context setup and the eNodeB acquires the UE AS capability directly from the UE as shown in Figure 6.23.

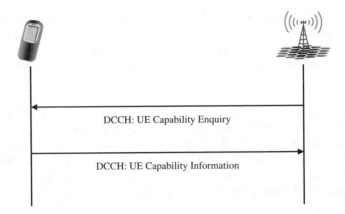

Figure 6.23   UE capability transfer procedure

6.6.2.13   Generic Protocol Error Handling

In UTRAN RRC, various error handling cases in DCCH messages have been specified thoroughly. However for LTE, there was no desire to add complexity by specifying the handling of network error cases, like ASN.1 errors. Thus only the error handling for the very important procedures is specified – the rest is unspecified. For example, if the RRC Connection Reconfiguration message has an ASN.1 error or integrity protection has failed, the UE starts the RRC Connection Re-establishment procedure, shown in Figure 6.24. But for other error cases where the failure response message is not defined, the UE may just ignore the message. Also RRC Connection Re-establishment procedure is used to recover from radio link failure.

6.6.3   Self Optimization – Minimization of Drive Tests

As part of the framework for self-configuration and self-optimization solutions for LTE systems, the Release 10 has been enhanced to include procedures to enable minimization

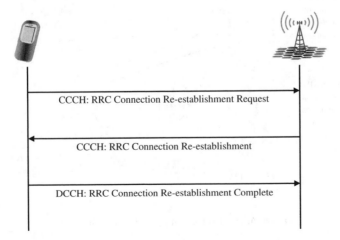

Figure 6.24   RRC connection re-establishment procedure

of drive tests (MDT) functionality. The key of the functionality is to enhance the measurement collection and reporting of the UEs in order to substitute collecting measurements data via traditional drive testing. Avoiding the drive tests is beneficial both in terms of costs and the resulting CO_2 emissions. The Release 10 functionalities are as follows [12].

- MDT-logged measurements. This refers to the measurements carried out in the idle-mode. The network will configure the UE to collect measurements during the idle-mode operation. The UE will collect the measurements and when connecting later the UE will then indicate that such measurements are available in order to enable the network to retrieve the data collected. The measurement quantities are then enhanced by time stamp and possible location information. An example of MDT idle-mode operation is shown in Figure 6.25.
- MDT immediate measurement. In this case the measurements are collected during the normal connected mode, enhanced with location information if available, while the time stamping itself comes from the network side receiving the information. The Radio Link Failure (RLF) reporting has been enhanced to include, as well as location information, optional information about, for example, the direction of movement when the RLF occurred. Some of the information will obviously depend on whether there a position location solution was activated when the event occurred.

The key issue in the design has been to allow an operator to define extra reporting in the areas where problems have been detected while not burdening the network in the areas where performance is detected to be at a decent level. The operation, as part of the control plane, also allows the smooth combination of information available both at the UE and the eNodeB. There are further procedures affecting the RRC specifications for the self-configuration and optimization, such as those supporting the measurements to support the automatic neighbor cell relation as covered in Chapter 7.

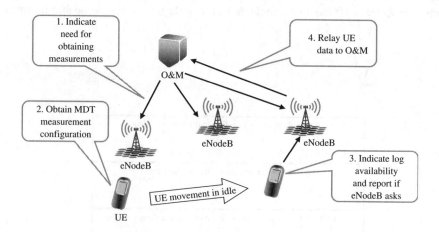

Figure 6.25   Logged mode MDT in idle mode

6.7   X2 Interface Protocols

As introduced in Chapter 3, the LTE has an X2 interface between the eNodeBs. The X2 is a logical interface and even though it is normally drawn as a direct connection between eNodeBs, it is usually routed via the same transport connection as the S1 interface to the site. The X2 interface control and user plane protocol stacks are shown in Figure 6.26, as presented in Chapter 3. X2 is an open interface, similar to the Iur interface in WCDMA. Normally the X2 interface is only for control plane information but in connection with the handover. It is temporary, used for user data forwarding. The key difference between the user plane and the control plane X2 interface protocol stacks is the use of Stream Control Transmission Protocol (SCTP) for control plane transmission between eNodeBs. Use of SCTP enables reliable delivery of control plane information between eNodeBs while, for data forwarding, the User Datagram Protocol (UDP) was seen as sufficient. The X2 Application Protocol (X2AP) covers the radio related signaling while the GPRS Tunneling Protocol, is used for user Plane (GTP-U). This is also the protocol used in the S1 interface for user data handling, as covered in Chapter 3.

The X2AP [13] functionalities are:

- Mobility management for Intra LTE mobility is covered in more detail in Chapter 6. The handover message between eNodeBs is transmitted on the X2 interface.
- Load management to enable inter-cell interference coordination by providing information of the resource status, overload and traffic situation between different eNodeBs.
- Setting up and resetting of the X2 interface.
- Error handling for covering specific or general error cases.

6.7.1   Handover on X2 Interface

The X2 interface has a key role in the intra-LTE handover operation. The source eNodeB will use the X2 interface to send the handover request message to the target eNodeB. If the X2 interface does not exist between the two eNodeBs in question, then procedures are needed to set one up before handover can be done, as explained in Chapter 6. The handover request message initiates the target eNodeB to reserve resources and it will send

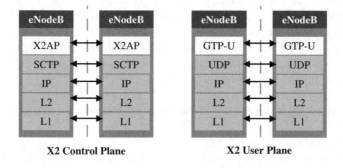

Figure 6.26   X2 interface User and Control Plane protocol stacks

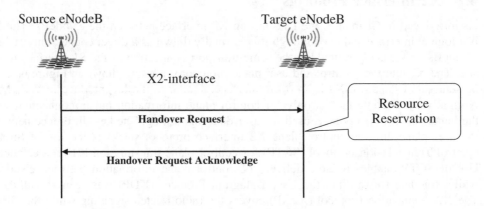

Figure 6.27   X2 Handover preparation over X2 interface

the handover request acknowledgement message assuming resources are found, as shown in Figure 6.27.

There are different information elements provided (some optional) on the handover Request message, such as:

- Requested SAE bearers to be handed over.
- Handover restrictions list, which may restrict the following handovers for the UE.
- Last visited cells the UE has been connected to if the UE historical information collection functionality is enabled. This has been considered to be useful in avoiding ping-pong effects between different cells when the target eNodeB is given information about how the serving eNodeB has been changing in the past, and thus in cases of frequent changes back and forth actions can be taken to limit that. Similar information is included in WCDMA Release 8 side as well for similar cases with flat architecture.

Upon sending the handover Request message, the eNodeB will start the timer and if no response is received the handover preparation is cancelled (and an indication is given to the target eNode that the timer has expired). There is also a Handover Cancel message to cancel an ongoing handover.

The Handover Request Acknowledge message contains more than just the information that the target eNodeB can accommodate (at least part) the requested SAE bearers to be handed over. The other information included is:

- GTP tunnel information for each SAE bearer to enable delivery of uplink and downlink PDUs.
- Possible SAE bearers not admitted.
- A transparent container having handover information (actual RRC Connection Reconfiguration message), which the source eNodeB will then sent to the UE as explained in section 6.6.2.8 and shown as handover commands in Chapter 7.

SN Status Transfer procedure is intended to provide information about the status of each SAE bearer that will be transferred to the eNodeB. The PDCP-SN and HFN are provided for both uplink and downlink direction for those SAE bearers where status presentation is applied. This will take place at the moment the source eNodeB stops assigning new PDCP SNs in the downlink and stops forwarding data over the S1 interface to EPC.

UE Context Release is signaling to the source eNodeB that 'the control point' for UE has now moved to target eNodeB and basically the source can now release all resources.

6.7.2 Load Management

Three different measurements are specified in Release 8 to support standardized operation for the load control and interference management over the X2-interface. The indications are:

- On the transmitter side there is an indication of the transmitted power level in the form of the Relative Narrowband Tx Power (RNTP) IE.
- On the receiver side there is both received interference level and interference sensitivity.

All these are measured/estimated by the Physical Resource Block (PRB) of 180 kHz. The received interference level measurement (or rather indication) on the X2 interface, is illustrated in Figure 6.28.

In the downlink direction the intention of the indication of the transmission power per PRB is to facilitate interference coordination between different eNodeBs. The eNodeB receiving the UL High Interference Indication IE should avoid scheduling cell-edge users on those PRBs where interference level is high in the neighboring eNodeB. The measurement of the TX power per PRB is illustrated in Figure 6.29.

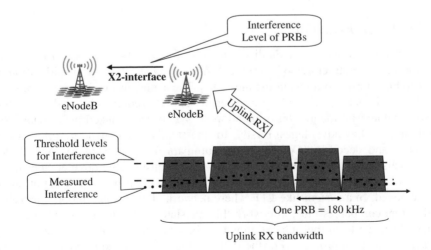

Figure 6.28 Interference level reporting over X2 interface

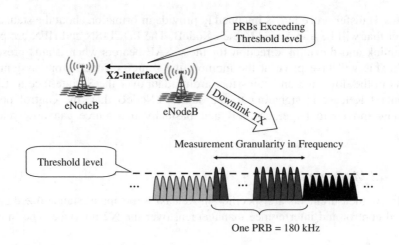

Figure 6.29 Transmission power versus threshold reporting over X2 interface

6.8 Understanding the RRC ASN.1 Protocol Definition[1]

In this section we will consider in detail the RRC protocol definition, which is defined using the Abstract Syntax Notation 1 (ASN.1) [14]. The key concepts regarding ASN.1 used in the RRC definition will be explained, and an overview will be given of the message structure. This section will also consider more advanced aspects such as the RRC extension mechanisms. This section is written in a tutorial style by taking examples from the 3GPP RRC protocol specification [11] and explaining the relevant ASN.1 concepts as they arise. It is worth noting that the ASN.1 is used also with other interfaces in LTE systems, namely in the X2 and S1 protocols.

6.8.1 ASN.1 Introduction

ASN.1 is an internationally standardized notation for specifying protocol definitions. It enables the message structures of a protocol to be designed without consideration of the actual encoding chosen to transmit the messages. The actual encoding (or bits transmitted) is then defined by selecting one of a number of standardized encoding rules [15–17] for the ASN.1-defined protocol. This separation of syntax from encoding provides ASN.1 with a number of key advantages. Firstly, for the standardized encoding rules, it enables the encoders and decoders (CoDec) to be automatically generated for ASN.1 defined protocols. This automatic generation not only saves considerable development time, but since the CoDecs are generated by a tool rather than written by hand a major source of error is removed. In a system like LTE where network elements and mobile devices from many different companies must interwork this translates into a major advantage. Another potential advantage from this separation is the ability to change the set of encoding rules used easily for a given protocol. For the RRC protocol the Packed Encoding Rules (PER) [16] are used. The main property of the PER is to provide a very compact encoding of the

---

[1] Reproduced by permission of © 2010 3GPP™.

messages. This compact encoding is very important because mobile telecommunications systems have limited air-interface bandwidth available. In the case of RRC, where we need to support legacy network elements and mobile devices using an earlier LTE release, the ability to change the encoding rules is problematic, however this need to ensure forward and backward compatibility highlights another major strength of ASN.1, namely, the built-in extension mechanism. Finally, the use of ASN.1 allows the protocol specification to be used directly within the TTCN-3 [18, 19] LTE test specification like those developed by the 3GPP [20]. The following sections should enable the reader to understand the RRC protocol definition including the issue of extension mechanisms with both critical and non-critical extensions.

6.8.2 *RRC Protocol Definition*

Normally an ASN.1 protocol specification is defined in one or more stand alone-textual modules. In the case of RRC [11], the specification is documented in a different way. Instead of defining the specification as one continuous module, the ASN.1 source is broken down into a series of fragments distributed though the document in sections 6.2, 6.3 and 6.4. This means that the RRC standard document cannot be directly used by standard ASN.1 tools.[2]

The first ASN.1 fragment from the RRC standard is shown in Table 6.2. The first thing to note from this example is how to distinguish the ASN.1 fragments. Each fragment in the standard has a highlighted background color and starts and ends with '-ASN1START' and '-ASN1STOP' respectively. In terms of ASN.1 this fragment simply defines the beginning of an ASN.1 module with the name EUTRA-RRC-Definitions. The actual ASN.1 definitions are then placed between the BEGIN and the eventual END keyword which denotes the end of the module. This small example also highlights several aspects of the ASN.1 language, firstly that ASN.1 keywords are always in capital letters, secondly that ASN.1 comments start with the '--' pattern and lastly that the symbol for assignment is ':: = '.

6.8.2.1 High-level Structure

At the highest level the structure of the RRC ASN.1 specification follows the logical channels introduced in section 6.3.1 and shown graphically in Figures 6.6 and 6.7. For each

Table 6.2 First code fragment form RRC ASN.1 protocol specification

```
--ASN1START

EUTRA-RRC-Definitions DEFINITIONS AUTOMATIC TAGS ::=

BEGIN

--ASN1STOP
```

---

[2] The following steps are necessary to use the RRC standard in a general ASN.1 tool. Firstly save the standard document as an ASCII file then strip out all text lines except the ASN.1 fragments.

logical channel – CCCH (uplink), DCCH(uplink), CCCH(downlink), DCCH(downlink), BCCH, and PCCH – there is a top-level data type defined, which represents the possible messages on that channel. For clarity, the BCCH messages are split into two data types; BCCH-BCH for broadcast channel messages and BCCH-DL-SCH for messages mapped to the downlink shared channel. For each of these logical channel types the structure is the same – an example is shown in Table 6.3 for the downlink CCCH channel. This ASN.1 fragment contains two type definitions. At the top-most level the type DL-CCCH-Message is defined. This type uses the SEQUENCE keyword, which is used to specify a sequential series of fields delimited by opening and closing curly brackets and separated (if necessary) with commas. In this case the definition only has a sequence of one field, with the name message. The type of this field is DL-CCCH-MessageType and it is this type that is specified in the second definition in this fragment. Note that, as shown in this example, there is a general rule within ASN.1 specifications that field names must start with a lower case letter and types must start with an upper case letter.

The DL-CCCH-MessageType definition makes use of the keyword CHOICE. This ASN.1 constructor is used to define alternatives. The alternatives are a set of fields specified within opening and closing curly brackets. When an actual message of this specification is sent or received it contains one of these alternatives. From this we can see that the DL-CCCH-MessageType can either contain the field c1 or messageClassExtension. If the branch c1 is taken then the type will contain one of the four message types defined in the nested CHOICE.

As stated earlier, all the logical channels types are defined in a similar way, a top-most type which is a simple SEQUENCE with one field, with the type of this field defined via CHOICE to resolve to one of the possible message types for that channel. Clearly the definitions have some other aspects, which have not been explained yet, for example the outer CHOICE and empty SEQUENCE. These constructs will be explained in the following section about extension mechanisms.

Table 6.3 Rel9 DL-CCCH logical channel message type

```
--ASN1START

DL-CCCH-Message ::= SEQUENCE {
   message              DL-CCCH-MessageType
}

DL-CCCH-MessageType ::= CHOICE {
   c1                   CHOICE {

      rrcConnectionReestablishment        RRCConnectionReestablishment,
      rrcConnectionReestablishmentReject   RRCConnectionReestablishmentReject,
      rrcConnectionReject                 RRCConnectionReject,
      rrcConnectionSetup                  RRCConnectionSetup
   },
   messageClassExtension SEQUENCE {}
}

--ASN1STOP
```

6.8.2.2   RRC Extension Mechanisms

The RRC protocol is constantly evolving with each new release through 3GPP standardization. To ensure backward and forward compatibility – that LTE equipment using an old release of the RRC protocol will work with LTE equipment using a later version of the protocol – we need to handle this evolution of the protocol carefully. Within the RRC protocol specification there are two distinct extension mechanisms used for this protocol evolution: critical extensions and non-critical extensions. Conceptually the difference between the two schemes is that, in a critical extension the existing definition is in effect completely replaced with a new definition, whereas, in the case of a non-critical extension only a part of the existing definition is replaced or the definition is simply extended. In both cases the 'hooks' in terms of specific ASN.1 constructs must be placed in the initial version of the protocol to allow extension at a given place in a future version.

Critical extensions

An example of critical extension can be seen in the DL-CCCH logical channel type definition in Table 6.3, where the outer CHOICE in the type definition gives two alternatives: c1 or messageClassExtension. In the current version of the RRC protocol this CHOICE will always take the c1 alternative. If, however, at some point in the future it is necessary to critically extend the channel type definition, the CHOICE will be set to message-ClassExtension. This in effect would mean from that version onwards all future LTE equipment would ignore the existing definitions in the c1 branch, except when communicating with legacy equipment – i.e. equipment still using the RRC protocol version with the CHOICE set to c1.

In the current specification where the c1 alternative is always chosen, the type for the possible future critical extension is given as an empty SEQUENCE. This can be considered simply as a placeholder for currently unknown future types. This, together with the field name, will be changed if critical extension takes place. To illustrate this, a possible future definition for the DL-CCCH-MessageType is shown in Table 6.4. In this hypothetical future version of the protocol the DL-CCCH-MessageType has been critical extended with the messageClassExtension field replaced with the r14-DL-CCCH-Messages definition. This definition also provides the possibility for further critical extension if needed with the c2 and messageClassExtension CHOICE construct.

The definition in Table 6.3 is from 3GPP Release 9 (Rel9). Let us assume that the possible future definition in Table 6.4 is from Release 14 (Rel14) and consider how LTE equipment using these two different versions of the RRC protocol can interwork. In the case of Rel9 device communicating with Rel9 network the situation is clear: the CHOICE in DL-CCCH-MessageType will always take the c1 alternative. Likewise in the case of a Rel14 device and network, the CHOICE will always take the r14-DL-CCCH-Messages alternative. When we have a Rel9 device communicating with a Rel14 network, all the messages must use the c1 alternative because the Rel9 device cannot understand or decode the other r14-DL-CCCH-Messages alternative. This highlights one aspect of critical extension: the communicating parties need to be aware of the protocol version of the entity they are communicating with.

Critical extension is always achieved using the CHOICE construct. The CHOICE may take the form discussed so far and shown in Table 6.3, or it may be based on the addition of 'spare' fields into an existing CHOICE as shown in Table 6.5. In this case the 'spare' field

Table 6.4 Possible future definition for `DL-CCCH-MessageType`

```
DL-CCCH-MessageType ::= CHOICE {
  c1                CHOICE {
    rrcConnectionReestablishment        RRCConnectionReestablishment,
    rrcConnectionReestablishmentReject   RRCConnectionReestablishmentReject,
    rrcConnectionReject                 RRCConnectionReject,
    rrcConnectionSetup                  RRCConnectionSetup
  },
  r14-DL-CCCH-Messages      CHOICE {
    c2                CHOICE {
      rrcConnectionReestablishment        RRCConnectionReestablishment-r14,
      rrcConnectionReestablishmentReject   RRCConnectionReestablishmentReject-
r14,
      rrcConnectionReject                 RRCConnectionReject-r14,
      rrcConnectionSetup                  RRCConnectionSetup-r14
    },
    messageClassExtension SEQUENCE {}

  }
}
```

Table 6.5 Definition of Rel9 RRC connection re-establishment message

```
--ASN1START

RRCConnectionReestablishment ::= SEQUENCE {
  rrc-TransactionIdentifier RRC-TransactionIdentifier,
  criticalExtensions      CHOICE {
    c1                    CHOICE{
      rrcConnectionReestablishment-r8   RRCConnectionReestablishment-r8-IEs,
      spare7 NULL,
      spare6 NULL, spare5 NULL, spare4  NULL,
      spare3 NULL, spare2 NULL, spare1  NULL
    },
    criticalExtensionsFuture  SEQUENCE {}
  }
}

RRCConnectionReestablishment-r8-IEs ::=  SEQUENCE {
  radioResourceConfigDedicated    RadioResourceConfigDedicated,
  nextHopChainingCount            NextHopChainingCount,
  nonCriticalExtension            RRCConnectionReestablishment-v8a0-IEs OPTIONAL
}

RRCConnectionReestablishment-v8a0-IEs ::=  SEQUENCE {
  lateR8NonCriticalExtension OCTET STRING      OPTIONAL,--Need OP
  nonCriticalExtension      SEQUENCE {}      OPTIONAL --Need OP
}

--ASN1STOP
```

has the type NULL. This is, in principle, just an empty place holder like the SEQUENCE{} in the previous example. However, formally NULL is a type with one unique value. The decision as to whether to add these 'spare' fields for critical extension and how many to add is governed by the perceived likelihood of the given element requiring critical extension in the future. In other words if we have a message definition where we believe that critical extension is unlikely in the future, then the simple c1 CHOICE mechanism shown in Table 6.3 is probably sufficient. In cases where there are likely to be many changes then one or more 'spare' fields can be added. Note that this decisions needs to be made in the first version of the protocol; 'spare' fields cannot be added later. If a definition needs to be critically extended and the definition has both 'spare' fields and the c1 CHOICE, the 'spare' fields are always used first.

Non-critical extensions

Non-critical extensions are a way to add new elements or information to the existing specification. Unlike critical extensions, LTE equipment based on previous versions of the protocol can skip over non-critical extensions even though they cannot understand the contents. The place holder for non-critical extension is defined in one of two ways. If one wishes to provide a non-critical extension at the end of a message or of a field contained within a BIT or OCTET STRING then an optional empty SEQUENCE construct is used. The OPTIONAL keyword in ASN.1 specifies that the field might or might not be in the message. An example of this construct can be seen in the RRCConnectionReestablishment-v8a0-IEs definition in Table 6.5 and will be explained in section 6.8.2.3. In all other cases the place holder for non-critical extension is provided by the built-in ASN.1 extension marker '...'. We take a closer look at this in section 6.8.2.4. Note that BIT STRING and OCTET STRING are two of the built-in types in ASN.1. As the name suggests, they are used to model an ordered set of bits or octets of arbitrary length. Within the 3GPP protocol specifications the types BIT STRING and OCTET STRING are sometimes used as a transparent container to transfer information without the need to understand the contained structure.

6.8.2.3   RRC Message Definitions

The definition for the message type RRCConnectionReestablishment is shown in Table 6.5. At the outer level, the defined message type is a SEQUENCE of rrc-TransactionIdentifier and criticalExtensions fields. Where the criticalExtensions field is defined to to be the classic nested CHOICE to support critical extension as described in 'Critical extensions section'. The actual information elements for the message are contained within the field rrcConnectionReestablishment-r8, whose type is RRCConnectionReestablishment-r8-IEs. This type shows an example of non-critical extension.

To understand this non-critical extension it is easiest to start with previous version of the RRCConnectionReestablishment-r8-IEs definition shown in Table 6.6. By comparing the release 8 and release 9 versions we can see that the optional empty sequence has been replaced by the type RRCConnectionReestablishment-v8a0-IEs. In effect this adds the field lateR8NonCriticalExtension to the existing (Rel8) message definition. Note that this field has been added to allow late corrections to release 8 even after release 9 has been frozen, with the OCTET STRING acting like a transparent container

Table 6.6 Definition of Rel8 `RRCConnectionReestablishment-r8-IEs`

```
RRCConnectionReestablishment-r8-IEs ::= SEQUENCE {
   radioResourceConfigDedicated RadioResourceConfigDedicated,
   nextHopChainingCount         NextHopChainingCount,
   nonCriticalExtension         SEQUENCE {}         OPTIONAL   -- Need OP
}
```

for any such additions. The release 8 and release 9 versions can interwork with each other because in release 8 the empty SEQUENCE for non-critical extensions was at the end of the message. This means when a release 8 device recieves an RRC connection reestablishment message from a release 9 network it will simply not try and decode the extra fields because, as far as it is concerned, the SEQUENCE is empty and it will ignore the extra bytes at the end of the message.

6.8.2.4 RRC Information Element Definitions

Having considered the channel and message type definitions we will now consider how the information elements of the RRC protocol are specified. Information elements are used to describe the fields in message types. As such they are not an ASN.1 concept, they are just a logical structuring concept in the RRC definition. An example for the `RadioResourceConfigDedicated` Information Element is shown in Table 6.7. The information element has a series of fields and at the end an example of non-critical extension using the built-in ASN.1 extension marker. To understand how this extension marker is used it is useful to look at the original version of the protocol shown in Table 6.8, where at the end of the SEQUENCE there is just an extension marker '...'. In the Rel9 version the field `rlf-TimersAndConstants-r9` has been added after the extension marker. This field is enclosed in double square brackets, which are used to denote an extension group. An extension group is just a way of keeping together all of the fields associated with one particular extension. When using the built-in ASN.1 extension mechanism, via the extension marker, all issues associated with interworking between different versions of the protocol are automatically handled. The associated machine-generated encoders and decoders handle the interaction between different versions of the protocol in a transparent and automated way. In this case, there is only a single extension group after the extension marker, but the ASN.1 extension mechanism allows far more. A single extension marker supports not just the addition of multiple fields and/or extension groups in a single version but also the development of an extension series though evolution of multiple versions of a protocol, each making one or more addition.

When considering the other type definition in Table 6.7 there are a number of ASN.1 constructs that are new. In the SRB-ToAddModList type definition there is a new variation on SEQUENCE, namely the SEQUENCE OF.

Whereas SEQUENCE is an ordered set of fields of arbitrary types, the SEQUENCE OF is an ordered list of the same type, in this case DRB-Identity. This definition also introduces the SIZE keyword. This keyword is used to constrain the size of the SEQUENCE OF – the SEQUENCE OF must have between 1 and maxDRB elements. Note that a SIZE constraint can also be specified with a single value instead of, as in this case, a range.

Table 6.7   Definition of Rel9 `RadioResourceConfigDedicated` Information Element

```
-- ASN1START

RadioResourceConfigDedicated ::= SEQUENCE {
    srb-ToAddModList              SRB-ToAddModList          OPTIONAL, -- Cond HO-Conn
    drb-ToAddModList              DRB-ToAddModList          OPTIONAL, -- Cond HO-
    drb-ToReleaseList             DRB-ToReleaseList         OPTIONAL, -- Need ON
    mac-MainConfig                CHOICE {
            explicitValue             MAC-MainConfig,
            defaultValue              NULL
    }       OPTIONAL,                                           -- Cond HO-toEUTRA2
    sps-Config                    SPS-Config                OPTIONAL, -- Need ON
    physicalConfigDedicated       PhysicalConfigDedicated   OPTIONAL, -- Need ON
    ...,
    [[ rlf-TimersAndConstants-r9  RLF-TimersAndConstants-r9 OPTIONAL -- Need ON
    ]]
}


SRB-ToAddModList ::= SEQUENCE (SIZE (1..2)) OF SRB-ToAddMod


SRB-ToAddMod ::= SEQUENCE {
    srb-Identity                  INTEGER (1..2),
    rlc-Config                    CHOICE {
        explicitValue                 RLC-Config,
        defaultValue                  NULL
    } OPTIONAL,                                                  -- Cond Setup
    logicalChannelConfig          CHOICE {
        explicitValue                 LogicalChannelConfig,
        defaultValue                  NULL
    } OPTIONAL,                                                  -- Cond Setup
    ...
}


DRB-ToAddModList ::= SEQUENCE (SIZE (1..maxDRB)) OF DRB-ToAddMod


DRB-ToAddMod ::= SEQUENCE {
    eps-BearerIdentity            INTEGER (0..15)     OPTIONAL,   -- Cond DRB-
    drb-Identity                  DRB-Identity,
    pdcp-Config                   PDCP-Config         OPTIONAL,   -- Cond PDCP
    rlc-Config                    RLC-Config          OPTIONAL,   -- Cond Setup
    logicalChannelIdentity        INTEGER (3..10)     OPTIONAL,   -- Cond DRB-
    logicalChannelConfig          LogicalChannelConfig OPTIONAL,  -- Cond Setup
    ...
}


DRB-ToReleaseList ::= SEQUENCE (SIZE (1..maxDRB)) OF DRB-Identity

-- ASN1STOP
```

Table 6.8 Definition of Rel8 `RadioResourceConfigDedicated` information element

```
RadioResourceConfigDedicated ::= SEQUENCE {
    srb-ToAddModList            SRB-ToAddModList          OPTIONAL, -- Cond HO-
    drb-ToAddModList            DRB-ToAddModList          OPTIONAL, -- Cond HO-
    drb-ToReleaseList           DRB-ToReleaseList         OPTIONAL, -- Need ON
    mac-MainConfig              CHOICE {
        explicitValue               MAC-MainConfig,
        defaultValue            NULL
    } OPTIONAL,                                           -- Cond HO-toEUTRA2
    sps-Config                  SPS-Config                OPTIONAL, -- Need ON
    physicalConfigDedicated     PhysicalConfigDedicated   OPTIONAL, -- Need ON
    ...
}
```

A `SIZE` constraint can also be used on the `BIT STRING` and `OCTET STRING` types that where introduced in 'Non-critical extensions section'. There is one more new type in the `DRB-ToAddMod` type definition: `INTEGER`. This will be considered in the next section.

6.8.2.5 Further ASN.1 Types and Constructs

Although in the previous sections we have considered all the main building blocks of the RRC protocol specification, the logical channel definitions, the RRC message definitions and the RRC information element definitions, there are still a number of ASN.1 types and constructs relevant for RRC that we have not yet come across. These will be covered in this section.

First let us consider the ASN.1 types `BOOLEAN`, `INTEGER` and `ENUMERATED`. The `BOOLEAN` type is used to model a field with two values: `TRUE` and `FALSE`. An example of the `BOOLEAN` type is the field `accumulationEnabled` shown in Table 6.9.

The `INTEGER` type is used to model a field that can be represented by an arbitrarily large positive or negative integer. To improve encoding efficiency – the number of bits or bytes needed to transfer the message – it is desirable to constrain `INTEGER` fields to the required range. The range constraint is specified in brackets after the key word as shown in Table 6.9. For example the type `p0-UE-PUSCH` may only take the values −8 to +7.

The `ENUMERATED` type allows the modeling of fields that contain named values. An example from Table 6.9 is the field `deltaMCS-Enabled` which can take the values 'en0'

Table 6.9 Definition of Rel9 `UplinkPowerControlDedicated` information element

```
UplinkPowerControlDedicated ::=      SEQUENCE {
    p0-UE-PUSCH                          INTEGER (-8..7),
    deltaMCS-Enabled                     ENUMERATED {en0, en1},
    accumulationEnabled                  BOOLEAN,
    p0-UE-PUCCH                          INTEGER (-8..7),
    pSRS-Offset                          INTEGER (0..15),
    filterCoefficient                    FilterCoefficient        DEFAULT fc4
}
```

Table 6.10   Definition of Rel9 FilterCoefficient information element

```
--ASN1START

FilterCoefficient ::=                    ENUMERATED {
                                            fc0, fc1, fc2, fc3, fc4, fc5,
                                            fc6, fc7, fc8, fc9, fc11, fc13,
                                            fc15, fc17, fc19, spare1,...}


--ASN1STOP
```

Table 6.11   Definition of named values

```
maxCellReport      INTEGER ::= 8  -- Maximum number of reported cells
maxDRB             INTEGER ::= 11 -- Maximum number of Data Radio Bearers
maxEARFCN          INTEGER ::= 65535 -- Maximum value of EUTRA carrier frequency
maxFreq            INTEGER ::= 8  -- Maximum number of EUTRA carrier frequencies
maxGERAN-SI        INTEGER ::=  10 -- Maximum number of GERAN SI blocks that can be
```

or 'en1'. A more complicated example of an ENUMERATED type including extension is shown in Table 6.10.

Up to this point we have concentrated on ASN.1 types but it is also possible to define ASN.1 values. Examples of such named values are shown in Table 6.11. These named values can be used for things like SIZE constraints and DEFAULT values (see later in this section). An example using such named values can be seen in Table 6.7, where the type DRB-ToAddModList has a SIZE constraint using the named value maxDRB which is defined in Table 6.11

As was mentioned in section 6.8.2, an ASN.1 definition is specified in one or more modules where the module starts with the module name followed by the DEFINITIONS keyword. All the module definitions are then placed between the BEGIN and END keywords. The next ASN.1 construct to consider is the IMPORT statement, which allows definitions to be imported from one module to another. An example of such an IMPORT statement is shown in Table 6.12. In this example a series of ASN.1 type and value definitions is imported from the EUTRA-RRC-Definitions module into the EUTRA-UE-Variables module. After importing the definitions can be used in exactly the same way as if they were directly specified in the importing module.

Finally let us consider the ASN.1 construct DEFAULT. DEFAULT allows a default value to be associated with a field. On the receiving side the decoder will add this field value to the message if no field value was received. On the sending side the field will not be present in the 'sent' message if the field value given is the default value. An example of the DEFAULT construct is the filterCoefficient field in Table 6.9, the type for this field is shown in Table 6.10. Like OPTIONAL, DEFAULT can only be specified on fields within a SEQUENCE type. The OPTIONAL and DEFAULT specifications are mutually exclusive – no single field can have both specified.

Table 6.12   Example of IMPORT definition

```
--ASN1START

EUTRA-UE-Variables DEFINITIONS AUTOMATIC TAGS ::=

BEGIN

IMPORTS
   CarrierFreqGERAN,
   CellIdentity,
   SpeedStateScaleFactors,
   C-RNTI,
   MeasId,
   MeasIdToAddModList,
   MeasObjectToAddModList,
   MobilityStateParameters,
   NeighCellConfig,
   PhysCellId,
   PhysCellIdCDMA2000,
   PhysCellIdGERAN,
   PhysCellIdUTRA-FDD,
   PhysCellIdUTRA-TDD,
   QuantityConfig,
   ReportConfigToAddModList,
   RSRP-Range,
   maxCellMeas,
   maxMeasId
FROM EUTRA-RRC-Definitions;

--ASN1STOP
```

6.9   Early UE Handling in LTE

In LTE the 3GPP has taken the approach to get feedback from the operators about the planned features for the first phase(s) of LTE deployment and then base the first phase test cases on the prioritization received. For the features that have lower priority in the testing area, UE capability signaling will basically be in place [11, 21] to indicate for the network whether the feature is fully implemented and tested or not as shown in Figure 6.30. This should avoid the problem that was faced with Release 99 deployment when many features that were mandatory for the UEs were not supported in the networks (and thus were not available for testing either) when the first networks were commercially opened. The Release 8-based test cases were prioritized, as described in [22]. The LTE test cases for signaling can be found from the recently created test specifications [23]. The completion of those signaling test cases naturally took longer than the actual radio protocol work in other groups as one needs a stable signaling specification before test cases can be finalized. In Release 9 the situation of the inter-operability tested features was reviewed and some of the features were declared 'expected to be supported', although, of course, they needed testing as well, with the change to the signaling reflected in [24] and reflected in the changes in the Release 9 September 2010 version of LTE RRC specification [11].

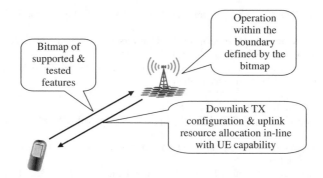

Figure 6.30 Signaling of the device capability at connection set-up

6.10 Summary

The LTE radio protocols follow a basic structure similar to WCDMA but there are obvious differences due to differences in the radio technology. From the functional split the distribution is also partly different with the most notable difference being the handling of ciphering in PDCP layer. The lack of Iub-like interface reduces the needed radio protocol signaling on the internal interfaces as the eNodeB takes care directly of the RRC signaling between the UE and the network. The X2 interface is thus a rather simple interface from the radio protocol perspective, with key functionalities being to take care of the mobility related signaling for intra-LTE mobility as well as handling the network-level information exchange for cell-level interference management. The same ASN.1 as was specified for WCDMA is used as the LTE radio protocol specification language to ensure straightforward extendibility with releases beyond LTE Release 8, as was covered in section 6.8 with the basic structure of the RRC ASN.1 protocol specification and the most important concepts from ASN.1. For UE capability, signaling the input from the operator community has shaped what the first phase commercial UEs will look like in terms of radio capability and respectively bitmap, with some updates made in Release 9 to reflect the expectations of feature availability in the actual networks.

References

[1] 3GPP Technical Specification, TS 36.321, 'Evolved Universal Terrestrial Radio Access (E-UTRA); Medium Access Control (MAC) Protocol Specification'.
[2] 3GPP Technical Specification, TS 36.322, 'Evolved Universal Terrestrial Radio Access (E-UTRA); Radio Link Control (RLC) protocol specification'
[3] H. Holma, A. Toskala 'WCDMA for UMTS', 5th edition, Wiley 2010.
[4] IETF RFC 4995, 'The RObust Header Compression (ROHC) Framework'.
[5] IETF RFC 4996, 'RObust Header Compression (ROHC): A Profile for TCP/IP (ROHC-TCP)'.
[6] IETF RFC 3095, 'RObust Header Compression (ROHC): Framework and four profiles: RTP, UDP, ESP and Uncompressed'.
[7] IETF RFC 3843, 'RObust Header Compression (ROHC): A Compression Profile for IP'.
[8] IETF RFC 4815, 'RObust Header Compression (ROHC): Corrections and Clarifications to RFC 3095'. © 2010. 3GPP™ TSs and TRs are the property of ARIB, ATIS, CCSA, ETSI, TTA and TTC who jointly own the copyright in them. They are subject to further modifications and are therefore provided to you 'as is' for information purposes only.

[9] IETF RFC 5225, 'RObust Header Compression (ROHC) Version 2: Profiles for RTP, UDP, IP, ESP and UDP Lite'.

[10] 3GPP Technical Specification, TS 36.323, 'Evolved Universal Terrestrial Radio Access (E-UTRA); Packet Data Convergence Protocol (PDCP) Specification'.

[11] 3GPP Technical Specification, TS 36.331, 'Evolved Universal Terrestrial Radio Access (E-UTRA); Radio Resource Control (RRC) Protocol Specification'.

[12] 3GPP Technical Specification 37.320 'Radio Measurement Collection for Minimization of Drive Tests (MDT), Overall description, Stage 2', August 2010.

[13] 3GPP Technical Specification, TS 36.423, 'X2 Application Protocol (XSAP)', Version 8.3.0, September 2008.

[14] ITU-T Recommendation X.680: 'Information Technology – Abstract Syntax Notation One (ASN.1): Specification of Basic Notation'.

[15] ITU-T X.690 (07/2002) 'Information Technology – ASN.1 Encoding Rules: Specification of Basic Encoding Rules (BER), Canonical Encoding Rules (CER) and Distinguished Encoding Rules (DER)'.

[16] ITU-T X.691 (07/2002) 'Information Technology – ASN.1 Encoding Rules: Specification of Packed Encoding Rules (PER)'.

[17] ITU-T X.693 (12/2001) 'Information Technology – ASN.1 Encoding Rules: XML Encoding Rules (XER)'.

[18] ETSI ES 201 873-1, 'Methods for Testing and Specification (MTS), The Testing and Test Control Notation', version 3; TTCN-3 Core Language'.

[19] ETSI ES 201 873-7, 'Methods for Testing and Specification (MTS), The Testing and Test Control Notation', version 3; Part 7: Using ASN.1 with TTCN-3.

[20] 3GPP TS 36.523-3 'Evolved Universal Terrestrial Radio Access (E-UTRA) and Evolved Packet Core (EPC);User Equipment (UE) Conformance Specification, Part 3: Test Suites'.

[21] 3GPP Technical Specification, TS 36.306, 'Evolved Universal Terrestrial Radio Access (E-UTRA); User Equipment (UE) Radio Access Capabilities'.

[22] 3GPP Tdoc R5-083674, 'LTE Features and Test Prioritisation', NTT DoCoMo.

[23] 3GPP Technical Specification, TS 36.523-1, 'Evolved Universal Terrestrial Radio Access (E-UTRA) and Evolved Packet Core User Equipment (UE) Conformance Specification; Part 1: Protocol Conformance Specification' (Release 8), Version 1.0.0, September 2008.

[24] 3GPP Tdoc RP-101008, 'FGI Ssetting in Release 9', NTT DoCoMo et al.

7

Mobility

Chris Callender, Harri Holma, Jarkko Koskela and Jussi Reunanen

7.1 Introduction

Mobility offers clear benefits to the end users: low delay services such as voice or real time video connections can be maintained while moving even in high speed trains. Mobility is beneficial also for nomadic services, such as laptop connectivity, since it allows a reliable connection to be maintained in the areas between two cells where the best serving cell is changing. This also implies a simple setup of the broadband connection without any location related configurations. Mobility typically has its price in the network complexity: the network algorithms and the network management get complex. The target of the LTE radio network is to provide seamless mobility while simultaneously keeping network management simple.

The mobility procedures can be divided into idle mode and connected mode for the attached UE (see Figure 7.1). Idle mode mobility is based on UE autonomous cell reselections according to the parameters provided by the network, thus this is quite similar to the one in WCDMA/HSPA today. The connected mode mobility in LTE, on the other hand, is quite different in LTE than in WCDMA/HSPA radio networks. The UE transition between idle and RRC connected mode is controlled by the network according to the UE activity and mobility. The algorithms for RRC connection management are described in Chapter 8.

This chapter explains the mobility principles and procedures in the idle and active state. Idle state mobility management is described in section 7.2. The intra-frequency handover is presented in section 7.3. The inter-RAT handovers are covered in section 7.4. The differences in mobility between UTRAN and E-UTRAN are summarized in section 7.8. The voice handover from E-UTRAN Voice over IP to 2G/3G circuit switched voice is discussed in Chapter 13. The performance requirements related to mobility are described in Chapter 14.

The general description of LTE mobility is presented in [1], idle mode mobility is specified in [2], the performance requirements for radio resource management are defined in [3] and the relevant Radio Resource Control specifications in [4]. The rest of this chapter uses the term E-UTRAN for LTE and UTRAN for WCDMA.

LTE for UMTS: Evolution to LTE-Advanced, Second Edition. Edited by Harri Holma and Antti Toskala.
© 2011 John Wiley & Sons, Ltd. Published 2011 by John Wiley & Sons, Ltd.

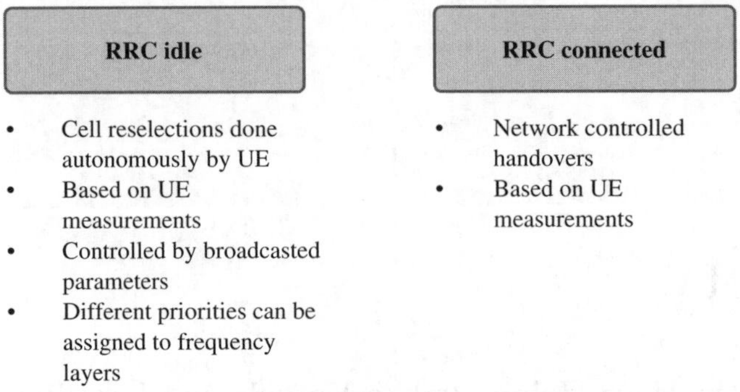

RRC idle	RRC connected
• Cell reselections done autonomously by UE • Based on UE measurements • Controlled by broadcasted parameters • Different priorities can be assigned to frequency layers	• Network controlled handovers • Based on UE measurements

Figure 7.1   Idle mode and RRC connected mode mobility

7.2   Mobility Management in Idle State

7.2.1   Overview of Idle Mode Mobility

The UE chooses a suitable cell of the selected Public Land Mobile Network (PLMN) based on radio measurements. This procedure is known as cell selection. The UE starts receiving the broadcast channels of that cell and finds out if the cell is suitable for camping, which requires that the cell is not barred and that radio quality is good enough. After cell selection, the UE must register itself to the network thus promoting the selected PLMN to the registered PLMN. If the UE is able to find a cell that is deemed a better candidate for reselection according to reselection criteria (section 7.2.2), it reselects onto that cell and camps on it and again checks if the cell is suitable for camping. If the cell where the UE camps does not belong to at least one tracking area to which the UE is registered, location registration needs to be performed. An overview is shown in Figure 7.2.

A priority value can be assigned to PLMNs. The UE searches for higher priority PLMNs at regular time intervals and searches for a suitable cell if another PLMN has been selected. For example, the operator may have configured preferred roaming operators to the Universal Subscriber Identity Module (USIM) card. When the UE is roaming and not camped to the preferred operators, it tries periodically to find the preferred operator.

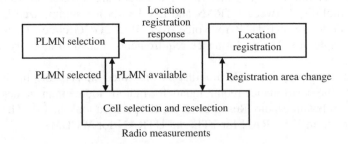

Figure 7.2   Idle mode overview

If the UE is unable to find a suitable cell to camp on or if the location registration fails, it attempts to camp on a cell irrespective of the PLMN identity, and enters a 'limited service' state which only allows emergency calls to be made.

The USIM must be inserted in the UE to perform the registration. UTRAN UE can use either the older SIM cards or the new USIM cards, while E-UTRAN uses only USIM. USIM can provide stronger protection against identity theft compared to SIM.

7.2.2 Cell Selection and Reselection Process

When UE is powered-on for the first time, it will start the Initial Cell Selection procedure. The UE scans all radio frequency (RF) channels in the E-UTRA bands according to its capabilities to find a suitable cell. On each carrier frequency, the UE needs only to search for the strongest cell. Once a suitable cell is found this cell is selected. Initial cell selection is used to ensure that the UE gets into service (or back to the service area) as fast as possible.

The UE may also have stored information about the available carrier frequencies and cells in the neighborhood. This information may be based on the system information or any other information the UE has acquired in the past – 3GPP specifications do not exactly define what kind of information the UE is required or allowed to use for Stored Information cell selection. If the UE does not find a suitable cell based on the stored information, the Initial Cell Selection procedure is started to ensure that a suitable cell is found.

For a cell to be suitable it has to fulfill S-criterion:

$$S_{rxlevel} > 0 \tag{7.1}$$

where

$$S_{rxlevel} > Q_{rxlevelmeas} - (Q_{rxlevmin} - Q_{rxlevelminoffset}) \tag{7.2}$$

$Q_{rxlevelmeas}$ is the measured cell received level (RSRP), $Q_{rxlevelmin}$ is the minimum required received level [dBm] and $Q_{rxlevelminoffset}$ is used when searching for a higher priority PLMN.

Whenever UE has camped to a cell, it will continue to find a better cell as a candidate for reselection according to the reselection criteria. Intra-frequency cell reselection is based on the cell ranking criterion. To do this the UE needs to measure neighbor cells, which are indicated in the neighbor cell list in the serving cell. The network may also ban the UE from considering some cells for reselection, also known as blacklisting of cells. To limit the need to carry out reselection measurements it has been defined that if $S_{ServingCell}$ is high enough, the UE does not need to make any intra-frequency, inter-frequency or inter-system measurements. The intra-frequency measurements must be started when $S_{ServingCell} \leq S_{intrasearch}$. The inter-frequency measurements must be started when $S_{ServingCell} \leq S_{nonintrasearch}$.

In case the UE moves fast it is possible for the network to adjust the cell reselection parameters. The high (or medium) mobility state is based on the number of cell reselections, N_{CR}, within a predefined time, T_{RCmax}. High mobility is typically characterized by different parameter values for the hysteresis and for the reselection timer. To avoid adjusting reselection parameters in case the UE is ping-ponging between two cells, these cell reselections are not counted in the mobility state calculations. As the 'speed' estimation is based on the count of reselections it does not give an exact speed, just a rough estimate of UE movement, but nevertheless provides a means for the network to control UE reselection behavior dependent on the UE movement. The speed dependent scaling method is also applied in the RRC_CONNECTED state for connected mode mobility parameters.

7.2.2.1 Intra-frequency and Equal Priority Reselections

Cell ranking is used to find the best cell for UE camping for intra-frequency reselection or on reselection to equal priority E-UTRAN frequency. The ranking is based on the criterion R_s for the serving cell and R_n for neighboring cells:

$$R_s = Q_{meas,s} + Q_{hyst}$$
$$R_n = Q_{meas,n} + Q_{offset} \qquad (7.3)$$

where Q_{meas} is the RSRP measurement quantity, Q_{hyst} is the power domain hysteresis to avoid ping-pong and Q_{offset} is an offset value to control different frequency specific characteristics (e.g. propagation properties of different carrier frequencies) or cell specific characteristics. In the time domain, $T_{reselection}$ is used to limit overly frequent reselections. The reselection occurs to the best ranked neighbor cell if it is better ranked than the serving cell for a longer time than $T_{reselection}$. The Q_{hyst} provides hysteresis by requiring any neighbor cell to be better than the serving cell by an RRC configurable amount before reselection can occur. The $Q_{offsets,n}$ and $Q_{offsetfrequency}$ make it possible to bias the reselection toward particular cells and/or frequencies. The cell reselection parameters are illustrated in Figure 7.3.

7.2.2.2 Inter-frequency/RAT reselections

In the UTRAN, inter-frequency and inter-RAT reselections were based on the same ranking as intra-frequency reselections. It was seen that this is very difficult for the network to control as the measurement quantities of different RATs are different and the network

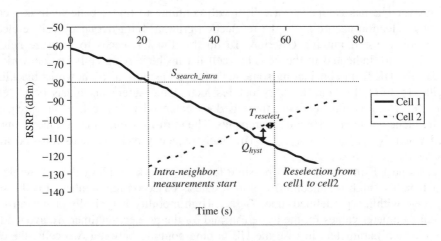

Figure 7.3   Idle mode intra-frequency cell reselection algorithm

needs to be able to control reselection between multiple 3GPP RATs (or even non-3GPP technologies). And as it was seen that operators would like to control how UE prioritizes camping on different RATs or frequencies of E-UTRAN, a new method for reselection handling between different RATs/frequencies (called layers from now on) was chosen. The method is known as absolute priority based reselection. Each layer is given a priority and based on this information the UE tries to camp on the highest priority frequency/RAT if it can provide decent service. In order for the UE to decide if decent service can be provided the network allocates each frequency/RAT a threshold ($Thresh_{x,high}$) which has to be fulfilled before reselection to such a layer is performed. A similar $T_{reselection}$ as in intra-frequency reselections is utilized, i.e. the new layer needs to fulfill the threshold for consecutive time of $T_{reselection}$ before reselection is performed. This is used to eliminate reselections if just temporary fading for the evaluated frequency occurs. To make reselection to a lower priority layer the UE will not reselect to that if the higher priority layer is still above the threshold or if the lower priority frequency is not above another threshold ($Thresh_{x,low}$).

The main parameters for controlling idle mode mobility are shown in Table 7.1.

7.2.3 Tracking Area Optimization

The UE location in RRC idle is known by the Mobility Management Entity (MME) with the accuracy of the tracking area. The size of the tracking area can be optimized in network planning. A larger tracking area is beneficial to avoid tracking area update signaling. On the other hand, a smaller tracking area is beneficial to reduce the paging

Table 7.1 Main parameters for idle mode mobility

Parameters	Description
Q_{hyst}	Specifies the hysteresis value for ranking criteria. The hysteresis is required to avoid ping-ponging between two cells.
$T_{reselection}$	Gives the cell reselection timer value. The cell reselection timer together with the hysteresis is applied to control the unnecessary cell reselections.
$Q_{rxlevmin}$	Specifies the minimum required received level in the cell in dBm. Used in the S-criterion, i.e. in the cell selection process.
$S_{intrasearch}$, $S_{nonintrasearch}$	Specifies the threshold (in dB) for intra-frequency, inter-frequency and inter-RAN measurements. UE is not required to make measurements if $S_{servingcell}$ is above the threshold.
N_{CR}, T_{RCmax}	Specifies when to enter the medium or high mobility state: if the number of cell reselections within time T_{RCmax} is higher than N_{CR}, the UE enters the medium or higher mobility state.
$Thresh_{x,high}$	This specifies the threshold used by the UE when reselecting towards the higher priority frequency X than currently serving frequency.
$Thresh_{x,low}$	This specifies the threshold used in reselection towards frequency X from a higher priority frequency.

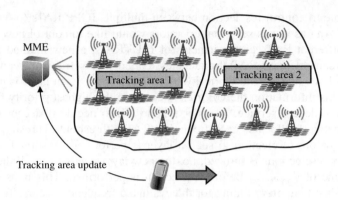

Figure 7.4   Tracking area concept

signaling load for the incoming packet calls. The corresponding concept in UTRAN is called a Routing area, which typically covers a few hundred base stations. The concept of the tracking area is illustrated in Figure 7.4.

UE can be assigned multiple tracking areas to avoid unnecessary tracking area updates at the tracking area borders, e.g. when the UE is ping-ponging between cells of two different tracking areas. UE can also be assigned both a LTE tracking area and a UTRAN routing area to avoid signaling when changing between the two systems.

7.3   Intra-LTE Handovers

UE mobility is solely controlled by the handovers when the RRC connection exists. Thus there is no UTRAN type of state as CELL_PCH where UE based mobility is possible while UE is in RRC_CONNECTED state. The handovers in E-UTRAN are based on the following principles:

1   The handovers are network controlled. E-UTRAN decides when to make the handover and what the target cell is.
2   The handovers are based on the UE measurements. The UE measurements and measurement reporting is controlled by parameters given by E-UTRAN.
3   The handovers in E-UTRAN are targeted to be lossless by using packet forwarding between the source and the target eNodeB.
4   The core network S1 connection is updated only when the radio handover has been completed. This approach is called Late path switch. The core network has no control on the handovers.

The E-UTRAN handover procedure, measurements, signaling and neighborlist control are described below.

7.3.1   Procedure

An overview of the intra-frequency handover procedure is shown in Figure 7.5. The UE is moving from left to right. In the initial phase the UE has user plane connection to the

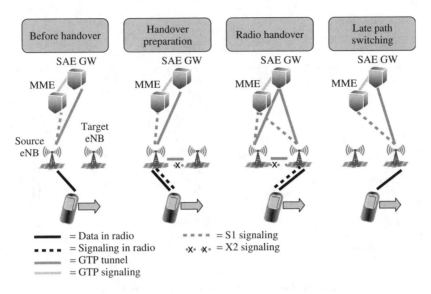

Figure 7.5   Intra-frequency handover procedure

source eNodeB and further to the System Architecture Evolution Gateway (SAE GW). The S1 signaling connection exists between eNodeB and MME. When the target cell fulfills the measurement reporting threshold, the UE sends the measurement report to the eNodeB. The eNodeB establishes the signaling connection and GTP (GPRS Tunneling Protocol) tunnel to the target cell. Once the target eNodeB has the resource available, the source eNodeB sends the handover command to the UE. The UE can switch the radio connection from the source to the target eNodeB. The core network is not aware of the handover at this point in time. The core network connection is finally updated and this procedure is called Late path switching.

The user plane switching is illustrated in Figure 7.6. The user plane packets in downlink are forwarded from the source eNodeB to the target eNodeB over the X2 interface before Late path switching. The X2 interface enables lossless handover. The source eNodeB forwards all downlink RLC SDUs (Service Data Units) that have not been acknowledged by the UE. The target eNodeB re-transmits and prioritizes all downlink RLC SDUs forwarded by the source eNodeB as soon as it obtains them. It may happen that some packets were already received by the UE from the source eNodeB but the source eNodeB did not receive the acknowledgement. The target eNodeB will then retransmit the packets unnecessarily and the UE receives the same packet twice. Therefore, the UE must be able to identify and remove duplicate packets.

In the uplink direction, the source eNodeB forwards all successfully received uplink RLC SDUs to the packet core. The UE re-transmits the uplink RLC SDUs that have not been acknowledged by the source eNodeB. It may happen also in uplink than some packets are sent twice. The reordering and the duplication avoidance in uplink are done in the packet core.

The packet forwarding takes place only for a short time until the late path switching has been completed. Therefore, the packet forwarding does not consume excessive transport resources and X2 interface requirements are not significant from the network dimensioning

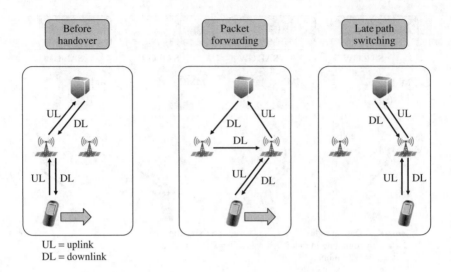

UL = uplink
DL = downlink

Figure 7.6   User plane switching in handover

point of view. The packets in the uplink direction can be sent directly from the target eNodeB to SAE GW.

7.3.2 Signaling

The detailed signaling messages during the handover procedure are described in this section. The procedure is divided into three parts: handover preparation (Figure 7.7), handover execution (Figure 7.8) and handover completion (Figure 7.9).

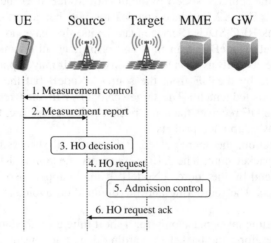

Figure 7.7   Handover preparation

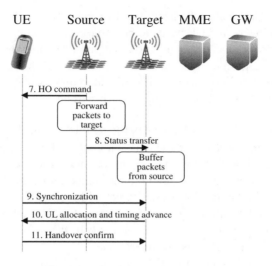

Figure 7.8 Handover execution

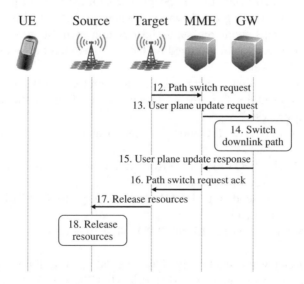

Figure 7.9 Handover completion

1   The source eNodeB configures the UE measurement procedures with a MEA-
    SUREMENT CONTROL message. The message defines the measurement reporting
    thresholds.
2   When the target cell fulfills the reporting threshold, the UE sends a MEASURE-
    MENT REPORT to the eNodeB. The typical reporting is event triggered where the
    UE sends measurements only when the reporting threshold has been fulfilled. It is
    also possible to configure periodic reporting.

3   The source eNodeB makes the handover decision based on the UE report. For intra-
     frequency handovers, the UE needs to be connected to the cell with the lowest path
     loss and the network has practically no freedom in deciding the handover target. For
     inter-frequency and inter-RAT handovers, the eNodeB can also take the load and
     service information into account. The operator may want to balance the loading
     between frequencies and may want to push certain services to certain frequency
     layers or systems.

4   The source eNodeB sends a HANDOVER REQUEST to the target eNodeB.

5   The target eNodeB performs the admission control. For intra-frequency handovers
     the network has little freedom in blocking the new connection since the UE trans-
     mission will anyway cause the uplink interference to the target cell even if the UE
     does not have the connection to the target cell. Actually, the uplink interference
     can be minimized by allowing the UE to connect to the cell with lowest path loss.
     If there are simply no resources in the target cell, the network may need to release
     the connection to avoid excessive interference.

6   The target eNodeB sends the HANDOVER REQUEST ACKNOWLEDGE to
     the source eNodeB. The target eNodeB is now ready to receive the incoming
     handover.

7   The source eNodeB sends the HANDOVER COMMAND to the UE. The source
     eNodeB starts forwarding the downlink packets to the target eNodeB.

8   The source eNodeB sends the status information to the target eNodeB indicating
     the packets that were acknowledged by the UE. The target eNodeB starts buffering
     the forwarded packets.

9   The UE makes the final synchronization to target eNodeB and accesses the cell
     via a RACH procedure. The pre-synchronization is already obtained during the cell
     identification process.

10   The target eNodeB gives the uplink allocation and timing advance information to
     the UE.

11   The UE sends HANDOVER CONFIRM to the target eNodeB. The target eNodeB
     can now begin to send data to the UE.

12   The target eNodeB sends a PATH SWITCH message to the MME to inform it that
     the UE has changed cell.

13   The MME sends a USER PLANE UPDATE REQUEST message to the Serving
     Gateway.

14   The Serving Gateway switches the downlink data path to the target eNodeB.

15   The Serving Gateway sends a USER PLANE UPDATE RESPONSE message to
     the MME.

16   The MME confirms the PATH SWITCH message with the PATH SWITCH ACK
     message.

17   The target eNodeB sends RELEASE RESOURCE to the source eNodeB, which
     allows the source eNodeB to release the resources.

18   The source eNodeB can release radio and control plane related resources associated
     with the UE context.

7.3.3 Handover Measurements

Before the UE is able to send the measurement report, it must identify the target cell. The UE identifies the cell using the synchronization signals (see Chapter 5). The UE measures the signal level using the reference symbols. There is no need for the E-UTRAN UE to read the Broadcast channel during the handover measurements. In UTRAN the UE needs to decode the Broadcast channel to find the system frame number, which is required to time-align the soft handover transmissions in downlink. There is no such requirement in E-UTRAN as there is no soft handover.

When the reporting threshold condition is fulfilled, the UE sends Handover measurements to the eNodeB.

7.3.4 Automatic Neighbor Relations

The neighborlist generation and maintenance has turned out to be heavy work in the existing mobile networks especially when the networks are expanded and new sites are being added. A missing neighbor is a common reason for the call drops. The UE in E-UTRAN can detect the intra-frequency neighbors without neighborlists, leading to simpler network management and better network quality. The automatic neighbor relation functionality is illustrated in Figure 7.10. The UE is moving towards a new cell and identifies the Physical Cell Identity (PCI) based on the synchronization signals. The UE sends a measurement report to eNodeB when the handover reporting threshold has been fulfilled. The eNodeB, however, does not have X2 connection to that cell. The physical cell identity (ID) is not enough to uniquely identify the cell since the number of physical cell IDs is just 504 while the large networks can have tens of thousands of cells. Therefore, the serving eNodeB requests the UE to decode the global cell ID from the broadcast channel of the target cell. The global cell ID identifies the cell uniquely. Based on the global cell ID the serving eNodeB can find the transport layer address of

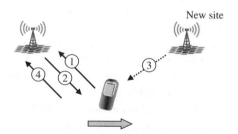

① = UE reports neighborcell signal including Physical Cell ID
② = Request for Global Cell ID reporting
③ = UE reads Global Cell ID from BCH
④ = UE reports Global Cell ID

Figure 7.10 Automatic intra-frequency neighbor identification. BCH, broadcast channel

the target cell using the information from the MME and set up a new X2 connection. The serving eNodeB can then proceed with the handover.

New X2 connections need to be created and some old unused connections can be removed when new cells are added to the network or when changes are done to the antenna systems.

The intra-frequency neighborlist generation is simple since the UE can easily identify all the cells within the same frequency. In order to create inter-RAT or inter-frequency neighbors the eNodeB must ask the UE to make inter-RAT or inter-frequency measurements. The eNodeB must also schedule gaps in the signal to allow the UE to perform the measurements.

7.3.5 Handover Frequency

The handover frequency in the network depends on a number of factors:

- The actual mobility of the users compared to the cell size. The statistics from HSPA networks can be used to estimate the impact of the mobility.
- The user activity. The typical voice activity is 30–50 mErl, which is 3–5% activity. Laptop users and smart phone users may have a RRC connection active all the time since the always-on applications keep sending packets at frequent intervals. User activity may therefore increase considerably in the future.
- The mobility solution of the system. The total number of handovers will likely be higher in LTE than in HSPA since part of the mobility in HSPA is carried out by the UE itself making cell reselections in Cell_PCH and Cell_FACH states, while LTE always uses handovers.

The impact of user mobility is illustrated by comparing the number of handovers for HSPA users, which are mostly laptop users today, and for WCDMA users, which are mostly voice users. An example of the WCDMA handover frequency is illustrated in Figure 7.11. The handover frequency is collected per cell per hour. The median active set update frequency is 6 s, corresponding to approximately 12 s handover frequency. Each HSPA cell change or LTE handover corresponds typically to two active set updates in WCDMA: add and drop. An example of the HSPA cell change frequency is shown in Figure 7.12. Both Figure 7.11 and Figure 7.12 were collected from the same network at the same time. The median number of HSPA serving cell changes is one serving cell change for every 120–140 s for each active connection. That indicates that current laptop HSPA users have clearly lower mobility than the WCDMA voice users. The HSPA cell change used 2 dB handover hysteresis where the target HSPA serving cell has to have 2 dB higher Ec/No than the current HSPA serving cell.

It is expected that the LTE mobility will initially be similar to HSPA mobility since LTE devices are likely to be USB modems. The introduction of handheld LTE devices will increase the handover rate per connection.

Figure 7.11 and Figure 7.12 show that the handover frequency varies considerably depending the environment, cell size and user mobility. For 20% of the cells the HSPA service cell change rate is below 40 s. There are, however, 15% of cells where the cell change rate is more than 500 s.

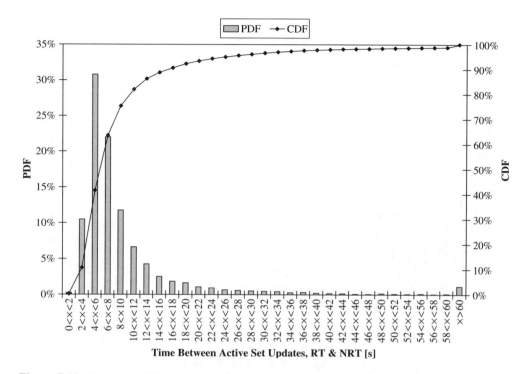

Figure 7.11   Example of WCDMA active set update frequency in 3G networks. PDF, probability density function; CDF, cumulative density function; RT, real time; NRT, non-real time

The most typical application today is voice with a normal activity of 3–5%. The application activity will increase with always on applications. Laptops keep sending small packets in the background implying that the RRC connection needs to be maintained 100% of the time. The same situation applies to the handheld devices that run always on applications such as push-email, push-to-talk or presence. Even if the end user does not use the application actively, the RRC connection is still active from the system perspective due to the transmission of the keep alive messages, reception of emails and presence updates. The user activity may increase from 5% to 100% due to the new applications – that is a 20 times higher activity than today. Therefore, the number of handovers may be considerably higher with new applications in the LTE network where all the mobility is carried by handovers for RRC connected users.

7.3.6   Handover Delay

The fast handover process is essential when the signal level from the source cell fades fast while the signal from the target cell increases. This may happen when the UE drives around the corner. The signal levels are illustrated in Figure 7.13. The critical point is that the UE must be able to receive the handover command from the source eNodeB before the signal-to-interference ratio gets too low. The handover reliability in this scenario can be improved by suitable window and averaging parameters as well as by minimizing the network delay in responding to the UE measurement report.

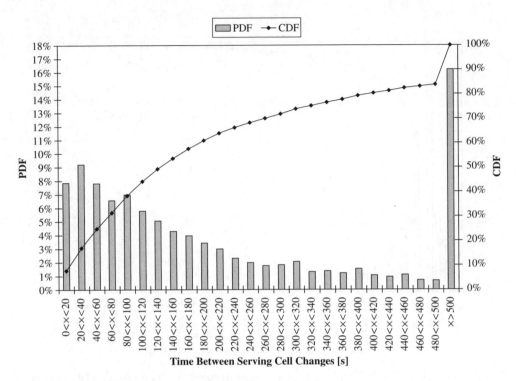

Figure 7.12   Example of HSPA Serving Cell Change frequency in 3G networks

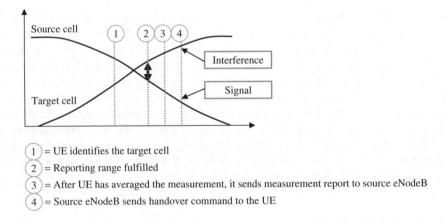

① = UE identifies the target cell
② = Reporting range fulfilled
③ = After UE has averaged the measurement, it sends measurement report to source eNodeB
④ = Source eNodeB sends handover command to the UE

Figure 7.13   Handover timings

7.4   Inter-system Handovers

The inter-RAT handovers refer to the inter-system handovers between E-UTRAN and GERAN, UTRAN or cdma2000® both for real time and for non-real time services. The inter-RAT handover is controlled by the source access system for starting the measurements and deciding the handover execution. The inter-RAT handover is a backwards

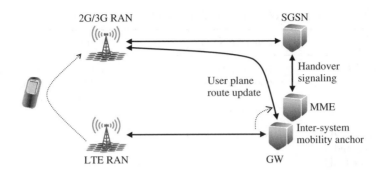

Figure 7.14   Overview of inter-RAT handover from E-UTRAN to UTRAN/GREAN

handover where the radio resources are reserved in the target system before the handover command is issued to the UE. As the GERAN system does not support Packet Switched Handover (PS HO), the resources are not reserved before the handover. The signaling is carried via the core network as there are no direct interfaces between the different radio access systems. The inter-RAT handover is similar to intra-LTE handover in the case where the packet core node is changed.

All the information from the target system is transported to the UE via the source system transparently. The user data can be forwarded from the source to the target system to avoid loss of user data. To speed up the handover procedure there is no need for the UE to have any signaling to the core network. The security and QoS context is transferred from the source to the target system.

The Serving GW can be used as the mobility anchor for the inter-system handovers.

The overview of the inter-system handover is shown in Figure 7.14. The interfaces are described in Chapter 3.

The interruption time in inter-system handovers can be very small. The interruption time from the UE point of view is defined as the time between the last TTI containing a transport block on the E-UTRAN side and the time the UE starts transmission of the new uplink DPCCH on the UTRAN side. This interruption time can be as low as 50 ms plus the frame alignments. When the UE enters UTRAN, it will initially be connected to one cell and the soft handover links can be added separately. The UE cannot go directly into soft handover in UTRAN since the UE is not required to read the system frame number (SFN) from UTRAN before the inter-system handover.

E-UTRAN also supports mobility to and from non-3GPP radio systems, such as cdma2000®.

7.5   Differences in E-UTRAN and UTRAN Mobility

The main differences between UTRAN and E-UTRAN mobility are summarized below and listed in Table 7.2. The idle mode mobility is similar in UTRAN and E-UTRAN for intra-frequency reselections. But a new priority based reselection method was chosen in E-UTRAN to ease network planning for the presence of multiple different RATs. The flat architecture in E-UTRAN brings some differences to the active mode mobility.

In UTRAN the UE must update the location both to the circuit switched core network (Location areas) and to the packet core network (Routing areas) while E-UTRAN only

Table 7.2   UTRAN and E-UTRAN differences in mobility

UTRAN	E-UTRAN	Notes
Location area (CS core)	Not relevant since no CS connections	For CS fallback handover, MME maps Tracking area to Location area
Routing area	Tracking area	
Soft handover is used for WCDMA uplink and downlink and for HSUPA uplink	No soft handovers	
Cell_FACH, Cell_PCH, URA_PCH	No similar RRC states	E-UTRAN always uses handovers for RRC connected users
RNC hides most of mobility	Core network sees every handover	Flat architecture HSPA is similar to E-UTRAN
Neighbor cell lists required	No need to provide cell specific information, i.e. only carrier frequency is required, but the network can provide cell specific reselection parameters (e.g. Q_{offset}) if desired	Also UTRAN UE can use detected cell reporting to identify a cell outside the neighborlist

uses Tracking areas (packet core). For so-called Circuit Switched Fall Back (CSFB) handover, the CS core network can send a paging message to the E-UTRAN UE. In that case the MME maps the Tracking area to the Location area. This procedure is explained in Chapter 13.

E-UTRAN uses handovers for all UEs that have RRC connection. The UE in UTRAN makes cell reselections without network control even if a RRC connection exists when the UE is in Cell_FACH, Cell_PCH or in URA_PCH states. Such states are not used in E-UTRAN. The UE location is known with UTRAN Registration area accuracy in URA_PCH state while the UE location is known with the accuracy of one cell in E-UTRAN whenever the RRC connection exists.

The WCDMA system uses soft handovers both in uplink and in downlink. The HSPA system uses soft handovers in uplink, but not in downlink. In the LTE system a need for soft handovers was no longer seen, thus simplifying the system significantly.

The Radio Network Controller (RNC) typically connects hundreds of base stations and hides the UE mobility from the core network. The packet core can only see Serving RNC (S-RNC) relocations when the SRNC is changed for the UE. For E-UTRAN, the packet core network can see every handover. 3GPP has specified the flat architecture option for UTRAN in Release 7. That architecture is known as Internet-HSPA. The RNC functionalities are embedded to the Node-B. Internet-HSPA mobility is similar to E-UTRAN mobility: serving RNC relocation over the Iur interface is used in the same way as E-UTRAN handover over the X2 interface.

7.6 Summary

3GPP E-UTRAN as well as the SAE standards are designed to support smooth mobility within E-UTRAN (LTE) and between E-UTRAN and GERAN/UTRAN (2G/3G RAN) and to cdma2000®. The idle mode mobility follows the same principles as in 2G/3G networks where the UE makes cell reselection autonomously. E-UTRAN adds the flexibility that the UE can be allocated multiple tracking areas. The network knows the UE location with the accuracy of the tracking area.

The handover is used in E-UTRAN RRC connected mode and is controlled by the network. The intra-frequency handover is based on fast hard handover in the radio followed by the core network path switching. The E-UTRAN flat architecture implies that the core network sees every handover while in UTRAN system most of the handovers are hidden by RNC. Additionally, the number of handovers in E-UTRAN will likely increase compared to UTRAN since the mobility is controlled by handovers for all RRC connected users.

The E-UTRAN system can be operated without pre-defined neighbor cell lists, which makes the network operation simpler than in GERAN or UTRAN networks.

References

[1] 3GPP Technical Specification 36.300 'E-UTRAN Overall Description; Stage 2', v. 8.6.0.
[2] 3GPP Technical Specification 36.304 'User Equipment (UE) procedures in idle mode', v. 8.3.0.
[3] 3GPP Technical Specification 36.133 'Requirements for support of radio resource management', v. 8.3.0.
[4] 3GPP Technical Specification 36.331 'Radio Resource Control (RRC); Protocol specification', v. 8.3.0.

8

Radio Resource Management

Harri Holma, Troels Kolding, Daniela Laselva, Klaus Pedersen, Claudio Rosa and Ingo Viering

8.1 Introduction

The role of Radio Resource Management (RRM) is to ensure that the radio resources are efficiently used, taking advantage of the available adaptation techniques, and to serve the users according to their configured Quality of Service (QoS) parameters. An overview of primary RRM functionalities is given in section 8.2, and the chapter then moves on to discuss QoS parameters and admission control issues in section 8.3. Next, details related to both downlink and uplink adaptation and scheduling methodologies are discussed in sections 8.4 and 8.5. Interference coordination is discussed in section 8.6, and a discussion of discontinuous reception is included in section 8.7. The Radio Resource Control (RRC) connection management is introduced in section 8.8, and the chapter is summarized in section 8.9.

8.2 Overview of RRM Algorithms

Figure 8.1 shows an overview of the user-plane and control-plane protocol stack at the eNodeB, as well as the corresponding mapping of the primary RRM related algorithms to the different layers. The family of RRM algorithms at the eNodeB exploits various functionalities from Layer 1 to Layer 3 as illustrated in Figure 8.1. The RRM functions at Layer 3, such as QoS management, admission control, and semi-persistent scheduling, are characterized as semi-dynamic mechanisms, since they are mainly executed during setup of new data flows. The RRM algorithms at Layer 1 and Layer 2, such as Hybrid Adaptive Repeat and Request (HARQ) management, dynamic packet scheduling, and link adaptation, are highly dynamic functions with new actions conducted every Transmission Time Interval (TTI) of 1 ms. The latter RRM functions are therefore characterized as *fast dynamic*.

The Channel Quality Indicator (CQI) manager at Layer 1 processes the received CQI reports (downlink) and Sounding Reference Signals (SRSs) (uplink) from active users in the cell. Each received CQI report and SRS is used by the eNodeB for scheduling

LTE for UMTS: Evolution to LTE-Advanced, Second Edition. Edited by Harri Holma and Antti Toskala.
© 2011 John Wiley & Sons, Ltd. Published 2011 by John Wiley & Sons, Ltd.

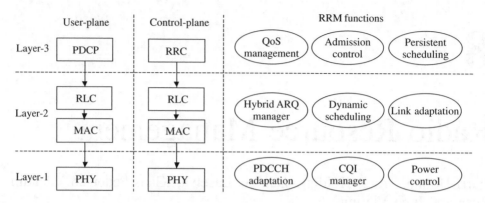

Figure 8.1 Overview of the eNodeB user plane and control plane protocol architecture, and the mapping of the primary RRM functionalities to the different layers. PHY = Physical layer; MAC = Medium access control; RLC = Radio link control; PDCP = Packet data convergence protocol; PDCCH = Physical downlink control channel

decisions and for link adaptation purposes in downlink and uplink, respectively. This chapter presents the Layer 3 and Layer 2 RRM functions except for the semi-persistent scheduling, which is part of the voice description in Chapter 13 as semi-persistent scheduling is typically used for voice service. The Layer 1 functions are covered in Chapter 5.

3GPP specifies the RRM related signaling but the actual RRM algorithms in the network are not defined in 3GPP – those algorithms can be vendor and operator dependent.

8.3　Admission Control and QoS Parameters

The eNodeB admission control algorithm decides whether the requests for new Evolved Packet System (EPS) bearers in the cell are granted or rejected. Admission control (AC) takes into account the resource situation in the cell, the QoS requirements for the new EPS bearer, as well as the priority levels, and the currently provided QoS to the active sessions in the cell. A new request is only granted if it is estimated that QoS for the new EPS bearer can be fulfilled, while still being able to provide acceptable service to the existing in-progress sessions in the cell having the same or higher priority. Thus, the admission control algorithm aims at only admitting new EPS bearers up to the point where the packet scheduler in the cell can converge to a feasible solution where the promised QoS requirements are fulfilled for at least all the bearers with high priority. The exact decision rules and algorithms for admission control are eNodeB vendor specific and are not specified by 3GPP. As an example, possible vendor specific admission control algorithms for OFDMA based systems are addressed in [1]. Similarly, the QoS-aware admission control algorithms in [2] and [3] can be extended to LTE.

Each LTE EPS bearer has a set of associated QoS parameters in the same way as GERAN and UTRAN radios. All the packets within the bearer have the same QoS treatment. It is possible to modify QoS parameters of the existing bearers dynamically. It is also possible to activate another parallel bearer to allow different QoS profiles for different services simultaneously. The new bearer can be initiated by the UE or by the packet core network.

The QoS profile of the EPS bearer consists of the following related parameters [4]:

- allocation retention priority (ARP);
- uplink and downlink guaranteed bit rate (GBR);
- QoS class identifier (QCI).

The GBR parameter is only specified for EPS GBR bearers. For non-GBR bearers, an aggregate MBR (AMBR) is specified. The ARP parameter, being an integer in the range 1–16, is primarily for prioritization when conducting admission control decisions. The QCI is a pointer to a more detailed set of QoS attributes. The QCI includes parameters like the Layer 2 packet delay budget, packet loss rate and scheduling priority. These parameters can be used by the eNodeB to configure the outer ARQ operation point for the RLC protocol, and the Layer 2 packet delay budget is used by the eNodeB packet scheduler, i.e. to prioritize certain queues in order to fulfill certain head-of-line packet delay targets.

3GPP specifications define a mapping table for nine different QCIs and their typical services, see Table 8.1. Further QCI values may be defined later depending on the need and on the new emerging services.

An additional QoS parameter called the prioritized bit rate (PBR) is specified for the uplink per bearer. The PBR is introduced to avoid the so-called uplink scheduling starvation problem that may occur for UE with multiple bearers. A simple rate control functionality per bearer is therefore introduced for sharing of uplink resources between radio bearers. RRC controls the uplink rate control function by giving each bearer a priority and a PBR. The PBR is not necessarily related to the GBR parameter signaled via S1 to the eNodeB, i.e. a PBR can also be defined for non-GBR bearers. The uplink rate control function ensures that the UE serves its radio bearer(s) in the following sequence [6]:

- all the radio bearer(s) in decreasing priority order up to their PBR;
- all the radio bearer(s) in decreasing priority order for the remaining resources.

When the PBR is set to zero for all RBs, the first step is skipped and the radio bearer(s) are served in strict priority order.

The QoS concept in LTE has been simplified compared to WCDMA/HSPA in which more than ten different QoS parameters are signaled over the Iu interface. LTE also enables

Table 8.1 QCI characteristics for the EPS bearer QoS profile [4] [5]

QCI#	Priority	L2 packet delay budget	L2 packet loss rate	Example services
1 (GBR)	2	100 ms	10^{-2}	Conversational voice
2 (GBR)	4	150 ms	10^{-3}	Conversational video
3 (GBR)	5	300 ms	10^{-6}	Buffered streaming
4 (GBR)	3	50 ms	10^{-3}	Real-time gaming
5 (non-GBR)	1	100 ms	10^{-6}	IMS signaling
6 (non-GBR)	7	100 ms	10^{-3}	Live streaming
7 (non-GBR)	6	300 ms	10^{-6}	Buffered streaming, email,
8 (non-GBR)	8	300 ms	10^{-6}	browsing, file download,
9 (non-GBR)	9	300 ms	10^{-6}	file sharing, etc.

a network activated GBR bearer without the need for the terminal application to initiate the request. There is, however, also more need for well performing QoS in LTE since circuit switched connections are not possible and voice is carried as Voice over IP (VoIP).

8.4 Downlink Dynamic Scheduling and Link Adaptation

Dynamic packet scheduling and link adaptation are key features to ensure high spectral efficiency while providing the required QoS in the cell. In this section, the general framework as well as specific algorithms and applications are presented.

8.4.1 Layer 2 Scheduling and Link Adaptation Framework

The controlling RRM entity at Layer 2 is the dynamic packet scheduler (PS), which performs scheduling decisions every TTI by allocating Physical Resource Blocks (PRBs) to the users, as well as transmission parameters including modulation and coding scheme. The latter is referred to as link adaptation. The allocated PRBs and selected modulation and coding scheme are signaled to the scheduled users on the PDCCH. The overall packet scheduling goal is to maximize the cell capacity, while making sure that the minimum QoS requirements for the EPS bearers are fulfilled and there are adequate resources also for best-effort bearers with no strict QoS requirements. The scheduling decisions are carried out on a per user basis even though a user has several data flows. Actually, each active user with an EPS bearer has at least two Layer 2 data flows; namely a control plane data flow for the RRC protocol and one or multiple user plane data flows for EPS bearers. Each of the data flows are uniquely identified with a 5-bit Logical Channel Identification (LCID) field. Given the scheduled Transport Block Size (TBS) for a particular user, the MAC protocol decides how many data are sent from each LCID.

As illustrated in Figure 8.2, the packet scheduler interacts closely with the HARQ manager as it is responsible for scheduling retransmissions. For downlink, asynchronous adaptive HARQ is supported with 8 stop-and-wait channels for each code-word, meaning that the scheduler has full flexibility to dynamically schedule pending HARQ retransmissions in the time and the frequency domain. For each TTI, the packet scheduler must decide between sending a new transmission or a pending HARQ transmission to each scheduled user, i.e. a combination is not allowed. The link adaptation provides information to the packet scheduler of the supported modulation and coding scheme for a user depending on the selected set of PRBs. The link adaptation unit primarily bases its decisions on CQI feedback from the users in the cell. An outer loop link adaptation unit may also be applied to control the block error rate of first transmissions, based on, for example, HARQ acknowledgements (positive or negative) from past transmissions [7]. A similar outer loop link adaptation algorithm is widely used for HSDPA [8].

8.4.2 Frequency Domain Packet Scheduling

Frequency Domain Packet Scheduling (FDPS) is a powerful technique for improving the LTE system capacity. The basic principle of FDPS is illustrated in Figure 8.3. The

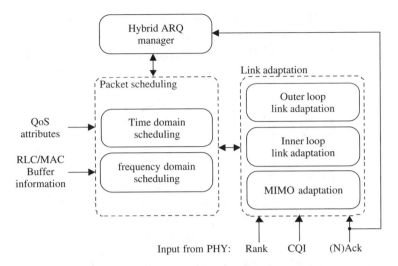

Figure 8.2 Layer 2 functionalities for dynamic packet scheduling, link adaptation, and HARQ management

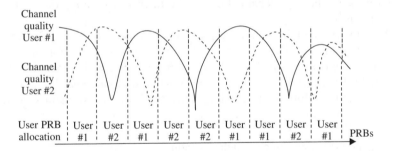

Figure 8.3 Frequency domain scheduling principle

FDPS principle exploits frequency selective power variations on either the desired signal (frequency selective fading) or the interference (fading or due to fractional other cell load) by only scheduling users on the PRBs with high channel quality, while avoiding the PRBs where a user experiences deep fades. A condition for achieving high FDPS gains is therefore that the radio channel's effective coherence bandwidth is less than the system bandwidth, which is typically the case for cellular macro and micro cell deployments with system bandwidths equal to or larger than 5 MHz.

As an example, Figure 8.4 shows the performance results from extensive macro cellular system level simulations versus the UE velocity. These results are obtained for a 10 MHz system bandwidth, a configuration with one transmit antenna and dual antenna UEs with interference rejection combining, Poisson arrival best effort traffic, and other assumptions in line with the 3GPP agreed simulation assumptions. The average cell throughput is reported. The results for simple wideband CQI correspond to cases with no FDPS, while the results with frequency selective CQI reporting are obtained with proportional FDPS. It is observed that a significant FDPS gain of approximately 40% is achievable for low

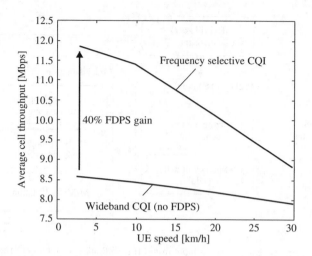

Figure 8.4   Capacity gain from Frequency Domain Packet Scheduling (FDPS)

to moderate UE speeds, while the FDPS gain decreases at higher UE speeds. The latter is observed because the radio channel cannot be tracked accurately due to the delay on uplink CQI reporting.

The time domain scheduling also can provide multi-user diversity gains when there is frequency flat fast fading. The gains depend on the amount of fading and on the speed of the fading. When there are deep fades, it gives more freedom for the scheduler to select optimally the user for the transmission. When the mobile speed is low enough, the scheduling is able to follow the fast fading. Therefore, the time domain scheduling gains generally get lower with:

- mobile antenna diversity since it reduces fading;
- base station transmit antenna diversity since it reduces fading;
- large bandwidth since it reduces fading due to frequency diversity;
- high mobile speed;
- multi-path propagation since it reduces fading due to multi-path diversity.

The time domain scheduling gains in LTE are relatively low since antenna diversity is a standard feature in the LTE terminal and also LTE uses typically large bandwidths.

Even though the spectral efficiency is maximized by operating with full frequency reuse [9], the system may start to operate in fractional load mode if there are only small data amounts in the eNodeB for the users in the cell. If such situations occur, the buffered data in the eNodeB are only transmitted on the required number of PRBs, while the rest of the PRBs are muted (i.e. no transmission). Under such fractional load conditions, the packet scheduler still aims at selecting the PRBs with the highest channel quality based on the CQI feedback. As illustrated in Figure 8.5 with a simple two-cell example, this allows convergence to a PRB allocation strategy between the two cells, where they aim at allocating complementary sets of PRBs to minimize the interference between the cells. As an example, it was shown in [10] that the average experienced Signal-to-Interference Noise Ratio (SINR) was improved by 10 dB compared to the full load case by

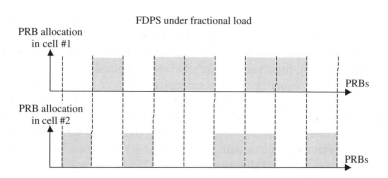

Figure 8.5   Frequency domain scheduling principle under fractional load conditions

operating a 3-sector network at 25% fractional load by using CQI assisted scheduling. The 10 dB improvement should be compared to the expected 6 dB improvement from a 25% fractional load (i.e. factor of 4 reduced load corresponds to 6 dB) with blind scheduling. The CQI assisted scheduling scheme under fractional load resembles the well-known carrier sense allocation scheme from the Ethernet protocol. No explicit load information was shared over the X2 interface in this case.

8.4.3   Combined Time and Frequency Domain Scheduling Algorithms

Figure 8.6 illustrates the three-step packet scheduling implementation as proposed in [7]. The first step consists of the time-domain (TD) scheduling algorithm, which selects up to N users for potential scheduling during the next TTI. The time domain scheduler only selects users with (i) pending data for transmission and (ii) users configured for scheduling in the following TTI, i.e. excluding users in Discontinuous Reception (DRX) mode. The time domain scheduler assigns a priority metric for each of the N selected users. Secondly, a control channel schedule check is performed to evaluate if there is sufficient control channel capacity to schedule all the users selected by the time domain scheduler. This implies evaluating if there are enough transmission resources within the first 3 OFDM symbols to have reliable transmission of the PDCCH for each of the selected users. If the latter is not the case, then only a sub-set of the N users are passed to the frequency domain (FD) scheduler. In cases with control channel congestion, the users with the lowest priority metric from the time domain scheduler are blocked. Users with pending HARQ retransmissions can be given priority. Finally, the frequency domain scheduler decides how to schedule the remaining users across the available PRBs. Compared to a fully joint time/frequency domain packet scheduling method, this method provides similar

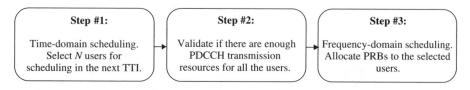

Figure 8.6   Illustration of the three-step packet scheduling algorithm framework

performance at a much lower complexity as the frequency domain scheduler only has to consider a sub-set of the users for multiplexing on the PRBs.

Assuming that the number of users scheduled per TTI is smaller than the number of schedulable users per cell, the time domain scheduler becomes the primary instrument for QoS control, while the frequency domain scheduler task is to benefit from radio channel aware multi-user frequency diversity scheduling. This implies that the overall scheduling QoS concept from HSDPA (see e.g. [11] and [12]) can be generalized to also apply for the LTE time domain scheduler part for most cases. Thus, the barrier function based scheduler in [13], the QoS-aware schedulers with required activity detection in [14] and [15], and the generalized delay aware schedulers in [16] and [17] are all relevant for the LTE time domain scheduler. Similarly, the well-known proportional fair scheduler (as previously studied in [18] and [19], among others) can be extended for OFDMA based systems as discussed in [20], [21] and [22].

The frequency domain scheduler allocates PRBs to the users selected by the time domain scheduler. The overall objective is to allocate the available PRBs among the selected users to maximize the benefit from multi-user FDPS, while still guaranteeing a certain fairness. The frequency domain scheduler is responsible for allocating PRBs to users with new transmissions and also to users with pending HARQ retransmissions. Note, however, that it is not possible to simultaneously schedule new data and pending HARQ retransmissions to the same user in the same TTI. Pending HARQ retransmissions will typically be scheduled on the same number of PRBs as the original transmission. As scheduling of HARQ retransmissions creates a combining gain it is typically not critical to have those retransmissions allocated on the best PRBs. The latter argument has led to the proposed frequency domain scheduling frame-work summarized in Figure 8.7 [7]. In Step #1 the frequency domain scheduler computes the number of required PRBs (denoted by Nre) for scheduling of pending HARQ retransmissions for the selected users by the time domain scheduler. Assuming Ntot available PRBs for PDSCH transmission, users with new data are allocated the best Ntot-Nre PRBs in Step #2. Finally, the remaining Nre PRBs are allocated for the users with pending HARQ retransmissions. This framework has the advantage that HARQ retransmissions are given priority (i.e. reducing HARQ delays) while maximizing the channel capacity by allocating the best PRBs for users with new data transmissions. The latter is an attractive solution for scenarios where the experienced channel quality is less critical for HARQ retransmissions due to the benefits from the HARQ combining gain. For highly coverage limited scenarios where it is important to maximize the channel quality for second transmissions, however, Steps #2 and #3 in Figure 8.7 can be exchanged. The exact algorithms for allocating PRBs to the considered users in Step #2 and Step #3 can be a simple diversity scheduler or a modified version of proportional fair or maximum throughput schedulers as discussed in [20], [7]

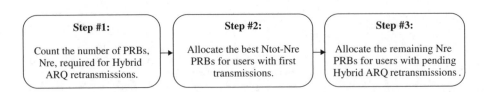

Figure 8.7   HARQ aware frequency domain packet scheduling

and [22], among others. The full benefit from FPDS requires that the considered users are configured to report frequency selective CQI reports such as full sub-band reporting or average best-M (see more details in Chapter 5 for the CQI schemes).

8.4.4 Packet Scheduling with MIMO

Multi-antenna transmission, Multiple Input Multiple Output (MIMO) with two transmit and two receive antennas can use either one transport block or two independently coded transport blocks to the user on virtual streams on the allocated PRBs. The coded transport blocks are also called code-words. The transmission of two independently coded transport blocks requires that the Rank of the MIMO radio channel is at least two. In the latter case, the CQI feedback from the user will include information that allows the eNodeB link adaptation unit to potentially select different modulation and coding schemes for the two independent transport blocks. Separate acknowledgements are transmitted for each code-word such that link adaptation and HARQ for the two streams are decoupled. The eNodeB can dynamically switch between transmission of one and two code-words depending on the CQI and Rank Indicator feedback from the scheduled user. As a simple rule of thumb, the average SINR has to be in excess of approximately 12 dB before transmission of two independent transport blocks becomes attractive on a 2×2 MIMO link with Rank 2. From a packet scheduling point of view, the single-user MIMO functionality is partially hidden as part of the link adaptation unit although supported TBS for the two streams is known by the packet scheduler. The different QoS-aware packet scheduling algorithms are equally applicable independent of whether advanced MIMO is being used or not. General discussions of MIMO schemes for OFDMA are available in [23].

LTE also supports multi-user MIMO transmission, where two users are scheduled on the exact same PRB(s), sending one independently coded transport block to each of the users by applying an appropriate set of antenna transmit weights. The two antenna transmission weights have to form orthogonal vectors while still being matched to achieve good received SINR for both users. To make multi-user MIMO attractive, i.e. maximizing the sum throughput served to the two users, the choice of the antenna transmission weights has to be accurate in both time and frequency domains. The latter also calls for coordinated packet scheduling actions, where users paired for transmission on the same PRB(s) are dynamically selected dependent on their joint short term radio channel characteristics. Hence, the usage of multi-user MIMO schemes also adds complexity to the packet scheduler and requires more control signaling (PDCCH).

8.4.5 Downlink Packet Scheduling Illustrations

This section presents performance results of LTE downlink packet scheduling. We first illustrate the impact of basic packet scheduling, mobile speed and antenna configuration on the cell and user throughput using a network emulation tool. Later in the section, we evaluate the QoS impact on the cell throughput and cell capacity by conducting dynamic system level simulations with multiple cells and multiple QoS UE classes.

Figure 8.8 and Figure 8.9 illustrate the impact of the packet scheduling, mobile speed and antenna configuration to the cell and user throughput. The plots are taken from an LTE network emulation tool. Figure 8.8 shows the cell throughput with proportional fair scheduler and with round robin scheduler with 3, 30 and 120 km/h. The highest cell

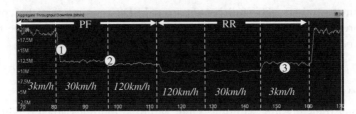

① = Significant impact to cell throughput when mobile speed increases from 3 to 30 km/h
    because the scheduling gain is reduced
② = Negligible impact when mobile speed increases from 30 to 120 km/h
③ = Gain from more accurate link adaptation at 3 km/h

Figure 8.8 Example of downlink cell throughput with different schedulers and mobile speeds

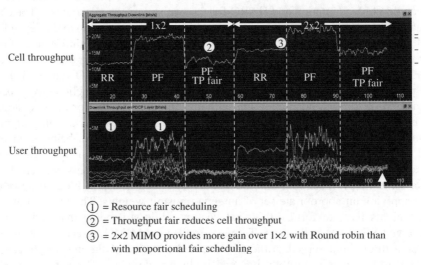

① = Resource fair scheduling
② = Throughput fair reduces cell throughput
③ = 2×2 MIMO provides more gain over 1×2 with Round robin than
    with proportional fair scheduling

Figure 8.9 Example of downlink cell throughputs and user throughputs with different antenna
configuration and different schedulers

throughput is obtained with proportional fair scheduler at 3 km/h. The throughput drops
considerably when the mobile speed increases to 30 km/h because the feedback is not able
to follow the fast fading, and the frequency domain scheduling gain is lost. Increasing
the mobile speed further to 120 km/h has only a marginal impact on the capacity. The
round robin scheduler has a lower cell throughput and less sensitivity to the mobile speed.
There is some capacity increase with the round robin scheduler at 3 km/h compared to 30
and 120 km/h because the link adaptation works more accurately at 3 km/h. The capacity
difference between proportional fair and round robin is small at 120 km/h, but increases
at low mobile speeds.

Figure 8.9 shows the cell and user throughput with 1 × 2 antenna configuration (1 transmission branch at eNodeB and 2 receiver branches at UE) and with 2 × 2 MIMO at 3 km/h.
Three different types of schedulers are presented: round robin, proportional fair with equal
resource scheduling, and proportional fair with equal throughput. The proportional fair

with equal resource scheduling increases both cell and user throughput with both antenna configurations. The relative gain from the proportional fair is higher with 1×2 since that antenna configuration has less diversity and experiences more fading than 2×2 MIMO. More fading gives more freedom for the scheduler to optimize the transmission. The throughput fair scheduling gives the same throughput for all users, but the cell through-put is clearly reduced compared to the resource fair scheduling. The cell edge users get higher throughput with throughput fair scheduling while those users in good locations get higher throughput with resource fair scheduling.

In the following we illustrate the design and the performance of a QoS-aware packet scheduler and the QoS impact on the cell throughput and cell capacity under full load. We consider the case of traffic mixes of best effort traffic (modeled as simple file download) with no QoS constraints and constant bit rate (CBR) applications with strict GBR QoS constraints. The results presented in the following are obtained with 1×2 interference rejection combining (IRC) antennas, for a system bandwidth of 10 MHz and for localized transmission.

The time domain (TD) packet scheduler is the primary entity to facilitate QoS differ-entiation, in terms of both service and user differentiation. An example of a QoS-aware packet scheduler is introduced in Table 8.2. The proposed scheduler is limited here to GBR-awareness for simplicity. It uses a GBR-aware time domain metric based on a mod-ified version of the proportional fair scheduler with adjustment according to the estimated required activity level for a certain user. Such a Required Activity Detection (RAD) sched-uler dynamically estimates the capacity used by serving the GBR users and it is able to share the remaining part, often called the excess capacity, according to any desired policy, e.g. by setting QoS class-specific shares. The frequency domain scheduler is based on a modification of the proportional fair metric, called proportional fair scheduler (PFsch). Similarly to the time domain proportional fair scheduler, the PFsch scheduler is designed to opportunistically benefit from radio channel aware multi-user frequency diversity and it aims at maximizing the spectral efficiency while providing resource fairness among the users. This scheduler decouples the behavior of the frequency domain scheduler from the time domain scheduler decisions, and vice versa [21]. In fact, the time–frequency

Table 8.2   Example of QoS-aware packet scheduler in LTE

Scheduler	Time domain scheduling metric, M_n	Frequency domain scheduling metric, $M_{n,k}$
(TD) QoS PS \	Required Activity Detection (RAD) [14], [15]	Proportional Fair scheduler (PFsch) [21]
	$\dfrac{D_n}{\overline{R}_n} \cdot \left(\dfrac{GBR_n}{\overline{R}_{sch,n}} + Share_n \cdot ExcessCap \right)$	$\dfrac{d_{n,k}}{\overline{R}_{sch,n}}$

D_n = Instantaneous wideband achievable throughput for user n.
$\overline{R}_n$ = Past average throughput of user n.
$d_{n,k}$ = Instantaneous achievable throughput for user n on PRB k.
$\overline{R}_{sch,n}$ = Past average throughput over the TTIs where user n is selected by the TD scheduler.
GBR_n = Guaranteed bit rate of user n.
$ExcessCap$ = Excess capacity (i.e. capacity left after minimum QoS is fulfilled).
$Share_n$ = Share of the excess capacity of user n.

domain coupling occurs whenever the time and the frequency domain metrics perform the allocation of the activity and the frequency resources based on the same feedback, e.g. the past average throughput. The coupling may lead to conflicting decisions when users are assigned a high activity rate but a low share of frequency resources, or vice versa.

Figure 8.10 shows the performance of the QoS packet scheduler as described in Table 8.2. The environment is 3GPP macro cell case #1. The results are obtained assuming different traffic mixes of best effort traffic and constant bit rate (CBR) applications with strict GBR QoS constraints of 256 kbps. As reference, results with the time/frequency domain proportional fair scheduler are also depicted. As expected, the proportional fair scheduler is not able to meet the GBR demands of a large percentage of CBR users as some users do not have sufficient performance with the proportional fair even-resource allocation principle. The figure illustrates that the QoS-aware scheduler strictly provides QoS to 100% of the CBR users and it has the ability to distribute the excess capacity among the best effort users. It is also observed that the average cell throughput with only best effort traffic decreases when more CBR users are present in the cell as shown with 12, 24 and 38 CBR users per cell, respectively, over the total of 40 users. The latter is observed because the CBR users are strictly served with their GBR requirement regardless of their location in the cell, hence reducing the freedom for the packet scheduler to optimize the cell throughput.

To tackle the need for supporting high GBRs compared to the cell throughput, the QoS aware time domain scheduler alone is not sufficient. To fully understand this limitation one should consider that, with a certain FDPS policy, a CBR user may not be capable of meeting its GBR requirement from a link budget perspective by only being assigned

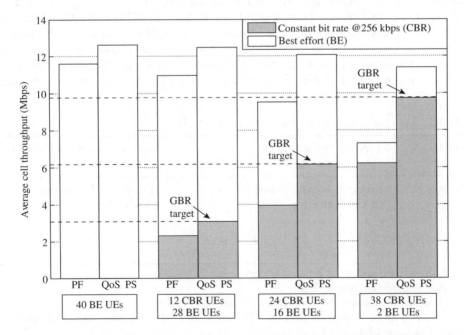

Figure 8.10 Average cell throughput for different traffic mixes of best effort and CBR traffic, for proportional fair scheduler and QoS-aware packet scheduler (see Table 8.2)

the maximum activity (i.e. the time domain scheduler selects the user every TTI). Note that the highest supportable GBR depends on the system bandwidth, the environment, the frequency domain packet scheduler, and the number of UEs multiplexed per TTI. Therefore, to support high GBR, QoS awareness is required to be further included in the frequency domain scheduler. An example of an adjustment to be applied to the frequency domain metric of GBR users is the following weight factor:

$$w_{FD_n} = \frac{GBR_n}{\overline{R}_{schPerPRB,n}} \tag{8.1}$$

where $\overline{R}_{schPerPRB,n}$ is the past average throughput per PRB over the TTIs where user n is allocated PRB resources. The adjustment estimates the share of frequency resources required by a GBR user to meet the GBR requirement, and it aims at allocating a higher share, when needed. For the sake of simplification, details are omitted on the weight lower bound and on the normalization of the non-GBR frequency domain metrics needed to account for the constraint of a limited number of frequency resources.

Figure 8.11 illustrates the performance of the QoS PS scheduler (see Table 8.2) with a mix of 12 best effort users with no QoS and 8 CBR users with strict GBR QoS constraints increasing from 512 kbps, to 768 kbps and up to 1024 kbps. It can be observed that the QoS differentiation uniquely enforced by QoS aware time domain is not

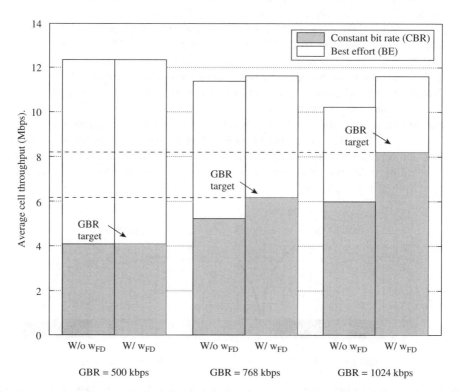

Figure 8.11 Average cell throughput for a traffic mix of 12 best effort and 8 CBR users per cell, when GBR equals 512, 768 and 1024 kbps. Results are obtained with the QoS aware packet scheduler as in Table 8.2, with and without the QoS-aware frequency domain weight

sufficient for supporting 100% of the CBR traffic when GBR equals 768 and 1024 kbps. When applying the frequency domain weight, w_{FD}, results show that a strict provision of the GBR requirements for all the CBR users is ensured with a limited cost in decreased FDPS gain.

8.5   Uplink Dynamic Scheduling and Link Adaptation

There are a few differences between uplink and downlink packet schedulers. The main differences are listed below.

1   eNodeB does not have full knowledge of the amount of buffered data at the UE. See section 8.5.1.2 for details on buffer status reports.

2   The uplink direction is more likely going to be power limited than the downlink due to low UE power compared to eNodeB. This means that especially in macro cell deployments users cannot always be allocated a high transmission bandwidth to compensate for poor SINR conditions.

3   Only adjacent physical resource blocks can be allocated to one user in uplink due to single carrier FDMA transmission. The fundamental difference between PRB allocation in LTE downlink and uplink stems from the uplink channel design and is illustrated with an example in Figure 8.12. The single-carrier constraint in LTE uplink limits both frequency and multi-user diversity.

4   The uplink is typically characterized by high interference variability [24] [25]. Interference variations from TTI to TTI in the order of 15–20 dB make it a hard task to estimate accurately the instantaneous uplink interference. Therefore in LTE uplink fast adaptive modulation and coding and channel-aware FDPS is typically based on channel state information about the own desired signal path rather than on the instantaneous interference level.

5   An uplink grant transmitted on PDCCH in TTI n refers to uplink transmission in TTI $n + 4$. This 4 ms delay is due to PDCCH decoding and processing time at the UE and represents a further limitation to link adaptation and channel-aware packet scheduling in uplink. For more details, please refer to the PDCCH physical layer description in Chapter 5.

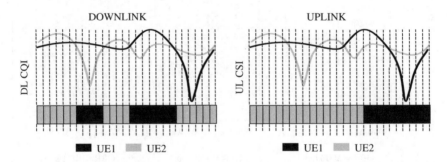

Figure 8.12   Example illustrating the single-carrier constraint to frequency domain packet scheduling in uplink

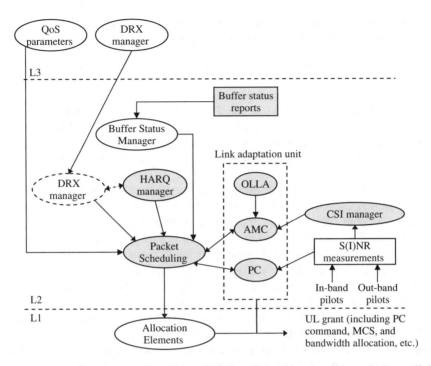

Figure 8.13 Inter-working between packet scheduling, link adaptation unit, and other uplink RRM functionalities

An overview of the uplink RRM functionalities and their interaction is given in Figure 8.13. As for the downlink, both uplink fast adaptive modulation and coding and frequency domain packet scheduler rely on frequency-selective channel state information (CSI). Uplink channel state information is estimated based on the SRS transmitted by the UE. The core of the uplink RRM functionality is the interaction between the packet scheduler, including fast Adaptive Transmission Bandwidth (ATB), and the so-called Link Adaptation (LA) unit, including Power Control (PC), adaptive modulation and coding and Outer-Loop Link Adaptation (OLLA). Buffer status reports and scheduling requests are also part of the core input to the uplink scheduler.

Interaction between the uplink dynamic packet scheduler and adaptive modulation and coding – The adaptive modulation and coding function is responsible for providing information to the packet scheduler on the channel state of a certain user in correspondence of a specific transmission bandwidth. In this sense the adaptive modulation and coding function is the link between the uplink packet scheduler and the uplink channel state information manager, which collects uplink channel state information based on the transmission of SRS in the uplink. Fast adaptive modulation and coding is also responsible for selecting the most appropriate MCS once the uplink packet scheduler has allocated a specific uplink band to the corresponding UE.

Interaction between packet scheduler and power control – The main scope of power control in LTE uplink is to limit inter-cell interference while respecting minimum SINR requirements based on QoS constraints, cell load and UE power capabilities. The uplink transmission power is set in the UE based on the following standardized formula [26]:

$$P_{\text{PUSCH}} = \min\{P_{\text{MAX}}, 10 \log_{10}(M_{\text{PUSCH}}) + P_{0\_\text{PUSCH}} + \alpha \cdot PL + \Delta_{\text{MCS}} + f(\Delta_i)\} \text{ [dBm]}$$
$$(8.2)$$

where P_{MAX} is the maximum allowed power and depends on the UE power class, M_{PUSCH} is the number of allocated PRBs on the Physical Uplink Shared Channel (PUSCH), PL is the downlink path loss measured by the UE (including distance-dependent path loss, shadowing and antenna gain), P_{0_PUSCH}, α and Δ_{MCS} are power control parameters, and $f(\Delta_i)$ is the closed loop power control correction transmitted by the eNodeB [26].

Figure 8.14 [27] is obtained assuming $\Delta_{\text{MCS}} = 0$ and $f(\Delta_i) = 0$ in equation (8.1), and clearly illustrates how the distribution of uplink SINR – and hence the system performance – strongly depends on the power control parameters, as well as on the propagation scenario. In general power control in uplink determines the average SINR region a user is operated at, which is similar to the G-factor in downlink. Since the UE performs power scaling depending on the allocated transmission bandwidth, the packet scheduler needs to have information on the UE transmit power spectral density so that the allocated transmission bandwidth does not cause the UE power capabilities to be exceeded. For such use, power headroom reports have been standardized for LTE uplink; see section 8.5.3.1 for more details.

Interaction between outer loop link adaptation and adaptive modulation and coding – As already pointed out, the adaptive modulation and coding functionality is responsible for two main tasks: (i) providing the link between the packet scheduler and the channel state information manager, and (ii) selecting the most appropriate MCS to be used for transmission on the selected uplink band by the uplink packet scheduler. In order to fulfill these tasks the adaptive modulation and coding functionality needs to map channel state information from the channel state information manager into SINR and/or

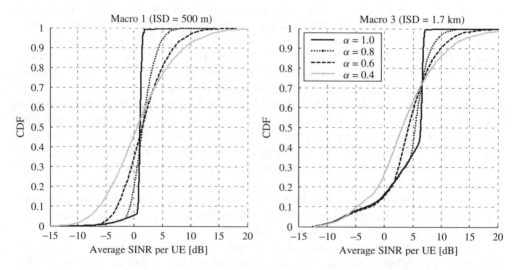

Figure 8.14 CDF of the average SINR per user for different propagation scenarios and values [27]

MCS. This function is similar to downlink; the main goal of OLLA is to compensate for systematic errors in the channel state information at the output of the channel state information manager so that packet scheduler and adaptive modulation and coding can perform correctly.

8.5.1 Signaling to Support Uplink Link Adaptation and Packet Scheduling

Fast link adaptation and channel-aware scheduling in LTE uplink strongly relies on the availability of frequency-selective channel state information based on SRS measurements. Moreover, since the uplink transmission buffers are located in the UE, the information on the buffer status needs to be transmitted in the uplink as well. Also, the user needs to report power headroom measurements in order to signal to the uplink packet scheduler (and to the fast ATB functionality in particular) how close it is operating to its maximum power capabilities. While the SRS physical layer details are covered in Chapter 5, in the following section RRM aspects of the LTE uplink channel sounding concept are briefly discussed. Next, the transmission of the so-called uplink scheduling information (i.e. buffer status and power headroom reports) is also addressed.

8.5.1.1 Sounding Reference Signals (SRS)

The uplink sounding concept basically determines 'when and how' an uplink SRS is transmitted. As a consequence, the sounding concept and parameters have a significant impact on the accuracy of uplink SRS measurements, and hence on the performance of uplink link adaptation and packet scheduling. From an uplink RRM perspective, some of the most important sounding parameters are [28]:

● SRS bandwidth: Indicates the transmission bandwidth of the uplink SRS. The SRS bandwidth is semi-statically signaled via RRC.
● SRS period and time offset: Indicates System Frame Number (SFN) modulo number and periodicity of SRS from one UE. The SRS period and time offset are semi-statically signaled via RRC. 3GPP has also standardized the possibility for the eNodeB to disable SRS transmission on a per-UE basis.
● SRS duration: Indicates for how long a UE must keep on transmitting the uplink SRS. SRS duration is also semi-statically signaled via RRC.
● Transmission combination, Constant Amplitude Zero Autocorrelation (CAZAC) sequence index and cyclic shift: Necessary to guarantee orthogonality among users transmitting uplink SRS using the same transmission bandwidth (see more details in Chapter 5). In LTE uplink up to 6 UEs/cell can share the same SRS bandwidth without interfering with each other.
● SRS sub-band hopping sequence: Determines the hopping sequence in case, for example, the SRS bandwidth is much narrower than the available bandwidth for scheduling.

It is an uplink RRM task to distribute the limited SRS resources among active users so that accurate and up-to-date channel state information is available. For instance, typically there is a tradeoff between measurement accuracy and SRS bandwidth (especially for power limited users): The narrower the SRS bandwidth, the higher the measurement accuracy [29]. On the other hand, several narrowband SRS transmissions are required to sound the entire schedulable bandwidth. Therefore, channel state information can become relatively outdated if the SRS periodicity and sub-band hopping are not configured properly. Another task of the uplink RRM functionality is to decide which users should be transmitting SRSs using the same time and frequency resources, since their orthogonality strongly depends on the received power level from the different UEs.

8.5.1.2   Buffer Status Reports (BSR)

The buffer status information is reported in uplink to inform the uplink packet scheduler about the amount of buffered data at the UE. LTE introduces a buffer status reporting mechanism that allows distinguishing between data with different scheduling priorities. The LTE buffer status reporting mechanism basically consists of two phases: (i) Triggering and (ii) Reporting.

Triggering
A Buffer Status Report (BSR) is triggered if any of the following events occur [30]:

- Uplink data arrive in the UE transmission buffer and the data belong to a radio bearer (logical channel) group with higher priority than those for which data already existed in the UE transmission buffer. This also covers the case of new data arriving in an empty buffer, The Buffer Status Report is referred to as 'regular BSR'.
- Uplink resources are allocated and number of padding bits is larger than the size of the BSR MAC control element, in which case the Buffer Status Report is referred to as 'padding BSR' (see below).
- A serving cell change occurs, in which case the Buffer Status Report is referred to as 'regular BSR' (see below).
- The periodic BSR timer expires, in which case the BSR is referred to as 'periodic BSR' (see below).

Reporting
The main uplink buffer status reporting mechanisms in LTE are the Scheduling Request (SR) and the Buffer Status Report (BSR) [30].
  Scheduling request (SR): The SR is typically used to request PUSCH resources and is transmitted when a reporting event has been triggered and the UE is not scheduled on PUSCH in the current TTI. The SR can be conveyed to the eNodeB in two ways:

- Using a dedicated 'one-bit' BSR on the Physical Uplink Control Channel (PUCCH), when available. The occurrence of SR resources on PUCCH is configured via RRC on a per UE basis. It might be possible that no resources for SR are allocated on PUCCH.
- Using the Random Access procedure. Random Access is used when neither PUSCH allocation nor SR resources are available on PUCCH.

According to the 3GPP specifications [30], a SR is only transmitted as a consequence of the triggering of a 'regular BSR'. The triggering of 'periodic BSR' and 'padding BSR' does not cause the transmission of a SR.

Buffer status report (BSR): BSRs are transmitted using a Medium Access Control (MAC) control element when the UE is allocated resources on PUSCH in the current TTI and a reporting event has been triggered. Basically the buffer status report is transmitted as a MAC-C PDU with only header, where the field length is omitted and replaced with buffer status information [30].

In summary, when a reporting event is triggered:

- If the UE has resources allocated on PUSCH, then a buffer status report is transmitted.
- If a 'regular BSR' is triggered and the UE has no PUSCH allocation in the current TTI but has SR resources allocated on PUCCH, then a SR is transmitted on PUCCH at the first opportunity.
- If a 'regular BSR' is triggered and the UE has neither SR resources allocated on PUCCH nor PUSCH allocation, then a SR is issued using the Random Access Procedure.

The design of buffer status reporting schemes and formats for UTRAN LTE uplink has been driven by two important factors [6]:

- Separate buffer status reports for data flows with different QoS characteristics/requirements are needed to support QoS-aware radio resource allocation.
- The overhead from the BSR needs to be minimized since it directly affects the uplink capacity.

Therefore within 3GPP it has been agreed that buffer status is reported on a per Radio Bearer Group (RBG) basis. A RBG is defined as a group of radio bearers with similar QoS requirements. The maximum number of radio bearer groups, however, has been fixed and is four. An example showing the mapping from radio bearers to RBG for buffer status reporting is shown in Figure 8.15.

Two BSR formats are used in LTE uplink [30]: Short BSR (only one radio bearer group is reported) and long BSR (all four radio bearer groups are reported), see Figure 8.16. In the first case 2 bits are required for radio bearer group identification, while in the latter case the 4 buffer size fields can be concatenated. In any case, the buffer size fields of the BSR are 6 bits long.

The basic idea is to transmit a short BSR if there are only data from one radio bearer group in the UE buffer, otherwise always transmit a long BSR. Some exceptions have

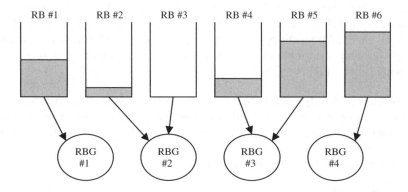

Figure 8.15 Example of mapping from RB to radio bearer group for buffer status reporting

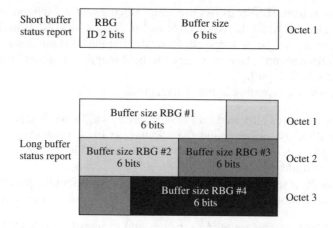

Figure 8.16 Short and long buffer status report types in LTE uplink [30]

been included in the standard in order to take into account the cases where, for example, a short BSR fits in the allocated transport block size, but a long BSR does not (see [30] for more details).

8.5.1.3 Power Headroom Reports

Due to the open loop component in the standardized power control formula (see Chapter 5) the eNodeB cannot always know the power spectral density level used at the UE. Information on the power spectral density is important for performing correct RRM decisions at the eNodeB, especially when allocating the transmission format including bandwidth and modulation and coding scheme. Not knowing the power spectral density used by a certain terminal could, for example, cause the allocation of a too high transmission bandwidth compared to the maximum UE power capabilities, thus resulting in a lower SINR than expected. Information on the power spectral density used at the UE can be obtained from power headroom reports, provided that the eNodeB knows the corresponding transmission bandwidth. Information on the power spectral density is primarily critical for the PUSCH as the transmission format for this channel is adaptively adjusted. For all these reasons power headroom reports have been standardized in 3GPP. The power headroom is calculated at the UE as the difference between the maximum UE transmit power and the 'nominal' PUSCH transmit power in the corresponding subframe set according to the PUSCH power control formula in [26]. A power headroom report is triggered if any of the following criteria are met [30]:

- An appositely defined prohibit timer expires or has expired and the path loss has changed by more than a pre-defined threshold since the last power headroom report. The prohibit timer is introduced to limit unnecessarily frequent transmissions of power headroom reports.
- An appositely defined periodic timer expires.

More details on the triggering and transmission of power headroom reports can be found in [30].

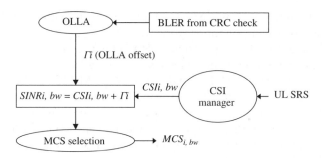

Figure 8.17   Schematic of the uplink fast adaptive modulation and coding functionality

8.5.2   *Uplink Link Adaptation*

Uplink link adaptation strongly relies on uplink SRS measurements. The uplink sounding concept and parameters basically determine how often and on which uplink physical resources the SRS measurements are available for a given UE, the measurement accuracy, etc. Then, similarly to downlink CQI, uplink SRS measurements are used by the channel state information manager to determine the most appropriate MCS for transmission on a specific uplink bandwidth. Also, uplink SRS measurements can be used by the uplink frequency domain scheduler when, for example, computing the scheduling priority of a specific user in correspondence of a specific uplink band. The uplink link adaptation functionality based on fast adaptive modulation and coding is schematically illustrated in Figure 8.17.

Uplink SRS measurements are eNodeB vendor specific and as a consequence are not strictly specified by 3GPP. However, it has been shown that uplink fast adaptive modulation and coding based on instantaneous SINR measurements can suffer from uplink interference variability [24]. On the other hand, selecting the MCS based on instantaneous channel conditions can still give a significant gain compared to slow AMC schemes [25]. Therefore one feasible implementation is to use uplink SRS to perform frequency-selective signal strength measurements with the aim of tracking fast fading on the signal component, while interference can be treated as a constant in order to minimize the effect of interference unpredictability.

8.5.3   *Uplink Packet Scheduling*

The main responsibility of the uplink packet scheduler is to share the available radio resources between users taking into account limitations and/or requirements imposed by other RRM functionalities.

- Packet scheduler and DRX/DTX. Users cannot be scheduled for transmission on PUSCH unless they are listening to the L1/L2 control channel.
- Packet scheduler and power control. UE transmission power capabilities must be considered when, for example, packet scheduler (ATB) allocates the uplink transmission bandwidth to a specific UE.
- Packet scheduler and QoS. The packet scheduler is responsible for fulfilling the user QoS requirements.
- Packet scheduler and BSRs. Users should be scheduled for data transmission only if they have data to transmit. Also, the prioritization between users can be carried out

based on the information conveyed by BSRs (e.g. users with high priority data are prioritized over users with low priority data).

- Packet scheduler and HARQ. Synchronous HARQ is used for LTE uplink. Hence, the UE must be scheduled if an earlier transmission has failed.
- Packet scheduler and MIMO. Uplink multi-user (MU-) MIMO can be used in Release 8 of the 3GPP specifications. With MU-MIMO the uplink packet scheduler can simultaneously allocate the same frequency resources to two users. The orthogonality between users is achieved by exploiting frequency-domain channel state information available via SRS measurements. No specific signaling (neither downlink nor uplink) is needed to support MU-MIMO in uplink.

As for the downlink (see Figure 8.6), the uplink packet scheduler is split in two sub-units: one that works in the time domain (TD) and one that works in the frequency domain (FD). The time domain metric is typically computed based on the experienced QoS vs QoS requirements and the estimated buffer occupancy. The channel-aware time domain scheduling metrics only take into account wideband channel state information. Basically, the time domain scheduler for uplink and downlink follows the same principle. The only difference is that the time domain scheduler in uplink must always prioritize users with pending retransmissions independently of other users' priority, QoS and channel conditions due to the uplink synchronous HARQ.

The allocation in the frequency domain is performed based on specific frequency domain scheduling metrics, which are typically computed based on frequency selective channel state information available from SRS measurements. QoS requirements and BSRs can also be considered when deriving frequency domain scheduling metrics. The main difference between uplink and downlink frequency domain PS is that in uplink the FDPS algorithm has to deal with the restrictions imposed by single-carrier transmission and limited UE power capabilities.

8.5.3.1 Example of Frequency Domain Packet Scheduler with Fast Adaptive Transmission Bandwidth

LTE supports fast Adaptive Transmission Bandwidth (ATB) in the uplink, i.e. the user transmission bandwidth can be modified on a TTI basis depending on channel conditions, traffic, fulfillment of QoS requirements, etc. With fast ATB the allocation of the user transmission bandwidth can be integrated in the FDPS algorithm. In this way, for instance, QoS-aware schedulers can be implemented in combination with channel-aware scheduling by simply introducing a QoS-based metric per UE depending on the level of fulfillment of its QoS requirements. In the time domain, users are ordered based on such a QoS-based metric, while in the frequency domain the frequency selective scheduling metric can potentially be weighted by the QoS-based metric. In this way fast ATB can guarantee automatic adaptation to variations in the cell load by, for example, differentiating resource allocation between different QoS users in the time or frequency domain depending on the temporary cell load.

Assuming that frequency-selective uplink channel state information is available at the uplink packet scheduler based on uplink SRS measurements (see section 8.5.1), a scheduling metric per user per PRB can then be defined based on uplink channel state information as well as on fulfillment of QoS requirements, buffers status information, etc.

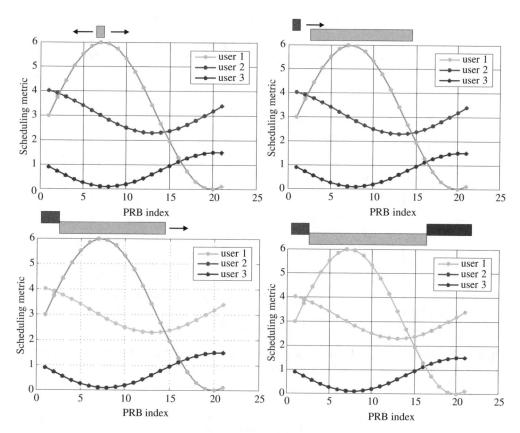

Figure 8.18 Example of combined fast adaptive transmission bandwidth and frequency domain packet scheduling algorithm in LTE uplink

An example of a FDPS algorithm, which is illustrated in Figure 8.18, starts allocating the user with the highest scheduling metric on the corresponding PRB. The user is then 'expanded' respecting the single-carrier transmission constraint of the LTE uplink until another user with a higher metric is met, or the UE transmission power constraints are exceeded (known via power headroom information reported by the UE), or the UE has no more data to transmit (known via buffer status information reported by the UE). Then a similar procedure is repeated until the entire bandwidth is used or there are no more users to schedule.

The fast ATB framework described in this example is able to achieve most of the gain compared to more complex FDPS algorithms, which try to derive near-optimum FDPS solutions using more exhaustive approaches [31] [32]. Due to its iterative nature the complexity of the algorithm remains very high, however, unless the time domain framework effectively limits the scheduling order of the system. Because of this, many simulations that have been conducted during the 3GPP work item phase have been based on the assumption of fixed bandwidth uplink packet scheduling [33].

Figure 8.19 and Figure 8.20 show some examples of gain numbers of uplink FDPS (proportional fair) compared to blind scheduling in the frequency domain.

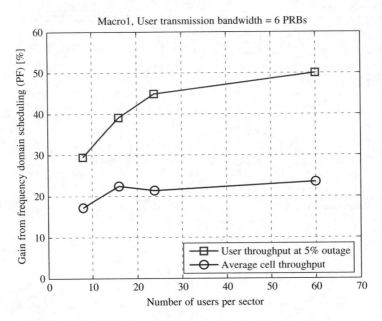

Figure 8.19 Gain from uplink frequency domain scheduling as a function of number of users for Macro 1 scenario and a fixed user transmission bandwidth of 6 PRBs

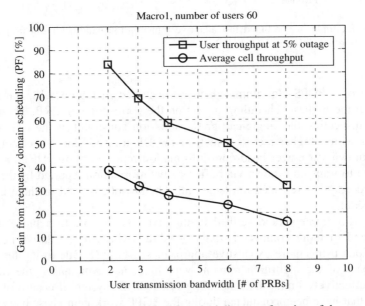

Figure 8.20 Gain from uplink frequency domain scheduling as a function of the user transmission bandwidth for Macro 1 propagation scenario and 60 simultaneous users

In general higher gain numbers can be observed when the capacity is measured with user throughput outage criteria. The reason is that the SINR gain from proportional fair scheduling typically maps into a higher throughput gain in the lower SINR regions. Also, note that for a given user transmission bandwidth, the cell throughput gain from frequency domain scheduling tends to saturate with 15 simultaneous users.

By reducing the user transmission bandwidth the gain from frequency domain scheduling increases (see Figure 8.20). Due to the single-carrier constraint, however, a high number of users needs to be scheduled each TTI to fully exploit the frequency selectivity of the channel. In practice due to PDCCH limitations approximately 8–10 users can be scheduled in uplink each TTI, which means that the user transmission bandwidth is about 5–6 PRBs. Therefore while the potential gain from uplink frequency scheduling is about 40% in average cell throughput and 85% in outage user throughput, single-carrier constraint and practical PDCCH implementation limit the actually achievable gain to 25% and 50%, respectively.

8.6 Interference Management and Power Settings

LTE includes a variety of mechanisms for controlling the interference between neighboring cells, also known as inter-cell interference control (ICIC). The standardized ICIC schemes for Release 8 primarily rely on frequency domain sharing between cells and adjustment of transmit powers. The X2 interface between eNodeBs includes standardized signaling for carrying interference information and scheduling information. The standardized ICIC methods are categorized as follows:

- Reactive schemes: Methods based on measurements of the past. Measurements are used to monitor the performance, and if the interference detected is too high, then appropriate actions are taken to reduce the interference to an acceptable level. Actions could include transmit power adjustments or packet scheduling actions to reduce the interference coupling between cells.
- Proactive schemes: Here an eNodeB informs its neighboring eNodeBs how it plans to schedule its users in the future (i.e. sending announcements), so that the neighboring eNodeB can take this information into account. Proactive schemes are supported via standardized signaling between eNodeBs over the X2 interface, e.g. by having eNodeBs sending frequency domain scheduling announcements to their neighbors.

3GPP Release 8 ICIC schemes are primarily designed for improving the performance of the uplink and downlink shared data channel (PDSCH and PUSCH). Hence, no explicit ICIC techniques are standardized for common channels like the BCCH, and control channels such as PDCCH and PUCCH (except for the use of power control). Note that for Release 8 time-synchronized networks (frame synchronized), the PDCCH and PUCCH always send at the same time and frequency.

8.6.1 Downlink Transmit Power Settings

Proactive downlink ICIC schemes are facilitated via the standardized Relative Narrowband Transmit Power (RNTP) indicator. The RNTP is an indicator (not a measurement) per PRB signaled to neighboring eNodeBs over the X2 interface. It indicates the maximum anticipated downlink transmit power level per PRB. Hence, from this information the neighboring eNodeBs will know at which PRBs a cell plans to use the most power, and the idea is thus that different power patterns can be used in those cells to improve the overall SINR conditions for the UEs. It is implementation specific how often a new RNTP indicator is sent from an eNodeB, and also the exact actions by an eNodeB receiving such messages are left unspecified by 3GPP.

By setting different downlink transmit powers per PRB, it is possible to dynamically configure different re-use patterns ranging from full frequency re-use (re-use one) with equal transmit power on all PRBs, to hard frequency re-use as shown in Figure 8.21, fractional frequency re-use as pictured in Figure 8.22, and soft frequency re-use as illustrated in Figure 8.23. The basic idea for fractional frequency re-use is that users close to their eNodeB are scheduled in the frequency band with frequency re-use one, while the cell-edge users are scheduled in the complementary frequency band with, for example, hard frequency re-use three in Figure 8.21. When using the soft frequency re-use configuration in Figure 8.23, users are scheduled over the entire bandwidth with different power levels. Cell-edge users are primarily scheduled in the bandwidth with the highest power level, while users closer to their serving eNodeB are scheduled on the PRBs with lower transmit power. Note that PRBs allocated to the same user should be transmitted with the same power level. Further information about performance results for different downlink transmit power profiles is presented in [9].

8.6.2   Uplink Interference Coordination

One proactive ICIC mechanism is standardized for the uplink, based on the High Interference Indicator (HII). The HII consists of sending a message over the X2 interface to the neighboring eNodeBs with a single bit per PRB, indicating whether the serving cell intends to schedule cell-edge UEs causing high inter-cell interference on those PRBs.

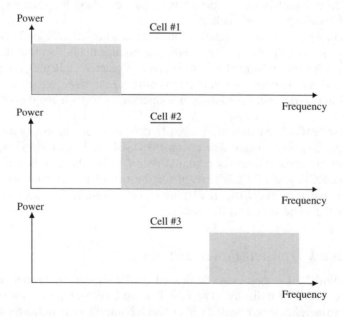

Figure 8.21   Example of downlink transmit power settings for hard frequency re-use three

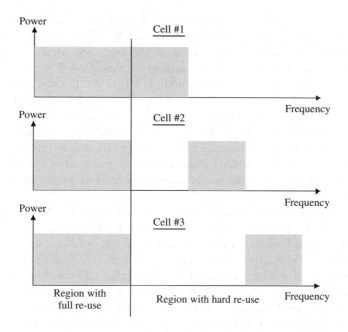

Figure 8.22   Example of downlink transmit power settings for a fractional frequency re-use

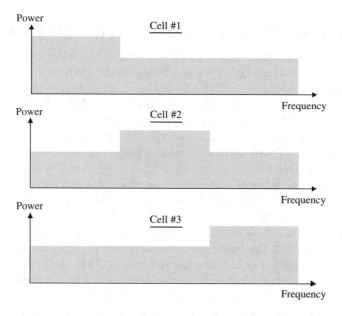

Figure 8.23   Example of downlink transmit power settings for a soft frequency re-use three

Different HII messages can be sent from the serving cell to different eNodeBs. There are no standardized handshake procedures between eNodeBs. The latter means that it is open for eNodeB vendors (i.e. not standardized by 3GPP) to decide when a new HII message is sent from an eNodeB, and the specific actions by the eNodeB receiving the message are also vendor specific. The basic idea behind the HII is that the serving cell will inform its neighboring eNodeBs at which PRBs it intends to schedule high interference users in the future. The neighboring eNodeBs should then subsequently aim at scheduling low interference users at those particular PRBs to avoid scheduling of cell-edge users at the same PRBs between two neighboring cells. Use of the HII mainly provides gain for fractional load cases, where there is only transmission on a sub-set of the PRBs per cell as in principle it can be used to dynamically achieve a hard/soft frequency re-use pattern in accordance with the offered traffic.

LTE also includes a reactive uplink ICIC scheme based on the Overload Indicator (OI). Here the basic idea is that the eNodeB measures the uplink interference + noise power, and creates OI reports based on this measurement. Low, medium, and high OI reports can be signaled over the X2 to neighboring cells. Hence, it is worth noting that the OI is primarily a function of the interference from other cells, and therefore does not include any information on the carried traffic or interference generated by the users in the serving cell. There is support for frequency selective OI, so that the aforementioned measurement and subsequent reporting to neighboring cells is per group of PRBs. One potential use is to dynamically adjust the uplink open loop power control parameters (e.g. Po) to maintain a certain maximum desirable uplink interference + noise level (or IoT level) based on the OI information exchanged between the eNodeBs.

8.7   Discontinuous Transmission and Reception (DTX/DRX)

LTE provides methods for the UE to micro-sleep even in the active state to reduce power consumption while providing high QoS and connectivity. DRX in the LTE sense means that the UE is not monitoring the PDCCH in the given subframe and is allowed to go to power saving mode. As uplink is scheduled in downlink PDCCH, the DRX parameter impacts both uplink and downlink performance for a UE.

The DRX concept contains different user-specific parameters that are configured via higher layer signaling. These parameters are described in Table 8.3 and illustrated in simplified form in Figure 8.24.

Basically, upon knowledge of the activity requirements in uplink and downlink for a certain UE, the regular DRX period including a certain planned on-time can be set. These two parameters are illustrated in Figure 8.24a. For highly predictable traffic (e.g. VoIP), the On Duration can be set to 1 subframe and the DRX Cycle to 20 ms or 40 ms if packet bundling is used. For traffic that is more dynamic and in bursts with tight delay requirements, it is possible to configure the user with a DRX Inactivity Timer where the packet scheduler can keep the UE awake by scheduling it within a certain time window. HARQ retransmissions are planned outside of the predefined DRX cycle to allow for a tighter DRX optimization without having to plan for worst-case retransmissions. There is a DRX retransmission timer defined so that a UE does not have to wait for a full DRX cycle for an expected retransmission that has been lost due to either ACK/NACK misinterpretation or PDCCH failed detection. Signaling of SRS and CQI is tied to the DRX parameters to reduce the use of PUCCH resources when signaling is not specifically

Table 8.3   DRX related parameters and examples of their use/setting [34]

DRX Parameter	Description	Example settings and purpose
DRX Cycle	Specifies the periodic repetition of the On Duration followed by a possible period of inactivity.	The overall DRX cycle is set from knowledge of the service requirements of a user (e.g. responsiveness/latency requirements) as well as requirements for other mobility/update related requirements in the cell.
On Duration timer	This parameter specifies the number of consecutive subframes the UE follows the short DRX cycle after the DRX Inactivity Timer has expired.	Depending on the user's relative scheduling priority, this parameter can be set by the network to achieve a tradeoff among multi-user scheduling performance and single-user power saving.
DRX Inactivity Timer	Specifies the number of consecutive PDCCH-subframe(s) after successfully decoding a PDCCH indicating an initial uplink or downlink user data transmission for this UE.	The parameter provides a means for the network to keep a user 'alive' when there is suddenly a large need for scheduling a UE. Hence, the mechanism allows for large power saving while facilitating high QoS for bursty traffic types. Setting is a tradeoff among scheduler freedom and UE power saving as UE is forced to continue monitoring the PDCCH whenever scheduled.
DRX Retransmission Timer	Specifies the maximum number of consecutive PDCCH-subframe(s) where a downlink retransmission is expected by the UE after the first available retransmission time.	This parameter can be set to allow for asynchronous HARQ in the downlink also for HARQ users. If disabled, the network will need to send its retransmission at the first available time. Hence, the parameter is set as a tradeoff among scheduling freedom for retransmissions and UE power saving.
DRX Short Cycle	Specifies the periodic repetition of the On Duration followed by a possible period of inactivity for the short DRX cycle.	Short DRX is basically like an inactivity timer with gaps that is more suitable for services where the incoming data become available in 'gaps'.
DRX Short Cycle Timer	This parameter specifies the number of consecutive subframe(s) the UE follows the short DRX cycle after the DRX Inactivity Timer has expired.	This parameter is set from knowledge of the service pattern. A larger setting allows for more distributed data transmission but increases UE power consumption.

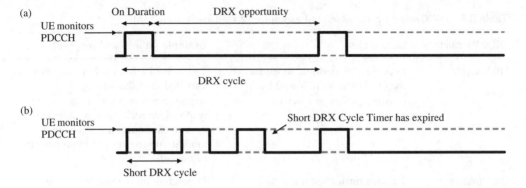

Figure 8.24   Simple illustration of DRX parameters

needed for link adaptation and scheduling purposes and thus increases the power saving
for the UE in long DRX periods. For very long DRX periods and a long time between SRS
transmissions, the network and the UE may lose the time alignment and so the UE has to
restart transmission to the network using RACH. Hence, this is also a consideration for
the RRM management function to properly set available DRX/SRS parameters according
to the planned load on the RACH channel.

Additionally, a short DRX cycle can be triggered that allows for periodic activity
within the regular DRX cycle if additional and time-distributed scheduling resources are
needed to facilitate the best power saving and QoS tradeoff for the given UE. The Short
DRX concept is shown in Figure 8.24b and if the UE is scheduled in the current DRX
window (e.g. regular On Duration window), new scheduling opportunities are created in
distributed form within the current DRX cycle. The availability of the scheduling resources
is controlled by the Short DRX Inactivity Timer and if it expires, the UE returns to the
normal DRX pattern immediately.

The impact of the parameter settings on web browsing experience was studied in [35].
It was shown that with a DRX cycle set to 100 ms and the On Duration set to 1 ms, the
DRX Inactivity timer could be set quite loosely from 25 to 50 ms while still achieving a
large UE power saving, large scheduling freedom for the network, and allow for multiple
web objects to be received within the active window thus improving the web browsing
performance. With such settings the UE saw a 95% reduction in the radio-related power
consumption while achieving 90% of the maximum achievable throughput.

Since DRX benefits UE power saving and thus is an important factor for ensuring
the success of LTE, it is important to integrate its use tightly with the RRM concept in
general. As the use of semi-static DRX takes the user out of the scheduling candidate
set, the gain from multi-user packet scheduling may be significantly reduced. Hence, it is
important that DRX parameters for a certain user are configured considering the multi-
user situation in the cell as well as the service requirements for that particular user; e.g.
activity needed to get the GBR requirement fulfilled or DRX period adjusted according to
latency requirements. By using the span of the different DRX parameters, it is possible to
effectively tune the tradeoff among scheduling flexibility for best cell-level performance
and the UE's performance/power tradeoff.

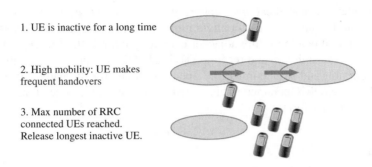

1. UE is inactive for a long time

2. High mobility: UE makes frequent handovers

3. Max number of RRC connected UEs reached. Release longest inactive UE.

Figure 8.25 Triggers for RRC connection release

8.8 RRC Connection Maintenance

The RRC connection maintenance is controlled by the eNodeB based on the RRM algorithms. When the UE has an RRC connection, the UE mobility is controlled by handovers. The handovers create some signaling traffic if the UE is moving. Therefore, if the UE is not transferring any data and is moving, it may be beneficial to release the RRC connection, while still maintaining the EPS bearer and the IP address. It may also be beneficial to release the RRC connection if the connection has been inactive for a long time or if the maximum number of RRC connections per base station is achieved. Example triggers for RRC connection release are illustrated in Figure 8.25. When the UE next transmits or receives data, the RRC connection is again re-established via RACH procedure.

8.9 Summary

The radio resource management algorithms are important to optimize the system capacity and end user performance. The network algorithms are not standardized but the network vendors and operators can design and tune the algorithms according to the needs. The main algorithms in LTE are packet scheduling, admission control, power control and interference control. The LTE radio gives a lot of freedom in the packet scheduling area since the scheduling can be done in both the time and the frequency domain. It is shown that frequency domain scheduling can provide a clear capacity benefit at low mobile speeds compared to random scheduling. The uplink scheduling has less freedom compared to the downlink because the signaling from the network to the terminal takes time and because SC-FDMA transmission in uplink must use adjacent resource blocks.

The signaling in 3GPP has been defined to support efficient scheduling including the downlink Channel Quality Information (CQI), uplink Sound Reference Signals (SRS), uplink Buffer Status Reports (BSR) and uplink Power Headroom Reports (PHR).

The QoS differentiation allows separate treatment for different services, applications, subscribers or depending on the used amounts of data. The QoS is relevant in LTE since all the services are packet based including voice. The QoS concept in LTE is fully network controlled and is simplified compared to the existing 2G/3G networks. The QoS priority is typically achieved in time domain scheduling.

The LTE interface specifications support the inter-cell interference coordination to achieve dynamic frequency re-use configurations. The co-ordination can be used to optimize the resource allocation in the adjacent cells to maximize the cell edge throughput.

Discontinuous transmission and reception (DTX/DRX) is important to minimize the UE power consumption and to maximize the operating times for the handheld devices. The tuning of the DTX/DRX parameters also needs to consider the scheduling performance and the end to end performance.

References

[1] D. Niyato, E. Hossain, 'Connection Admission Control Algorithms for OFDM Wireless Networks', IEEE Proc. Globecomm, pp. 2455–2459, September 2005.

[2] P. Hosein, 'A Class-Based Admission Control Algorithm for Shared Wireless Channels Supporting QoS Services', in Proceedings of the Fifth IFIP TC6 International Conference on Mobile and Wireless Communications Networks, Singapore, October 2003.

[3] K.I. Pedersen, 'Quality Based HSDPA Access Algorithms', in IEEE Proc. Vehicular Technology Conference Fall, September 2005.

[4] 3GPP 23.401, 'Technical Specification Group Services And System Aspects; GPRS enhancements for E-UTRAN access (Release 8)', August 2007.

[5] 3GPP 23.203, 'Technical Specification Group Services and System Aspects; Policy and charging control architecture', v. 8.3.1.

[6] 3GPP 36.300 'Technical Specification Group Radio Access Network; Evolved Universal Terrestrial Radio Access (E-UTRA) and Evolved Universal Terrestrial Radio Access Network (E-UTRAN); Overall description; Stage 2 (Release 8)', October 2007.

[7] A. Pokhariyal, et al., 'HARQ Aware Frequency Domain Packet Scheduler with Different Degrees of Fairness for the UTRAN Long Term Evolution', IEEE Proc. Vehicular Technology Conference, pp. 2761–2765, April 2007.

[8] K.I. Pedersen, F. Frederiksen, T.E. Kolding, T.F. Lootsma, P.E. Mogensen, 'Performance of High Speed Downlink Packet Access in Co-existence with Dedicated Channels', IEEE Trans. Vehicular Technology, Vol. 56, No. 3, pp. 1261–1271, May 2007.

[9] A. Simonsson, 'Frequency Reuse and Intercell Interference Co-ordination in E-UTRA', IEEE Proc. Vehicular Technology Conference, pp. 3091–3095, April 2007.

[10] A. Pokhariyal, et al., 'Frequency Domain Packet Scheduling Under Fractional Load for the UTRAN LTE Downlink', IEEE Proc. Vehicular Technology Conference, pp. 699–703, April 2007.

[11] H. Holma, A. Toskala, 'WCDMA for UMTS – HSPA Evolution and LTE', 4th edition, Wiley, 2007.

[12] K.I. Pedersen, P.E. Mogensen, Troels E. Kolding, 'Overview of QoS Options for HSDPA', IEEE Communications Magazine, Vol. 44, No. 7, pp. 100–105, July 2006.

[13] P.A. Hosein, 'QoS Control for WCDMA High Speed Packet Data', IEEE Proc. Vehicular Technology Conference 2002.

[14] D. Laselva, et al., 'Optimization of QoS-aware Packet Schedulers in Multi-Service Scenarios over HSDPA', 4th International Symposium on Wireless Communications Systems, pp. 123 – 127, ISWCS 2007, 17–19 October 2007.

[15] T.E. Kolding, 'QoS-Aware Proportional Fair Packet Scheduling with Required Activity Detection', IEEE Proc. Vehicular Technology Conference, September 2006.

[16] G. Barriac, J. Holtzman, Introducing Delay Sensitivity into the Proportional Fair algorithm for CDMA Downlink Scheduling, IEEE Proc. ISSSTA, pp. 652–656, September 2002.

[17] M. Andrews, K. Kumaran, K. Ramanan, A. Stolyar, P. Whiting, 'Providing Quality of Service over a Shared Wireless Link,' IEEE Communications Magazine, Vol. 39, No. 2, pp. 150–154, February 2001.

[18] F. Kelly, 'Charging and rate control for elastic traffic', European Trans. On Telecommunications, No. 8, pp. 33–37, 1997.

[19] J.M. Holtzman, Asymptotic Analysis of Proportional Fair Algorithm, IEEE Proc. PIMRC, Personal Indoor and Mobile Radio Communication Conference, pp. F33–F37, September 2000.

[20] G. Song, Y. Li, 'Utility-Based Resource Allocation and Scheduling in OFDM-Based Wireless Broadband Networks', IEEE Communications Magazine, pp. 127–134, December 2005.

[21] G. Monghal, *et al.*, 'QoS Oriented Time and Frequency Domain Packet Schedulers for The UTRAN Long Term Evolution', in IEEE Proc. VTC-2008 Spring, May 2008.

[22] C. Wengerter, J. Ohlhorst, A.G.E. v. ElbWart, 'Fairness and throughput analysis for Generalized Proportional Fair Frequency Scheduling in OFDMA', IEEE Proc. of Vehicular Technology Conference, pp. 1903–1907, Sweden, May 2005.

[23] H. Yang, 'A Road to Future Broadband Wireless Access: MIMO-OFDM-Based Air Interface', IEEE Communication Magazine, pp. 553–560, January 2005.

[24] I. Viering, A. Klein, M. Ivrlac, M. Castaneda, J.A. Nossek, 'On Uplink Intercell Interference in a Cellular System', IEEE International Conference on Communications (ICC), Vol. 5, pp. 2095–2100, Istanbul, Turkey, June 2006.

[25] C. Rosa, D. López Villa, C. Úbeda Castellanos, F.D. Calabrese, P. Michaelsen, K.I. Pedersen, Peter Skov, 'Performance of Fast AMC in E-UTRAN Uplink', IEEE International Conference on Communications (ICC 2008), May 2008.

[26] 3GPP TS 36.213, 'Technical Specification Group Radio Access Network; Physical layer procedures (Release 8)', November 2007.

[27] C. Úbeda Castellanos, D. López Villa, C. Rosa, K.I. Pedersen, F.D. Calabrese, P. Michaelsen, J. Michel, 'Performance of Uplink Fractional Power Control in UTRAN LTE', IEEE Trans. on Vehicular Technology, May 2008.

[28] R1-080900, 'Physical-layer parameters to be configured by RRC', TSG RAN WG1 #52, Sorrento, Italy, 11–15 February 2008.

[29] R1-060048, 'Channel-Dependent Packet Scheduling for Single-Carrier FDMA in E-UTRA Uplink', 3GPP TSG RAN WG1 LTE Ad Hoc Meeting, Helsinki, Finland, 23–25 January 2006.

[30] 3GPP TS 36.321, 'Technical Specification Group Radio Access Network; Medium Access Control (MAC) protocol specification (Release 8)', March 2008.

[31] J. Lim, H.G. Myung, D.J. Goodman, 'Proportional Fair Scheduling of Uplink Single Carrier FDMA Systems', IEEE Personal Indoor and Mobile Radio Communication Conference (PIMRC), Helsinki, Finland, September 2006.

[32] F.D. Calabrese, P.H. Michaelsen, C. Rosa, M. Anas, C. Úbeda Castellanos, D. López Villa, K.I. Pedersen, P.E. Mogensen, 'Search-Tree Based Uplink Channel Aware Packet Scheduling for UTRAN LTE', in IEEE Proc. Vehicular Technology Conference, 11–14 May 2008.

[33] R1-060188, 'Frequency-domain user multiplexing for the E-UTRAN downlink', 3GPP TSG RAN WG1 LTE Ad Hoc Meeting, Helsinki, Finland, 23–25 January 2006.

[34] 3GPP 36.321 Technical Specifications 'Medium Access Control (MAC) protocol specification', v. 8.3.0.

[35] T. Kolding, J. Wigard, L. Dalsgaard, 'Balancing Power Saving and Single-User Experience with Discontinuous Reception in LTE', Proc. of IEEE International Symposium on Wireless Communication Systems (ISWCS), Reykjavik, Iceland, October 2008.

9

Self Organizing Networks (SON)

Krzysztof Kordybach, Seppo Hamalainen, Cinzia Sartori and Ingo Viering

9.1 Introduction

The introduction of LTE means additional network integration and operation challenges. LTE will run in parallel with existing 2G and 3G networks (Multi-Radio Access Technology RAT). Network complexity will also increase because of the large number of smaller cells (Heterogeneous Networks), base stations and related parameters that need to be configured and optimized in such a scenario. It is in the interest of the operators to minimize operational effort and cost; especially in the early deployment phase. Attempts to set up and optimize the network are significant and traditionally take lengthy periods of time. In order to minimize the network operating expenses, the provision of *Self Organizing Networks* (SON) is one of the key objectives of LTE. This chapter introduces SON architecture, standardization, its main uses, and example algorithms.

In the early GSM deployments, maintenance was carried out on site. Any updates had to be done physically next to the base station. Today, the cellular networks are managed by centralized remote applications in the network management systems, with nearly all processes, like planning, configuration, and optimization, supported by computer-aided tools. The SON paradigm will allow automation, often transferring these functions to the network, in such a way that the network will automatically and dynamically adapt to varying situations.

The 3GPP standardization defines the necessary measurements, procedures and open interfaces to support better operability in a multi-vendor environment. These elements together form the standardized SON framework. The Next Generation Mobile Network (NGMN) alliance and 3GPP have pushed the cost reduction requirements by defining a number of SON Use Cases, which may be placed either in eNodeB (Distributed) or in the Operations, Administration and Maintenance (OAM) system (Centralized). Different kinds of Use Cases coexist, so a Hybrid SON architecture is needed. In summary, SON aims at minimizing the manual configuration and optimization of the network. It achieves

LTE for UMTS: Evolution to LTE-Advanced, Second Edition. Edited by Harri Holma and Antti Toskala.
© 2011 John Wiley & Sons, Ltd. Published 2011 by John Wiley & Sons, Ltd.

this goal by automating, to a great extent, a large number of everyday tasks making them easily repeatable and quickly implementable.

The main SON functionalities are separated into Self-Configuration, Self-Optimization and Self-Healing categories.

Self-Configuration is the process that occurs in pre-operational phase before the base station is switched on and the signal is sent on the air. During self-configuration a new base station is automatically integrated in the network by secure auto-connection, secure software and data base download. With respect to current processes, the aim of self-configuration is to simplify and accelerate the network integration of the new node reducing the manual interaction, especially onsite commissioning.

The purpose of Self-Optimization is to maximize network performance and provide cost savings, greater than in case of manual optimization. Self-optimization is made by auto-tuning the network automatically with the help of UE and eNodeB measurements and performance measurements.

Self-healing runs in parallel to Self-Optimization and aims to automatically detect and localize faults such as site and cell malfunctions and to heal possible coverage and capacity losses. In order to recover service in the malfunctioning cell, self-healing appropriately adjusts the parameters and algorithms in surrounding cells. Self-healing functionality, as such, can be seen as a combination of automatic troubleshooting and consequent network self-optimization.

9.2   SON Architecture

SON architectures can be divided into three classes, according to the location of optimization algorithms: Centralized SON, Distributed SON and Localized SON. Figure 9.1 depicts the applicability of centralized, distributed and localized SON, depending on the number of involved cells and network reactions times.

Centralized SON is associated with the mechanisms of slow update rates and it is based on long-term statistics that are collected from many cells involved in the optimization process. Centralized SON functions are mainly implemented into network management or element management systems, depending on requirements regarding multi-vendor capabilities.

The Coverage and Capacity Optimization (CCO) is a case that involves a large number of cells; in fact, the effects of changing the antenna tilt or the transmit power of a base station may cause interference to remote eNodeBs that are not the neighbors of cells. In the case of centralized SON, optimization algorithms are executed in the OAM system, which controls the configuration of many eNodeBs.

Distributed SON is applicable to processes that require a fast reaction time and that affect only a few cells, or processes where changed parameters have only local impact but information about configuration or status of neighbor cells is required. Information between involved cells can therefore be exchanged using X2 interfaces. In case of distributed SON, optimization algorithms are executed in eNodeB.

Localized SON applies to processes in which other cells are not involved. The cases for localized SON require fast response time and have single cell scope – they do not affect their neighbors. In this case optimization may be implemented by Radio Resource Management (RRM) functions, which are executed in eNodeB.

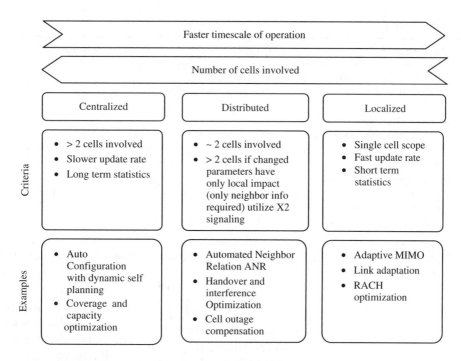

Figure 9.1   Applicability of centralized, distributed and localized SON

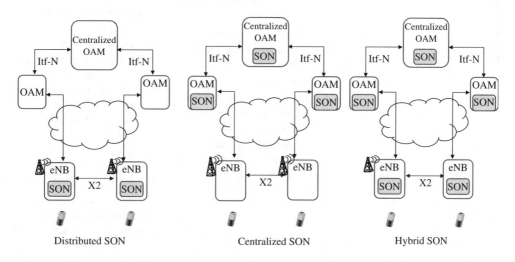

Figure 9.2   SON architecture options

In practice, a *Hybrid SON* is required, in which all three kinds of approaches are in use simultaneously, coping with different needs for different use cases. In Hybrid SON, part of the optimization algorithms are executed in the OAM system, while others are executed in eNodeBs.

The SON architecture options are shown in Figure 9.2.

Table 9.1   SON use cases

	LTE Release 8	LTE Release 9	LTE Release 10
Self-Configuration	Automated configuration of Physical Cell Identity (PCI) Automatic Neighbor Relation (ANR) function: setting up a neighbor relation based on UE reports Self-configuration of eNodeBs Automatic software management	Automatic radio configuration function	
Self-Optimization	Mobility load balancing: load reporting between eNodeBs over X2	Mobility load balancing: composite load information and mobility setting negotiation; also, basic inter-RAT load information exchange Mobility robustness optimization: basic radio link failure detection and reporting RACH optimization: reporting RACH statistics from UE Energy savings for LTE: cells status information and wake-up request Management of load balancing optimization management of handover optimization	Minimization of drive test Enhancements to mobility load balancing Enhancements to mobility robustness optimization The main focus of Release 10 enhancements is Multi-RAT environment Enhancements to energy savings Enhanced inter-cell interference coordination for non-carrier aggregation based deployments of heterogeneous networks for LTE Coordination of SON functions Energy savings management
Self-Healing			Capacity and coverage optimization: detection of capacity problems Minimization of drive tests: logging and reporting of various measurement data by the UE and collection of data in a server to minimize the number of drive tests run by operators Cell outage compensation/ mitigation

9.3 SON Functions

Multiple SON use cases have been specified by 3GPP for three areas: self-configuration (covering pre-operational phases of network deployment, such as planning and initial configuration), self-optimization (optimization during the operational phase) and self-healing (maintenance, recovery from faults). Different SON use cases are relevant at different times of network operation, for example during roll-out, early phases of operation, or in high loaded and mature networks. In general the use cases related to self-configuration and coverage are most important in earlier phases, whereas quality and capacity-based use cases become more relevant when network usage increases.

While self-configuration features are mainly specified in 3GPP Release 8 and include Automated Configuration of Physical Cell ID and Automatic Neighbor Relation Function, self-optimization is mainly addressed in 3GPP Release 9, where the focus is on Mobility Load Balancing (MLB), Mobility Robustness and RACH optimization. Self-healing features, such as Capacity and Coverage Optimization (CCO) are part of Release 10. Release 10 work started in 2010 and provides CCO, MRO and MLB enhancements, Minimization of Drive Test (MDT) and Energy Saving.

Release 10 LTE includes major architectural features such as Heterogeneous Networks, Relays and Carrier Aggregation. SON will also have to face this network evolution, in particular, if the uncoordinated deployment of HeNodeBs (home eNodeBs) is considered. HeNodeBs are known also as femto cells. In such a network environment, self-configuration with minimum involvement of centralized entity and mobility of users between macro and HeNodeB is important.

SON features across releases are represented in Table 9.1.

9.4 Self-Configuration

The Self-Configuration process allows the automated network integration of new eNodeB by auto-connection and auto-configuration, core connectivity (S1) and automated neighbor site configuration (X2).

Although most of the Self-Configuration process is vendor specific, including mechanisms to connect to OAM and to secure the connection, 3GPP support is necessary to enable inter-vendor operability. The Self-Configuration process works in the pre-operational state and covers the on-site installation and the initial set-up of an eNodeB. It consists of the following steps:

1   The eNodeB is delivered with initial basic software to allow the basic bootstrapping process and software for the basic connectivity.
2   An IP address is allocated to the eNodeB by the DHCP (Dynamic Host Configuration Protocol) server and a connection to the IP backhaul is established.
3   The eNodeB and Authentication Center mutually authenticate and establish a secure IP connection.
4   Software and configuration data are downloaded from the OAM subsystem.
5   The eNodeB establishes S1 connection to the Core Network.
6   The eNodeB establishes X2 connections to neighbor eNodeBs.

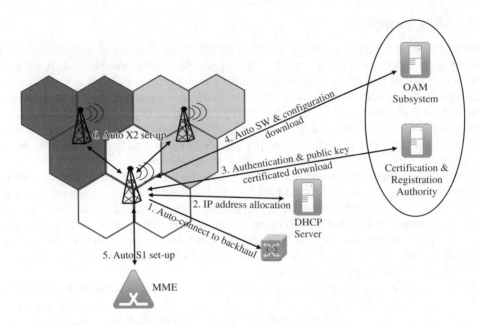

Figure 9.3   Self-configuration procedure

The self-configuration procedure is illustrated in Figure 9.3.

In addition to the above steps, the eNodeB needs the Physical Cell ID and the Neighbor Cell list. The Automated Configuration of Physical Cell ID and Automatic Neighbor Relation will allow eNodeB to operate fully. These two functionalities are the two main use cases addressed in 3GPP Release 8.

The functional description of these use cases is presented in TS 36.300 [1].

9.4.1   Configuration of Physical Cell ID

Dynamic configuration includes the Layer 1 identifier, called Physical Cell Identity (PHY-CID), management and its mapping to Cell Global ID (CGID) and IP address. Dynamic configuration is needed to a newly deployed eNodeB to make it completely operational. There are 504 Physical Cell IDs in LTE. Physical Cell ID corresponds to the scrambling code in WCDMA.

When a new eNodeB is brought into the field, a Physical Cell ID needs to be selected for each of its supported cells, avoiding collisions with respective neighboring cells. If an identical Physical Cell ID is used by two cells that can be received simultaneously by a UE, the UE is not able to distinguish the two cells. Traditionally, the proper Physical Cell ID is derived from radio network planning and is part of the initial configuration of the node. The Physical Cell ID assignment fulfills the following conditions:

- It is collision free. The Physical Cell ID is unique in the area that the cell covers.
- It is confusion free. The cell does not have neighboring cells with identical Physical Cell ID.

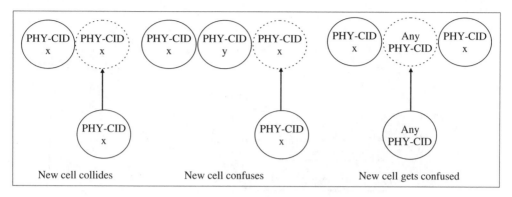

Figure 9.4 Possibilities of conflicts associated with the insertion of a new Physical Cell ID (PHY-CID)

The possible conflicts are shown in Figure 9.4.

The Physical Cell ID assignment may be either centralized or distributed. The natural approach for the Physical Cell ID assignment is the centralized one because the OAM subsystem has a complete knowledge and control of the assignment of Physical Cell IDs. The OAM can select Physical Cell ID as a part of the process.

Another approach is the distributed Physical Cell ID assignment. The OAM subsystem signals a list of Physical Cell IDs values. The eNodeB further limits the scope by removing from the list the Physical Cell ID values detected from UE measurement reports, or using a downlink receiver to identify its surrounding cells by hearing their Physical Cell IDs over the air, or via the inter eNodeB interface. Then, it should randomly select its ID from the remaining values.

9.4.2 Automatic Neighbor Relations (ANR)

Automatic Neighbor Relations aims to avoid the manual work required for configuring neighbor relations in a newly deployed eNodeB or to optimize its configuration during operation. Network performance will also benefit from such optimized and up-to-date lists. Proper neighbors setting will increase the number of successful handovers and minimize the number of call drops due to missing neighbor relations.

The LTE UE can detect Physical Cell IDs without any advertised neighbor lists. The Physical Cell ID that UE has reported must be mapped to the Cell Global Cell ID before the handover can be executed. That mapping can be done in the network by using the intelligence and the location information in the OAM system. The mapping can also be done based on the UE decoding the target cell Cell Global ID from the broadcast channel. The capability for decoding the Cell Global ID is an optional UE feature. When one UE has done the Cell Global ID decoding, the information can be stored in eNodeB and there is no need for Cell Global ID decoding for other UEs anymore. The neighbor list updating based on UE measurements is illustrated in Figure 9.5.

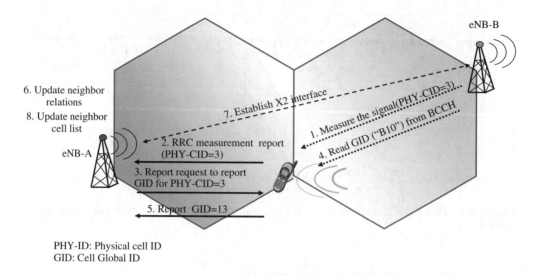

PHY-ID: Physical cell ID
GID: Cell Global ID

Figure 9.5   ANR in LTE

1   The UE is instructed to report neighbors to the serving eNodeB while in RRC
      connected mode.
2   The UE measures the Physical Cell ID of a cell that it has detected but not in a
      predefined neighbor list and reports it to the eNodeB.
3   If the eNodeB receives a report by the UE that indicates an unknown Physical
      Cell ID, it orders the UE to read Global Cell ID from the broadcast channel of the
      detected cell.
4   The eNodeB requests the IP address corresponding to this Cell Global ID from
      Mobility Management Entity (MME), updates its neighbor list, establishes a X2
      connection with the peer eNodeB and exchanges configuration data over X2.

   The above is according to distributed ANR procedure. However, network operator
can configure black list/white list cells for ANR in the network management system. A
blacklisted cell will never be subject to ANR, even if reported as a strong neighbor by
a UE. The operator may want to block certain handover candidates, for example, from
indoor cells to outdoor cells in high-rise buildings.
   In addition to LTE intra-frequency cases, 3GPP specifies ANR for LTE Inter-Frequency
and Inter-RAT. In this case the eNodeB can instruct a UE to perform measurements and
detect cells on other frequencies as well as on other Radio Access Technologies (RAT).
The inter-frequency and Inter-RAT measurements need scheduling of the measurements
gaps for UE.
   The ANR may also remove redundant neighbor relations. This functionality is fully
vendor specific – for example, a time stamp may be added to every added neighbor
relation. Then, if the cell is not reported for a given period of time, the corresponding
relation is deleted.

9.5   Self-Optimization and Self-Healing Use Cases

SON use cases are illustrated in 3GPP documents TR 36.902 [2] and TS 36.300 [1] and
summarized below.

9.5.1 Mobility Load Balancing (MLB)

The idea of the MLB use case is to enable cells that suffer congestion to transfer load to other cells, which have spare resources. Such transfer must usually be forced against radio conditions – some terminals have to be connected to non-optimal cells. New mechanisms therefore had to be standardized.

The solution for MLB consists of three elements:

- Load reporting. The reporting is different in the case of intra-LTE and in case of inter-RAT.
- Load balancing action. The load balancing in Release 8 and Release 9 concerns the active user and is therefore based on handovers.
- Amending handovers so that the load remains balanced.

Load reporting, or load information exchange, enables cell controllers (eNodeBs in case of LTE) to exchange information about load level in cells and about available capacity. In case of load information exchange between LTE cells, the reporting is initialized by one eNodeB toward a peer eNodeB with the Resource Status Reporting Initiation X2 procedure. The request passed in RESOURCE STATUS REQUEST X2AP message contains the information to be reported and reporting periodicity. If the peer eNodeB accepts the request, it starts reporting using Resource Status Reporting X2 procedure – the RESOURCE STATUS UPDATE X2AP message is sent with requested periodicity and contains the requested information.

The periodicity of the reporting can be requested in the range of 1 through 10 s. The information to be reported can be:

1   Hardware load: four levels reported separately for uplink and downlink.
2   S1 transport network load: 4 levels reported separately for uplink and downlink.
3   Radio resource status: percentage of allocated Physical Resource Blocks (PRB), separately for uplink and downlink and split into:
    - total allocation;
    - guaranteed bit-rate traffic;
    - non-guaranteed bit-rate traffic;
    - composite available capacity information: percentage of PRBs available for load balancing, separately in the uplink and downlink.

In the above list the composite available capacity, in particular, deserves some attention. The idea is that this information combines all the limitations that a cell experiences. Furthermore, the information is expressed in the form of generally understandable radio resources – PRBs. Even though it is coded as percentage, the receiver knows the bandwidth of the reported cell from X2 set-up phase. It is therefore very straightforward to recalculate the percentage to the actual number of PRBs offered for load balancing purposes.

For example, if a cell is not much loaded on the radio interface but for some reason suffers high load on the hardware, it should report only as much PRBs to be available as its congested hardware is able to handle. Other sources of such limitations may also be backbone capacity (for example, the capacity of the connection between a relay station and a donor eNodeB) or operator's policy (OAM configuration). It is important to note that the standards do not mandate reporting of the actually available PRBs – this can be calculated anyway from the radio resource status information. As the composite available capacity is not linked with the actual status of the cell, the admission control in a cell may further modify the reported value: offer more resources than there are actually available at the moment, if it knows some ongoing services may be downgraded, or less, if it wants to keep some resources booked in order to protect existing traffic. The implementation of this is vendor specific though.

In case of inter-RAT load reporting, the information cannot be reported periodically and may concern only a single cell. Since different RATs use different load information, the load is reported in a special container that can bear either UTRAN/GERAN load information or E-UTRAN composite available capacity. The container is transported using RIM (RAN Information Management) protocol [3]. The composite available capacity reported from LTE follows the same principles as in case of intra-LTE load reporting. The load information sent from UMTS or GSM contains the following data, separately for uplink and downlink:

- load experienced by a cell as a percentage;
- load in the cell corresponding to real-time traffic in percentage;
- load in the cell corresponding to non-real time traffic with four levels.

Additionally, in order to enable comparisons of percentages in different RATs, cell capacity class values are added to both, LTE composite available capacity and UMTS/GSM load information. The cell capacity class value is set arbitrarily by the operator per cell but usually is the same for the same bandwidth of the same RAT. The cell capacity class can be used to compare total capacities of RATs. For example, if UTRAN has the class value of 20 and the LTE is set to be 50, then when a UMTS cell reports it is loaded in 40% and therefore may be assumed to have 60% of resources free, the LTE cell that receives it can calculate the corresponding load that can be transferred to that UMTS cell:

$$Load_{LTE} = Load_{UMTS} \frac{Class_{UMTS}}{Class_{LTE}} = 60\% \frac{20}{50} = 24\%$$

This means, the 60% of available capacity in the UMTS cell corresponds to 24% of the capacity in the LTE cell. Similarly, in the other direction, 60% of available resources reported as the composite available capacity would mean 150% of the capacity of a single UMTS cell.

The load reporting is needed to assess how much traffic can be transferred to the reporting cell. In the intra-LTE scenario, the eNodeB that controls the congested cell must select users that can technically be handed over (those that report a good enough radio connection to the potential target) and that can fit to the available space reported from that target. That means the eNodeB must estimate the resource allocation in the target, which does not have to be identical to that in the congested cell – radio conditions will be different, and scheduler implementation and configuration may be different, so the number of PRBs needed for the same service may change. This lies, however, in the realm of vendor-specific implementation of the load-balancing algorithms.

Once the eNodeB finds a user that can be handed over to a cell that most likely has enough resources, the actual load-balancing action is executed. In practice this is a regular handover, but triggered not because of radio conditions but because of the load situation. To differentiate it from a regular handover, the eNodeB that initializes it may set an appropriate cause value. An example value for a load balancing handover may be 'Reduce Load in Serving Cell' or 'Load Balancing'.

Once the user is handed over, it may be necessary to amend the handover parameters so that the user does not come back to the congested cell due to radio reasons. The amendment must be performed in both cells, so that handover settings between the two cells remain coherent. For this purpose the Mobility Settings Change X2 procedure is defined. The eNodeB that detects the change needs to estimate how much the cell border would need to be shifted to avoid quick return of the user. This shift is expressed in dBs and informed to the other cell in MOBILITY CHANGE REQUEST X2AP message. The eNodeB may also include the information about the change of its own handover threshold but, in case of load balancing, it is assumed that the change is usually symmetrical: if the congested cell shifts its handover border by, for example, $-2\,dB$, the other cell should extend its own by $+2\,dB$. The exact interpretation of the informed change depends on the implemented handover algorithm and therefore can not be precisely specified. Similarly, it is up to the implementation of the eNodeB, how the change is reflected in the idle mode mobility settings. It is also assumed that the eNodeB that requested the change of the mobility setting should not execute it until it is acknowledged from the target eNodeB. The mobility load balancing is illustrated in Figure 9.6.

It is important to mention that the cell that receives the MOBILITY CHANGE REQUEST message may reject it, if it finds the proposed change impossible to execute. In that case, it may provide information about the allowed change range.

An example message exchange between two LTE cells involved in load balancing is presented in Figure 9.7.

Details of the X2 procedures used for Mobility Load Balancing solution can be found in TS 36.423 [4]. The LTE part that concerns inter-RAT mechanisms is defined in the S1 specification, i.e. the TS 36.413.

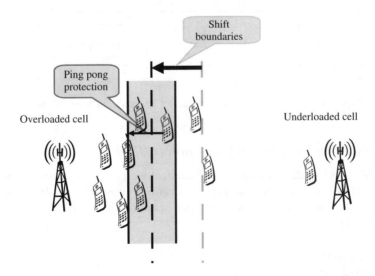

Figure 9.6 Mobility Load Balancing principles

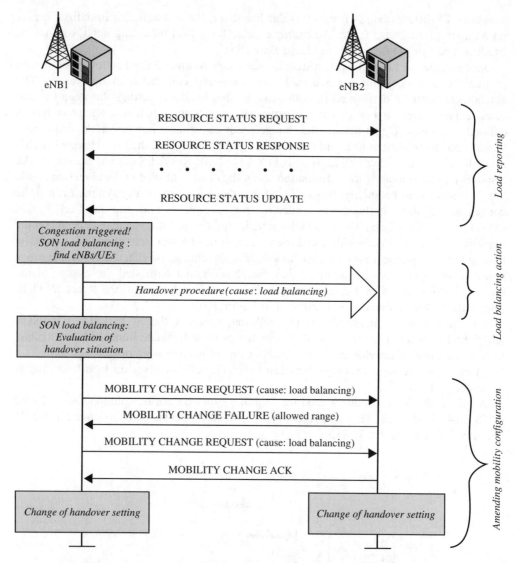

Figure 9.7 Example of a X2 procedure for mobility load balancing

9.5.2 Mobility Robustness Optimization (MRO)

The idea of MRO is to detect and correct automatically errors in the mobility configuration. Such errors can be very different and can have various consequences. In the Release 9 of LTE the use case was limited to the errors that lead to a Radio Link Failure (RLF) and that are caused by:

- too late handover;
- too early handover;
- handover to wrong cell.

When users' mobility is considered, the main problem is the one of wrong parameter settings at eNodeBs. The detection is based on the RRC message reception from the UE requesting re-establishment of the connection. The serving cell detects a broken connection but has no means of finding out that the problem is due to user's mobility.

The detection of a late handover is simple: the connection re-establishment request contains the Physical Cell ID of the last serving cell and the Common Radio Network Temporary Identifier (C-RNTI) of the UE in that cell. If the Physical Cell ID is different from that of the cell that received the request, it may be assumed that there should have been a handover and that it was not initialized, or was initialized too late, and eventually the connection was broken. The eNodeB that controls the cell where the re-establishment is attempted can therefore inform the eNodeB that controls the source cell about the Radio Link Failure. For this purpose the RLF INDICATION X2AP message is used. This is presented in Figure 9.8.

It is slightly more difficult to detect a handover that is too early. This scenario assumes that there is a successful handover to the target, but it is triggered too soon, when radio conditions to the target are still poor. Then, even though the handover is completed successfully, the connection is dropped almost immediately and the UE tries to re-establish the connection at the source cell. The source cell cannot recognize the UE though – the C-RNTI and Physical Cell ID are of the target. It therefore considers the Radio Link Failure as too late handover from the target and sends the Radio Link Failure indication. However, from the UE identifiers included in the indication, the target can recognize that the indication concerns a UE that, just before the Radio Link Failure occurred, had been handed over from the cell that now sends the indication. The target should then send back a report of that handover to let the source know that that the first handover was too early. For this purpose HO REPORT X2AP message is used. It is also necessary to define a timer running internally in an eNodeB that defines what the 'immediately after successful handover' means. A complete too early handover scenario is presented in Figure 9.9.

The scenario of a handover to wrong cell is similar to the handover that is too early: there is a successful handover from source to target and a Radio Link Failure happens at the target immediately after the handover is completed. The difference is that the UE tries to re-establish the connection at a different cell than the source or target – a third cell. It is assumed the handover executed to the target should instead have been executed to that third cell. The detection is therefore the same, as in case of the handover that was

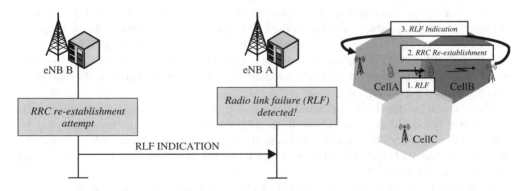

Figure 9.8   Detection of a late handover

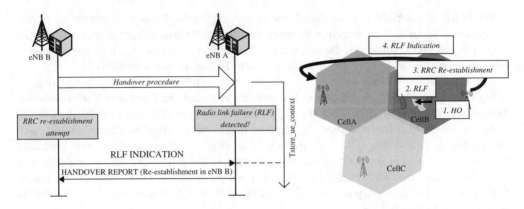

Figure 9.9   Detection of a handover that is too early

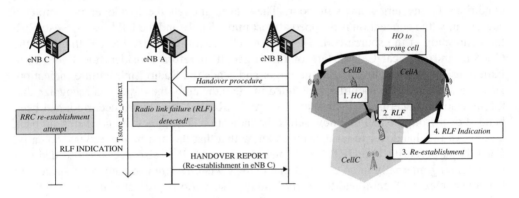

Figure 9.10   Detection of handover to wrong cell

too early. However, when the target cell receives RLF INDICATION message from the third cell, it forwards the HO REPORT to the source cell, not back to the third one. The complete detection procedure of handover to wrong cell is presented in Figure 9.10.

One might wonder how the eNodeB that receives the HO REPORT message can figure out whether it concerns a handover that was too early or a handover to a wrong cell. This is based on the three values included in the HO REPORT message: ID of the cell that the UE arrived from (the source cell), ID of the cell where the Radio Link Failure occurred (the target cell) and ID of the cell that the re-establishment was attempted at. If the third and first IDs are the same, the initial handover was triggered too early; if they differ, it was directed to the wrong cell.

These scenarios may have additional flavors and, depending on the assignment of cells (i.e. if they belong to the same eNodeB, or different eNodeBs), some messages may be substituted with eNodeB internal communication. All of them are described in the TS 36.300 [1]. The RLF INDICATION and HO REPORT messages are defined in the X2 specification, TS 36.423 [4].

In order to distinguish between a handover that is too late and a coverage hole, the UE may be requested to provide last set of measurements, as stored before the Radio Link

Failure. This can be done if the re-establishment was successful. This measurement report may then be forwarded in the RLF INDICATION message.

If the SON algorithm is triggered due to a mobility setting error and is able to correct it, it may be necessary to inform the neighbors about the change. In some cases the correction may actually require a change also on the side of the neighbor. For this purpose the procedure for amending mobility settings described in the MLB section can be used. In order to inform the target cell about the reason for change, the cause value should be set appropriately, for example, to 'Handover Optimization'.

9.5.3 RACH Optimization

The use case addresses the problem of suboptimal RACH configuration that leads to sending many RACH preambles and therefore higher interference caused by RACH access. In order to enable eNodeB to detect the problem and possibly to correct it automatically, a UE may be polled for RACH statistics after it successfully connects. The reported statistics are:

- number of RACH preambles sent until the successful RACH completion;
- contention resolution failure indication.

A second element of the solution is the possibility to exchange PRACH configuration among eNodeBs. This can be done when X2 is set up or when eNodeB configuration is updated. The exchanged PRACH information contains:

- root sequence index;
- zero correlation zone configuration;
- high speed flag;
- PRACH frequency offset;
- PRACH configuration index (for TDD only).

9.5.4 Energy Saving

The energy saving use case originates from the idea of switching off capacity-booster cells when the extra capacity is not needed. Part of the energy consumed by a base station depends on the load but many parts require power even when there is no user to serve. The cost of the power has motivated operators to switch off cells when extra capacity is not needed. This may happen, for example, in shopping malls' indoor cells on closing days. The method of switching off cells can be manual but attempts have been made to find a solution for an automated process.

There is one important assumption to note, derived rather from the legal requirements of operator licenses: the coverage must be provided at all times. The solution must not create holes in coverage for basic services.

Switching off a cell that is fully backed-up in terms of coverage can be considered as fully localized action: the cell knows when it is empty (at least if only active mode users are considered). Even if neighbor relations are not deleted, the risk of handover failure is low because no UE will report a switched-off cell as a handover candidate. However, there is a problem in defining a moment for automatic wake up when there is enough traffic in its serving area to justify the energy cost of extra capacity. This is known in

the cells that remain on to provide the coverage – they can wake up suspended cells, if they know they are suspended. The SON solution for the energy-saving use case relies, therefore, on two elements:

- cell status information that can be sent to neighbors;
- a cell wake-up call that a neighbor may issue when it detects that more capacity is needed.

Since the method for switching off a cell is not specified, it is possible that it may rely on a gradual power-down. The suspension may also be delayed until the last active user leaves the cell. In both cases, all incoming handovers will probably be rejected. In order to enable precise feedback information to the eNodeB initializing such handovers, but also for other possible failure or rejection messages due to cell switch off procedure, additional cause value was defined: "Switch Off Ongoing".

9.5.5 Summary of the Available SON Procedures

In the previous chapters a set of new functions defined for LTE was presented. These functions form the standardized SON framework. Figure 9.11 summarizes their mapping on SON use cases solved in LTE Release 9.

9.5.6 SON Management

Management of SON functions described above comprises three functions:

- monitoring;
- corrective actions;
- operator control.

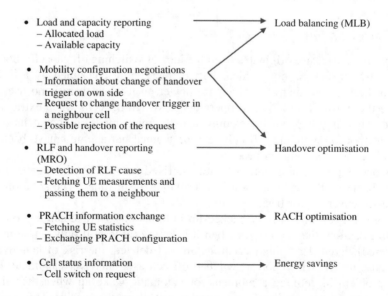

Figure 9.11   SON functionality and its mapping on use cases

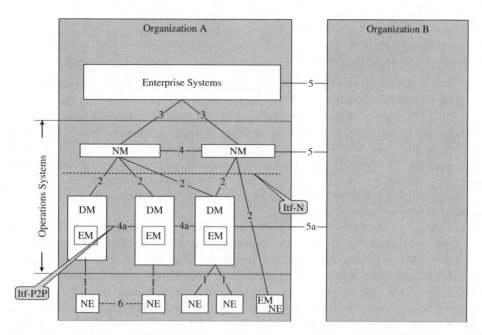

Figure 9.12   3GPP Management Reference Model. NM = Network Management, DM = Domain Management, EM = Element Management

The role of monitoring is to collect and enable reporting of measurements of the key performance indicators (KPIs), alarms and events monitoring. The information provided by the monitoring function allows the execution of corrective actions, which, in the end, involve changing configuration or other parameters. The ultimate control over complete SON functionality is given to the operator. The management reference model is shown in Figure 9.12.

Management architecture is split into standardized 'northbound' (Itf-N) and proprietary 'southbound' (Itf-S) interfaces. The Itf-N enables the above functionalities in a multi-vendor environment, while each vendor may enhance the standardized SON mechanisms with proprietary features, managed through the Itf-S. The mix of 'centralized-standardized' and 'distributed-supported by proprietary centralized mechanisms' leads to a hybrid SON architecture, where some functionalities are present in eNodeBs and some others in the central OAM.

Standardization of SON management goes in parallel to the development of SON mechanisms that are to be managed. For example, the Itf-N may be used for automatic neighbor relations (ANR), the automatic radio configuration function (ARCF), handover optimization and load-balancing optimization (LBO). The overall SON management concepts are defined in TS 32.500 [5].

9.6   3GPP Release 10 Use Cases

Release 10 introduces a number of additional use cases:

- Coverage and Capacity Optimization (CCO);
- Enhanced Mobility Robustness Optimization;

- Enhanced Mobility Load Balancing;
- Enhanced Energy Saving;
- Minimization of Drive Tests (MDT).

Coverage and Capacity Optimization addresses mainly coverage problems but can be extended to capacity. It will enable automatic correction of coverage and capacity problems caused by slowly changing environment. As an example, seasonal variations can be followed much more precisely than on manual or static adaptation. For example, trees with leaves attenuate signals differently that when they are bare in winter – thus the coverage situation changes and optimization is needed. In addition, since the amendment in coverage configuration requires good knowledge of the whole network, it is very likely that the intelligence will be implemented in a central entity that will collect necessary statistics from controlled eNodeBs.

The enhanced Mobility Robustness Optimization and enhanced Mobility Load Balancing are meant mainly to extend or improve functionalities defined already in Release 9. One extension concerns the possibility that UE could send measurements even if the connection re-establishment fails and UE sets up a new connection. Release 10 will also extend the Use Cases to inter-RAT, as Release 9 focused on LTE.

Energy Saving, already defined in Release 9, is also a possible candidate to be enhanced. Like Mobility Robustness Optimization, Release 10 would extend the Release 9 LTE solution to Multi-RAT. Having overlapping service areas of different RATs, operator may want to switch off one RAT and keeping the other for coverage. In addition other more sophisticated options are considered, including compensating for switched off cells with extended coverage of remaining cells, which enables energy saving in single RAT – single layer deployment.

One of the most interesting and important Use Case developed in Release 10 is Minimization of Drive Tests. The origin of the idea is the fact that drive tests are expensive and inefficient. However, if normal UEs are enabled to provide same information, operators will get a powerful method to analyze and optimize their networks. Even indoor measurements can be retrieved which is typically not possible with current drive tests. This must however be enabled in such a way that users' privacy and experienced QoS are not endangered. Because the MDT has been defined as RRC functionality, the work has mainly impact on the RRC specification in TS 36.331.

9.7 Summary

Self Organizing Networks (SON) is a set of network algorithms for simplifying the network configuration and optimization. SON targets to provide lower operating expenditures for the operators and better performance for the customers. The need for SON has emerged especially together with LTE because the network complexity increases with more Radio Access Technologies (RATs) and with more base stations. SON algorithms can be centralized based on long-term statistics and targeting for multi-cell optimization, or distributed for fast and local optimization. SON algorithms can be categorized into self-configuration like automatic neighbor relation management, self-optimization like load balancing and self-healing, like coverage and capacity optimization. SON algorithms are included in 3GPP Releases 8, 9 and 10, and further work continues in upcoming 3GPP releases. SON is not limited to LTE radio only but many of these SON algorithms are applied to 3G HSPA networks as well.

References

[1] 3GPP TS 36.300 V9.3.0, '3rd Generation Partnership Project; Technical Specification Group Radio Access Network; Evolved Universal Terrestrial Radio Access (E-UTRA) and Evolved Universal Terrestrial Radio Access Network (E-UTRAN); Overall description; Stage 2,' Release 9, March 2010.

[2] 3GPP TR 36.902 V9.1.0, '3rd Generation Partnership Project; Technical Specification Group Radio Access Network; Evolved Universal Terrestrial Radio Access Network (E-UTRAN); Self-Configuring and Self-Optimizing Network (SON) Use Cases and Solutions', Release 9, March 2010.

[3] 3GPP TS 48.018 V9.1.0, '3rd Generation Partnership Project; Technical Specification Group GSM/EDGE Radio Access Network; General Packet Radio Service (GPRS); Base Station System (BSS) – Serving GPRS Support Node (SGSN); BSS GPRS Protocol (BSSGP)', Release 9, March 2010.

[4] 3GPP TS 36.423 V9.2.0, '3rd Generation Partnership Project; Technical Specification Group Radio Access Network; Evolved Universal Terrestrial Radio Access Network (E-UTRAN); X2 Application Protocol (X2AP)', Release 9, March 2010.

[5] 3GPP TS 32.500 V9.0.0, '3rd Generation Partnership Project; Technical Specification Group Services and System Aspects; Telecommunication Management; Self-Organizing Networks (SON); Concepts and requirements,' Release 9, December 2009.

10

Performance

Harri Holma, Pasi Kinnunen, István Z. Kovács, Kari Pajukoski,
Klaus Pedersen and Jussi Reunanen

10.1 Introduction

This chapter illustrates LTE capabilities from the end user's and from the operator's point
of view. Radio performance has a direct impact on the cost of deploying the network
in terms of the required number of base station sites and in terms of the transceivers
required. The operator is interested in the network efficiency: how many customers can be
served, how much data can be provided and how many base station sites are required. The
efficiency is considered in the link budget calculations and in the capacity simulations. The
end user application performance depends on the available bit rate, latency and seamless
mobility. The radio performance defines what applications can be used and how these
applications perform.

The link level studies in this chapter illustrate the main factors affecting LTE performance including mobile speed and transmission bandwidth. The LTE link performance is
benchmarked with the theoretical Shannon limit. The capacity studies present the impact
of the environment and show the bit rate distributions in typical network deployments.
The relative capacity gains compared to HSPA networks are shown analytically and with
simulations. The chapter presents also general dimensioning guidelines and the specific
aspects of refarming LTE to the GSM spectrum. The network capacity management is
illustrated with examples from High Speed Packet Access (HSPA) networks.

Practical performance is also impacted by the performance of the commercial UE
and eNodeBs. To guarantee consistent performance, 3GPP has defined a set of radio
performance requirements. 3GPP performance requirements are explained in detail in
Chapter 11.

10.2 Layer 1 Peak Bit Rates

LTE provides high peak bit rates by using a large bandwidth up to 20 MHz, high order
64QAM modulation and multistream MIMO transmission. Quadrature Phase Shift Keying

LTE for UMTS: Evolution to LTE-Advanced, Second Edition. Edited by Harri Holma and Antti Toskala.
© 2011 John Wiley & Sons, Ltd. Published 2011 by John Wiley & Sons, Ltd.

(QPSK) modulation carries 2 bits per symbol, 16QAM 4 bits and 64QAM 6 bits. 2 × 2 MIMO further doubles the peak bit rate up to 12 bits per symbol. Therefore, QPSK 1/2 rate coding carries 1 bps/Hz while 64QAM without any coding and with 2 × 2 Multiple Input Multiple Output (MIMO) carries 12 bps/Hz. The bandwidth is included in the calculation by taking the corresponding number of resource blocks for each bandwidth option: 6 with 1.4 MHz and 15 with 3 MHz bandwidth. The number of resource blocks for the bandwidths 5, 10, 15 and 20 MHz are 25, 50, 75 and 100 respectively. We assume the following control and reference signal overheads:

- Physical Downlink Control Channel (PDCCH) takes one symbol out of 14 symbols. That is the minimum possible PDCCH allocation. It is enough when considering single user peak bit rate. The resulting control overhead is 7.1% (= 1/14).
- Downlink Reference Signals (RS) depend on the antenna configuration. Single stream transmission uses 2 RS out of 14 in every 3rd sub-carrier, 2 × 2 MIMO 4 symbols and 4 × 4 MIMO 6 symbols. The overhead varies between 4.8% and 14.3%. The RS partly overlap with PDCCH and the overlapping is taken into account.
- Other downlink symbols are subtracted: synchronization signal, Physical Broadcast Channel (PBCH), Physical Control Format Indicator Channel (PCFICH) and one group of Physical Hybrid Automatic Repeat Request Indicator Channel (PHICH). The overhead depends on the bandwidth ranging from below 1% at 20 MHz to approximately 9% at 1.4 MHz.
- No Physical Uplink Control Channel (PUCCH) is included in the calculation. PUCCH would slightly reduce the uplink data rate.
- Uplink reference signals take 1 symbol out of 7 symbols resulting in an overhead of 14.3% (= 1/7).

The achievable peak bit rates are shown in Table 10.1 The highest theoretical data rate is approximately 172 Mbps. If 4 × 4 MIMO option is applied, the theoretical peak data rate increases to 325 Mbps. The bit rate scales down according to the bandwidth. The 5-MHz peak bit rate is 42.5 Mbps and 1.4-MHz 8.8 Mbps with 2 × 2 MIMO.

Table 10.1 Downlink peak bit rates (Mbps)

Resource blocks			1.4 MHz	3.0 MHz	5.0 MHz	10 MHz	15 MHz	20 MHz
			6	15	25	50	75	100
Modulation and coding	Bits/ symbol	MIMO usage						
QPSK 1/2	1.0	Single stream	0.8	2.2	3.7	7.4	11.2	14.9
16QAM 1/2	2.0	Single stream	1.5	4.4	7.4	14.9	22.4	29.9
16QAM 3/4	3.0	Single stream	2.3	6.6	11.1	22.3	33.6	44.8
64QAM 3/4	4.5	Single stream	3.5	9.9	16.6	33.5	50.4	67.2
64QAM 1/1	6.0	Single stream	4.6	13.2	22.2	44.7	67.2	89.7
64QAM 3/4	9.0	2 × 2 MIMO	6.6	18.9	31.9	64.3	96.7	129.1
64QAM 1/1	12.0	2 × 2 MIMO	8.8	25.3	42.5	85.7	128.9	172.1
64QAM 1/1	24.0	4 × 4 MIMO	16.6	47.7	80.3	161.9	243.5	325.1

Table 10.2 Uplink peak bit rates (Mbps)

Resource blocks			1.4 MHz	3.0 MHz	5.0 MHz	10 MHz	15 MHz	20 MHz
Modulation and coding	Bits/ symbol	MIMO usage	6	15	25	50	75	100
QPSK 1/2	1.0	Single stream	0.9	2.2	3.6	7.2	10.8	14.4
16QAM 1/2	2.0	Single stream	1.7	4.3	7.2	14.4	21.6	28.8
16QAM 3/4	3.0	Single stream	2.6	6.5	10.8	21.6	32.4	43.2
16QAM 1/1	4.0	Single stream	3.5	8.6	14.4	28.8	43.2	57.6
64QAM 3/4	4.5	Single stream	3.9	9.7	16.2	32.4	48.6	64.8
64QAM 1/1	6.0	Single stream	5.2	13.0	21.6	43.2	64.8	86.4

The uplink peak data rates are shown in Table 10.2: up to 86 Mbps with 64QAM and up to 57 Mbps with 16QAM with 20 MHz. The peak rates are lower in uplink than in downlink since single user MIMO is not specified in uplink in 3GPP Release 8. The single user MIMO in uplink would require two power amplifiers in the terminal. MIMO can be used in Release 8 uplink as well to increase the aggregate cell capacity, but not single user peak data rates. The cell level uplink MIMO is called Virtual MIMO (V-MIMO) where the transmission from two terminals, each with single antenna, is organized so that the cell level peak throughput can be doubled.

The transport block sizes in [1] have been defined so that uncoded transmission is not possible. The maximum achievable bit rates taking the transport block sizes into account are shown in Table 10.3 for downlink and in Table 10.4 for uplink for the different

Table 10.3 Downlink peak bit rates with transport block size considered (Mbps)

Resource blocks		1.4 MHz	3.0 MHz	5.0 MHz	10 MHz	15 MHz	20 MHz
Modulation and coding	MIMO usage	6	15	25	50	75	100
QPSK	Single stream	0.9	2.3	4.0	8.0	11.8	15.8
16QAM	Single stream	1.8	4.6	7.7	15.3	22.9	30.6
64QAM	Single stream	4.4	11.1	18.3	36.7	55.1	75.4
64QAM	2 × 2 MIMO	8.8	22.2	36.7	73.7	110.1	149.8

Table 10.4 Uplink peak bit rates with transport block size considered (Mbps)

Resource blocks		1.4 MHz	3.0 MHz	5.0 MHz	10 MHz	15 MHz	20 MHz
Modulation and coding	MIMO usage	6	15	25	50	75	100
QPSK	Single stream	1.0	2.7	4.4	8.8	13.0	17.6
16QAM	Single stream	3.0	7.5	12.6	25.5	37.9	51.0
64QAM	Single stream	4.4	11.1	18.3	36.7	55.1	75.4

modulation schemes. The downlink peak rate with 2×2 MIMO goes up to 150 Mbps and the uplink rate up to 75 Mbps. The calculations assume that uplink 16QAM uses Transport Block Size (TBS) index 21, uplink QPSK uses TBS index 10, downlink 16QAM uses TBS index 15 and downlink QPSK uses TBS index 9.

The original LTE target was peak data rates of 100 Mbps in downlink and 50 Mbps in uplink, which are clearly met with the 3GPP Release 8 physical layer.

10.3   Terminal Categories

3GPP Release 8 has defined five terminal categories having different bit rate capabilities. Category 1 is the lowest capability with maximum bit rates of 10 Mbps downlink and 5 Mbps uplink while Category 5 is the highest capability with data rates of 300 Mbps in downlink and 75 Mbps uplink. The bit rate capability in practice is defined as the maximum transport block size that the terminal is able to process in 1 ms.

All categories must support all RF bandwidth options from 1.4 to 20 MHz, 64QAM modulation in downlink and 1–4 transmission branches at eNodeB. The receive antenna diversity is mandated via performance requirements. The support of MIMO transmission depends on the category. Category 1 does not need to support any MIMO while categories 2–4 support 2×2 MIMO and Category 5 support 4×4 MIMO. The uplink modulation is up to 16QAM in categories 1–4 while 64QAM is required in category 5.

The terminal categories are shown in Table 10.5. Further terminal categories may be defined in later 3GPP releases. The initial LTE deployment phase is expected to have terminal categories 2, 3 and 4 available providing downlink data rates up to 150 Mbps and supporting 2×2 MIMO.

Table 10.5   Terminal categories [2]

	Category 1	Category 2	Category 3	Category 4	Category 5
Peak rate downlink (approximately)	10 Mbps	50 Mbps	100 Mbps	150 Mbps	300 Mbps
Peak rate uplink (approximately)	5 Mbps	25 Mbps	50 Mbps	50 Mbps	75 Mbps
Max bits received within TTI	10296	51024	102048	149776	299552
Max bits transmitted within TTI	5160	25456	51024	51024	75376
RF bandwidth	20 MHz	20 MHz	20 MHz	20 MHz	20 MHz
Modulation downlink	64QAM	64QAM	64QAM	64QAM	64QAM
Modulation uplink	16QAM	16QAM	16QAM	16QAM	64QAM
Receiver diversity	Yes	Yes	Yes	Yes	Yes
eNodeB diversity	1–4 tx	1–4 tx	1–4 tx	1–4 tx	1–4 tx
MIMO downlink	Optional	2×2	2×2	2×2	4×4

10.4 Link Level Performance

10.4.1 Downlink Link Performance

The peak data rates are available only in extremely good channel conditions. The practical data rate is limited by the amount of interference and noise in the network. The maximum theoretical data rate with single antenna transmission in static channel can be derived using the Shannon formula. The formula gives the data rate as a function of two parameters: bandwidth and the received Signal-to-Noise Ratio (SNR).

$$\text{Bit rate [Mbps]} = \text{Bandwidth [MHz]} \cdot \log_2(1 + SNR) \qquad (10.1)$$

The Shannon capacity bound in equation (10.1) cannot be reached in practice due to several implementation issues. To represent these loss mechanisms accurately, we use a modified Shannon capacity formula [3]:

$$\text{Bit rate [Mbps]} = BW\_eff \cdot \text{Bandwidth [MHz]} \cdot \log_2(1 + SNR/SNR\_eff) \qquad (10.2)$$

where BW_eff accounts for the system bandwidth efficiency of LTE and SNR_eff accounts for the SNR implementation efficiency of LTE.

The bandwidth efficiency of LTE is reduced by several issues, as listed in Table 10.6. Due to the requirements of the Adjacent Channel Leakage Ratio (ACLR) and practical filter implementation, the bandwidth occupancy is reduced to 0.9. The overhead of the cyclic prefix is approximately 7% and the overhead of pilot assisted channel estimation is approximately 6% for single antenna transmission. For dual antenna transmission the overhead is approximately doubled to 11%. Note that here ideal channel estimation is used, which is the reason why the pilot overhead is not included in the link performance bandwidth efficiency but must be included in the system bandwidth level efficiency [3]. This issue also impacts the SNR_eff. The overall link-level bandwidth efficiency is therefore approximately 83%.

When fitting Equation 10.2 to the simulated performance curve in Additive White Gaussian Noise (AWGN) channel conditions, we extract the best value for SNR_eff using the setting for BW_eff of 0.83 from Table 10.6 and the fitting parameters are indicated in parentheses as (BW_eff, SNR_eff). The results are presented in Figure 10.1. We can observe that LTE is performing less than 1.6~2 dB off from the Shannon capacity bound. There is nevertheless a minor discrepancy in both ends of the G-factor dynamic range. This is because the SNR_eff is not constant but changes with the G-factor. It is shown in [3] that this dependency can be accounted for using the fudge factor, η multiplying

Table 10.6 Link bandwidth efficiency for LTE downlink with a 10 MHz system

Impairment	Link BW_eff	System BW_eff
BW efficiency	0.90	0.90
Cyclic prefix	0.93	0.93
Pilot overhead	–	0.94
Dedicated and common control channels	–	0.715
Total	0.83	0.57

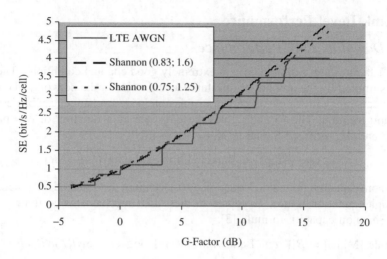

Figure 10.1 LTE spectral efficiency (SE) as a function of G-factor (indB) including curves for best Shannon fit. The steps are due to the limited number of modulation and coding schemes in the simulation [3]. © 2007 IEEE

the *BW_eff* parameter. For AWGN, $\eta = 0.9$ (*BW_eff*$^*\eta = 0.75$) and *SNR_eff* $= 1.0$. The best fit to the link adaptation curve [3] is 1.25 dB.

In Figure 10.2 we show the spectral efficiency results versus G-factor from LTE link level studies and the best Shannon fit for a 3 km/h typical urban fading channel. Different antenna configurations are presented: single antenna transmission and reception Single Input Single Output (SISO) (1×1), single antenna transmission and antenna receive diversity Single Input Multiple Output (SIMO) (1×2) and Closed Loop MIMO (2×2). Using the G-factor dependent *SNR_eff* we achieve a visibly almost perfect fit to the link simulation results. For SISO it can be observed that the best Shannon fit parameters are significantly worsened compared to AWGN: the equivalent *BW_eff* has reduced from 0.83 to 0.56 (corresponding to fudge factor $\eta = 0.6$) and the *SNR_eff* parameter is increased from 1.6~2 dB to 2~3 dB. The match between the fitted Shannon curves and the actual link-level results is not perfect but sufficiently close for practical purposes.

These link-level results show that LTE downlink with ideal channel estimation performs approximately 2 dB from Shannon Capacity in AWGN, whereas the deviation between Shannon and LTE becomes much larger for a fading channel. This fading loss can, however, be compensated with the multi-user diversity gain when using frequency domain packet scheduling and the fitted Shannon curves can also include the system level scheduling gain [3]. With these adjustments, cell capacity results can be accurately estimated from the suggested modified Shannon formula and a G-Factor distribution according to a certain cellular scenario.

10.4.2   Uplink Link Performance

10.4.2.1   Impact of Transmission Bandwidth

The uplink coverage can be optimized by bandwidth selection in LTE. The bandwidth adaptation allows the UE power to be emitted in an optimum bandwidth, which extends

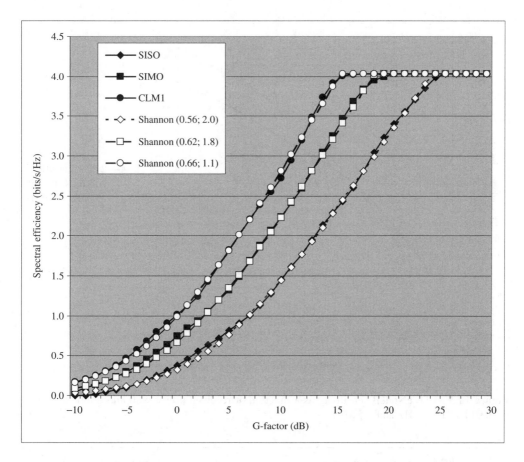

Figure 10.2 Spectral efficiency for SISO (1 × 1), SIMO (1 × 2) and Closed loop MIMO (2 × 2), as a function of G-factor. The best Shannon fit curves are plotted with parameters (*BW_eff*, *SNR_eff*) using Equation 10.2 [3]

the coverage compared to having a fixed transmission bandwidth. A larger number of channel parameters need to be estimated for the wideband transmissions, which reduces the accuracy of the channel estimation due to noise. The optimized transmission bandwidth in LTE enables more accurate channel estimation for the lower data rates compared to WCDMA/HSUPA.

In Figure 10.3, the throughput as a function of SNR with different bandwidth allocation is shown. The smallest bandwidth allocation, 360 kHz, optimizes the coverage for bit rates below 100 kbps. The moderate bandwidth allocation, 1.08 MHz, gives the best coverage for bit rates from 100 kbps to 360 kbps. The assumed base station noise figure is 2 dB and no interference margin is included.

10.4.2.2 Impact of Mobile Speed

Figure 10.4 shows the link adaptation curve of LTE uplink for the UE speeds of 3, 50, 120 and 250 km/h. The received pilot signals are utilized for channel estimation. The channel

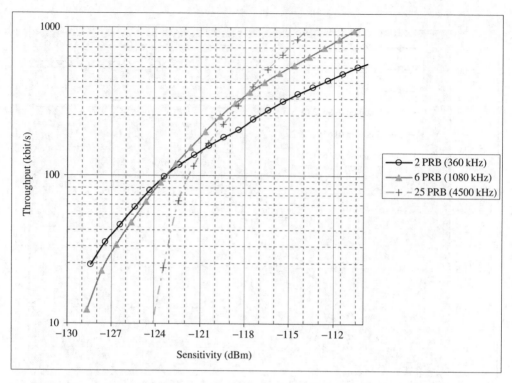

Figure 10.3 LTE eNodeB sensitivity as a function of received power with allocation bandwidths of 360 kHz, 1.08 MHz and 4.5 MHz

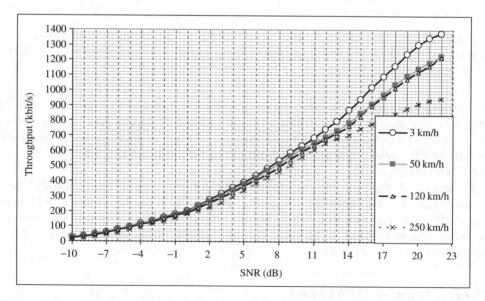

Figure 10.4 LTE eNodeB throughput as a function of SNR with UE speeds of 3 km/h, 50 km/h, 120 km/h and 250 km/h

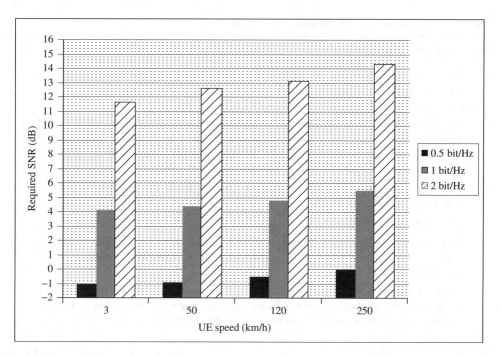

Figure 10.5   Required SNR values for various spectral efficiencies with UE speeds of 3 km/h, 50 km/h, 120 km/h and 250 km/h

estimation is based on Wiener filter in both time and frequency domain. The impact of the mobile speed is small at low data rates while at high data rates the very high mobile speed of 250 km/h shows 30% lower throughput than at 3 km/h.

The required SNR values for various link efficiencies are given in Figure 10.5. In the efficiency range of QPSK (0.1–1 bits/Hz), the LTE uplink performance is affected only by 1–1.5 dB, when increasing UE speed from 3 to 250 km/h. For higher order modulations, 16QAM and 64QAM, the impact of the UE speed is higher – up to 3 dB for the efficiency of 2 bps/Hz/cell. In general, LTE uplink is robust against Doppler frequency shifts. The UE speed affects the throughput and SNR requirements mainly due to channel estimation performance: when the channel is changing fast, it is not possible to use a long averaging time and the accuracy of the channel estimation is impacted.

10.5   Link Budgets

The link budget calculations estimate the maximum allowed signal attenuation, called path loss, between the mobile and the base station antenna. The maximum path loss allows the maximum cell range to be estimated with a suitable propagation model, such as Okumura–Hata. The cell range gives the number of base station sites required to cover the target geographical area. The link budget calculation can also be used to compare the relative coverage of the different systems. The relative link budget indicates how well the new LTE radio system will perform when it is deployed on the existing base station sites that are designed for GSM and WCDMA.

Table 10.7 Uplink link budget parameters for LTE

Field	Description	Typical value
a	UE maximum transmission power for power class 3. Different power classes would have different power levels. The power can be reduced depending on the modulation, see Chapter 14 for details.	23 dBm
b	UE antenna gain depends on the type of device and on the frequency band. Small handheld terminal at a low frequency band (like Band VIII) can have an antenna gain of −5 dBi while a fixed wireless terminal with directive antenna can have a gain of up to 10 dBi.	−5 to 10 dBi
c	Body loss is typically included for voice link budget where the terminal is held close to the user's head.	3 to 5 dB for voice
d	Calculated as a + b − c.	
e	Base station RF noise figure. Depends on the implementation design. The minimum performance requirement is approximately 5 dB but the practical performance can be better.	2 dB
f	Terminal noise can be calculated as k (Boltzmann constant) $\times T$ (290 K) $\times$ bandwidth. The bandwidth depends on bit rate, which defines the number of resource blocks. We assume two resource blocks for 64 kbps uplink.	−118.4 dBm for two resource blocks (360 kHz)
g	Calculated as e + f.	
h	Signal-to-noise ratio from link simulations or measurements. The value depends on the modulation and coding schemes, which again depend on the data rate and on the number of resource blocks allocated.	−7 dB for 64 kbps and two resource blocks
i	Calculated as g + h.	
j	Interference margin accounts for the increase in the terminal noise level caused by the interference from other users. Since LTE uplink is orthogonal, there is no intra-cell interference but we still need a margin for the other cell interference. The interference margin in practice depends heavily on the planned capacity – there is a tradeoff between capacity and coverage. The LTE interference margin can be smaller than in WCDMA/HSUPA where the intra-cell users are not orthogonal. In other words, the cell breathing will be smaller in LTE than in CDMA based systems.	1 to 10 dB
k	Cable loss between the base station antenna and the low noise amplifier. The cable loss value depends on the cable length, cable type and frequency band. Many installations today use RF heads where the RF parts are close to the antenna making the cable loss very small. The cable loss can also be compensated by using mast head amplifiers.	1 to 6 dB
l	Base station antenna gain depends on the antenna size and the number of sectors. Typical 3-sector antenna 1.3 m high at 2 GHz band gives 18 dBi gain. The same size antenna at 900 MHz gives smaller gain.	15 to 21 dBi for sectorized base station
m	Fast fading margin is typically used with WCDMA due to fast power control to allow headroom for the power control operation. LTE does not use fast power control and the fast fading margin is not necessary in LTE.	0 dB
n	Soft handover is not used in LTE.	0 dB

Table 10.8 Uplink link budgets

	Uplink	GSM voice	HSPA	LTE
	Data rate (kbps)	12.2	64	64
	Transmitter – UE			
a	Max tx power (dBm)	33.0	23.0	23.0
b	Tx antenna gain (dBi)	0.0	0.0	0.0
c	Body loss (dB)	3.0	0.0	0.0
d	EIRP (dBm)	30.0	23.0	23.0
	Receiver – Node B			
e	Node B noise figure (dB)	–	2.0	2.0
f	Thermal noise (dB)	−119.7	−108.2	−118.4
g	Receiver noise (dBm)	–	−106.2	−116.4
h	SINR (dB)	–	−17.3	−7.0
i	Receiver sensitivity	−114.0	−123.4	−123.4
j	Interference margin (dB)	0.0	3.0	1.0
k	Cable loss (dB)	0.0	0.0	0.0
l	Rx antenna gain (dBi)	18.0	18.0	18.0
m	Fast fade margin (dB)	0.0	1.8	0.0
n	Soft handover gain (dB)	0.0	2.0	0.0
	Maximum path loss	162.0	161.6	163.4

The parameters for the LTE link budget for uplink are introduced in Table 10.7, and the link budget is presented in Table 10.8. The corresponding LTE downlink link budget parameters are presented in Table 10.9, and the link budget in Table 10.10. The link budgets are calculated for 64 kbps uplink with 2-antenna base station receive diversity and 1 Mbps downlink with 2-antenna mobile receive diversity. For reference, the link budgets for GSM voice and for HSPA data are illustrated as well.

The LTE link budget in downlink has several similarities with HSPA and the maximum path loss is similar. The uplink part has some differences: smaller interference margin in LTE, no macro diversity gain in LTE and no fast fading margin in LTE. The maximum path loss values are summarized in Figure 10.6. The link budgets show that LTE can be deployed using existing GSM and HSPA sites assuming that the same frequency is used for LTE as for GSM and HSPA. LTE itself does not provide any major boost in the coverage. That is because the transmission power levels and the RF noise figures are also similar in GSM and HSPA technologies, and the link performance at low data rates is not much different in LTE than in HSPA.

The link budget was calculated for 64 kbps uplink, which is likely not a high enough data rate for true broadband service. If we want to guarantee higher data rates in LTE, we may need low frequency deployment, additional sites, active antenna solutions or local area solutions.

The coverage can be boosted by using a lower frequency since a low frequency propagates better than a higher frequency. The benefit of the lower frequency depends on the environment and on its use. The difference between 900 MHz and 2600 MHz is illustrated in Table 10.11. Part of the benefit of the lower frequency is lost since the antenna gains tend to get smaller at a lower frequency band. To maintain the antenna gain at lower frequency would require a physically larger antenna which is not always feasible at base station sites and in small terminals. The greatest benefits from low frequencies can be

Table 10.9 Downlink link budget parameters for LTE

Field	Description	Typical value
a	Base station maximum transmission power. A typical value for macro cell base station is 20–60 W at the antenna connector.	43–48 dBm
b	Base station antenna gain. See uplink link budget.	
c	Cable loss between the base station antenna connector and the antenna. The cable loss value depends on the cable length, cable thickness and frequency band. Many installations today use RF heads where the power amplifiers are close to the antenna making the cable loss very small.	1–6 dB
d	Calculated as A + B − C.	
e	UE RF noise figure. Depends on the frequency band, Duplex separation and on the allocated bandwidth. For details, see Chapter 12.	6–11 dB
f	Terminal noise can be calculated as k (Boltzmann constant) $\times T$ (290 K) $\times$ bandwidth. The bandwidth depends on bit rate, which defines the number of resource blocks. We assume 50 resource blocks, equal to 9 MHz, transmission for 1 Mbps downlink.	−104.5 dBm for 50 resource blocks (9 MHz)
g	Calculated as E + F.	
h	Signal-to-noise ratio from link simulations or measurements. The value depends on the modulation and coding schemes, which again depend on the data rate and on the number of resource blocks allocated.	−9 dB for 1 Mbps and 50 resource blocks
i	Calculated as G + H.	
j	Interference margin accounts for the increase in the terminal noise level caused by the other cell. If we assume a minimum G-factor of −4 dB, that corresponds to $10*\log10(1+10^{(4/10)}) = 5.5$ dB interference margin.	3–8 dB
k	Control channel overhead includes the overhead from reference signals, PBCH, PDCCH and PHICH.	10–25% = 0.4–1.0 dB
l	UE antenna gain. See uplink link budget.	−5–10 dBi
m	Body loss. See uplink link budget.	3.5 dB for voice

obtained when the base station site can use large antennas 2.5 m high and where the external antenna can be used in the terminal. This is fixed wireless deployment.

Example cell ranges are shown in Figure 10.7. Note that the y-axis has a logarithmic scale. The cell range is shown for 900 MHz, 1800 MHz, 2100 MHz and 2600 MHz frequency variants. These frequencies do not cover all possible LTE frequency variants. Typical US frequencies would be 700 MHz and 1700/2100 MHz, but the end result would be very similar to the values in Figure 10.7. The cell ranges are calculated using the Okumura–Hata propagation model with the parameters shown in Table 10.12. The urban cell range varies from 0.6 km to 1.4 km and suburban from 1.5 km to 3.4 km. Such cell ranges are also typically found in existing GSM and UMTS networks. The rural case shows clearly higher cell ranges: 26 km for the outdoor mobile coverage and even up to 50 km for the rural fixed installation at 900 MHz.

Note that the earth's curvature limits the maximum cell range to approximately 40 km with an 80 m high base station antenna assuming that the terminal is at ground level.

Table 10.10 Downlink link budgets

Downlink		GSM voice	HSPA	LTE
	Data rate (kbps)	12.2	1024	1024
	Transmitter – Node B			
a	Tx power (dBm)	44.5	46.0	46.0
b	Tx antenna gain (dBi)	18.0	18.0	18.0
c	Cable loss (dB)	2.0	2.0	2.0
d	EIRP (dBm)	60.5	62.0	62.0
	Receiver – UE			
e	UE noise figure (dB)	–	7.0	7.0
f	Thermal noise (dB)	−119.7	−108.2	−104.5
g	Receiver noise floor (dBm)	–	−101.2	−97.5
h	SINR (dB)	–	−5.2	−9.0
i	Receiver sensitivity (dBm)	−104.0	−106.4	−106.5
j	Interference margin (dB)	0.0	4.0	4.0
k	Control channel overhead (%)	0.0	20.0	20.0
l	Rx antenna gain (dBi)	0.0	0.0	0.0
m	Body loss (dB)	3.0	0.0	0.0
	Maximum path loss	161.5	163.4	163.5

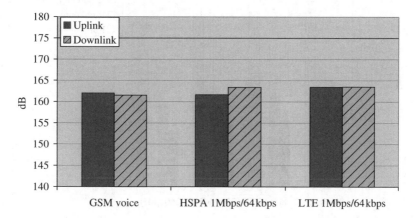

Figure 10.6 Maximum path loss values for GSM voice and HSPA and LTE data

The maximum cell range can be calculated with Equation 10.3, where R is the effective earth radius of 8650 km, h is the base station antenna height and d is the distance from the terminal to the base station. The calculation is illustrated in Figure 10.8. To achieve 100 km cell range, the required antenna height is 580 m, which in practice is feasible if the base station antenna is located on a mountain pointing towards the sea or other flat terrain.

$$R^2 + d^2 = (R + h)^2$$
$$h = \sqrt{R^2 + d^2} - R$$

(10.3)

Table 10.11   Benefit of 900 MHz compared to 2600 MHz

	Urban	Rural	Rural fixed wireless
Propagation loss[a]	+14 dB	+14 dB	+14 dB
BTS antenna gain	−3 dB[b]	0 dB[c]	0 dB[c]
BTS cable loss[d]	+1 dB	+3 dB	+3 dB
UE antenna gain	−5 dB[e]	−5 dB[e]	0 dB[f]
UE sensitivity[g]	−1 dB	−1 dB	−1 dB
Total	+6 dB	+11 dB	+16 dB

[a]According to Okumura–Hata.
[b]Shared 1.3 m antenna for 900/2600 giving 15 and 18 dBi gain at 900 vs 2600 MHz.
[c]2.5 m antenna giving 18 dBi gain at 900 MHz.
[d]Cable 1/2'. Urban 30 m and rural 100 m.
[e]Based on 3GPP RAN4 contributions.
[f]External fixed antenna assumed.
[g]UE sensitivity can be up to 3 dB worse at 900 MHz but the difference in practice is less.

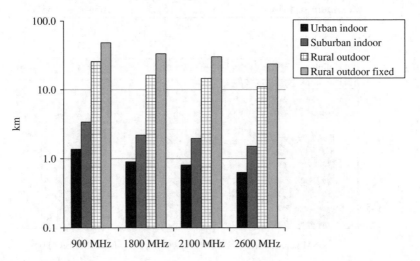

Figure 10.7   Cell ranges with Okumura–Hata propagation model

10.6   Spectral Efficiency

This section presents the typical spectral efficiency of LTE downlink and uplink in terms ofbps/Hz/cell. The impact of the radio resource management algorithms on performance is discussed in Chapter 8.

10.6.1   System Deployment Scenarios

The specified LTA evaluation and test scenarios correspond to various radio propagation conditions and environments, similarly to HSPA [4]. The assumed macro and micro

Table 10.12   Parameters for Okumura–Hata propagation model

	Urban indoor	Suburban indoor	Rural outdoor	Rural outdoor fixed
Base station antenna height (m)	30	50	80	80
Mobile antenna height (m)	1.5	1.5	1.5	5
Mobile antenna gain (dBi)	0.0	0.0	0.0	0.0
Slow fading standard deviation (dB)	8.0	8.0	8.0	8.0
Location probability	95%	95%	95%	95%
Correction factor (dB)	0	−5	−15	−15
Indoor loss (dB)	20	15	0	0
Slow fading margin (dB)	8.8	8.8	8.8	8.8
Max path loss at 1800/2100/2600 (dB)	163	163	163	163
Max path loss at 900 (dB)	160	160	160	160

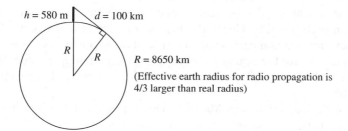

Figure 10.8   Maximum cell range limitation caused by earth curvature

Table 10.13   LTE reference system deployment cases – macro cell scenarios

Simulation cases (macro)	Frequency (GHz)	Inter-site distance (m)	Bandwidth (MHz)	UE Speed (kmph)	Penetration loss (dB)
1	2.0	500	10	3	20
3	2.0	1732	10	3	20

scenarios are shown in Table 10.13 and Table 10.14, and both assume a hexagonal-grid cellular layout.

The typical evaluation scenarios used for LTE are macro case 1 and macro case 3 with 10 MHz bandwidth and low UE mobility. The propagation models for macro cell scenario are based on the Okamura–Hata model presented in section 10.5.

The micro cell scenarios assume an outdoor base station location and an outdoor and/or indoor mobile station location. The propagation model for these micro cell scenarios is similar to the macro cell models, but with different parameter settings, specific to the dense urban deployment scenario. The assumed micro cell base station antenna configuration is omni-directional (one-sector) as opposed to the directional (three sectors) deployments for macro cell scenarios. For the micro cell outdoor-to-outdoor scenario a dual-slope distance

Table 10.14 LTE reference system deployment cases – micro cell scenarios. The indoor:outdoor UE location ratio used is 50:50

Simulation cases (macro)	Frequency (GHz)	Inter-site distance (m)	Bandwidth (MHz)	UE Speed (kmph)	Penetration loss (dB)
Outdoor-to-outdoor	2.0	130	10	3	0
Outdoor-to-indoor	2.0	130	10	3	Included in the path loss model

dependent path loss model is used in order to account for the Line-of-Sight (LOS) and Non-Line-of-Sight (NLOS) propagation conditions. For the micro cell outdoor-to-indoor scenario a single-slope distance dependent path loss model is used with a higher path loss exponent compared to the macro cell Okamura–Hata and micro cell outdoor-to-outdoor models. Typically, for system evaluation both outdoor and indoor mobile stations are assumed simultaneously with a 50:50 distribution ratio.

The differences between the reference system deployment cases, as expected, yield different system performances. Generally, the macro cell scenarios provide the baseline LTE system performance numbers, while the special deployment cases, such as the micro cell scenario, are used for the evaluation of adaptive spatial multiplexing MIMO enhancement techniques, which are able to better exploit the high Signal-to-Interference-and-Noise (SINR) conditions.

The results with Spatial Channel Model in Urban macro (SCM-C) radio channels and with a mixed velocity are also shown, as defined in reference [5].

The main simulation assumptions are shown in Table 10.15.

Table 10.15 Simulation assumptions for simulation scenarios

Description	Settings
Number of active UEs per cell	10 UEs, uniformly distributed
Bandwidth	10 MHz
Channel	3GPP TU at 3 km/h (6 tap) SCM-C at 3 km/h and with correlations for spatial channels ITU modified Pedestrian B and Vehicular A with correlation Mixed velocity case: −60% of UEs with ITU modified Pedestrian B at 3 km/h, −30% of UEs with ITU modified Vehicular A at 30 km/h and −10% of UEs with ITU modified Vehicular A at 120 km/h
Cell selection	Best cell selected with 0 dB margin
Transmission power	Uplink: Max 24 dBm (latest 3GPP specifications use 23 dBm)
Antenna configuration	Downlink: 1×2, 2×2 Uplink: 1×2, 1×4
Receiver algorithms	Downlink: LMMSE (Linear Minimum Mean Square Error) receiver Uplink: LMMSE receiver with Maximal Ratio Combining

10.6.2 Downlink System Performance

As a typical comparison between the different deployment environments, Figure 10.9 shows the distribution of the average wideband channel SINR also known as Geometry factor or G-factor. The corresponding downlink system performance numbers for LTE SIMO and 2×2 MIMO transmission schemes are presented in Figure 10.10 and Figure 10.11, respectively.

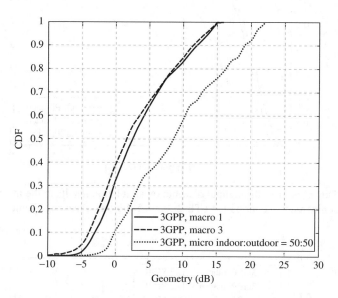

Figure 10.9 Distribution of the average wide-band channel SINR (geometry factor) macro case 1, case 3 and micro

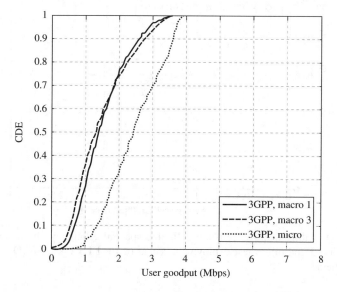

Figure 10.10 Downlink user throughput distribution simulation results in macro case 1, case 3 and micro test deployment scenarios for LTE 1×2 SIMO transmission scheme

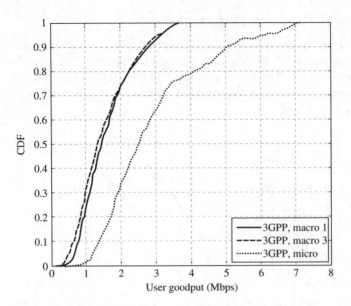

Figure 10.11   Downlink user throughput distribution simulation results in the macro case 1, case 3 and micro test deployment scenarios for LTE 2 × 2 MIMO transmission scheme

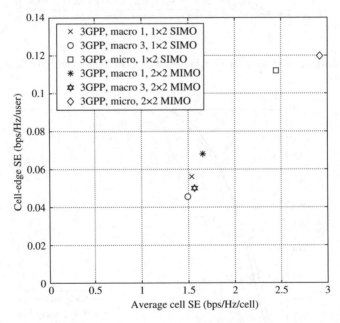

Figure 10.12   Downlink average cell and cell-edge spectral efficiency simulation results in the macro case 1, case 3 and micro test deployment scenarios for LTE 1 × 2 SIMO and 2 × 2 MIMO transmission scheme

Macro scenarios have median G-factors of 2–3 dB while the micro scenario has a G-factor above 8 dB. The macro cell has practically no values above 15 dB while the micro cell has 25% of the samples above 15 dB. The higher G-factor in micro cells turns into a higher throughput compared to the macro cells.

Figure 10.10 shows the user throughput with 10 users without MIMO. The median macro cell throughput is 1.4 Mbps while the micro cell provides 2.5 Mbps. The maximum values are 3–4 Mbps. Figure 10.11 presents the same case but with 2×2 MIMO. The median data rates are similar to the non-MIMO case but the highest data rates in micro cells are increased from 4 Mbps to 7 Mbps.

Figure 10.12 shows the summary of the average cell spectral efficiency (SE) and the cell edge efficiency. Micro cell scenarios show considerably higher efficiency than macro scenarios. MIMO increases the average efficiency in micro cells but provides only marginal gains in the macro cell environment.

10.6.3 Uplink System Performance

Uplink baseline results with Best Effort data are shown for a two and four antenna receiver with Round Robin (RR) and Proportional Fair (PF) schedulers. In these results Interference over Thermal (IoT) probability at 20 dB was limited to 5%.

Figure 10.13 shows cell throughput for two types of scenarios for Inter-Site Distances (ISD) of 500 m and 1500 m. The cell throughput is presented with full Sounding Reference Signal (SRS) usage, thus giving a lower bound of system performance. From these results the PF scheduler gives about a 15–20% gain for mean cell throughput over the RR scheduler. Most of this gain comes from using SRS for better frequency domain scheduling. With higher UE speeds (e.g. 120 km/h) this gain will decrease towards the corresponding RR scheduler results. Furthermore, the cell edge throughput with the same cell mean throughput is about 125% higher in ISD 500 m than in a larger cell size. UE power limitation is seen at ISD 1500 m with lower cell edge performance. Note that the impact of the cell size is larger in uplink than in downlink due to the lower UE power compared to the eNodeB transmission power.

The 95% limit for IoT distribution is set to 20 dB. Smaller cell sizes like ISD 500 m are affected by this constraint because higher IoT values would be possible from the UE power point of view. The mean IoT is 13 dB in Figure 10.14. For a larger cell size (like ISD 1500 m) IoT is not a limiting factor and the mean IoT is 2 dB.

The uplink power control range in ISD 500 m is seen in Figure 10.15. The power control range can be used well at a smaller cell size. The median UE transmission power is 15 dBm. This power is relatively high compared to voice users, but it is expected for the data connection. In a larger cell a higher proportion of users are limited by maximum power, thus moving the remaining distribution towards the right side.

Some spectral efficiency figures are collected into Table 10.16 for comparison with HSUPA. From these figures we observe that uplink performance is doubled from the cell mean spectral efficiency point of view with RR scheduler and 2-branch reception. Furthermore, at the same time cell edge performance is tripled. With PF and 4-branch reception, the gain over HSUPA can be further improved.

The average spectral efficiency in different channel models and cell sizes is summarized in Figure 10.16 and the cell edge efficiency in Figure 10.17. The efficiency is higher in small cells than in large cells since the UE runs out of power in large cells. The impact of

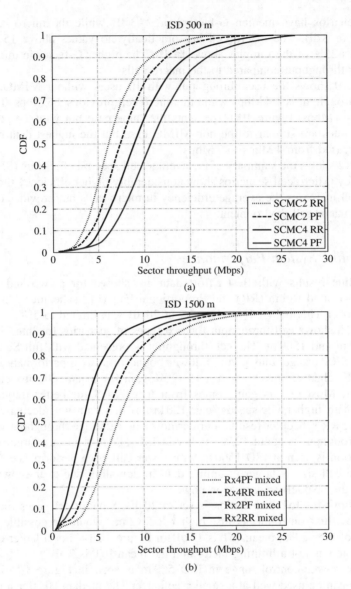

Figure 10.13 Uplink cell throughput for (a) ISD 500 m; (b) ISD 1500 m

the cell size is most clearly visible in cell edge efficiency. We can also see that 4-branch reception and PF scheduling provide higher capacity compared to 2-branch reception and RR scheduling.

10.6.4 Multi-antenna MIMO Evolution Beyond 2 × 2

The downlink LTE MIMO transmission schemes are specified to support up to a four transmit antenna port configuration. A brief summary of these schemes is given in Table 10.17, along with the employed terminology.

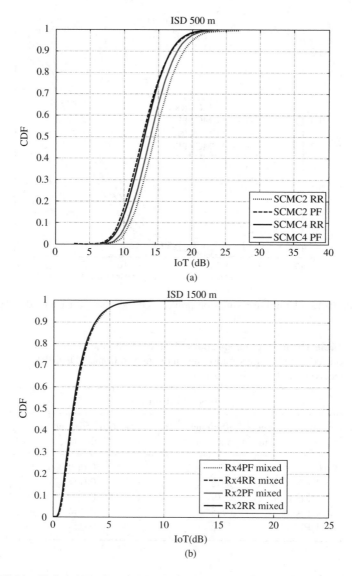

Figure 10.14 Uplink instantaneous Noise Raise for (a) ISD 500 m; (b) ISD 1500 m

The MIMO rank identifies the number of spatial layers (streams) while the MIMO codeword is used to jointly encode up to two spatial layers. In this context the closed-loop MIMO operation is implemented using codebook based pre-coding, i.e. the antenna weights (vectors or matrices) to be applied at the base station side are selected from a fixed number of possibilities, which form a quantized codebook set. These codebook sizes are summarized in Table 10.17. Furthermore, for all the single-user spatial multiplexing MIMO schemes it is assumed that the multiplexed spatial layers are transmitted to the same mobile station.

In a practical system, to ensure proper performance of these MIMO schemes, the overall radio Layer 1 (L1) and Layer 2 (L2) resource management has to include a minimum set

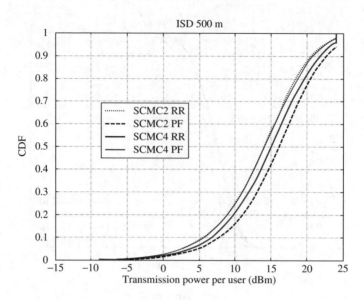

Figure 10.15 Uplink transmission power CDF per user in case of ISD 500 m

Table 10.16 Uplink spectral efficiency at the mean and at the
cell edge for various multi-antenna configurations, FDE/MRC
receivers and RR/PF schedulers

	Best effort traffic	
Features	Spectral efficiency cell mean (b/s/Hz/cell)	Spectral efficiency cell edge (5%-ile) (b/s/Hz/cell)
HSUPA	0.33	0.009
1 × 2 RR	0.66	0.027
1 × 2 PF	0.79	0.033
1 × 4 RR	0.95	0.044
1 × 4 PF	1.12	0.053

of MIMO-aware mechanisms at both eNodeB and UE side. The minimum set of required
feedback information from the UE to support downlink adaptive spatial multiplexing
MIMO transmission comprises:

- channel state/quality information (CQI): direct (e.g. SINR) or indirect (e.g. MCS) information on the average channel conditions estimated on the physical resources and
MIMO codewords;
- MIMO rank information: indicates the optimum number of spatial layers (streams) to
be used and the corresponding pre-coding vector/matrix index (PMI);
- HARQ information: indicates the reception status (ACKnowledged/Not ACKnowledged) of the transmitted data packets on the scheduled MIMO spatial layers.

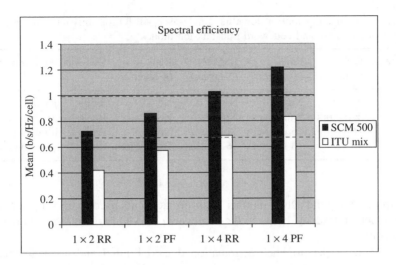

Figure 10.16   Uplink mean spectral efficiency

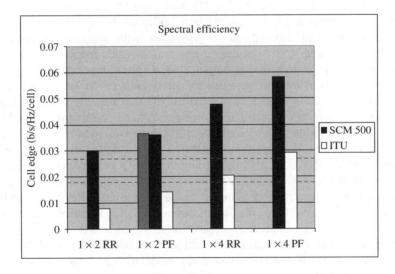

Figure 10.17   Uplink cell edge spectral efficiency

The first two items above, CQI and PMI, are estimated by the mobile station based on the downlink pilot measurements and are transmitted back to the base station in a quantized form. The HARQ mechanism has an impact on the freedom of the radio resource management blocks, when link adaptation and optimized packet scheduling has to be performed for single/dual codeword MIMO transmission.

The mobile station feedback information is heavily dependent on the channel/signal estimation performance, which has an upper bound due to the practical system design limitations (reference signals, measurement/estimation time, propagation delays, etc.). Furthermore, all this feedback information requires uplink control channel capacity; thus it is

Table 10.17 LTE specification for downlink single-user MIMO transmission schemes. MIMO pre-coding codebook size and codeword transmission modes

MIMO scheme	Rank 1 (1 layer)	Rank 2 (2 layers)	Rank 3 (3 layers)	Rank 4 (4 layers)	Codeword transmission mode
2 × 1	4 vectors	No	No	No	Single codeword
2 × 2	4 vectors	3 matrices	No	No	Rank 2 with dual codeword
4 × 2	16 vectors	16 matrices	No	No	Rank 2 with dual codeword
4 × 4	16 vectors	16 matrices	16 matrices	16 matrices	Rank 2–4 with dual codeword

desirable to minimize the actual number of informationbits used to encode them and converge to a practical tradeoff between the achievable MIMO performance and the required uplink capacity.

The system level performance degradation due to limited and noisy feedback from the terminals has been studied and presented in the literature under practical LTE-like assumptions, e.g. [6, 7]. In the following the main concepts related to these performance tradeoffs in simple reference MIMO configuration cases are summarized. The radio resource management aspects are discussed in more detail in Chapter 8.

In theory, the optimal link adaptation operation with spatial-multiplexing MIMO schemes requires separate CQI measures estimated for each separately encoded spatial layer. Without significant loss of performance the CQI feedback has been reduced by grouping the spatial layers into two groups with a maximum of two layers each, so that two CQI measures are sufficient for the considered MIMO transmission schemes. Each group of spatial layers is separately encoded and corresponds to one codeword (see Table 10.17).

The optimized operation of the adaptive MIMO transmission schemes requires, in addition to the classical link adaptation, a MIMO rank adaptation mechanism. A straightforward and simple scheme is with the optimum MIMO rank selected at the terminal side such that it maximizes the instantaneous downlink user throughput. The optimum pre-coding vector/matrix can be determined based on various criteria: maximum beamforming gain, maximum SINR or maximum throughput. For simplicity, here we use the first criterion because this implies the computation of signal power levels only. The latter options are more complex and require information about the number of allocated physical resources blocks, the interference levels and the modulation and coding used.

Based on the instantaneous channel conditions the terminal estimates the optimum MIMO rank, and the corresponding optimum pre-coding vector/matrix. Ideally, both the MIMO rank and the PMI should be estimated for each physical resource block, i.e. assuming frequency-selective MIMO adaptation. For large system bandwidths of 5 MHz and above, in practice this approach leads to a very large signaling overhead introduced in the uplink control channel. A good compromise can be achieved by using MIMO feedback per group of several consecutive physical resource blocks. This approach is further motivated by the choice made for the CQI feedback in the LTE systems. The CQI feedback is described in Chapter 5. The MIMO feedback information is assumed to be transmitted to the serving base station together with, or embedded in, the CQI reports.

In the context of the LTE system, because the same modulation and coding scheme is used on all the allocated physical resource blocks, it is also reasonable to assume that the same MIMO rank can be used for all the allocated physical resource blocks. Furthermore, given that the actual resource allocation is determined by the packet scheduler at the base station, and not at the terminal side, a practical solution is to use one single MIMO rank determined for the entire monitored system bandwidth, i.e. frequency non-selective MIMO rank information is fed back to the base station.

The minimal system-level restrictions described above, still allow for quite a large degree of freedom in the MIMO adaptation mechanisms, in both the time and frequency domains. Some relevant examples are presented in the following.

The frequency granularity of the PMI feedback can be different from the granularity of the MIMO rank. Two reference PMI selection schemes can be assumed combined with the above described full-bandwidth MIMO rank selection:

- Frequency Non-Selective (FNS) PMI: one PMI is determined for the entire effective system bandwidth;
- Frequency Selective (FS) PMI: one PMI is determined for each group of two consecutive physical resource blocks, i.e. using the same $2 \times$ PRB granularity as the CQI measures.

Using the MIMO and CQI feedback information from all the active terminals in the serving cell, the RRM algorithm in the serving base station performs the actual resource allocation and scheduling for each terminal. Previous investigations with low mobility scenarios have disclosed the influence on the overall system performance of the rate at which the MIMO adaptation is performed. The results show that a semi-adaptive scheme with slow update rate ($\sim$100 ms) based on the average channel conditions yields only small cell throughput losses of the order of 5% compared to the case with a fast-adaptive scheme with MIMO rank selected in each scheduling period. Thus, there are two reference schemes:

- G-factor based (GF): with rank update based on the average wideband channel SINR conditions;
- quasi-dynamic (QD): with rank selected only when a new (1st HARQ) transmission is scheduled.

The quasi-dynamic MIMO scheme is a tradeoff solution between the fast-adaptive (per TTI) and the G-factor based (slow, per 5 ms to 10 ms) adaptation schemes, thus it can be successfully used in both medium and low mobility scenarios.

In combination with these two L1–L2 mechanisms – MIMO rank adaptation, and MIMO pre-coding information feedback frequency domain granularity – the downlink adaptive 2×2 and 4×4 MIMO transmission schemes have been simulated. These evaluations assume limited, imperfect and delayed channel feedback from the terminals. The PMI and rank information feedback is assumed to be error free, while the CQI information is modeled with errors introduced by the channel estimation.

Figure 10.18 and Figure 10.19 show representative simulation results for the spectral efficiency in the macro cell case 1 and micro cell deployment scenarios (see section 10.6.1), respectively. The number of mobile stations is 20 simulated per cell/sector. Time–frequency proportional fair packet scheduling is employed.

For both the 2×2 and 4×4 MIMO transmission schemes, the macro cell investigations show gains in the average cell spectral efficiency and the cell-edge spectral

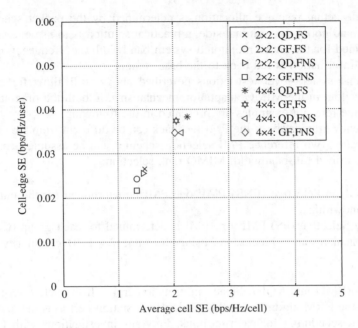

Figure 10.18 Average cell vs cell-edge user (coverage) spectral efficiency simulation results for MIMO schemes operating in the macro cell case 1 scenario

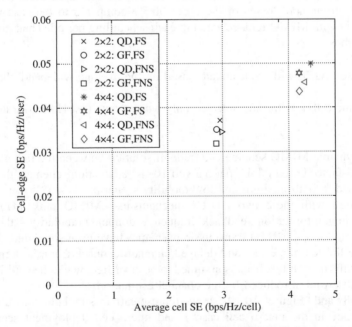

Figure 10.19 Average cell vs cell-edge user (coverage) spectral efficiency simulation results for MIMO schemes operating in the micro cell scenario

efficiency of the order of 10% and 18%, respectively, when using quasi-dynamic MIMO rank adaptation and frequency selective closed-loop information feedback compared to the G-factor based and frequency non-selective cases. The performance in micro cell scenarios is less impacted, with gains of up to 4% and 15%, respectively. The latter results confirm the suitability and high potential of the downlink MIMO schemes in microcellular environments without introducing significant overhead in the uplink signaling and control channels payload.

10.6.5 Higher Order Sectorization (Six Sectors)

One of the methods for increasing the performance by using more antennas at the base station is to use higher order sectorization. The use of higher order sectorization is especially considered to be an option for macro cell installations, where antennas are mounted above rooftops. Typically, three-sector sites are assumed in most of the LTE macro cell performance evaluations, by using three separate panel antennas per site, each with a 3 dB beamwidth of 65 or 70 degrees. A first step could therefore be to increase the sectorization to six sectors per site by simply using six panel antennas with a narrower beamwidth of, for example, 35 degrees. This provides a simple method for increasing the capacity of the network at the sites with high offered traffic. The six sector antennas could also be implemented with three panel antennas and digital fixed beamforming.

The performance improvements of six sectors vs three sectors have been evaluated by means of system simulations in a typical macro cell environment with 500 m site-to-site distance, i.e. similar to 3GPP macro cell case 1 with LTE 10 MHz bandwidth. The following four cases are considered:

- all sites have three sectors with antenna beamwidth of 70 degrees;
- all sites have six sectors with antenna beamwidth of 35 degrees;
- only one centre site has six sectors, while other sites have three sectors (denoted Mode-1);
- a cluster of seven sites has six sectors, while all other sites have three sectors (denoted Mode-2).

Performance results for these four cases are presented in Figure 10.20, where the average throughput per site is reported. These results were obtained with a simple full buffer traffic model, assuming two antennas at the terminals, and standard proportional fair time-frequency packet scheduling. A significant performance gain is achieved by increasing the sectorization from three sectors to six sectors: the site capacity is increased by 88%. If only one site is upgraded to have six sectors (Mode-1), then the capacity of that particular site is increased by 98% compared to using three sectors. The latter result demonstrates that upgrading sites to using higher order sectorization is a possible solution for solving potential capacity bottlenecks that may arise in certain hotspot areas. Assuming that the number of users per site is the same, independent of whether three sectors or six sectors are used, then similar improvements are found in terms of user experienced throughput. However, the use of higher order sectorization does not increase the peak user throughput as is the case for the use of MIMO schemes. Similar capacity gains of six sectors vs three sectors have also been reported in [9] for WCDMA networks. The simulated capacity has considered the downlink direction. It is expected that the gains are also similar in the uplink direction.

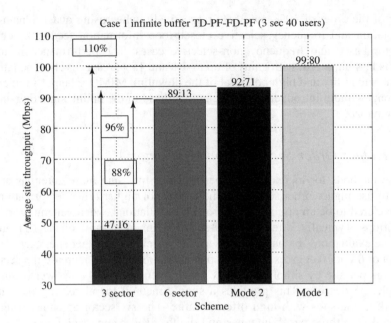

Figure 10.20   Average throughput per macro cell site for different sectorization configurations [8].
© 2008 IEEE

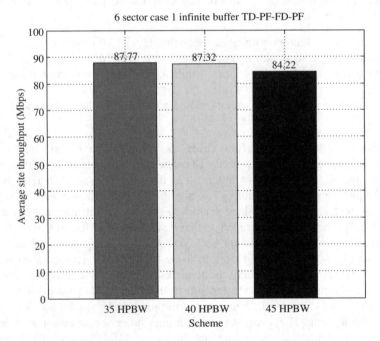

Figure 10.21   Comparison of site throughput for different degrees of azimuth spread for six-sector-site deployment. © 2008 IEEE

The impact of the beamwidth with a six-sector antenna was also studied and the results are shown in Figure 10.21. The motivation for presenting these results is that even though the raw antenna beamwidth is 35 degrees, the effective antenna beamwidth is larger due to the radio channels' azimuthal dispersion [10]. The azimuthal dispersion is found to be of the order of 5–10 degrees for typical urban macro cells. The simulation results show that the average site throughput with antenna beamwidth of 40 degrees is 1% lower compared to 35 degrees, and with 45 degrees it is 4% lower.

The six-sector deployment also improves the network coverage since the typical antenna gain is increased from 18 dBi in three-sector 70-degree antennas to 21 dBi in six-sector 35-degree antennas.

10.6.6 Spectral Efficiency as a Function of LTE Bandwidth

Most LTE system simulations assume 10 MHz bandwidth. The system bandwidth has some impact on the efficiency mainly due to the following factors:

- The frequency domain scheduling gain is higher with larger bandwidth since there is more freedom for the scheduler to optimize the transmission. For 1.4 MHz bandwidth the frequency domain scheduling gain is very low since the whole channel bandwidth likely has flat fading.
- The relative overhead from common channels and control channel – PBCH, Synchronization signal and PUCCH – is lower with larger bandwidths.
- The relative guard band is higher with a 1.4 MHz bandwidth: six resource blocks at 180 kHz equal to 1.08 MHz corresponding to 77% utilization compared to the channel spacing of 1.4 MHz. The utilization is 90% for other bandwidths.

The calculations for the relative efficiency of the bandwidths are shown in Table 10.18 for downlink and in Table 10.19 for uplink. The frequency domain scheduling gain depends on several factors including environment, mobile speed, services and antenna structure. We assume that the scheduling gain in uplink is half of the downlink gain for all bandwidths.

The results are summarized in Figure 10.22. The efficiency is quite similar for the bandwidths of 5 to 20 MHz while there is a 15% difference for 3 MHz. The efficiency with 1.4 MHz bandwidth is approximately 35–40% lower than for 10 MHz. The typical downlink cell throughput for 20 MHz is 1.74 bps/Hz/cell × 20 MHz = 35 Mbps, while for 1.4 MHz the cell throughput is 1.74 bps/Hz/cell × 60% × 1.4 MHz = 1.5 Mbps. The

Table 10.18 Relative efficiency of different LTE bandwidths in downlink

	LTE 1.4 MHz	LTE 3 MHz	LTE 5 MHz	LTE 10 MHz	LTE 20 MHz
Resource blocks	6	15	25	50	100
Guard band overhead	23%	10%	10%	10%	10%
BCH overhead	2.9%	1.1%	0.7%	0.3%	0.2%
SCH overhead	2.5%	1.0%	0.6%	0.3%	0.2%
Frequency domain scheduling gain	5%	20%	35%	40%	45%

Table 10.19 Relative efficiency of different LTE bandwidths in uplink

	LTE 1.4 MHz	LTE 3 MHz	LTE 5 MHz	LTE 10 MHz	LTE 20 MHz
Guard band overhead	23%	10%	10%	10%	10%
PUCCH overhead	16.7%	13.3%	8.0%	8.0%	8.0%
Frequency domain scheduling gain	2.5%	10%	17.5%	20%	22.5%

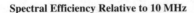

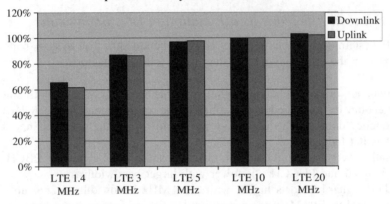

Figure 10.22 Relative spectral efficiency compared to 10 MHz bandwidth in macro cells

conclusion is that LTE should be deployed using as large a bandwidth as possible. The main motivation is to maximize the LTE data rates, but the second motivation is to optimize the spectral efficiency as well. The narrowband options are still useful for refarming purposes if it is not possible to allocate larger bandwidths initially. Even if LTE 1.4 MHz cell throughput of 1.5 Mbps seems relatively low, it is still more than what can be offered by narrowband 2G GPRS/EDGE networks.

10.6.7 Spectral Efficiency Evaluation in 3GPP

The LTE target was to provide spectral efficiency at least three times higher than HSDPA Release 6 in downlink and two times higher than HSUPA Release 6 in uplink. The relative efficiency was studied by system simulations in 3GPP during 2007. The results of the different company contributions are illustrated in Figure 10.23 for downlink and in Figure 10.24 for uplink. These results are for Case 1, which represents interference limited small urban macro cells. The average downlink efficiency increases from HSDPA 0.55 bps/Hz/cell to LTE 1.75 bps/Hz/cell, corresponding to three times higher efficiency. The uplink efficiencies are 0.33 for HSUPA and 0.75 for LTE, corresponding to more than two times higher efficiency. The performance evaluation showed that LTE can fulfill those targets that have been defined for the system in the early phase.

HSPA Release 6 with two-antenna Rake receiver terminals was used as the reference case since it was the latest 3GPP release available at the time when LTE work started.

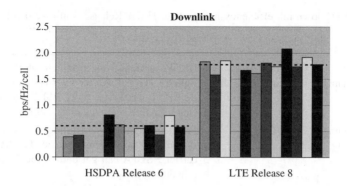

Figure 10.23   Downlink spectral efficiency results from different company contributions [11]

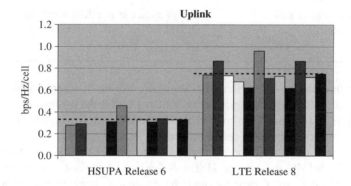

Figure 10.24   Uplink spectral efficiency results from different company contributions [12]

HSPA Release 6 is based on 1×2 maximum ratio combining. LTE Release 8 uses 2×2 MIMO transmission.

10.6.8   Benchmarking LTE to HSPA

The spectral efficiency gain of LTE over HSDPA can be explained by a few characteristics in the LTE system design. Those factors are shown in Table 10.20 for the downlink. LTE uses orthogonal modulation to avoid intra-cell interference providing a major boost in capacity compared to HSDPA with Rake receiver. The CDMA transmission in HSDPA suffers from intra-cell interference caused by multi-path propagation. The CDMA codes are orthogonal but only in the single path channel. The LTE benefit over HSDPA depends on the amount of multi-path propagation.

Another major benefit in LTE is the frequency domain scheduling, which is not possible in a CDMA based system. The CDMA systems transmit the signal using the full bandwidth. MIMO provides some efficiency benefit in LTE since MIMO was not included in HSDPA Release 6. The inter-cell interference rejection combining is used in LTE simulation results and it gives some further capacity boost for LTE. Adding these factors together gives the result that LTE is indeed expected to provide approximately three times higher efficiency than HSDPA Release 6. The LTE performance gain over HSDPA Release 6

Table 10.20 LTE downlink efficiency benefit over HSPA Release 6 in macro cells

LTE benefit	Gain	Explanation
OFDM with frequency domain equalization	Up to +70% depending on the multi-path profile	HSDPA suffers from intra-cell interference for the Rake receiver. Rake receiver is assumed in Release 6. However, most HSDPA terminals have an equalizer that removes most intra-cell interference.
Frequency domain packet scheduling	+40%	Frequency domain scheduling is possible in OFDM system, but not in single carrier HSDPA. The dual carrier HSDPA can get part of the frequency domain scheduling gain.
MIMO	+15%	No MIMO defined in HSDPA Release 6. The gain is relative to single antenna base station transmission. HSDPA Release 7 includes MIMO.
Inter-cell interference rejection combining	+10%	The interference rejection combining works better in OFDM system with long symbols.
Total difference	= 3.0×	$1.7 \times 1.4 \times 1.15 \times 1.1$

will be larger in small micro or indoor cells where the multistream MIMO transmission can be used more often for taking advantage of high LTE data rates.

The LTE uplink efficiency gain over HSUPA mainly comes from the orthogonal uplink while HSUPA suffers from intra-cell interference.

HSPA evolution brings some performance enhancements to HSPA capacity; for details see Chapter 13. MIMO is part of HSPA Release 7; terminal equalizers can remove intra-cell interference in downlink and base station interference cancellation can remove intra-cell interference in uplink. Dual carrier HSDPA transmission is included in Release 8, which enables some frequency domain scheduling gain. Even if all the latest enhancements from HSPA evolution are included, LTE still shows an efficiency benefit over HSPA.

The efficiencies from system simulations are illustrated in Figure 10.25. HSPA Release 7 includes here a UE equalizer and 2×2 MIMO in downlink and base station interference cancellation in uplink. HSPA Release 8 includes dual carrier HSDPA with frequency domain packet scheduling. HSPA Release 9 is expected to allow the combination of MIMO and DC-HSDPA.

10.7 Latency

10.7.1 User Plane Latency

User plane latency is relevant for the performance of many applications. There are several applications that do not require a very high data rate, but they do require very low latency. Such applications are, for example, voice, real time gaming and other interactive applications. The latency can be measured by the time it takes for a small IP packet to travel from the terminal through the network to the internet server, and back. That measure is called round trip time and is illustrated in Figure 10.26.

The end-to-end delay budget is calculated in Table 10.21 and illustrated in Figure 10.27. The 1-ms frame size allows a very low transmission time. On average, the packet needs to

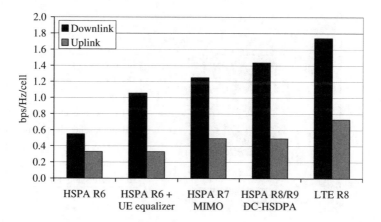

Figure 10.25   Spectral efficiency of HSPA and LTE

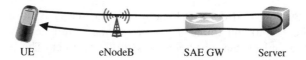

Figure 10.26   Round trip time measurement

Table 10.21   Latency components

Delay component	Delay value
Transmission time uplink + downlink	2 ms
Buffering time (0.5 × transmission time)	$2 \times 0.5 \times 1\,\text{ms} = 1\,\text{ms}$
Retransmissions 10%	$2 \times 0.1 \times 8\,\text{ms} = 1.6\,\text{ms}$
Uplink scheduling request	$0.5 \times 5\,\text{ms} = 2.5\,\text{ms}$
Uplink scheduling grant	4 ms
UE delay estimated	4 ms
eNodeB delay estimated	4 ms
Core network	1 ms
Total delay with pre-allocated resources	13.6 ms
Total delay with scheduling	20.1 ms

wait for 0.5 ms for the start of the next frame. The retransmissions take 8 ms at best and the assumed retransmission probability is 10%. The average delay for sending the scheduling request is 2.5 ms and the scheduling grant 4 ms. We further assume a UE processing delay of 4 ms, an eNodeB processing delay of 4 ms and a core network delay of 1 ms.

The average round trip including retransmission can be clearly below 15 ms if there are pre-allocated resources. If the scheduling delay is included, the delay round trip time will be approximately 20 ms. These round trip time values are low enough even for applications with very tough delay requirements. The practical round trip time in the field

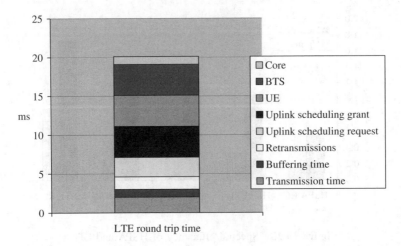

Figure 10.27   End-to-end round trip time including scheduling latency

may be higher if the transport delay is longer, or if the server is far away from the core network. Often the end-to-end round trip time can be dominated by non-radio delays, e.g. by the distance and by the other elements in the internet. The propagation time of 5000 km is more than 20 ms.

10.8   LTE Refarming to GSM Spectrum

LTE will be deployed in the existing GSM spectrum like 900 MHz or 1800 MHz. The flexible LTE bandwidth makes refarming easier than with WCDMA because LTE can start with 1.4 MHz or 3.0 MHz bandwidths and then grow later when the GSM traffic has decreased. The required separation of the LTE carrier to the closest GSM carrier is shown in Table 10.22. The required total spectrum for LTE can be calculated based on the carrier spacing. The coordinated case assumes that LTE and GSM use the same sites while the uncoordinated case assumes that different sites are used for LTE and GSM. The uncoordinated case causes larger power differences between the systems and leads to a larger guard band requirement. The coordinated case values are based on the GSM UE emissions and the uncoordinated values on LTE UE blocking requirements. It may be possible to push the LTE spectrum requirements down further for coordinated deployment depending on the GSM UE power levels and the allowed LTE uplink interference levels. The limiting factor is the maximum allowed interference to the PUCCH RBs that are located at the edge of the carrier.

Table 10.22   Spectrum requirements for LTE refarming

	LTE–GSM carrier spacing		LTE total spectrum requirement	
	Coordinated	Uncoordinated	Coordinated	Uncoordinated
5 MHz LTE (25 RBs)	2.5 MHz	2.7 MHz	4.8 MHz	5.2 MHz
3 MHz LTE (125 RBs)	1.6 MHz	1.7 MHz	3.0 MHz	3.2 MHz
1.4 MHz LTE (6 RBs)	0.8 MHz	0.9 MHz	1.4 MHz	1.6 MHz

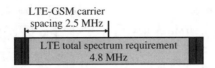

Figure 10.28   LTE 5-MHz refarming example

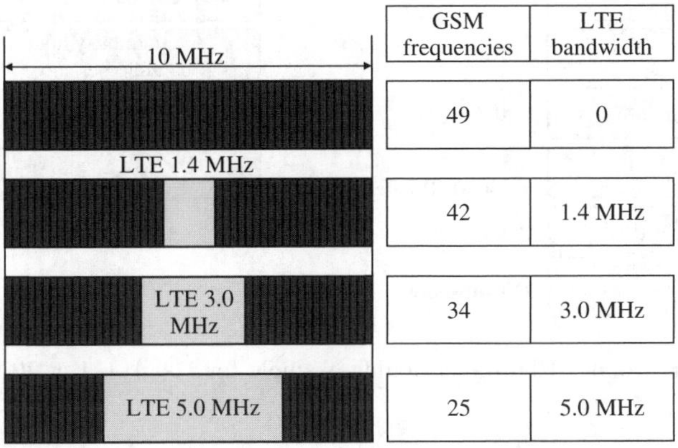

10 MHz	GSM frequencies	LTE bandwidth
	49	0
LTE 1.4 MHz	42	1.4 MHz
LTE 3.0 MHz	34	3.0 MHz
LTE 5.0 MHz	25	5.0 MHz

Figure 10.29   LTE refarming to GSM spectrum

The carrier spacing definition is illustrated in Figure 10.28. Figure 10.29 shows the expansion of the LTE carrier bandwidth when the GSM traffic decreases. Only seven GSM carriers need to be removed to make room for LTE 1.4 MHz and 15 GSM carriers for LTE 3.0 MHz.

10.9   Dimensioning

This section presents examples on how to convert the cell throughput values to the maximum number of broadband subscribers. Figure 10.30 shows two methods: a traffic volume based approach and a data rate based approach. The traffic volume based approach estimates the maximum traffic volume in gigabytes that can be carried by LTE 20 MHz 1 + 1 + 1 configuration. The spectral efficiency is assumed to be 1.74 bps/Hz/cell using 2 × 2 MIMO. The busy hour is assumed to carry 15% of the daily traffic according to Figure 10.30 and the busy hour average loading is 50%. The loading depends on the targeted data rates during the busy hour: the higher the loading, the lower are the data rates. The maximum loading also depends on the applied QoS differentiation strategy: QoS differentiation pushes the loading closer to 100% while still maintaining the data rates for more important connections.

The calculation shows that the total site throughput per month is 4600 GB. To offer 5 GB data for every subscriber per month, the number of subscribers per site will be 920.

Another approach assumes a target of 1 Mbps per subscriber. Since only some of the subscribers are downloading data simultaneously, we can apply an overbooking factor, for example 20. This essentially means that the average busy hour data rate is 50 kbps per subscriber. The number of subscribers per site using this approach is 1050.

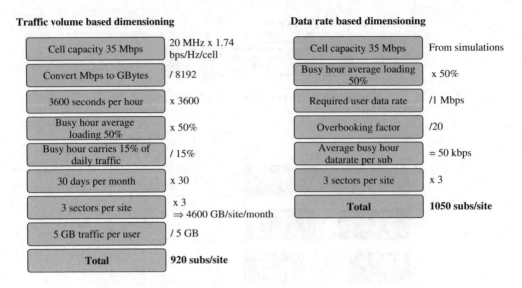

Figure 10.30   LTE dimensioning example for $1 + 1 + 1$ at 20 MHz

The calculation illustrates that LTE has the capability to support a large number of broadband data subscribers.

Figure 10.31 illustrates the technology and spectrum limits for the traffic growth assuming that HSPA and LTE use the existing GSM sites. The starting point is voice only traffic in a GSM network with $12 + 12 + 12$ configuration, which corresponds to a high capacity GSM site found in busy urban areas. This corresponds to $12 \times 8 \times 0.016 = 1.5$ Mbps sector throughput assuming that each time slot carries 16 kbps voice rate. The voice traffic was the dominant part of the network traffic before flat rate HSDPA was launched. The data traffic has already exceeded the voice traffic in many networks in data volume. The second scenario assumes that the total traffic has increased 10 times compared to the voice only case. The sector throughput would then be 15 Mbps, which can be carried with three HSPA carriers using a 15 MHz spectrum.

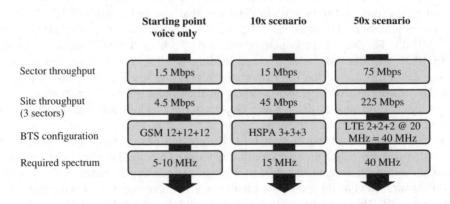

Figure 10.31   Traffic growth scenarios with 10 times and 50 times more traffic

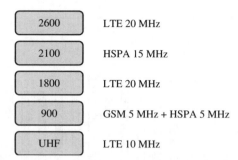

Figure 10.32   Example of a European operator with good spectrum resources

The last scenario assumes 50 times more traffic compared to voice only, which leads to 75 Mbps and can be carried with two LTE carriers each of 20 MHz with a total 40 MHz of spectrum. The site throughput will be beyond 200 Mbps, setting corresponding requirements for the transport network capacity also.

An example of the availability of the spectrum resources by a European operator within a few years is illustrated in Figure 10.32: GSM with 5 MHz, HSPA with 20 MHz and LTE with 50 MHz. Such an amount of spectrum would allow the traffic to increase more than 50 times compared to the voice only scenario. There are many operators in Asia and Latin America with less spectrum resources, which makes it more difficult to provide the high broadband wireless capacities.

10.10   Capacity Management Examples from HSPA Networks

In this section some of the HSDPA traffic analysis in RNC and at the cell level is shown and the implications discussed. It is expected that the analysis from broadband HSDPA networks will also be useful for the dimensioning of broadband LTE networks. All the analysis is based on the statistics of a single RNC and up to 200 NodeBs. The NodeBs were equipped with a 5-code round robin type shared HSDPA scheduler where five codes of HSDPA are shared among three cells and the transport solution was 2*E1 per NodeB. The maximum power for HSDPA was limited to 6 W. This area was selected to be analyzed as in this RNC the RF capability and transport capability are in line with each other, i.e. the transport solution can deliver the same throughput as the shared 5-code HSDPA scheduler. First it is necessary to evaluate the cell level data volume fluctuations and contributions to RNC level total data volume so that the RNC throughput capacity can be checked. Then the cell level data volume and throughput limits are evaluated for when the new throughput improving features (proportional fair scheduler, cell dedicated scheduler, more codes, code multiplexing and so on) for the radio interface are introduced.

10.10.1   Data Volume Analysis

Figure 10.33 shows the data volume distribution over 24h for the RNC on a typical working day. It can be seen that the single hour data volume share is a maximum of 6% from the total daily traffic volume and the fluctuation is just 3% to 6%. Also the hourly data volume share from the busy hour data volume share is 50% during the early morning hours

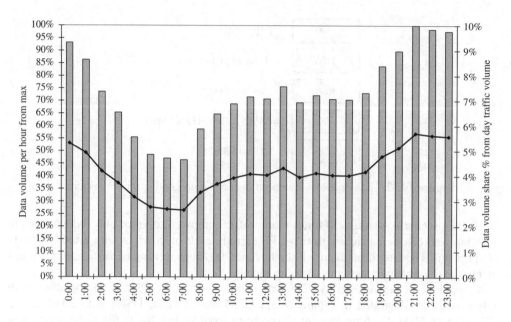

Figure 10.33 Daily RNC level hourly data volume deviation

and steadily increases towards the busiest hours from 8 pm to 1 am. The 3 hours from 9 pm to midnight are the busiest hours during the day. The usage increases heavily after about 6 pm, which indicates that as the working day ends then is the time for internet usage.

Looking into the individual cell contribution to the total RNC level data volume in Figure 10.34, it can be seen that during the night when the data volume is low and

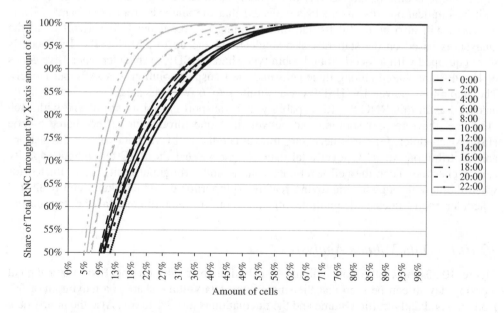

Figure 10.34 Cells' data volume contribution to total RNC data volume

mobility is low, the traffic is also heavily concentrated on certain areas (cells) and during the day the share of cells contributing to the total data volume also increases.

As can be seen from Figure 10.34 during the early morning hours, 10% of the cells contribute 70–85% of the total RNC level data volume whereas during the busiest hours the same 70–85% data volume is contributed by 19–25% of the cells. During the early morning hours the data volume is very concentrated on just a couple of cells, which means that cells covering the residential areas should have very high data volume during the night time and early morning and due to low mobility the channel reservation times should be extremely long. This is shown in Figure 10.35, which indicates that during the early morning hours the average HS-DSCH allocation duration under the cell is a lot longer than during the highest usage. The lack of mobility and potentially some file sharing dominating usage during night time means that certain cells have extremely high data volumes at certain hours only and some cells have a fairly high usage and low data volume variation throughout the whole day. This makes efficient cell dimensioning a challenging task as if capacity is given to a cell according to the busy hour needs, then some cells are totally empty during most times of the day.

In Figure 10.36, the data volume fluctuation per hour is shown as the cumulative distribution of cells having a certain percentage of data volume from the highest data volume during the specific hour. From the graphs it can be seen that during the early morning and low data volume times there is also the highest number of cells having the lowest data volume. This indicates a very high fluctuation of data volume between all the cells.

Figure 10.37 shows the busy hour share of the total daily traffic on the cell level. The cells are sorted according to the data volume: the cells on the left have the highest traffic

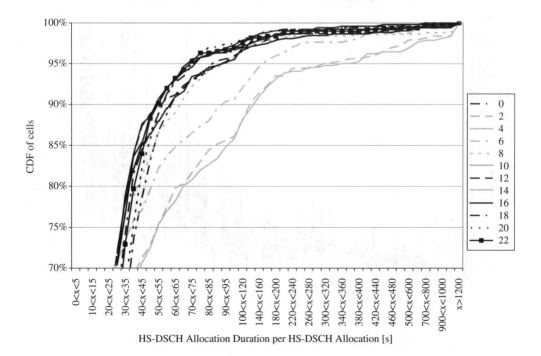

Figure 10.35   Average HS-DSCH allocation duration CDF of cells

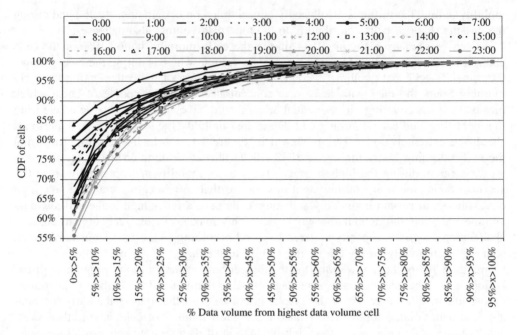

Figure 10.36 Percentage data volume per cell from highest data volume cell per hour

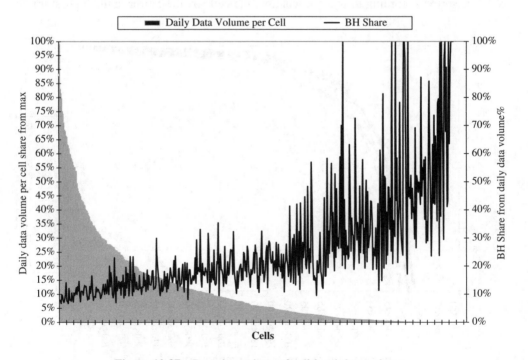

Figure 10.37 Busy hour share of cell level data volume

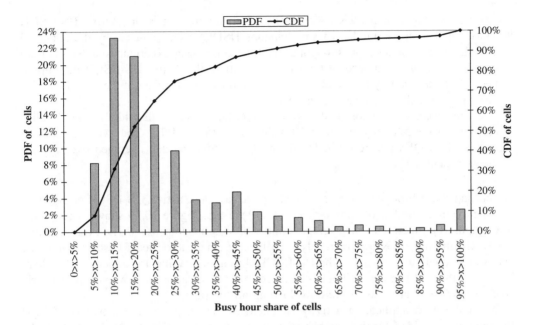

Figure 10.38 CDF and PDF of data volume busy hour share

volume. The busy hour carries 10–20% of the daily traffic in busy cells. Figure 10.38 shows the distribution of the cells depending on the percentage of the data carried by the busy hour. The most typical values are 10–15% and 15–20%. We need to note that the busy hour share on the RNC level was only 6%. The difference between the cell and RNC level traffic distribution is explained by the trunking gain provided by RNC since the traffic is averaged over hundreds of cell.

10.10.2 Cell Performance Analysis

The cell performance analysis is carried out using the key indicators below:

- Active HS-DSCH Throughput – typically given in this analysis as kbps, it is the throughput under a cell when data have been sent in TTIs. Put simply it is the amount of data (kbit)/number of active TTIs (s) averaged over a 1 h measurement period.
- HSDPA Data Volume – typically given in this analysis as Mbyte, kbit or Mbit, it is the amount of data sent per cell during the 1 h measurement period.
- Average number of simultaneous HSDPA users, during HSDPA usage – the amount of simultaneous users during the active TTIs, i.e. how many users are being scheduled during active TTIs per cell (the maximum amount depends on operator purchased feature set). When taking the used application into account, the average number of simultaneous users during HSDPA usage needs to be replaced by the number of active users who have data in the NodeB buffers. Averaged over the 1 h measurement period.
- Allocated TTI share – the average number of active TTIs during the measurement period (i.e. when there are data to send, the TTI is active) over all the possible TTIs during the 1 h measurement period.

- Average Throughput per User – typically given as kbps, it is the Active HS-DSCH Throughput/Average number of simultaneous HSDPA users adjusted by the allocated TTI share; it is the average throughput that one single user experiences under a certain cell. When taking into account the used application, the Active HS-DSCH throughput needs to be divided by the number of active users who have data in the NodeB buffer. Averaged over the 1 h measurement period.
- Average reported Channel Quality Indicator (CQI) – every UE with HS-DSCH allocated measures and reports the CQI value back to the BTS. The average reported CQI is the CQI value reported by all the UEs under the cell averaged over the 1 h measurement period.

The end user throughput depends then on the average active HS-DSCH throughput and the average number of simultaneous HSDPA users. Figure 10.39 shows the average throughput as a function of the average number of simultaneous users. The average active HS-DSCH throughput is approximately 1.2 Mbps. When the allocated TTI share is low, the end user throughput approaches the HS-DSCH throughput. When the allocated TTI share increases, the end user throughput is reduced. This calculation is, however, pessimistic since it assumes that all users are downloading all the time.

When the used application activity is taken into account, i.e. the active HS-DSCH throughput is divided by the number of users who have data in the NodeB buffer, the average throughput per user is quite different from the formula used above. Figure 10.40

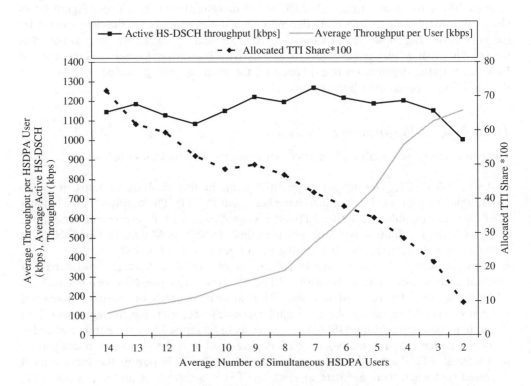

Figure 10.39   Average throughput per user

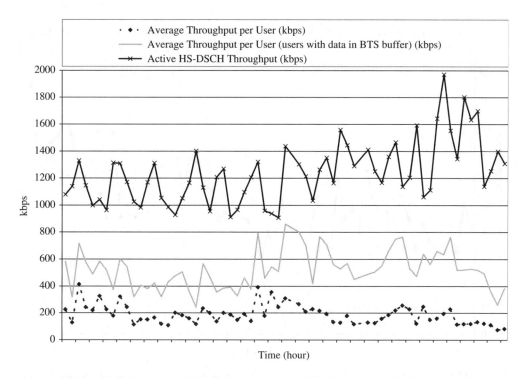

Figure 10.40 Average throughput per user, comparisons of different formulas

shows the comparison of the two different average throughput per user formulas and the Active HS-DSCH throughput. It should be noted that Figure 10.39 and Figure 10.40 are not from exactly the same time. The end user throughput is 400–800 kbps when only those users are considered that have data in the buffer. This can be explained when analyzing the difference between the number of simultaneous users and the number of simultaneous users who have data in the NodeB buffer as shown in Figure 10.41.

Figure 10.41 shows that the used application plays a significant role in the end user throughput evaluation and it should not be ignored. Therefore, the average user throughput may be low because the application simply does not need a high average data rate. The network performance analysis needs to separate the data rate limitations caused by the radio network and the actual used application data rate.

10.11 Summary

3GPP LTE Release 8 enables a peak bit rate of 150 Mbps in downlink and 50 Mbps in uplink with 2×2 MIMO antenna configuration in downlink and 16QAM modulation in uplink. The bit rates can be pushed to 300 Mbps in downlink with 4×4 MIMO and 75 Mbps in uplink with 64QAM. It is expected that the first LTE deployments will provide bit rates up to 150 Mbps.

The LTE link level performance can be modeled with the theoretical Shannon limit when suitable correction factors are included. LTE link level performance was shown to be robust with high mobile speeds, and the uplink LTE performance can be optimized

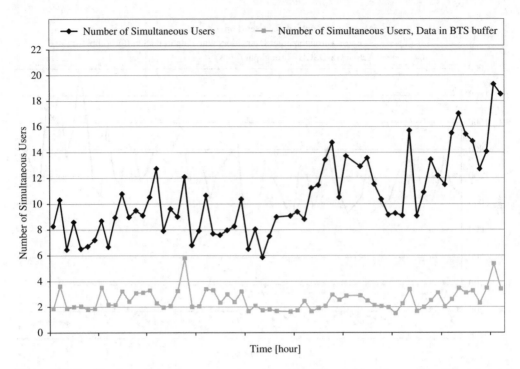

Figure 10.41 Number of simultaneous users with and without taking the used application into account

by using adaptive transmission bandwidth. The link level simulations are used in the link budget calculations, which indicate that the LTE link budget is similar to the HSPA link budget with the same data rate and same spectrum.

The LTE system performance is optimized by orthogonal transmission schemes, by MIMO transmission and by frequency domain scheduling. The spectral efficiency can be further enhanced with multi-antenna transmission and higher order sectorization. The high efficiency can be maintained for different LTE bandwidths between 5 and 20 MHz, while the spectral efficiency is slightly lower with the narrowband LTE carriers 1.4 and 3.0 MHz. It was shown that LTE provides higher spectral efficiency compared to HSPA and HSPA evolution especially due to the frequency domain packet scheduling.

The user plane latency in LTE can be as low as 10–20 ms. The low latency is relevant for improving the end user performance since many applications and protocols benefit from low latency. The low latency is enabled by the short sub-frame size of 1 ms.

The flexible refarming of LTE to the GSM spectrum is enabled by the narrowband options: the refarming can be started with 1.4 or 3.0 MHz and later expanded when the GSM traffic has decreased. All UEs need to support all bandwidths between 1.4 and 20 MHz.

The dimensioning of the broadband wireless networks is different from voice networks. The examples from HSPA networks illustrate that the traffic distribution over the day, over the geographical area and the user mobility need to be considered.

References

[1] 3GPP Technical Specification 36.213 'Physical layer procedures', v. 8.3.0.

[2] 3GPP Technical Specification 36.306 v8.2.0: 'User Equipment (UE) radio access capabilities', August 2008.

[3] P. Mogensen *et al.* 'LTE Capacity compared to the Shannon Bound', *IEEE Proc. Vehicular Technology Conference*, pp. 699–703, April 2007.

[4] 3GPP Technical Specification 25.942 'Radio Frequency (RF) system scenarios', v. 7.0.0.

[5] 3GPP Technical Report 25.996 'Spatial channel model for Multiple Input Multiple Output (MIMO) simulations', V.7.0.0.

[6] I.Z. Kovács *et al.*, 'Effects of Non-Ideal Channel Feedback on Dual-Stream MIMO OFDMA System Performance', *IEEE Proc. Veh. Technol. Conf.*, Oct. 2007

[7] I.Z. Kovács *et al.*, 'Performance of MIMO Aware RRM in Downlink OFDMA', *IEEE Proc. Veh. Technol. Conf.*, pp. 1171–1175, May 2008.

[8] S. Kumar, *et al.*, 'Performance Evaluation of 6-Sector-Site Deployment for Downlink UTRAN Long Term Evolution', *IEEE Proc. Vehicular Technology Conference*, September 2008.

[9] B. Hagerman, D. Imbeni, J. Barta, A. Pollard, R. Wohlmuth, P. Cosimini, 'WCDMA 6 Sector Deployment-Case study of a real installed UMTS-FDD Network', *Vehicular Technology Conference*, Vol. 2, pp. 703–707, Spring 2006.

[10] K.I. Pedersen, P.E. Mogensen, B. Fleury, 'Spatial Channel Characteristics in Outdoor Environments and Their Impact on BS Antenna System Performance', *IEEE Proc. Vehicular Technology Conference*, pp. 719–723, May 1998.

[11] 3GPP TSG RAN R1-072444 'Summary of Downlink Performance Evaluation', May 2007.

[12] 3GPP TSG RAN R1-072261 'LTE Performance Evaluation – Uplink Summary', May 2007.

11

LTE Measurements

Marilynn P. Wylie-Green, Harri Holma, Jussi Reunanen
and Antti Toskala

11.1 Introduction

This chapter presents measurement results using the LTE air interface with 5 to 20 MHz bandwidths in the laboratory and in the field in Europe, the USA and Asia. The carrier frequencies used ranged from 700 MHz to 2600 MHz. The measurements were made during field trials as well as in commercial networks. The aims of this chapter are to introduce measurement procedures and methods of analysis and to illustrate LTE's performance compared to the expected results. The analysis of the results focuses on the peak data rates, link adaptation, multi-user capacity, mobility and latency.

11.2 Theoretical Peak Data Rates

First, we will discuss the methodology used to determine the 3GPP layer 1 reference peak data rates in order to compare the values with the actual layer 1 performance observed in trials.

The achievable peak throughput in downlink and uplink transmissions is limited by the maximum number of bits of a Shared Channel (SCH) transport block that can be accommodated by the LTE UE device during each Transmission Time Interval (TTI). The transport block sizes are summarized in Table 11.1. As defined in 3GPP TS 36.306 [1], the maximum data rate for a Category 2 LTE device is 51.0 Mbps in the downlink and 25.5 Mbps in the uplink. The corresponding data rates for Category 3 and 4 are 102.0 Mbps and 150.8 Mbps in the downlink. The uplink data rate for Categories 3 and 4 is 51.0 Mbps.

The 3GPP peak layer 1 data rate, which serves as the reference for system performance, is determined according to information contained in 3GPP TS 36.213 [1]. That specification document provides uplink and downlink Modulation and Transport Block Size (TBS) index tables, which map each Modulation and Coding Scheme (MCS) to a corresponding TBS index. In addition, TS 36.213 contains a Transport Block Size (TBS) table that can

LTE for UMTS: Evolution to LTE-Advanced, Second Edition. Edited by Harri Holma and Antti Toskala.
© 2011 John Wiley & Sons, Ltd. Published 2011 by John Wiley & Sons, Ltd.

Table 11.1 Downlink physical layer parameters values per UE category (3GPP TS 36.306)

UE Category	Direction	Maximum number of SCH transport block bits received within a TTI	Number of supported layers for spatial multiplexing
Category 2	Downlink	51 024	2
	Uplink	25 456	–
Category 3	Downlink	102 048	2
	Uplink	51 024	–
Category 4	Downlink	150 752	2
	Uplink	51 024	–

be used in order to map the TBS index and the number of Physical Downlink Shared Channel (PDSCH) Physical Resource Blocks (PRB) allocated to a user to the actual TBS per TTI for each codeword. The anticipated peak layer 1 data rate can be calculated as the scaled sum of the total number of bits per TTI. We also add the contributions from both codewords if the dual stream is active. Consequently, the peak data rate, R, in Mbps, is given by

$$R[Mbps] = \frac{(\text{Transport Block Size}[bits/TTI]) \cdot \left(\frac{TTI}{ms}\right) \cdot \left(\frac{1000\,ms}{\sec}\right)}{10^6}$$

The process of mapping MCS ID and the number of PRBs to a peak data rate is summarized in Figure 11.1.

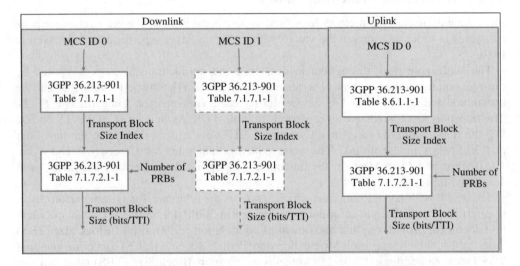

Figure 11.1 Generic procedure for determining expected L1 throughput from (3GPP TS 36.213)

Table 11.2 Layer 1 peak data rates

Description	5 MHz	10 MHz	20 MHz
Maximum resource blocks	DL 25	DL 50	DL 100
	UL 20	UL 45	UL 95
Maximum data rate with Category 4 UE	DL 36.7 Mbps	DL 73.4 Mbps	DL 150.8 Mbps
	UL 10.7 Mbps	UL 19.1 Mbps	UL 40.6 Mbps
Maximum data rate with Category 3 UE	DL 36.7 Mbps	DL 73.4 Mbps	DL 102.0 Mbps
	UL 10.7 Mbps	UL 19.1 Mbps	UL 40.6 Mbps
Maximum data rate with Category 2 UE	DL 36.7 Mbps	DL 51.0 Mbps	DL 51.0 Mbps
	UL 10.7 Mbps	UL 19.1 Mbps	UL 25.5 Mbps

The total number of PRBs in 20 MHz bandwidth is 100 and in 10 MHz bandwidth is 50. The maximum MSC ID was 28 in the downlink in the measurements. The resource manager has assigned the MCS ID in accordance with the radio conditions and the device capability. This flexibility in the assignment of modulation and coding schemes results in a variety of ways in which the system can achieve near peak performance. For example, if 50 PRBs are allocated to a single user on the PDSCH, and the scheduler assigns an MCS ID of 23 to both codewords then, according to TS 36.213, the system can achieve 49.04 Mbps. In a second example, if the prevailing radio conditions or system load warrants the allocation of only 39 PRBs for the PDSCH, then an assignment of MCS ID 27 to both codewords can achieve the same 49.04 Mbps peak layer 1 throughput as in the former example.

In the uplink, the maximum number of PRBs available for a single user on the PUSCH has been 95 with 20 MHz and 45 with 10 MHz. A proportion of the remaining five PRBs have been used for the control channel and random access purposes. Hence, for a 95 PRB PUSCH allocation, the resource manager could allocate a maximum MCS ID of 20 for the uplink transmissions, which would result in a 3GPP layer 1 peak throughput of 40.576 Mbps. The corresponding peak layer 1 throughput with 10 MHz is 19.080 Mbps.

The layer 1 peak data rates are shown in Table 11.2.

11.3 Laboratory Measurements

The configuration used in the laboratory testing has facilitated an analysis of LTE radio link performance over a 20 MHz LTE bandwidth using a Category 4 LTE terminal under static radio conditions. The tests, conducted over the 2 GHz band, have been used in order to demonstrate the peak throughput with 2×2 MIMO that can be achieved under good channel conditions. The configuration settings used in the lab are described in Table 11.3.

The duration of this example test was 30 minutes. Figure 11.2 captures 1 minute of the logged measurements where the peak layer 1 throughput is 149.4 Mbps which is very close to the maximum attainable 3GPP layer 1 upper limit.

Table 11.3 Laboratory settings for single-user throughput testing in the downlink

Description	Settings
Downlink Modulation Schemes	OFDM (64-QAM, 16-QAM, QPSK)
Terminal	LTE UE Category 4 150 Mbps
LTE Bandwidth	20 MHz
Maximum Number of PDSCH PRBs over 20 MHz	100
Maximum MCS ID over 20 MHz	28
Reference Peak Throughput over 20 MHz (Dual Stream MIMO)	150.8 Mbps
MIMO	Open Loop Adaptive
Antenna Configuration	2 × 2

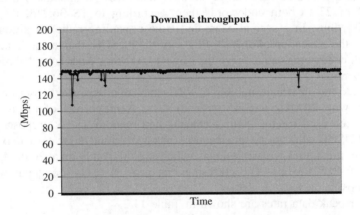

Figure 11.2 Data rate as a function of time with a Category 4 device

11.4 Field Measurement Setups

The field measurements in this chapter are based on trial networks on Bands 3 and 17 with 10 MHz bandwidth in the USA and on commercial network on Band 7 with 20 MHz in Europe. The main parameter settings are show in Table 11.4.

In order to test the air interface capability, the transport network capacity needs to be sufficiently higher in order to avoid transport network limiting the end-to-end performance. Figure 11.3 illustrates the transport network architecture in the field trials. The transport network allows the data from each eNodeB to be routed through the Metro Ethernet MPLS (Multiprotocol Label Switching) network by using a single 100 Mbps/GigE link. The MME, HSS, P-GW and S-GW are located close to the trial area to minimize the latency.

Table 11.4 Parameter settings for field performance testing

Description	Settings
Frequency Bands	Band 4 (trial) Downlink: 2110 – 2155 MHz Uplink: 1710 – 1755 MHz Band 17 (trial) Downlink: 734 – 746 MHz Uplink: 704 – 716 MHz Band 7 (commercial) Downlink: 2620 – 2690 MHz Uplink: 2500 – 2570 MHz
Maximum Number of PDSCH PRBs over 20 MHz	Downlink: 100 Uplink: 95
Maximum Number of PDSCH PRBs over 10 MHz	Downlink: 50 Uplink: 45
Maximum Number of PDSCH PRBs over 5 MHz	Downlink: 25 Uplink: 20
Maximum MCS ID	Downlink: 28 Uplink: 20
Loading level in trial network	Downlink load at 50% Uplink load at 3 dB noise rise level
MIMO	Open Loop Adaptive
Target BLER	10%
Antenna Configuration	Downlink: 2 × 2 Uplink: 1 × 2

11.5 Artificial Load Generation

Network performance needs to be validated both in unloaded and in loaded cases. In order to simplify the practical testing, the load can be generated without real traffic and large number of terminals. In the downlink tests, interference has been created by introducing the desired level of power over the designated percentage of the physical resource blocks. The downlink load has been defined as the percentage of PRBs occupied by the artificial load. The downlink load generation is illustrated in Figure 11.4.

For the uplink loaded scenarios, the solution has been to use a vector signal generator to transmit a pseudo-random sequence on the PUSCH with a determined amount of transmit power. The total amount of transmit power sent by the UE is adjusted according to the pathloss of the simulator location and the victim cell in order to generate the desired noise level at the victim cell. The uplink load generation is illustrated in Figure 11.5.

During the technology trial, both 0% and 50% cell loading has been used for both uplink and downlink test cases.

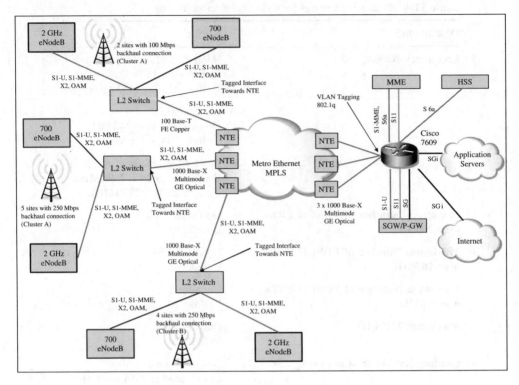

Figure 11.3 Transport Network Architecture (HSS = Home Subscriber Server, MME = Mobility Management Entity, NTE = Network Termination Equipment, OAM = Operation Administration Maintenance, VLAN = Virtual LAN)

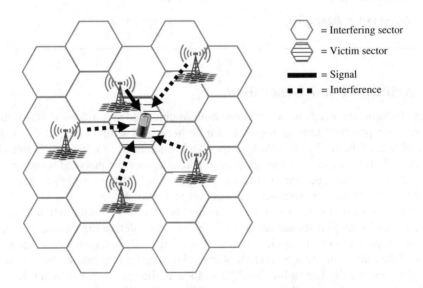

Figure 11.4 Artificial load generation in the downlink

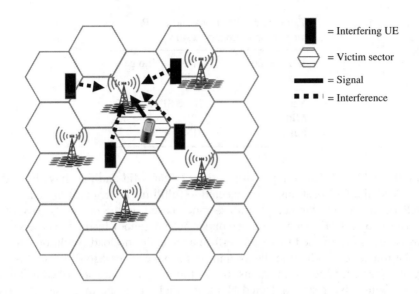

Figure 11.5 Artificial load generation in the uplink

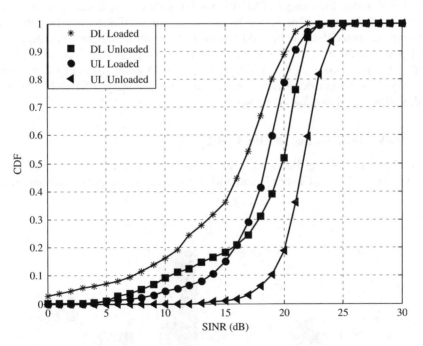

Figure 11.6 Uplink and downlink SINRs measured in the loaded and unloaded scenarios along the drive route in Cluster A

Table 11.5 Positions and SINR ranges for stationary testing

UE Position	SINR Range (dB)
Near	20 < SINR < 25
Mid	10 < SINR < 15
Far	SINR < 5

During the field trials, three types of locations (Near, Mid and Far) have been defined. The Near, Mid and Far locations were selected to fulfill the criterion of high, medium and low SINR (signal-to-interference plus noise ratio) based on the downlink SINR that has been measured by a LTE scanning device under 50% neighbor loading. These same three locations have also been used to assess performance under no load conditions, in order to analyze the impact of loading on the system performance. The SINR thresholds used for defining the Near, Mid and Far locations for all stationary tests are described in Table 11.5.

The data from drive tests conducted along the mobility route in Cluster A have been used in order to calculate the cumulative distribution of the measured SINRs within the two victim sites. The results for the data measured in Cluster A are displayed in Figure 11.6. A rather high range of SINRs is seen within the test area with median values above 15 dB in all cases. This is due in part to the predominance of the suburban-rural propagation characteristics of the field test area. In addition, the network deploys a high transmit power of 60 W at the eNodeB as well as a mast head amplifier, which introduces a significant improvement on the uplink signal power. A comparison of the average SINR in the loaded versus unloaded cases reveals an approximately 3 dB shift of the median for both uplink and downlink scenarios.

11.6   Peak Data Rates in the Field

Figure 11.7 provides a snapshot comparison of the 3GPP reference peak throughput to that achieved with 10 MHz in Cluster B and with 20 MHz in a European commercial network

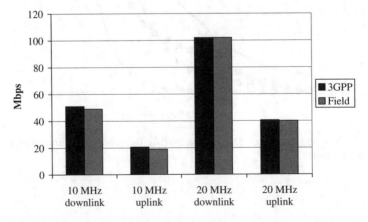

Figure 11.7   Comparison of 3GPP reference L1 peak throughputs to L1 peak throughputs achieved in the field with 10 MHz and 20 MHz bandwidths

during downlink and uplink mobility tests using a Category 3 device. The measured throughput values in the field match the expected peak rates very well.

11.7　Link Adaptation and MIMO Utilization

Link adaptation is a dynamic layer 2 radio resource management operation that determines the best modulation and coding schemes per TTI (1 ms). Based on CQI feedback from the users and QoS requirements, the link adaptation functionality provides information to the packet scheduler regarding the supported modulation and coding schemes for each for the allocated PRBs. However, as discussed in [2], an outer loop link adaptation may also be used in order to control the block error rate of the first transmission based on the positive or negative HARQ acknowledgements from past transmissions. Link adaptation, packet scheduling and HARQ operation are illustrated in Figure 11.8.

Figure 11.9 shows an example of how the link adaptation is used in order to maintain a constant bit rate of 1 Mbps during a drive route in which variable SINR is experienced during the entire test. As shown, the number of resource blocks that are being allocated varies with the radio conditions, as reflected in the fluctuation of the CQI. As the radio conditions become poorer, CQI drops and we observe an increase in the number of PRBs allocated because more robust coding schemes are used. Conversely, as the radio conditions improve and CQI increases, the system allocates fewer PRBs to maintain the desired bit rate. The variations in the transport block size also reflect the system's ability to adapt the data rate based on the prevailing radio conditions during each TTI.

A second example of link adaptation is found for a test case in which the LTE UE is taken along the drive route in Cluster A using full buffer download. Figure 11.10 shows the variation of the throughput as a function of time as well as the corresponding SINR as measured at the antenna connector of the LTE UE. The SINR varies between 4 dB and 23 dB along the drive route. The minimum throughput during the test run is

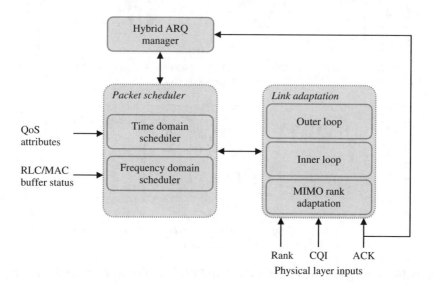

Figure 11.8　Block diagram of the layer 2 functionalities for link adaptation, packet scheduling and Hybrid ARQ management

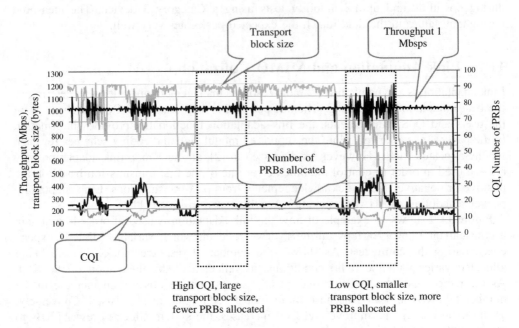

Figure 11.9 Link adaptation in order to maintain a constant data rate of 1 Mbps over a drive route with variable SINR

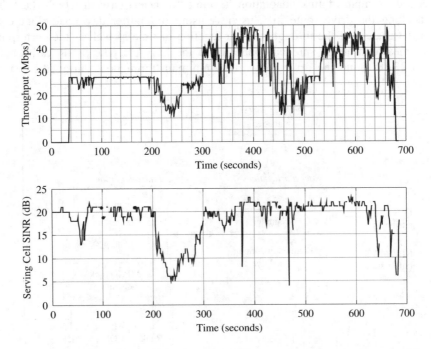

Figure 11.10 Throughput (top) and Serving Cell SINR variation (bottom) along the drive route

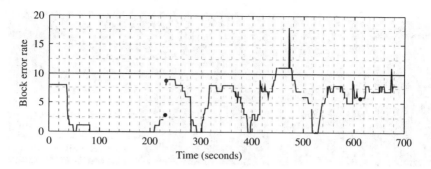

Figure 11.11   Block error rate along the drive route

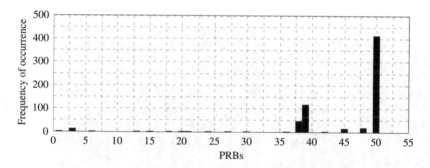

Figure 11.12   Probability distribution of the number of PRBs allocated to the PDSCH

12 Mbps while the maximum is 49 Mbps, which is close to the theoretical peak value with a 10 MHz bandwidth. As shown in Figure 11.11, for the majority of the testing, the BLER is maintained below the 10% target, with a minor excursion above the target when the SINR falls to its minimum value of 4 dB. As the radio conditions deteriorate, the system increases the coding rate and reduces the allocated transport block size. As shown in Figure 11.12, the tendency to assign a similar number of PRBs for the PDSCH (of 50) regardless of the radio conditions is an indication that the eNodeB is attempting to maximize the throughput.

11.8   Handover Performance

This section discusses the handover functionality and performance during the field measurements. These tests have been conducted over both the 2 GHz and 700 MHz bands. The objective of these tests is to study the inter-sector and inter-eNodeB handovers including Radio Resource Control (RRC) messaging and layer 1 and Application Layer throughput before, during and after handover completion.

The handover reporting is typically configured as event triggered. The UE will only send a measurement report when there is a need to make a handover. In this example, Event A3 was used to trigger the handover. Event A3 refers to a neighbor cell becoming offset better than a serving cell. The offset value was defined as 3 dB. When the reporting

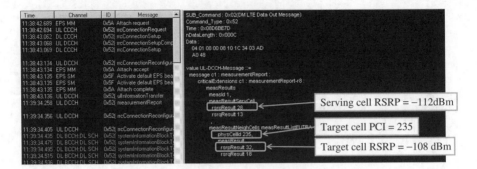

Figure 11.13 Measurement reporting from UE to eNodeB: PCI = Physical Cell Identify; RSRP = Reference Symbol Received Power

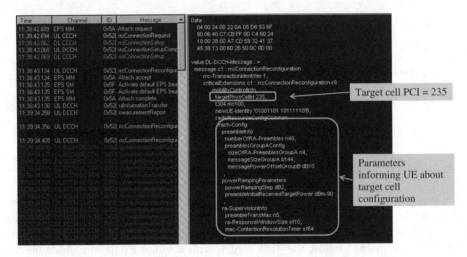

Figure 11.14 RRC Reconfiguration message from eNodeB to UE

threshold is fulfilled, the UE sends a measurement report, which is shown in Figure 11.13. The measurement report gives the RSRP (Reference Symbol Received Power) value and the target cell PCI (Physical Cell Identify). The eNodeB responses with a reconfiguration message, which gives the target PCI and the target cell configuration parameters including RACH information. The reconfiguration message is shown in Figure 11.14. After the UE has completed the radio handover successfully, the UE sends a reconfiguration complete message back to eNodeB. The interruption time between the reconfiguration and the reconfiguration complete message has typically been below 25 ms. An example case with 23 ms delay is shown in Figure 11.15.

The handover performance was tested using a 10 Mbps UDP application and then observing the average data rate during mobility. The average layer 1 data rate was very close to 10 Mbps, except for the high band (Band 4) in the uplink where UE was running

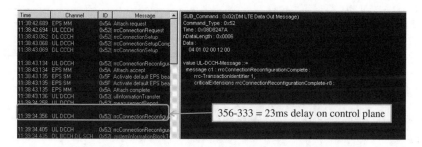

Figure 11.15   Layer 1 interruption time in the handover

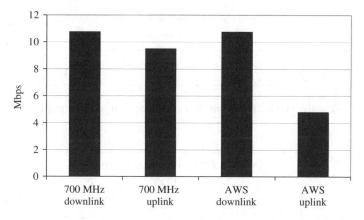

Figure 11.16   Average throughput during mobility and handovers with a 10 Mbps UDP application

out of power and was not able to maintain the target data rate. The summary of layer 1 throughputs achieved during the handover tests is shown in Figure 11.16.

11.9   Data Rates in Drive Tests

An example downlink layer 1 throughput with 20 MHz bandwidth and a Category 3 UE is illustrated in Figure 11.17 and the corresponding SIR in Figure 11.18. When SIR values are higher than 20 dB, the data rate varies between 80 and 100 Mbps. When UE moves further away from the site, SIR gets lower and the data rate is reduced. At SIR of 10 dB, the data rate is 25-40 Mbps. At the very end of the curve SIR is approximately 0 dB which still gives a 15 Mbps throughput.

Figure 11.19 illustrates the average data rates achieved during mobility tests. The average downlink rate in the 50% loaded case with 10 MHz was 29 Mbps and in uplink rate 13 Mbps. The average rate with 20 MHz commercial network was 44 Mbps downlink and 30 Mbps uplink. The fully loaded system simulations in interference limited case with

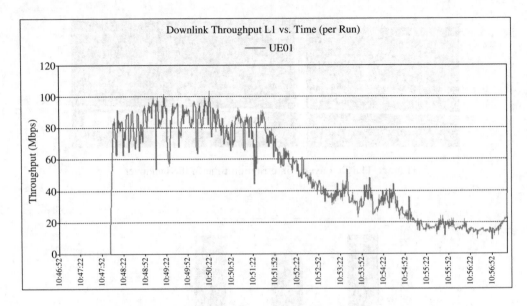

Figure 11.17   Downlink layer 1 throughput in drive testing

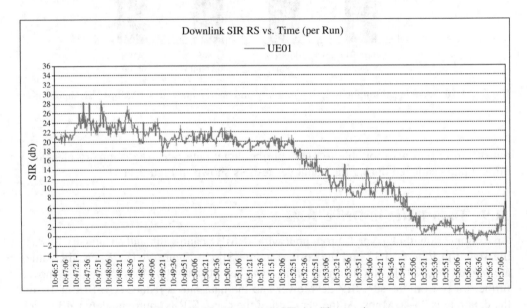

Figure 11.18   Downlink SIR in drive testing

multi-user case typically give 17 Mbps average cell throughput in downlink and 8 Mbps in uplink in 10 MHz bandwidth, see Chapter 10. The throughputs in these measurements are clearly higher because the other cells are not 100% loaded.

The MCS distributions for the downlink along the drive route are shown in Figure 11.20. The most common MCS has been 27, which illustrates the availability of high SNR. Low

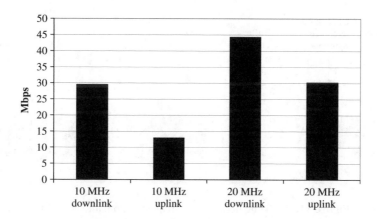

Figure 11.19   Average throughputs in mobility testing with 10 and 20 MHz bandwidths

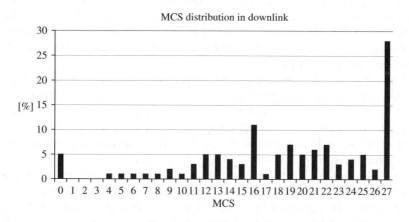

Figure 11.20   Downlink MCS distribution during the drive route in Cluster A

MCS values, below 10, are rare in this case. A MCS of 10 would correspond to 16 Mbps data rate with 10% retransmissions and 100 PRBs. The lower data rates were very rare during the measurements, which matches the MSC distribution.

11.10   Multi-user Packet Scheduling

In the previous sections we reported the performance of the single user throughput test cases. In this section, we focus on multi-user scenarios, in which four LTE UEs are actively transmitting within the cell.

The downlink scheduler provisions for frequency domain scheduling based on channel quality feedback to exploit frequency domain gain. This algorithm is mainly responsible for allocating those PRBs that have the best channel quality for the UE. This is a multiple-step optimization algorithm whereby, at each step, the feasible set of the UEs decreases while the computational complexity increases.

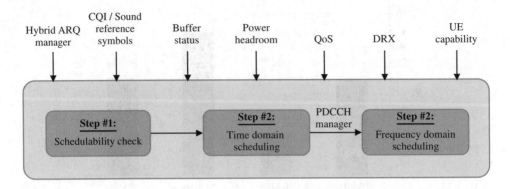

Figure 11.21 Time domain and frequency domain scheduling

The three-step process used to schedule users is generically shown in Figure 11.21. The algorithm begins with a pre-scheduling step. During this step, there is an evaluation to determine if a UE can be scheduled in the sub-frame under consideration. The pre-scheduling step considers data availability, HARQ retransmission, control channel resources and measurement gaps.

In the second phase, the downlink scheduler performs time-domain scheduling. Signaling and HARQ retransmissions will receive higher priority in the time domain scheduling but otherwise the simultaneous users had the same priority in these measurements.

Finally, in the last phase, the scheduler applies frequency domain scheduling based on channel quality feedback. This portion of the algorithm is mainly responsible for selecting those PRBs that have the best channel quality for the UE.

We provide an example of the performance of the downlink and uplink channel aware scheduling algorithms based on a review of the logs collected during this test. As mentioned previously, the downlink resource blocks are scheduled based on the UE's CQI feedback for different PRBs. In what follows, we demonstrate how the eNodeB assigns the PRBs for which the UE has reported the best CQI in order to maximize the downlink cell capacity. In the uplink, the eNB scheduler is unaware of the channel conditions. Consequently, the scheduler assigns the PRBs using a random frequency hopping algorithm.

Figure 11.22 illustrates sub-band and wideband CQI reporting from the UE's measurements. The UE report can indicate CQI values both for the wideband signal and for sub-bands separately. The higher the CQI value, the better is the channel quality and the higher data rate can be supported. The reporting period in this example has been 10 ms.

While the wideband CQI is 15 (its maximum value) in Figure 11.22, the CQI value reported for sub-band 3 is 13. Hence, the scheduler prioritizes other sub-bands ahead of sub-band 3 during this time frame. The prioritization of other sub-bands is seen in Figure 11.23, where we show that there are no PRB assignments made over sub-band 3 for this reporting period.

The uplink direction in these measurements uses channel-unaware pseudorandom hopping. Uplink channel-aware frequency domain scheduling is more difficult than in downlink. The downlink direction can use CQI reporting from the reference symbols but the uplink direction must rely on the sounding reference symbols, which is one symbol wideband transmission from each UE separately. For more details, see Chapter 8.

| Time | Frame | Subframe | Wideband | | Subband 0 | | Subband 1 | | Subband 2 | | Subband 3 | |
			Periodic Report Type	Wideband CQI 1	Subband PMI	Subband CQI	Subband PMI	Subband CQI	Subband PMI	Subband CQI	Subband PMI	Subband CQI
10:14:58:846	158	1	-	-	-	-	-	-	-	-	-	13
10:14:58:856	159	1	-	-	-	-	-	-	-	-	-	-
10:14:58:866	160	1	15	-	-	-	-	-	-	-	-	-
10:14:58:876	161	1	-	-	-	15	-	-	-	-	-	-
10:14:58:886	162	1	-	-	-	-	-	-	-	-	-	13
10:14:58:896	163	1	-	-	-	-	-	-	-	-	-	-
10:14:58:906	164	1	15	-	-	-	-	-	-	-	-	-
10:14:58:916	165	1	-	-	-	15	-	-	-	-	-	-
10:14:58:926	166	1	-	-	-	-	-	-	-	-	-	13
10:14:58:936	167	1	-	-	-	-	-	-	-	-	-	-
10:14:58:946	168	1	15	-	-	-	-	-	-	-	-	-
10:14:58:956	169	1	-	-	-	15	-	-	-	-	-	-

Figure 11.22 Wideband and sub-band CQI reporting from UE

| | | PRB# | 49 | 48 | 47 | 46 | 45 | 44 | 43 | 42 | 41 | 40 | 39 | 38 | 37 | 36 | 35 | 34 | 33 | 32 | 31 | 30 | 29 | 28 | 27 | 26 | 25 | 24 | 23 | 22 | 21 | 20 | 19 | 18 | 17 | 16 | 15 | 14 | 13 |
SFN	Subframe	Subbands	8					7					6					5					4					3					2						
191	9	PRB's"Assinged	0	0	0	0	0	1	1	1	0	0	0	1	1	1	1	1	1	1	1	1	0	0	0	1	1	1	0	0	0	0	0	0	1	1	1	1	
192	1	PRB's"Assinged	0	0	0	0	0	1	1	1	0	0	0	1	1	1	1	1	1	1	1	1	0	0	0	1	1	1	0	0	0	0	0	0	1	1	1	1	
192	3	PRB's"Assinged	0	0	0	0	0	1	1	1	0	0	0	1	1	1	1	1	1	1	1	1	0	0	0	1	1	1	0	0	0	0	0	0	1	1	1	1	
192	5	PRB's"Assinged	0	0	0	0	0	1	1	1	0	0	0	1	1	1	1	1	1	1	1	1	0	0	0	1	1	0	0	0	0	0	0	1	1	1	0	0	

Figure 11.23 Due to the low CQI reported in sub-band 3 during the current period, the eNodeB does not schedule any resource blocks over this sub-band

#State	Time	SFN	Subframe	UL Hard	UL RB H	CQI rec	UL RB Origin	UL RB Number	Grant R	UL-SCH	UL-SCH	UL-SCH	D
0	10:28:15:1	938	6	0	-	-		20	1	-	QPSK	-	17 -
0	10:28:15:1	939	3	7	-	-		10	1	-	QPSK	-	17 -
0	10:28:15:1	940	0	6	-	-		10	1	-	QPSK	-	17 -
0	10:28:15:2	948	6	4	-	-		20	1	-	QPSK	-	17 -
0	10:28:15:2	949	3	3	-	-		11	1	-	QPSK	-	17 -
0	10:28:15:2	950	0	2	-	-	Randomly	17	1	-	QPSK	-	17 -
0	10:28:15:2	950	7	1	-	-	Generated	24	1	-	QPSK	-	17 -
0	10:28:15:3	951	4	0	-	-		32	1	-	QPSK	-	17 -
0	10:28:15:3	960	0	6	-	-		26	1	-	QPSK	-	17 -
0	10:28:15:3	960	7	5	-	-		47	1	-	QPSK	-	17 -
0	10:28:15:4	961	4	4	-	-		31	1	-	QPSK	-	17 -
0	10:28:15:4	962	1	3	-	-		9	1	-	QPSK	-	17 -
0	10:28:15:4	962	8	2	-	-		8	1	-	QPSK	-	17 -
0	10:28:15:5	971	4	0	-	-		30	1	-	QPSK	-	17 -
0	10:28:15:5	972	1	7	-	-		37	1	-	QPSK	-	17 -
0	10:28:15:5	972	8	6	-	-		46	1	-	QPSK	-	17 -
0	10:28:15:5	973	5	5	-	-		45	1	-	QPSK	-	18 -
0	10:28:15:5	974	2	4	-	-		21	1	-	QPSK	-	18 -
0	10:28:15:6	982	8	2	-	-		42	1	-	QPSK	-	18 -

Figure 11.24 A single PRB is allocated to the UE in the uplink, however, the PRB number is allowed to vary pseudorandomly

Figure 11.24 illustrates the uplink allocation. The UE is allocated one PRB and the allocation varies randomly from sub-frame to another.

11.11   Latency

Ping tests measure latency by determining the time it takes a given packet to travel from laptop via UE, LTE radio network and core network to the server and back, measuring, for example, the speed of information traveling from a computer to a game server and back. The expected latency is shown in Figure 11.25: approximately 20 ms for small packet without any retransmissions. We assume a 4 ms processing delay in UE, 4 ms in eNodeB, 1 ms in the core network and a 10 ms scheduling period.

The layer 1 retransmission takes 8 ms. The average BLER of 10% leads to an increase of 0.8 ms average additional latency both in the uplink and in the downlink, and a total of 1.6 ms, which makes the expected ping latency 22 ms including the retransmissions.

The latency will be reduced if the network has a pre-allocated resource for the UE, which then avoids the latency caused by scheduling request and uplink grant. Pre-allocation reduces latency by 9 ms, which makes the best ping latency approximately 10 ms without retransmissions. The LTE latency is discussed in more detail in Chapter 10.

Example ping measurements with pre-allocated resources are shown in Figure 11.26, which gives an example of round-trip time measurements showing retransmissions. The end-to-end latency is 11 ms without retransmissions. The retransmission adds 8 ms more latency. The minimum latency was 9.4 ms and the average 11.4 ms.

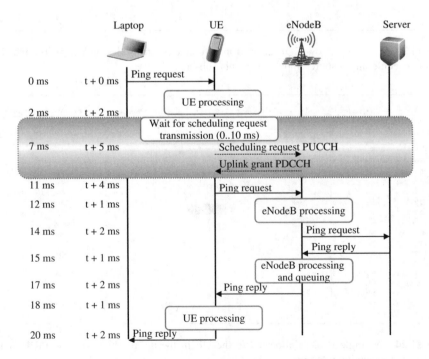

Figure 11.25   User plane latency for a small ping without retransmissions

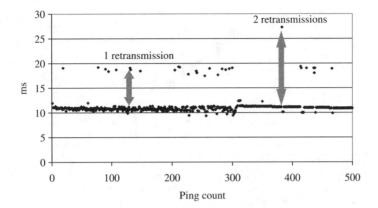

Figure 11.26   Example round trip time measurements showing retransmissions

```
Reply from 10.21.128.123: bytes=32 time=10ms TTL=63
Reply from 10.21.128.123: bytes=32 time=10ms TTL=63
Reply from 10.21.128.123: bytes=32 time=10ms TTL=63
Reply from 10.21.128.123: bytes=32 time=10ms TTL=63
Reply from 10.21.128.123: bytes=32 time=9ms TTL=63
Reply from 10.21.128.123: bytes=32 time=9ms TTL=63
Reply from 10.21.128.123: bytes=32 time=9ms TTL=63
Reply from 10.21.128.123: bytes=32 time=10ms TTL=63
Reply from 10.21.128.123: bytes=32 time=10ms TTL=63
Reply from 10.21.128.123: bytes=32 time=10ms TTL=63
Reply from 10.21.128.123: bytes=32 time=10ms TTL=63
Reply from 10.21.128.123: bytes=32 time=11ms TTL=63
Reply from 10.21.128.123: bytes=32 time=10ms TTL=63
Reply from 10.21.128.123: bytes=32 time=10ms TTL=63
Reply from 10.21.128.123: bytes=32 time=10ms TTL=63
Reply from 10.21.128.123: bytes=32 time=10ms TTL=63
Reply from 10.21.128.123: bytes=32 time=10ms TTL=63
Reply from 10.21.128.123: bytes=32 time=10ms TTL=63
Reply from 10.21.128.123: bytes=32 time=10ms TTL=63
Reply from 10.21.128.123: bytes=32 time=10ms TTL=63
Reply from 10.21.128.123: bytes=32 time=10ms TTL=63
Reply from 10.21.128.123: bytes=32 time=10ms TTL=63
Reply from 10.21.128.123: bytes=32 time=10ms TTL=63
Reply from 10.21.128.123: bytes=32 time=10ms TTL=63
```

Figure 11.27   Example of round-trip time measurements

A screen shot from ping measurements is illustrated in Figure 11.27, which shows an example of round-trip time measurements. The ping values vary between 9 and 10 ms without retransmissions. Overall, we can conclude that the measured end-to-end latency matches expectations.

11.12   Very Large Cell Size

LTE radio can be efficiently utilized to provide broadband wireless services in rural areas. The cell radius can be very large if the propagation is favorable by having high mounted base station antennas, by using external antennas in UE and by using low frequency. A test case is illustrated in this section showing the link performance with a 75 km cell range. The map of the test area in rural Australia is shown in Figure 11.28. The test was

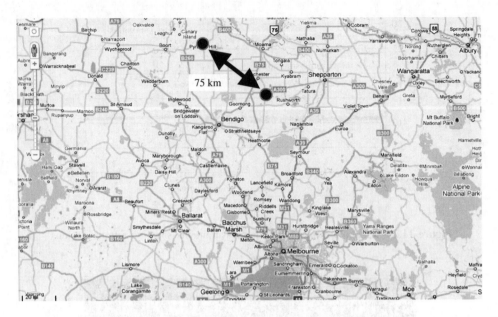

Figure 11.28   Measurement area for testing 75 km cell range

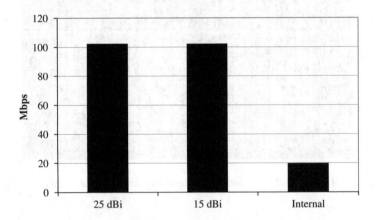

Figure 11.29   Downlink peak throughput with 75 km cell range with different UE antennas

done using 2600 MHz frequency. The propagation would naturally be better if using lower frequency. The base station was configured with 2×60 W power amplifiers and 18 dBi antennas. The measurements were done by using different antennas in the terminal: 25 dBi high gain directive antenna, 15 dBi directive antenna and UE internal antenna. The peak data rates are shown in Figure 11.29. The directive antennas provided the peak rate of 100 Mbps while the UE internal antenna allowed data rate of 20 Mbps. The corresponding uplink data rates in the measurements have been 15-30 Mbps.

11.13 Summary

This chapter illustrated LTE measurement procedures and tools as well as measurement results. The measurement results match nicely with theory and simulations. The peak rate of 150 Mbps was achieved with Category 4 UE with a 20 MHz bandwidth. The measured end-to-end latency is 10 ms. The multi-cell drive tests show close 30 Mbps average data rate with a 10 MHz bandwidth and 40–50 Mbps with 20 MHz. The packet scheduler allocation was also studied in the measurements and it was shown that the UE CQI based frequency domain scheduler can improve the system efficiency. The measurements also illustrated a 75 km connection with 100 Mbps by using external antennas.

References

[1] 3GPP, Technical Specification, TS 36.213-901, 'Group Radio Access Network; Evolved Universal Terrestrial Radio Access (E-UTRA); Physical Layer Procedures' (Release 9), Version 9.0.1, December 2009.

[2] K. Pedersen, T. Kolding, F. Frederiksen, I. Kovacs, D. Laselva and P. Mogensen, 'An overview of downlink radio resource management for UTRAN Long-Term Evolution', IEEE Communications Magazine, pp. 86–93, July 2009.

12

Transport

Torsten Musiol

12.1 Introduction

Radio Transport, also known as Mobile Backhaul, refers to solutions for transferring data between base stations and the core network. As air interface data rates and capacities are increasing with LTE, the requirements on the transport network are also increasing. Transport solutions may cause a substantial part of network operating expenses and in some cases the transport network may be the bottleneck from an end-to-end point of view. So, transport optimization may be even more important than air-interface capacity optimization. The transport network can utilize a number of different physical media. Globally, more than 50% of the base stations are connected using microwave radio technology. Fiber, today, constitutes only a small part of transport connections but the share is increasing, especially in urban areas with high LTE capacity requirements. In turn, the share of solutions based on copper lines for connecting macro base stations is decreasing. However, advanced DSL technologies will have their position mainly for connecting femto base stations. Whereas the Radio Network Layer (RNL) is being specified by 3GPP, as shown in Chapter 3 with the presentation of the protocol stacks for different interfaces, the Transport Network Layer (TNL) for LTE is nowhere defined consistently. Mostly reference is made to protocols outside 3GPP. Several standardization bodies are contributing, among them IETF (Internet Engineering Task Force), IEEE (Institute of Electrical and Electronics Engineers), ITU-T (International Telecommunication Union Telecommunication) and MEF (Metro Ethernet Forum). This chapter provides an overview of generic LTE transport concepts and issues, with special focus on the E-UTRAN, i.e. mobile backhaul solutions for LTE base stations.

12.2 Protocol Stacks and Interfaces

12.2.1 Functional Planes

All protocol stacks for user (U), control (C), management (M) and (optionally) synchronization (S) planes are based on IP. A common IP transport infrastructure can be used

LTE for UMTS: Evolution to LTE-Advanced, Second Edition. Edited by Harri Holma and Antti Toskala.
© 2011 John Wiley & Sons, Ltd. Published 2011 by John Wiley & Sons, Ltd.

for all functional planes but some specific requirements need to be considered for the network design.

User plane
The U-plane service layer is IP. UE mobility is achieved through tunnels, which originate from the Packet Data Network Gateway (PDN-GW) and are anchored at the Serving Gateway (S-GW). Within the E-UTRAN, the GPRS Tunneling Protocol (GTP-U) is used as tunneling protocol. See Figure 12.1.

Control plane
Within the E-UTRAN, SCTP (Stream Control Transmission Protocol) carries the control plane protocols ('Application Protocols') S1-AP and X2-AP. SCTP has been used previously to carry NBAP (Node B Application Part) over IP (Iub/IP). See Figure 12.2.

Management plane
O&M traffic, also known as Management Plane (M-plane) is usually vendor proprietary. See Figure 12.3.

Synchronization plane
A fourth functional plane will be introduced here: the Synchronization Plane (S-plane), see Figure 12.4. Terminated by respective applications in the network (synchronization

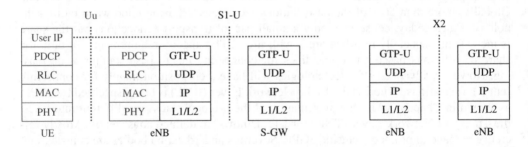

Figure 12.1   User-plane protocol stacks

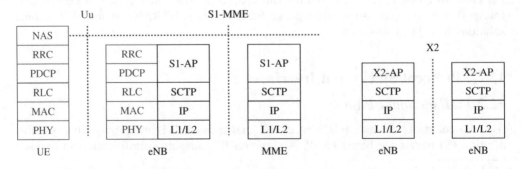

Figure 12.2   Control-plane protocol stacks

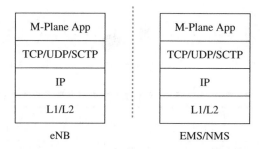

Figure 12.3 Management-plane protocol stack

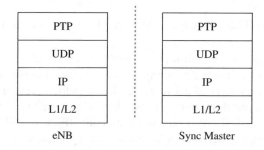

Figure 12.4 Synchronization-plane protocol stack

grandmaster) and in the base station (synchronization slave), this concept can easily be understood because synchronization traffic flowing over the IP transport network has its own set of quality of service (QoS) requirements and should be distinguished from U/C/M-plane traffic correspondingly (see section 12.8.1).

12.2.2 Network Layer (L3) – IP

According to [1] and [2], the eNB Radio Network Layer (RNL) protocols are running on top of IP, which is part of the Transport Network Layer (TNL). This layer should be called TNL IP layer in the following. In contrast to that, user applications are hooked up on the User IP layer, which is independent from the TNL IP layer. GTP-U/UDP is used as a tunneling mechanism in between. This traffic may be put into yet another tunnel if IPsec (tunnel mode) is used for transport security purposes (see section 12.7). So, there may be up to three IP layers, which are all independent from each other, see Figure 12.5:

- the User IP layer, terminated by user application at the UE and corresponding peer;
- the TNL IP layer (LTE application IP layer), terminated by LTE/SAE network element applications;
- IPsec tunnel IP layer (optional), terminated by Security Gateway (SEG) functions (integrated in the eNodeB, external device at the other end).

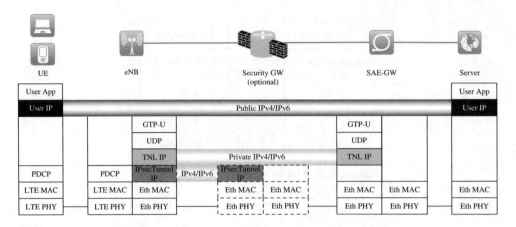

Figure 12.5   User IPv4/IPv6 over LTE transport IPv4/IPv6

The GTP-U tunneling concept allows the use of private IP addresses in the LTE transport network (eNodeB backhaul) while user applications require a public IP address. There are more than 16 million private IPv4 addresses available, which are even reusable in distinct parts of the operator network if routing in between is not needed. This is particularly the case with respect to the TNL IP layer (and optionally IPsec tunnel IP layer). Sooner or later, IPv6 in the transport network will become an operational cost-optimization topic for network elements that anyway have to have IPv6 addresses due to support for IPv6 in the User IP layer (e.g. PDN-GW). It can be expected that these network elements would be simpler to manage if network protocols are harmonized. So, if availability of private IPv4 addresses is no issue, migration towards IPv6 will be driven from the core network later anyway.

12.2.3   Data Link Layer (L2) – Ethernet

Due to highly scalable capacity and favorable economics, Ethernet based on the [3] standard has become the *de facto* standard for eNodeB interfaces. It should be noted that Ethernet connectivity services, based on Ethernet carrier grade equipment, have been originally developed for fixed network applications, where the amount of traffic is much higher than the mobile traffic.

The concept of Virtual LAN's (VLAN) according to [4] is commonly used to separate (differentiate) traffic terminated at a single physical Ethernet interface. This could be applied, for example, to separate traffic from/to different base stations at an aggregation point (edge router), or to separate different types of base station traffic (U/C/M/S-plane). Apart from traffic separation, the VLAN header also supports traffic prioritization using the Ethernet p-bits as defined in [5].

Physical Ethernet interfaces may support data rates (100 Mbps or 1000 Mbps) and burst sizes that are higher than the capacity of the Ethernet backhaul link (for example, leased line or capacity limited microwave radio link). Traffic shaping is required to ensure conformance to the link capacity in order to minimize packet loss. This applies in particular to Ethernet leased lines where the link capacity is governed by a Service Level Agreement (SLA).

Traffic shaping reduces the burstiness of the traffic in conformance with a given maximum average bit rate and maximum burst size. If packets need to be dropped despite shaping, this should be done based on priorities (QoS aware). Shaping can be applied per L2 interface, i.e. per physical Ethernet interface or per VLAN. This means that traffic could be shaped per traffic type (U/C/M/S-plane) if VLAN differentiation is used accordingly.

12.2.4 Physical Layer (L1) – Ethernet Over Any Media

Usually, LTE base stations support various physical interface types. Examples are Fast Ethernet (FE) 100Base-Tx and Gigabit Ethernet (GE) 100/1000Base-T for electrical transmission, or 1000Base-SX/LX/ZX for optical transmission via fiber.

A variety of physical media (with many sub-variants) are available to carry Ethernet frames from/to LTE base stations. With direct fiber access to the eNodeB site, capacities are virtually unlimited. Due to high deployment costs, fiber penetration is still rather low, but will be growing as capacity demands increase. Microwave radio (MWR)-based transmission has played a major role in 2G/3G mobile backhaul, and will continue to do so for LTE. Current MWR capacities are typically up to 300 Mbps and use licensed bands from 6 GHz to 38 GHz. Hop-lengths are in the range of 20–50 km for low-MW-band links and 5–15 km for high-frequency links. The hop length also depends on capacity, geographical location (rain zone) and antenna size. Capacities even up to 10 Gbps can be provided in E-band (70–90 GHz). However, hop-lengths are limited to 1–3 km. An example microwave product is shown in Figure 12.6. Copper based transmission (xDSL) has inherent capacity limitations, but can be used if the lines are short.

For media adaptation, but also for network demarcation purposes, a network terminating device (xDSL CPE, MWR IDU, leased line termination equipment, and so forth) is often present at the eNodeB site, which connects to the eNodeB via Ethernet.

Traditionally used PDH (E1/T1) based transport (IP over PDH or Ethernet over PDH) is too capacity limited for LTE, so will be used only in exceptional cases, in particular if only limited radio interface bandwidth is available.

Figure 12.6   An example 1 Gbps microwave link implementation

12.2.5   Maximum Transmission Unit Size Issues

Both GTP/UDP/IP and IPsec tunneling (optional) add overheads to the user IP packet. If this has been generated with an Maximum Transmission Unit (MTU) size of 1500 bytes, corresponding to the maximum SDU size of a standard Ethernet frame between internet routers, it will not fit to the Ethernet interface SDU of the mobile backhaul network. There are multiple ways to solve this problem:

- enforcement of smaller MTU size at the user IP layer;
- IP fragmentation and reassembly at the TNL IP layer;
- enlargement of Ethernet SDU using Jumbo Frames under the TNL IP layer (or IPsec tunnel IP layer, respectively);
- MSS Clamping for TCP user traffic (TCP Maximum Segment Size (MSS) adjusted by a router, effectively similar to enforcing a smaller MTU size at the user IP layer).

All solutions have their advantages and disadvantages. Based on existing standards, a smaller MTU size cannot be enforced in all circumstances. Ethernet Jumbo Frames would solve the problem nicely but are not available everywhere. IP fragmentation and reassembly would always work but has performance issues.

Enforcement of smaller MTU size at the user IP layer

The MTU size used for packets at the user IP layer can be enforced through static or dynamic configuration of the IP protocol stack SW of the UE. With one approach, the MTU size of the IP protocol stack in the UE could be configured to a lower value. There are UE vendor-specific settings that support this idea but this is not mandated by any standard. Figure 12.7 illustrates that the MTU size at the user IP layer needs to be reduced to 1394 bytes with respect to IPv4 in the transport layers (1358 bytes with IPv6 transport).

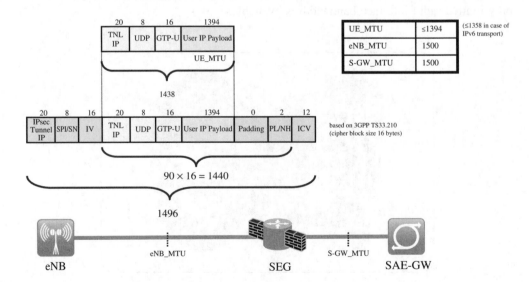

Figure 12.7   Lower MTU size at the user IP layer

In case of IPv6 in the user IP layer, the MTU size at the UE could be set with the Router Advertisement message that the PDN-GW sends after PDN connection establishment ([6], section 4.6.4).

Alternatively, the user IP layer MTU could also be enforced through Path MTU discovery with a lower MTU properly set in the PDN-GW. Unfortunately, Path MTU discovery (IPv4: RFC1191, MTU $\geq$ 68, IPv6: RFC1981, MTU $\geq$ 1280) is not mandated for either IPv4 or IPv6, although it is strongly recommended. It is usually performed for TCP/SCTP, but not for UDP-based applications.

The impact of a lower MTU size on the transport efficiency is negligible (<1%).

IP fragmentation and reassembly

Without using IPsec, Fragmentation and Reassembly of TNL IP layer packets are terminated at the eNodeB at one end and the SAE-GW at the other end. With IPsec, the same approach should be applied in order to avoid multiple fragmentation. In this case IPsec packets are carrying fragments, but are not fragmented themselves. The MTU size (S-GW_MTU) would need to be set up to a maximum of 1438 bytes in order not to exceed a value of 1500 for the Ethernet SDU at the eNodeB. For an example with IPv4 S1-U over IPv4 IPsec tunnel, see Figure 12.8. While this is a well standardized and generic solution, there are potential drawbacks, such as additional processing delay and throughput degradation due to processing load at the end points. The impact on transport efficiency is minor ($\sim$2.5%). Large packets ($\sim$25%) will be fragmented – they will cause double transport overheads ($\sim$10% per fragment).

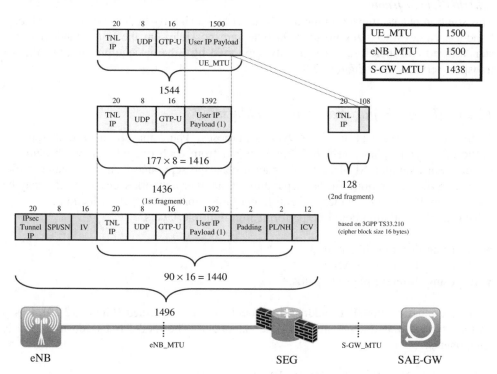

Figure 12.8 Fragmentation and Reassembly with IPv4

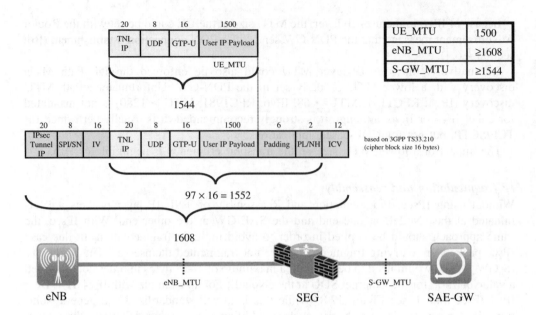

Figure 12.9 Ethernet Jumbo Frames with IPv4

Ethernet jumbo frames

This solution has no impact on the user IP layer and a slightly higher transport efficiency. Vendor-specific SDU sizes up to 9000 bytes are possible. On the other hand, it is not standardized and therefore not generally supported by Ethernet equipment and Ethernet transport services. See Figure 12.9.

12.2.6 Traffic Separation and IP Addressing

Traffic separation can be used for various purposes. There may be a need to separate traffic of different planes (eNodeB applications) from each other for network planning reasons. For example, M-plane traffic may need to be separated from U/C-plane traffic. S-plane traffic may require further separation. In another network scenario, traffic may be split into multiple paths, which have different QoS capabilities (path selection).

Traffic differentiation could be performed on multiple layers:

- L3: using different IP sub-nets;
- L2: using different VLANs;
- L1: using different physical paths.

First, a generic eNodeB IP addressing model shall be explained (Figure 12.10). Based on that, typical network scenarios will be elaborated. The model is based on two basic components:

- binding of eNodeB applications to IP addresses;
- assignment of interface IP addresses to eNodeB interfaces.

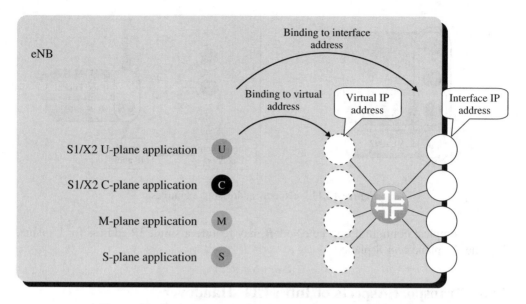

Figure 12.10 Binding of eNodeB applications to IP addresses

Binding of eNodeB applications to IP addresses
The eNodeB applications (S1/X2 U-plane, S1/X2 C-plane, M-plane, S-plane) may be arbitrarily bound to

- eNodeB interface address(es); or
- eNodeB virtual address(es).

Interface addresses are directly visible at eNodeB interfaces, whereas virtual addresses appear behind an eNodeB internal IP router. Configurations where eNodeB applications are bound to virtual addresses are typically used in scenarios where the transport link (VLAN, IPsec tunnel) is terminated with one IP address while application separation on L3 is also required.

Assignment of interface IP addresses to eNodeB interfaces
eNodeB interface IP address(es) may be assigned to different types of data link layer interfaces, which are provided by

- physical interface(s); or
- logical interface(s)

A physical interface is provided by an Ethernet port, whereas a logical interface is provided by a VLAN termination.

This generic model supports various network scenarios. Two examples are depicted in Figure 12.11. In the configuration on the left side, functional planes (eNodeB applications) are separated on L2. In contrast to that, IPsec is used with a single tunnel while functional separation is performed on L3 in the other example.

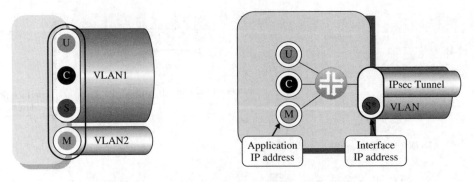

Figure 12.11   eNodeB addressing examples

In the simplest configuration, the eNodeB may feature a single IP address for U-plane, C-plane, M-plane and S-plane.

12.3   Transport Aspects of Intra-LTE Handover

Intra-LTE handover (HO) can be performed over the S1 interface ('S1 handover') or X2 interface ('X2 handover'). With X2 handover, downlink data forwarding between Source eNodeB and Target eNodeB will be optimized. Implications for the Mobile Backhaul Network will now be elaborated. In principle, similar considerations apply to both S1 and X2 handover. Since the X2 interface is expected to be present in many cases, this will be analyzed in more detail. X2 is required for Self Optimized Network (SON) functionalities, see Chapter 9.

During the X2 handover procedure the radio link is interrupted for a short moment (30–50 ms). Downlink packets arriving at the Source eNodeB will be forwarded to the Target eNodeB via the (logical) X2 interface until the S1 path has been switched to the Target eNodeB.

In the first phase (1) of the handover (see Figure 12.12, left side), the downlink traffic destined to the Source eNodeB fills the normally lightly used Source eNodeB uplink path

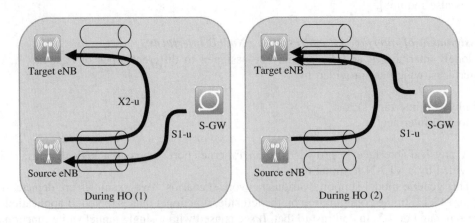

Figure 12.12   X2 Handover

(asymmetric user traffic and symmetric backhaul capacity downlink/uplink assumed). The duration of phase (1) T_1 is determined by C-plane processing (including transport latency) and may be significantly longer than the air interface interruption time.

In the second phase (2) (see Figure 12.12, right side), the Target eNodeB connection may be temporarily congested if many handovers are taking place simultaneously and downlink packets are already arriving through the S1 path to the Target eNodeB while X2 packets are still in transit. This 'overlap time' T_2 is mainly given by the X2 transport latency, assuming S1 latency toward Source eNodeB and Target eNodeB are almost equal. If congestion occurs packets may be queued (and potentially QoS aware dropped) at the affected transport node(s). But even if the X2 traffic goes completely along with S1 traffic ($T_2 \sim$ S1 uplink latency + S1 downlink latency), the probability and duration of such bursts will be very limited.

As a conclusion, the capacity need for X2 is very small compared to S1. Analysis of the mobility model resulted in 2%...3% extra capacity.

12.4 Transport Performance Requirements

12.4.1 Throughput (Capacity)

Maximum cell peak rates of 150 Mbps downlink (64QAM 2×2 MIMO) and 75 Mbps uplink (64QAM single stream) are available for a 20 MHz configuration. Cell peak rates are achievable only under ideal air interface conditions – very close to the base station and without interference from other cells. Average downlink cell spectral efficiency has been determined as $\sim$1.5–1.7 bit/s/Hz in macro cell deployment (three sectors) and 2.4–2.9 bit/s/Hz in micro cells (one sector). It should be noted that this describes the cell peak capacity under average conditions and is not an averaging over time. For further consideration in this chapter, macro cell configurations with a spectral efficiency of 1.7 bit/s/Hz downlink and 0.7 bit/s/Hz uplink are assumed, which is a good approximation for 10 MHz and 20 MHz bandwidth.

The required transport capacity can be dimensioned with two different approaches:

- dimensioning based on user traffic profile;
- dimensioning based on air interface capabilities.

Dimensioning based on the user traffic profile requires planning data from the operator. It can be tailored to actual needs and allows QoS parameters to be taken into account. On the other hand, dimensioning based on air interface capabilities provides a simple and straightforward alternative if the traffic profile is not available and should be sufficient for initial planning considerations. This approach will be elaborated further below. Note that the considerations herein refer to the eNodeB backhaul interface only. Depending on the traffic profile, statistical multiplexing gain could reduce the total capacity requirements at aggregation points.

To 'translate' the capacity at the air interface into an equivalent capacity need at the transport (backhaul) interface of the eNodeB, air interface overhead has to be subtracted, while transport overhead has to be added.

For a typical traffic profile with 50% small (60 bytes), 25% medium-size (600 bytes) and 25% large (1500 bytes) packets, the transport overhead can be estimated as $\sim$25%

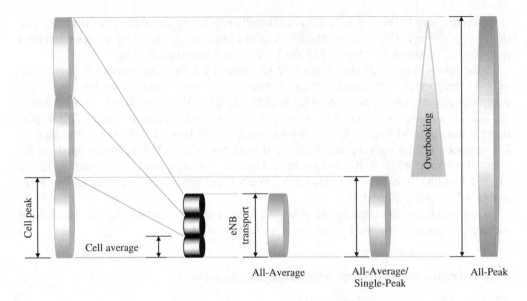

Figure 12.13   Transport capacity dimensioning scenarios

with IPv4 and IPsec and ~15% without IPsec. After subtracting the air interface overhead (PDCP/RLC), 20% with IPsec and 10% without IPsec has to be added to the data rate at the air interfaces to calculate the corresponding transport capacity.

If N is the number of cells of one eNodeB site and C_{peak} and C_{avrg} denote the peak and average capacity of one cell, the following main concepts can be distinguished (example in Figure 12.13 with 3 cells):

- 'All-Average.' The backhaul connection supports the aggregated average capacity of all cells.

$$C_{trs} = N \times C_{avrg}$$

- 'All-Average/Single-Peak.' The backhaul connection supports the aggregated average capacity of all cells and the peak capacity of one cell, whatever value is greater.

$$C_{trs} = MAX (N \times C_{avrg}; \ C_{peak})$$

- 'All-Peak.' The backhaul connection supports the aggregated peak capacity of all cells ('non-blocking').

$$C_{trs} = N \times C_{peak}$$

In most cases, 'All-Average/Single-Peak' is a reasonable trade-off between user service capabilities and transport capacity requirements, which may have a great impact on the operating costs. With that approach, advertised user service peak rates up to the cell peak rate can be momentarily supported in one cell. The advertised user service rate, however, will be only a fraction of the cell peak rate. It is mainly driven by user expectations derived from fixed broadband (xDSL) service rates and respective applications and will be enforced by LTE QoS parameters. This also means that multiple users can be supported with that rate simultaneously. With service differentiation, different maximum rates could be applied to different users – premium users could be served with highest rates. If the number of users is low, transport costs could be reduced even further if the user service rate will be applied for dimensioning instead of the cell peak rate. In summary, 'All-Average/Single Peak' is a good trade-off in general but it may be under-dimensioned for hot spots where users are located close to the antenna in multiple cells of the same eNodeB, and it may be over-dimensioned for sites with low utilization.

The 'All-Peak' concept will always lead to over-dimensioning, thus usually extra costs. However, if fiber is available at eNodeB sites, the incremental cost impact may be tolerable. There will be cases where the operator requires that the transport subsystem will have no bottleneck at all.

As an example for the 'All-Average/Single-Peak' concept, a typical $1 + 1 + 1$ 10 MHz downlink 64QAM 2×2 MIMO uplink 16QAM configuration is shown in Figure 12.14. This configuration can be supported with a Fast Ethernet backhaul line.

The 'All-Peak' scenario with a $2 + 2 + 2$ 20 MHz downlink 64QAM 2×2 MIMO uplink 16QAM configuration is shown in Figure 12.15. It should be noted that with a

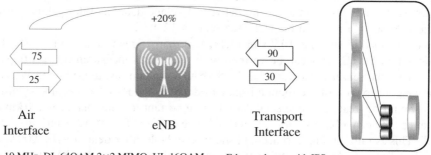

3 cells, 10 MHz, DL 64QAM 2×2 MIMO, UL 16QAM   Ethernet layer, with IPSec
DL 17 Mbit/s net PHY average rate per cell
UL 7 Mbit/s net PHY average rate per cell
DL 75 Mbit/s net PHY peak rate per cell
UL 25 Mbit/s net PHY peak rate per cell

Figure 12.14   Example: 'All-Average/Single-Peak' $1 + 1 + 1$ 10 MHz

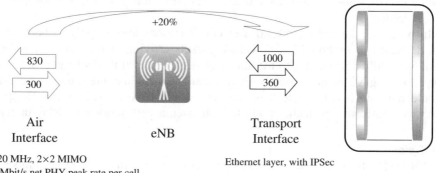

6 cells, 20 MHz, 2×2 MIMO   Ethernet layer, with IPSec
DL 150 Mbit/s net PHY peak rate per cell
UL 50 Mbit/s net PHY peak rate per cell

Figure 12.15   Example: 'All-Peak' $2 + 2 + 2$ 20 MHz

single Gigabit Ethernet interface at the eNodeB the downlink transport capacity is limited to 1 Gbps, so the theoretically possible air interface capacity cannot be utilized fully.

As stated in section 12.3, the capacity need for Intra-LTE handover (over S1 or X2 U-plane) is very small compared to S1 U-plane traffic (2%–3%).

M-plane, C-plane and S-plane capacity requirements are negligible compared to S1 U-plane.

12.4.2 Delay (Latency), Delay Variation (Jitter)

Delay requirements originate from both user services (U-plane end-to-end delay/delay variation for VoIP, web surfing, file transfer, gaming, streaming, email, and so on), radio network layer protocols (C-plane) and synchronization (S-plane).

U-plane

U-plane delay requirements are determined by user service quality expectations. Obviously, delay requirements are important for interactive services (impact on response time). However, there is also a relationship between U-plane delay and throughput performance if the user service is based on TCP (see section 12.4.3).

As explained in sections 10.7 and 11.11, less than 15 ms round-trip time (RTT) can be achieved with LTE. This can be verified with ping delay measurements starting from the UE, where the server is located near the SAE-GW. In contrast to GSM and WCDMA, the relative contribution of the mobile backhaul delay to the RTT has become significant and thus should be treated with special care. It is important to note that each 500 km one-way distance between a base station and a usually centralized SAE-GW contribute 5 ms to the RTT. This is only taking into account the unavoidable fiber delay ($\sim$200 000 km/s is the speed of light in glass). Including the delay introduced by intermediate transport nodes, the transport part of the RTT can easily exceed the radio part.

Delay variation (jitter) has to be considered in addition to delay (latency). Both the LTE air interface scheduler, retransmissions and the transport network contribute to the end-to-end jitter. Real-time end user applications including VoIP and audio/video streaming are usually designed so that they can tolerate jitter in the order of 10–20 ms, using a properly dimensioned jitter buffer. Network dimensioning and in-built QoS mechanisms have to assure that the end-to-end delay and delay variation budget of supported applications are not exceeded.

In principle, the service-related user considerations above apply to both S1 and X2. As described in section 12.3, downlink packets arriving at the Source eNodeB during handover will be forwarded to the Target eNodeB, i.e. will take a longer route. Basically, implementing X2 latency significantly less than the radio link interruption time (30–50 ms) would have no benefit since those packets would have to wait at the Target eNodeB anyway. Most user applications will tolerate such a very short-term delay increase.

C/M-plane

In WCDMA the RNL related latency requirements are imposed by a number of RAN functions over Iub/Iur including Macro-Diversity Combining, Outer Loop Power Control, Frame Synchronization and Packet Scheduler. In contrast in LTE the transport latency requirements are mainly driven by user services (U-plane). In particular, most of the handover messaging is performed in the latency-insensitive preparation phase.

S-plane

A specific requirement for delay variation (jitter) applies if Precision Time Protocol (PTP) based on IEEE1588-2008 is used for eNodeB synchronization (see section 12.8.1).

12.4.3 TCP Issues

The Transmission Control Protocol (TCP) is designed to provide reliable transport for packet data. Today TCP is used for 80–90% of all packet traffic on the internet. TCP is typically used for applications and their respective protocols where reliability is important, for example web browsing (HTTP), email (SMTP) and file transfer (FTP). Due to its flow control mechanism, a single TCP connection is rate limited by the ratio between the TCP window size and the RTT. Given a standard 64 KB TCP receive window size (RFC793), 20 ms RTT would limit the achievable service data rate at ∼25 Mbps for a single TCP connection in the steady state due to TCP flow control, see Figure 12.16. This limitation could be mitigated with multiple concurrent TCP connections, which are typically the case for web browsing, and/or enlarging the receive window through TCP Window Scaling (RFC1323). This is supported by many web servers today. In particular large file transfers, like music, video, SW download, will benefit from these measures, if high service rates would be offered by the operator.

Apart from the steady state, TCP connection start-up ('TCP slow start') and behavior after packet loss ('TCP congestion avoidance') are also affected by the RTT. With today's browsers and web sites and with average page size below 1 MB, it can be assumed that most web traffic is transferred within the slow start phase, so TCP Window Scaling will not help that much here. In any case, the transport network should be designed for low latency both for higher TCP data rate in the steady state and for faster TCP slow start.

Content Distribution Networks (CDN) reduce the network load but also reduce the end-to-end RTT, thus having a positive impact on the file transfer performance. For CDN supported services, the mobile network related RTT will dominate.

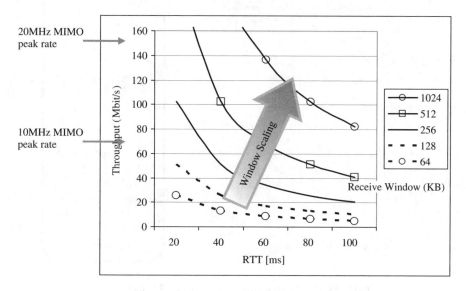

Figure 12.16   TCP data rate versus RTT

12.5 Transport Network Architecture for LTE

12.5.1 Implementation Examples

Mobile networks are commonly structured and dimensioned with respect to user densities (metropolitan areas versus rural areas) and topological requirements (highways, railways, and so forth). From a transport perspective, the network is usually partitioned into a Backbone Network and a Mobile Backhaul Network. The LTE Backbone Networks are typically IP/MPLS based. Much greater variety can be found in the Mobile Backhaul Network.

The eNodeB supports an IP-over-Ethernet interface so both Ethernet (L2VPN – Carrier Ethernet or IP/MPLS-based) and IP (L3VPN – IP/MPLS-based) transport services would be adequate in the Mobile Backhaul Network. Such a transport service can be delivered by the mobile operator itself (self-built transport network) or by another operator (leased line service). In both cases, service attributes and parameters are usually stated in a Service Level Specification (SLS), which in the latter case is part of a Service-Level Agreement (SLA). While this makes a big difference with respect to the operator business case, it will be treated here conceptually as equal.

In many cases, the demarcation point between the Mobile Backhaul Network and the Backbone Network, implemented by an Edge Router, coincides with the borderline between the unsecure and secure parts of the network. If IPsec is applied for network security purposes (see section 12.7), the Security Gateway (SEG) terminates the IPsec tunnel IP layer with the eNodeB at the other end. Thus, the SEG may be collocated with the Edge Router.

Figure 12.17 illustrates the physical view where the Mobile Backhaul Network is further subdivided into the Access Network and the Aggregation Network. In many cases, the

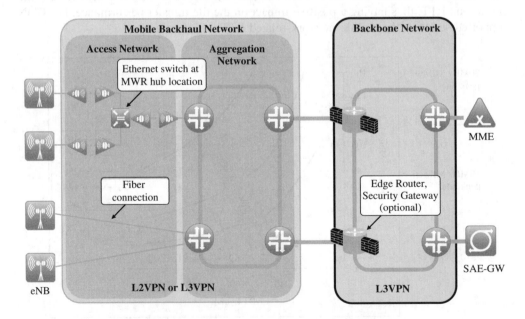

Figure 12.17   Implementation example (physical view)

Access Network is built upon a fiber and/or microwave radio (MWR) based tree or chain topology, whereas the Aggregation Network uses fiber rings.

The Access Network may be implemented using Ethernet technology natively. For traffic separation and scalability reasons, the network should be partitioned into VLANs where bridging takes place at intermediate nodes (for example, microwave radio hubs).

12.5.2 X2 Connectivity Requirements

X2 handover principles are explained in section 12.3. There is acommon misconception with respect to the requirements on the transport network. At a first glance, it seems obvious that it would be beneficial if the X2 'turning point' could be located close to the base stations. See Figure 12.18 for Near-End X2 connectivity. This can be implemented with Ethernet switching or IP routing functions. X2 latency would be low and X2 traffic wouldn't load higher parts of the Mobile Backhaul Network. On the other hand, the S1 transport path should be optimized for low latency anyhow, see section 12.4.2. X2 latency significantly less than radio link interruption time (30–50 ms) doesn't add value. Furthermore, the amount of X2 traffic is marginal compared to S1 traffic, so the potential for savings is very limited.

Far-End X2 connectivity in Figure 12.19 can be built on existing topologies with hub-and-spoke. It requires IP routing at the far end, which could be implemented with the Backbone Edge Router. Also this architecture meets the latency and capacity requirements analyzed earlier.

It can be concluded that Near-End X2 connectivity is not generally advantageous for 3GPP Release 8/9/10 LTE network architectures. In particular, direct physical connections between adjacent eNodeBs are not required.

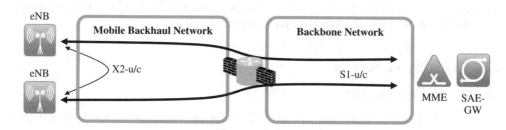

Figure 12.18   Near-End X2 connectivity

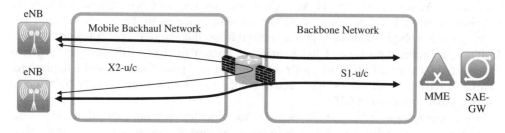

Figure 12.19   Far-End X2 connectivity

Table 12.1 Transport service attributes (recommendations)

Service Attribute	Value
L3: Packet Delay (PD)	≤ 20 ms
L2: Frame Delay (FD)	
L3: Packet Delay Variation (PDV)	$\leq +/- 5.0$ ms
L2: Frame Delay Variation (FDV)	
L3: Packet Loss Ratio (PLR)	$\leq 10e-6$
L2: Frame Loss Ratio (FLR)	

12.5.3   Transport Service Attributes

Transport service attributes are used to describe the requirements on the transport service for eNodeB backhaul. It should be noted that these attributes are defined end-to-end (between eNodeB and core network elements) on the IP layer (L3) and have to be allocated in general to multiple parts of the network. In particular, if the transport service is (partially) provided by Ethernet Virtual Connections (EVC), these attributes have to be translated into their L2 equivalents.

Ethernet service attributes are defined by Metro Ethernet Forum [7]. Note that these attributes are optional and a subset can be fully sufficient to define an LTE transport service. Ethernet capacity requirements are described by the service attributes

- Committed Information Rate (CIR)/Committed Burst Size (CBS); and
- Excess Information Rate (EIR)/Excess Burst Size (EBS) (optional).

As explained in section 12.4.2, performance requirements on the transport service are primarily driven by user service aspects. The values given in Table 12.1 should be understood as recommendations rather than hard requirements.

12.6   Quality of Service

12.6.1   End-to-End QoS

In order to facilitate a certain end-to-end QoS for a user service, radio and transport QoS have to be considered consistently, see Figure 12.20.

Radio interface resources are scarce, so strict control has to be applied. This means that capacity has to be reserved for Guaranteed Bit Rate (GBR)-type services through Connection Admission Control (CAC). Controlled overbooking with prioritization is the solution for non-GBR type of services. The resource situation in the access section ('last mile') of the mobile backhaul network is similar, so the solution principles are similar: Committed Information Rate (CIR) guarantees continuation of GBR services, whereas Extended Information Rate (EIR) with overbooking and prioritization can be applied for non-GBR traffic. Nevertheless, it is recommended that a minimum capacity be reserved for non-GBR traffic using CIR.

In the aggregation section of the network the situation is more relaxed, so less strict control is required. Still, there might be a need for capacity reservation and traffic prioritization. If sufficient resources are available, this section may be ignored by the CAC.

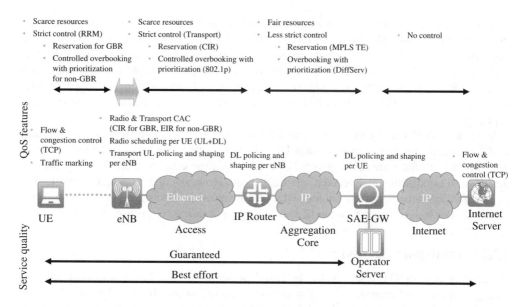

Figure 12.20 End-to-end QoS

The mobile network up to the SAE-GW is operator controlled, so with proper network dimensioning the service quality can be guaranteed. Naturally, CAC makes most sense here. This is not the case if the actual user service is provided by a server located anywhere in the internet. Only Best Effort can be expected. If the user service is based on TCP (80%–90% of the traffic), TCP congestion control mechanisms will assure a fair distribution of capacity.

12.6.2 Transport QoS

DiffServ is the most common way of traffic prioritization on IP layer [8, 9]. Traffic end points (eNodeB, S-GW, MME, NMS) will set DiffServ Code Point (DSCP) values, while intermediate network elements have to handle the packets accordingly. If IPsec tunnel mode is used (see section 12.7), this will still work if the tunnel terminating network elements (eNodeB and SEG) copy the application layer ('inner') DSCP into the IPsec tunnel ('outer') DSCP.

If traffic aggregation is performed by Ethernet switching, not IP routing, the transport network may not be IP QoS (DiffServ) aware. In this case, traffic prioritization on the Ethernet layer using frame marking methods (IEEE 802.1p Ethernet priority bits [5]) have to be used.

Quality of service has to be applied consistently to both the radio and the transport interface. As per [10], each U-plane Service Data Flow (SDF) is associated with a QoS Class Identifier (QCI). This has to be mapped into the appropriate transport layer QoS code points.

In contrast to the Iub interface in HSDPA, there are no standardized means for congestion control at the S1 interface in LTE. This has to be taken into account when designing the network. GBR traffic should be governed by Connection Admission Control (CAC) and performance should be assured by SLA/SLS (CIR/CBS attributes) and traffic prioritization. Only non-GBR traffic should be affected by network congestion. Most of the non-GBR traffic is TCP based (web browsing, file transfer, email, and so forth). End-to-end TCP congestion control in conjunction with IP transport features like Random Early Discard (RED, in order to de-synchronize multiple TCP connections) as used in fixed networks for a long time, will sufficiently guarantee fair and efficient usage of transport capacity.

Note that TCP cannot efficiently participate in HSDPA congestion control because RLC retransmission between RNC and UE will hide congestion situations both in the air interface and in the transport network. The HSDPA flow control algorithm was initially designed to control the air interface scheduler only. However, the transport network (Iub) is the often bottleneck and inefficient RLC retransmission is to be minimized, so it was recognized that the transport network would require its own congestion control.

12.7 Transport Security

As compared with TDM and ATM, IP-based protocols enable lower transport cost and easier planning and configuration. On the other hand, mobile backhaul traffic becomes more vulnerable to hacker attacks. Attack methods have evolved rapidly with cheap HW providing high processing power and better tools, which are widely available. In addition, base stations that were traditionally located in secure, locked sites are increasingly set up in public places or homes.

Furthermore, there are two major differences compared to WCDMA based on RNC based architecture with respect to transport security (see Figure 12.21):

- Air interface security of user plane and control plane (RRC) traffic is terminated at the eNodeB. Traffic in the LTE mobile backhaul network is not secured by Radio Network Layer protocols.
- Since the LTE network architecture is flat, adjacent base stations and core nodes (MME, S-GW) become directly IP-reachable from base station sites. If physical access to the site cannot be prohibited, a hacker could connect his device to the network port and attack aforementioned network elements.

Transport security features are seen as mandatory so that both the mobile backhaul network *and* the eNodeB site can be regarded as secure. IPsec provides a comprehensive set of security features like data origin authentication, encryption, integrity protection, which solve both problems. The 3GPP security architecture, based on IPsec and Public Key Infrastructure (PKI), is outlined in following specifications: TS33.210 [11], TS33.310 [12], and TS33.401 [13].

IPsec is applied between Security Gateways (SEG), see Figure 12.22. Each eNodeB site instantiates one SEG function. One or more Security Gateways should be located at the edge of the operators 'Security Domain' as per 3GPP Network Domain Security. Typically, the Security Domain includes the backbone network. Connectivity provided by a third party (leased lines) is usually treated as outside the Security Domain. The SEG is effectively hiding the core nodes (MME, SAE-GW, Network Management System) from the mobile backhaul network. Since the traffic itself is transparent to the SEG, various core network configurations can be supported, for example MME/SAE-GW pooling and S1-flex.

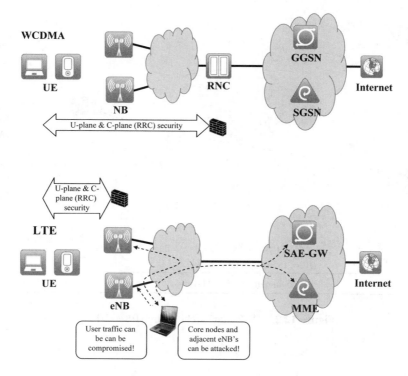

Figure 12.21   Security with WCDMA and LTE

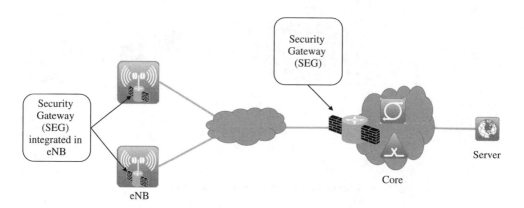

Figure 12.22   Security Gateway functions in the network

In the simplest scenario, all S1 and X2 traffic of an eNodeB could pass through a single IPsec tunnel. That means that X2 traffic would need to be routed back through an SEG – passing this SEG function twice, see 'X2 Star' architecture in Figure 12.23.

Alternatively, X2 traffic (U/C-plane) may be secured directly between eNodeB-integrated SEGs directly ('X2 Mesh' architecture, Figure 12.24). There is a common misconception that the flat LTE architecture implies 'X2 Mesh'. In reality, the advantages are minor. The amount of backhaul traffic that can be saved in the aggregation network

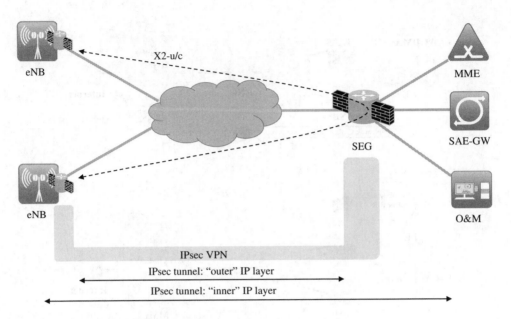

Figure 12.23   'X2 Star' architecture (IPsec VPN)

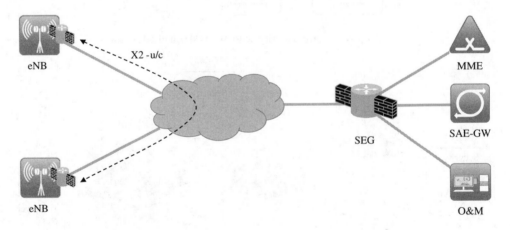

Figure 12.24   'X2 Mesh' architecture

is very small. Even though X2 latency will be reduced, the effect will be insignificant since the transport path used for S1 should be designed for low latency anyhow.

It should be noted that the 3GPP Release 8 and 9 X2 self-configuration as per [14] does not fully support the 'X2 Mesh' architecture. The TNL Address Discovery procedure determines the C-plane application address only, which is not sufficient if the eNodeB application IP addresses are different from the interface IP address that terminates the IPsec tunnel layer.

With IPsec Tunnel Mode, the payload IP frame is encapsulated with an extra IP header. IPsec Transport Mode doesn't need that extra IP header, so the transport efficiency is slightly better. However, Transport Mode is only applicable between two IP hosts (RFC2401). So, if a centralized SEG is applied, Tunnel Mode is the only choice. In turn, IPsec Transport Mode would only be applicable to direct X2 connections. Due to the very low amount of X2 traffic, the transport efficiency impact is negligible. If IPsec Tunnel Mode is used, the transport QoS mechanism (DiffServ) will not be affected if application layer ('inner') DSCP values are copied into the IPsec tunnel ('outer') DSCP's.

12.8 Synchronization from Transport Network

As per 3GPP requirement, the air interface at an eNodeB in FDD mode needs to be frequency synchronized with an accuracy of ±50 ppb. In addition, phase (time) synchronization is required for LTE TDD mode to enable frame timing and uplink-downlink configuration synchronization across the coverage area, as discussed in Chapter 15. In the future synchronization will also be needed for Single Frequency Network (SFN) Multimedia Broadcast Multicast Service (MBMS) and location-based services.

Traditionally, synchronization information for 2G and 3G base stations has been carried over TDM-based transport networks or derived from satellites (GPS). One of the development steps with HSPA networks was also to use E1/T1 line(s) to enable synchronization (and other time-critical communications) and provide HSPA data over packet-based transport with less stringent timing requirements. Since transport networks for LTE are packet based only, alternative technologies have been developed that will be introduced here briefly.

12.8.1 Precision Time Protocol

Precision Time Protocol (PTP), based on [15], also known as Timing-over-Packet (ToP), is a field-proven method to support synchronization through an IP transport network. It should be noted that the network has to support low delay variation (jitter) for PTP packets. Most robust PTP implementations can tolerate jitter up to ±5 ms.

The PTP solution consists of a PTP Grandmaster (server) at the core site and PTP Slaves (clients) implemented in the eNodeBs, see Figure 12.25. The master and slaves

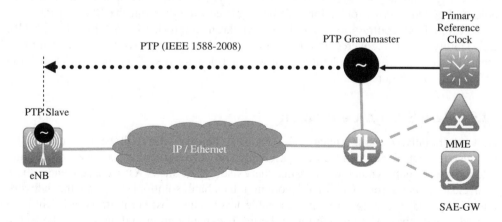

Figure 12.25 Synchronization with PTP (IEEE1588-2008 [15])

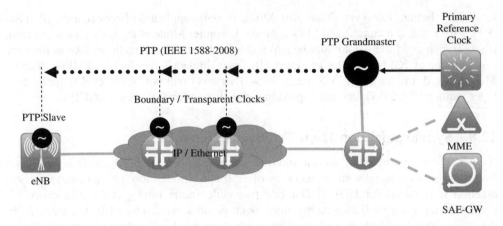

Figure 12.26 PTP for Time/Phase Synchronization

communicate through the bidirectional protocol containing time stamps. The slaves use these timestamps to synchronize themselves to the grandmaster's clock. Accurate frequency can be derived from that clock.

LTE TDD mode requires accurate phase synchronization ($\pm1.5\,\mu$s) in addition to frequency synchronization (±50 ppb). While frequency synchronization can tolerate typical network jitter in the order of several milliseconds through long-time averaging, uplink/downlink delay asymmetry cannot be controlled with the required accuracy. In order to solve this problem, PTP has to be implemented at all intermediate nodes on the synchronization traffic path so that outgoing timing packets will be (re-)marked based on the local node clock (on-path support with IEEE1588-2008 [15] boundary clock or transparent clock). See Figure 12.26.

12.8.2 Synchronous Ethernet

Synchronous Ethernet (SyncE), as per G.8261/8262/8264, is an SDH-like mechanism for distributing frequency at layer 1. In contrast to PTP, which is essentially a layer-3 technology, frequency synchronization will be extracted directly from the Ethernet interface at the eNodeB. In contrary to packet based synchronization (IEEE1588-2008 [15], NTP), the stability of the recovered frequency does not depend on network load and impairments. SyncE has to be implemented at all intermediate nodes on the synchronization traffic path. See Figure 12.27.

12.9 Base Station Co-location

At least initially, existing (macro) 2G/3G base stations sites will be reused for LTE in order to keep deployment costs reasonably low. Similarly that has been done quite often with 3G deployment when using the existing 2G sites. As a consequence, there is a clear demand for 2G/3G/LTE common backhaul solutions to keep the network complexity and operating costs at reasonable level. The most straightforward way is to upgrade legacy base stations with an Ethernet backhaul option and aggregate the traffic using a QoS aware Ethernet switch or IP router, which may be an external device or

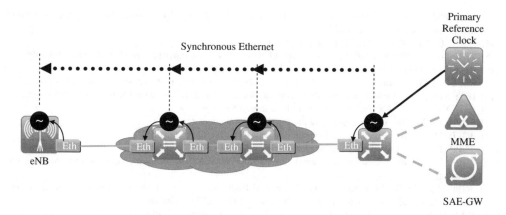

Figure 12.27 Synchronous Ethernet

integrated in the eNodeB. If a co-located base station doesn't support Ethernet, circuit emulation (pseudo-wire) technology can be applied as an alternative. In addition to traffic aggregation, propagation and distribution of synchronization information is required.

12.10 Summary

This chapter provided an overview of transport (mobile backhaul)-related issues and solutions for S1 and X2 logical interfaces in the LTE network. While air interface specifications are defined by 3GPP, specifications from a number of different standardization bodies need to be combined for the design of the transport network. E1/T1 connections were typically sufficient for carrying traffic for 2G voice networks. HSPA pushed the requirements to tens of Mbps and now LTE beyond 100 Mbps, implying major changes to the transport network. The transport capacity could be dimensioned according to the very highest LTE radio peak data rate but it may be more cost efficient to base the dimensioning on the average and typical radio capacities instead. The optimization of transport network is highly relevant because transport solutions may cause a substantial part of network operating expenses. Quality of Service (QoS) differentiation should be used in a harmonized way to enhance the network efficiency both in the radio interface and in the transport network. LTE supports not only higher peak data rates but also reduces the latency at the radio interface. In order to provide the LTE benefits to the end users, the transport network latency and jitter should be minimized as well.

The flat network architecture in LTE also has an impact in terms of transport security and transport congestion control algorithms. In the hierarchical 3G architecture, the RNC provided ciphering and congestion control while different solutions are applied in LTE. The transport network is also used to provide the base-station synchronization, which sets specific requirements for the network design.

References

[1] 3GPP, TS 36.410, 'S1 General Aspects and Principles'.
[2] 3GPP, TS 36.420, 'Evolved Universal Terrestrial Radio Access Network (E-UTRAN); X2 General Aspects and Principles'.

[3] IEEE 802.3 – 2004, 'Carrier sense multiple access with collision detection (CSMA/CD) access method and physical layer specifications'.

[4] IEEE 802.1q - 2005, 'Virtual Bridged Local Area Networks'.

[5] IEEE 802.1p, 'Traffic Class Expediting and Dynamic Multicast Filtering' (published in 802.1D-1998).

[6] IETF RFC 4861, 'Neighbor Discovery for IP version 6 (IPv6)'.

[7] [MEF6.1] Metro Ethernet Forum, Technical Specification 6.1 'Ethernet Services Definitions – Phase 2'.

[8] IETF RFC 2474, 'Definition of the Differentiated Services Field (DS Field) in the IPv4 and IPv6 Headers'.

[9] IETF RFC 2475, 'An Architecture for Differentiated Services'.

[10] 3GPP, TS 23.401, 'GPRS enhancements for E-UTRAN access'.

[11] 3GPP, TS 33.210, '3G Network Domain Security (NDS), IPlayer Security'.

[12] 3GPP, TS 33.310, 'Authentication Framework'.

[13] 3GPP, TS 33.401, 'Evolved Universal Terrestrial Radio Access Network (E-UTRAN); Architecture Description'.

[14] 3GPP, TS 36.300, 'Evolved Universal Terrestrial Radio Access (E-UTRA) and Evolved Universal Terrestrial Radio Access Network (E-UTRAN); Stage 2 Overall Description'.

[15] IEEE 1588 – 2008, 'IEEE Standard for a Precision Clock Synchronization Protocol for Networked Measurement and Control Systems'.

13

Voice over IP (VoIP)

Harri Holma, Juha Kallio, Markku Kuusela, Petteri Lundén,
Esa Malkamäki, Jussi Ojala and Haiming Wang

13.1 Introduction

While the data traffic and the data revenues are increasing, the voice service still makes
the majority of operators' revenue. Therefore, LTE is designed to support not only data
services efficiently, but also good quality voice service with high efficiency. As LTE radio
only supports packet services, the voice service will also be Voice over IP (VoIP), not
Circuit Switched (CS) voice. This chapter presents the LTE voice solution including voice
delay, system performance, coverage and inter-working with the CS networks. First, the
general requirements for voice and the typical voice codecs are introduced. The LTE voice
delay budget calculation is presented. The packet scheduling options are presented and
the resulting system capacities are discussed. The voice uplink coverage challenges and
solutions are also presented. Finally, the LTE VoIP inter-working with the existing CS
networks is presented.

13.2 VoIP Codecs

GSM networks started with Full rate (FR) speech codec and evolved to Enhanced Full
Rate (EFR). The Adaptive Multi-Rate (AMR) codec was added to 3GPP Release 98 for
GSM to enable codec rate adaptation to the radio conditions. AMR data rates range from
4.75 kbps to 12.2 kbps. The highest AMR rate is equal to the EFR. AMR uses a sampling
rate of 8 kHz, which provides 300–3400 Hz audio bandwidth. The same AMR codec was
included also for WCDMA in Release 99 and is also used for running the voice service
on top of HSPA. The AMR codec can also be used in LTE.

The AMR-Wideband (AMR-WB) codec was added to 3GPP Release 5. AMR-WB uses
a sampling rate of 16 kHz, which provides 50–7000 Hz audio bandwidth and substantially
better voice quality and mean opinion score (MOS). As the sampling rate of AMR-WB is
double the sampling rate of AMR, AMR is often referred to as AMR-NB (narrowband).
AMR-WB data rates range from 6.6 kbps to 23.85 kbps. The typical rate is 12.65 kbps,
which is similar to the normal AMR of 12.2 kbps. AMR-WB offers clearly better voice

LTE for UMTS: Evolution to LTE-Advanced, Second Edition. Edited by Harri Holma and Antti Toskala.
© 2011 John Wiley & Sons, Ltd. Published 2011 by John Wiley & Sons, Ltd.

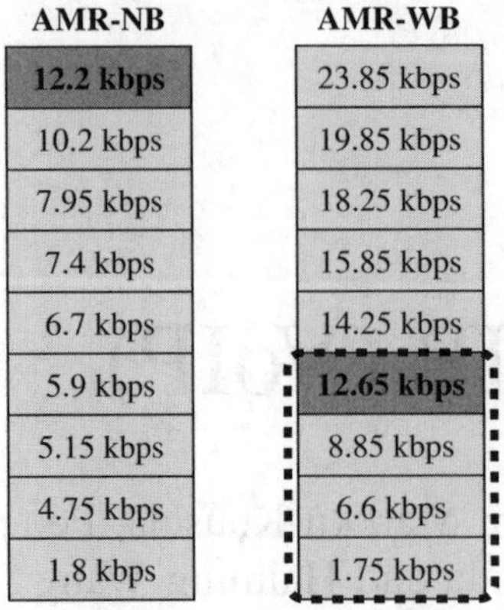

AMR-NB	AMR-WB
12.2 kbps	23.85 kbps
10.2 kbps	19.85 kbps
7.95 kbps	18.25 kbps
7.4 kbps	15.85 kbps
6.7 kbps	14.25 kbps
5.9 kbps	**12.65 kbps**
5.15 kbps	8.85 kbps
4.75 kbps	6.6 kbps
1.8 kbps	1.75 kbps

Figure 13.1   Adaptive Multirate (AMR) Voice Codec radio bandwidth

quality than AMR-NB with the same data rate and can be called wideband audio with narrowband radio transmission. The radio bandwidth is illustrated in Figure 13.1 and the audio bandwidth in Figure 13.2. The smallest bit rates, 1.8 and 1.75 kbps, are used for the transmission of Silence Indicator Frames (SID).

This chapter considers AMR codec rates of 12.2, 7.95 and 5.9 kbps. The resulting capacity of 12.2 kbps would also be approximately valid for AMR-WB 12.65 kbps.

When calling outside mobile networks, voice transcoding is typically required to 64 kbps Pulse Code Modulation (PCM) in links using ITU G.711 coding. For UE-to-UE calls, the transcoding can be omitted with transcoder free or tandem free operation [1]. Transcoding generally degrades the voice quality and is not desirable within the network.

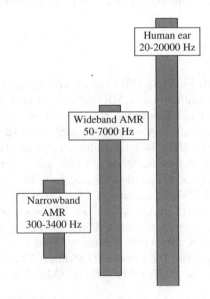

Figure 13.2   Adaptive Multirate (AMR) Voice Codec audio bandwidth

There are also a number of other voice codecs that are generally used for VoIP. A few examples are G.729, using an 8 kbps coding rate, and Internet Low Bit Rate Codec (iLBC), using 13 kbps which is used, for example, in Skype and in Googletalk.

13.3 VoIP Requirements

There is a high requirement set for the radio network to provide a reliable and good quality voice service. Some of the main requirements are considered below.

The impact of the mouth-to-ear latency on user satisfaction is illustrated in Figure 13.3. The delay preferably should be below 200 ms, which is similar to the delay in GSM or WCDMA voice calls. The maximum allowed delay for a satisfactory voice service is 280 ms.

IP Multimedia Subsystem (IMS) can be deployed to control VoIP. IMS is described in Chapter 3. IMS provides the information about the required Quality of Service (QoS) to the radio network by using 3GPP standardized Policy and Charging Control (PCC) [3]. The radio network must be able to have the algorithms to offer the required QoS better than Best Effort. QoS includes mainly delay, error rate and bandwidth requirements. QoS in LTE is described in Chapter 8.

The voice call drop rates are very low in the optimized GSM/WCDMA networks today – in the best case below 0.3%. VoIP in LTE must offer similar retainability including smooth inter-working between GSM/WCDMA circuit switched (CS) voice calls. The handover functionality from VoIP in LTE to GSM/WCDMA CS voice is called Single radio Voice Call Continuity (SR-VCC) and described in detail in section 13.10.

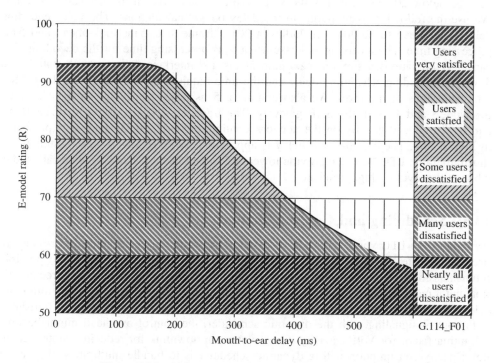

Figure 13.3   Voice mouth-to-ear delay requirements [2]

The AMR 12.2 kbps packet size is 31 bytes while the IP header is 40–60 bytes. IP header compression is a mandatory requirement for an efficient VoIP solution. IP header compression is required both in UE and in eNodeB. All the VoIP simulations in this chapter assume IP header compression.

The IP connectivity requires keep alive messages when the UE does not have a phone call running. The frequency of the keep alive messages depends on the VoIP solution: operator IMS VoIP can use fairly infrequent keep alive messages since IMS is within the operator's own network and no firewalls or Network Address Tables (NAT) are required in between. The internet VoIP requires very frequent keep alive message to keep the connectivity open through firewalls and NATs. The frequent keep alive message can affect UE power consumption and network efficiency.

VoIP roaming cases need further attention especially if there are some LTE networks designed for VoIP and data, while some networks are designed for data only transmission without the required voice features. VoIP roaming also requires IMS and GPRS roaming agreements and the use of visited GGSN/MME model. One option is to use circuit switched (CS) calls whenever roaming with CS Fallback for LTE procedures. Similarly CS calls can also be used for emergency calls since 3GPP Release 8 LTE specifications do not provide the priority information from the radio to the core network nor a specific emergency bearer.

13.4 Delay Budget

The end-to-end delay budget for LTE VoIP is considered here. The delay should preferably be below 200 ms, which is the value typically achieved in the CS network today. We use the following assumptions in the delay budget calculations. The voice encoding delay is assumed to be 30 ms including a 20 ms frame size, 5 ms look-ahead and 5 ms processing time. The receiving end assumes a 5 ms processing time for the decoding. The capacity simulations assume a maximum 50 ms air–interface delay in both uplink and downlink including scheduling delay and the time required for the initial and the HARQ retransmissions of a packet. We also assume a 5 ms processing delay in the UE, 5 ms in eNodeB and 1 ms in the SAE gateway. The transport delay is assumed to be 10 ms and it will depend heavily on the network configuration. The delay budget is presented in Figure 13.4. The mouth-to-ear delay is approximately 160 ms with these assumptions, illustrating that LTE VoIP can provide lower end-to-end latency than CS voice calls today while still providing high efficiency.

13.5 Scheduling and Control Channels

By default the LTE voice service uses dynamic scheduling, where control channels are used to allocate the resources for each voice packet transmission and for the possible retransmissions. Dynamic scheduling gives full flexibility for optimizing resource allocation, but it requires control channel capacity. Multi-user channel sharing plays an important role when optimizing the air interface for VoIP traffic. Because each user is scheduled with control signaling with the dynamic scheduler, the control overhead might become a limiting factor for VoIP capacity. One solution in downlink for reducing control channel capacity consumption with a dynamic scheduler is to bundle multiple VoIP packets together at Layer 1 into one transmission of a user. The packet bundling is CQI based

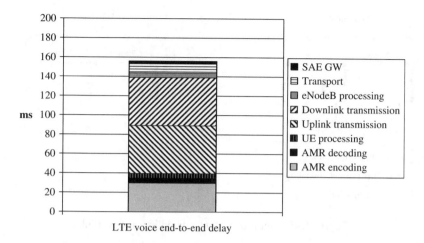

Figure 13.4 LTE voice end-to-end delay budget

and is applied only for users whose channel conditions favor the use of bundling. The main benefit from the bundling is that more users could be fitted to the network with the same control channel overhead as users in good channel conditions are scheduled less often. The drawback from the bundling is that bundled packets experience a tighter delay budget, but the negative impact of this to VoIP capacity can be kept to a minimum by applying bundling for good enough users that are not relying on HARQ retransmissions. From the voice quality perspective it is also important to minimize the probability of losing consecutive packets. This can be achieved by making the link adaptation for the TTIs carrying bundled packets in a more conservative way leading to a reduced packet error rate for the first transmission. Due to the low UE transmission power and non-CQI based scheduling, packet bundling is not seen as an attractive technique in LTE uplink.

The 3GPP solution to avoid control channel limitation for VoIP capacity is the Semi-Persistent Scheduling (SPS) method [4, 5], where the initial transmissions of VoIP packets are sent without associated scheduling control information by using persistently allocated transmission resources instead. The semi-persistent scheduling is configured by higher layers (Radio Resource Control, RRC), and the periodicity of the semi-persistent scheduling is signaled by RRC. At the beginning of a talk spurt, the semi-persistent scheduling is activated by sending the allocated transmission resources by Physical Downlink Control Channel (PDCCH) and the UE stores the allocation and uses it periodically according to the periodicity. With semi-persistent scheduling only retransmissions and SID frames are scheduled dynamically, implying that the control channel capacity is not a problem for the semi-persistent scheduler. On the other hand, only limited time and frequency domain scheduling gains are available for semi-persistent scheduling. The semi-persistent allocation is released during the silence periods. Semi-persistent scheduling is adopted for 3GPP Release 8. The downlink operation of the semi-persistent scheduler is illustrated in Figure 13.5.

In the following the impact of control channel capacity to VoIP maximum capacity with dynamic scheduler is illustrated with theoretical calculations.

It is assumed that the number of Control Channel Elements (CCE) is approximately 20 per 5 MHz bandwidth. Two, four or eight CCEs can be aggregated per user depending

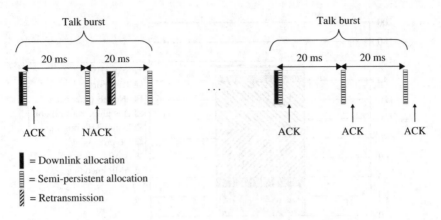

Figure 13.5 Semi-persistent scheduling in downlink

on the channel conditions in low signal-to-noise ratios. We further apply the following assumptions for the calculations: voice activity 50%, voice packet arrival period 20 ms, downlink share of the traffic 50% and number of retransmissions 20%. The results are calculated without packet bundling and with two packet bundling. To simplify the calculations it is further assumed that SID frames are not taken into account. VoIP maximum capacity can be calculated by using Equation 13.1.

$$Max\_users = \frac{\#CCEs}{\#CCEs\_per\_user} \cdot \frac{Packet\_period[ms]}{Voice\_activity}$$

$$\times Packet\_bundling \cdot Downlink\_share \cdot \frac{1}{1 + BLER} \quad (13.1)$$

The calculation results are shown in Figure 13.6 for a 5 MHz channel. As an example, the maximum capacity would be 330 users without CCE aggregation and without packet bundling. Assuming an average CCE aggregation of three brings the maximum capacity to

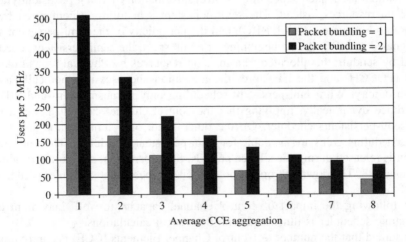

Figure 13.6 Maximum control channel limited capacity with fully dynamic scheduling

110 without packet bundling and 220 with packet bundling. These theoretical calculations illustrate the importance that multi-user channel sharing has for VoIP system performance, and furthermore, the maximum gain in capacity that packet bundling may provide with the dynamic scheduler. According to the system level simulations, the average size of CCE aggregation is approximately 1.5 for a scheduled user. Comparison of the simulated downlink capacities[1] for a dynamic scheduler (presented in section 13.6) with the theoretical upper limits shows that the simulated capacities without packet bundling match rather nicely the theoretical calculations, being approximately 5% lower than the theoretical maximum. With packet bundling the simulated capacities are notably below the theoretical upper limits, as in practice the probability for using bundling is clearly limited below 100%.

13.6   LTE Voice Capacity

This section presents the system level performance of VoIP traffic in LTE at a 5 MHz system bandwidth. VoIP capacity is given as the maximum number of users that could be supported within a cell without exceeding 5% outage. A user is defined to be in an outage if at least 2% of its VoIP packets are lost (i.e. either erroneous or discarded) during the call. VoIP capacity numbers are obtained from system level simulations in a macro cell scenario 1 [6], the main system simulation parameters being aligned with [7]. An illustration of a VoIP capacity curve is presented in Figure 13.7, which shows how the outage goes up steeply when the maximum capacity is approached. This enables running

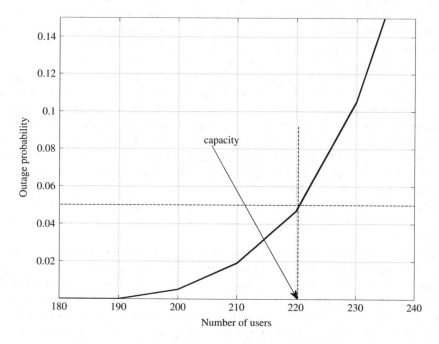

Figure 13.7   An example of VoIP capacity curve

[1] Simulated capacities should be up-scaled by 12.5% to remove the impact of SID transmissions on the capacity.

Table 13.1 VoIP capacity in LTE at 5 MHz

VoIP codec	AMR 5.9	AMR 7.95	AMR 12.2
Downlink capacity			
Dynamic scheduler, without bundling	210	210	210
Dynamic scheduler, with bundling	410	400	370
Semi-persistent scheduler	470	430	320
Uplink capacity			
Dynamic scheduler	230	230	210
Semi-persistent scheduler	410	320	240

the system with a relatively high load (compared to the maximum capacity) while still maintaining low outage.

The results of VoIP capacity simulations are summarized in Table 13.1 for three different AMR codecs (AMR 5.9, AMR 7.95 and AMR 12.2) and for both dynamic and semi-persistent schedulers. The capacity is 210–470 users in downlink and 210–410 users in uplink, which corresponds to 42–94 users per MHz per cell in downlink and 42–82 users per MHz per cell in uplink. The lower AMR rates provide higher capacity than the higher AMR rates. The AMR codec data rate can be defined by the operator allowing a tradeoff between the capacity and the voice quality. The lower AMR rates do not increase the capacity with dynamic scheduling due to the control channel limitation.

The results also show that voice capacity is uplink limited, which can be beneficial when there is asymmetric data transmission on the same carrier taking more downlink capacity. The downlink offers higher capacity than uplink because the downlink scheduling uses a point-to-multipoint approach and can be optimized compared to the uplink.

In the following, the simulated capacities are analyzed in more detail. All the supporting statistics presented in the sequel are assuming load as close as possible to the 5% outage.

As described in section 13.5, VoIP system performance with the dynamic scheduler is limited by the available PDCCH resources, and therefore the dynamic scheduler cannot fully exploit the Physical Downlink Shared Channel (PDSCH) air interface capacity as there are not enough CCEs to schedule the unused Physical Resource Blocks (PRBs). This is illustrated in Figure 13.8 for downlink, which contains a cumulative distribution function for the scheduled PRBs per Transmission Time Interval (TTI) with AMR 12.2. The total number of PRBs on 5 MHz bandwidth is 25. As can be seen from Figure 13.8, the average utilization rate for the dynamic scheduler is only 40% if packet bundling is not allowed.

Due to the control channel limitations, the savings in VoIP packet payload size with lower AMR rates are not mapped to capacity gains with the dynamic scheduler, if packet bundling is not used. The different codecs provide almost identical performance in downlink whereas there are small gains in capacity in uplink with lower AMR rates. This uplink gain originates from the improved coverage as more robust MCS could be used when transmitting a VoIP packet in uplink – this gain does not exist in the downlink direction as it is not coverage limited due to higher eNodeB transmission power.

When packet bundling is enabled, the average size of an allocation per scheduled user is increased and hence the air interface capacity of PDSCH can be more efficiently

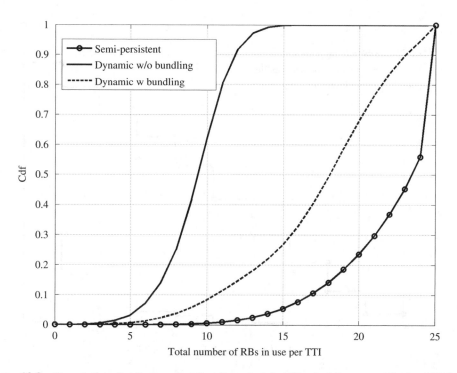

Figure 13.8 Cumulative distribution function for scheduled Physical Resource Blocks (PRB) per TTI in downlink (out of a total of 25 PRBs on 5MHz bandwidth)

exploited, which leads to an increased average PRB utilization rate of 70%. Gains of 75–90% are achieved for capacity when packet bundling is enabled. The probability of using packet bundling in a macro cell scenario is approximately 70% for AMR 12.2 codec, and it slightly increases when the VoIP packet payload decreases. This explains the small (<10%) gains in capacity achieved by reducing VoIP packet payload.

The performance of the semi-persistent scheduler does not suffer from control channel limitations, as initial transmissions are scheduled without associated control information by using persistently allocated time and frequency resources instead. The difference in control channel consumption for simulated packet scheduling methods is presented for the downlink direction in Figure 13.9, which contains the cumulative distribution function for the total number of consumed CCEs per TTI. With the dynamic scheduler all control channel elements are in use 70% of the time, if bundling is not allowed. With packet bundling, load is higher at 5% outage and hence control channel capacity could be more efficiently exploited implying that all CCEs are used almost 100% of the time. With the semi-persistent scheduler, the control channel overhead is clearly lower, the probability of using all CCEs being only slightly higher than 10%. Similar observations can be made from the distribution of consumed CCEs per TTI in the uplink direction, which is presented in Figure 13.10. As expected, the CCE consumption for semi-persistent scheduling is much lower than that for dynamic scheduling. Here we assume that the total number of CCEs reserved for downlink/uplink scheduling grants is 10.

The performance of the semi-persistent scheduler is not limited by the control channel resources, but it is limited by the available PDSCH bandwidth, which is illustrated

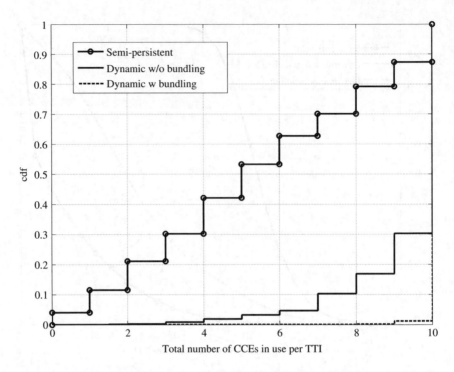

Figure 13.9   Cumulative distribution function for the total number of CCEs per TTI in downlink

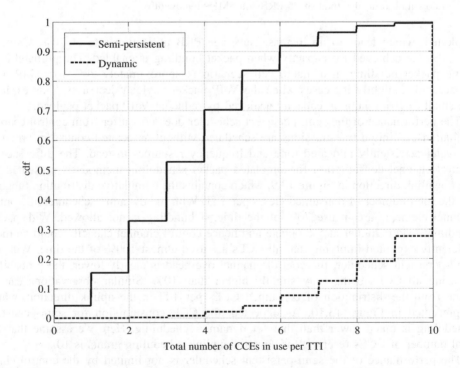

Figure 13.10   Cumulative distribution function for the total number of CCEs per TTI in uplink

in Figure 13.8, showing that the average PDSCH utilization rate for a semi-persistent scheduler is more than 90%. As the performance of the semi-persistent scheduler is user plane limited, the size of a VoIP packet has a significant impact on the capacity: an approximately 35% higher capacity is obtained if AMR 7.95 is used instead of AMR 12.2, whereas the use of AMR 5.9 instead of AMR 7.95 provides capacity gains of only 10%. The reason for the reduced gain for AMR 5.9 is presented in Figure 13.11, which shows the distribution of the size of the persistent resource allocation in terms of PRBs for all simulated codecs in the downlink direction. According to the distribution, the size of the persistent allocation is one (1) PRB for almost 50% of the users with AMR 7.95, and hence no savings in allocated bandwidth can be achieved for those users when using AMR 5.9 instead of AMR 7.95. In the downlink the size of the persistent allocation is calculated dynamically from wideband CQI for each talk spurt separately, whereas in the uplink, due to practical constraints, the size of the persistent resource allocation is decided from the path loss measurements of a user. The reason for the uplink solution is that the sounding is not accurate enough with a large number of users. The size of the persistent resource allocation has the following distribution during the uplink simulation: with AMR 12.2 all the users have a persistent allocation of size two (2) PRBs, whereas with AMR 7.95 half of the users have a persistent allocation of size one (1) PRB, and half of the users have a persistent allocation of size two (2) PRBs. With AMR 5.9, 90% of the users have a persistent allocation of size one (1) PRB, and others have a persistent allocation of size two (2) PRB. An approximately 33% higher capacity is obtained if AMR 7.95 is used instead of AMR 12.2. Furthermore, AMR 5.9 provides approximately

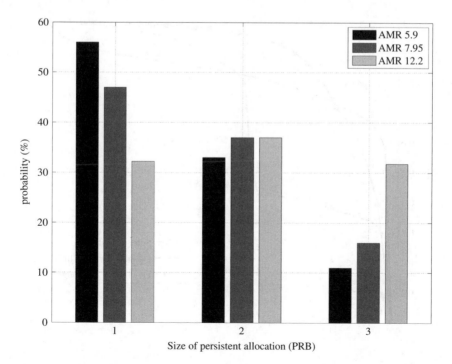

Figure 13.11 Probability distribution function for the size of the persistent resource allocation for different codecs in downlink

28% gains in capacity over AMR 7.95. The reduced gain is mainly due to a slightly increased number of retransmissions for AMR 5.9.

When comparing the schedulers in the downlink direction for experienced air interface delay, it is observed that due to the control channel limitations the relative amount of packets experiencing delay close to the used delay bound (50 ms) is clearly higher for the dynamic scheduler. With packet bundling, control channel limitations can be relaxed, and therefore delay critical packets can be scheduled earlier. Bundling itself adds delays in steps of 20 ms. For the semi-persistent scheduler, the time the packet has to wait in the packet scheduler buffer before initial scheduling takes place depends on the location of the persistent resource allocation in the time domain, which has a rather even distribution with a full load. Therefore the packet delay distribution for the semi-persistent scheduler is smooth. In all these schemes, retransmissions are scheduled at the earliest convenience, after an 8 ms HARQ cycle. Due to the first HARQ retransmission the distributions plateau at 10 ms. A cumulative distribution function for the packet delay is presented in Figure 13.12.

When summarizing VoIP system performance in downlink, it is concluded that the performance of the dynamic scheduler is control channel limited without packet bundling and hence the semi-persistent scheduler is able to have 50–125% higher capacities than the dynamic scheduler. The performance of the semi-persistent scheduler is user plane limited and the gains in capacity over the dynamic scheduler are smallest for AMR 12.2, which has the highest data rate amongst the simulated codecs and hence the biggest VoIP packet size. When the packet bundling is used with AMR 12.2, the control channel limitation

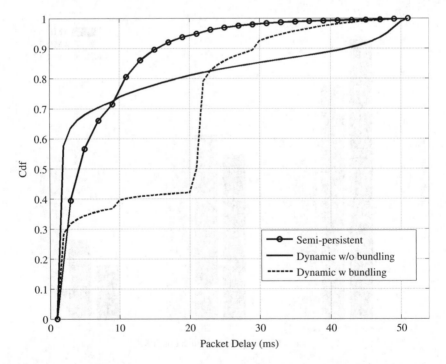

Figure 13.12 Cumulative distribution function for the experienced air interface delay in downlink with AMR 12.2

for the performance of the dynamic scheduler is less significant compared with lower rate codecs as the number of supported users at 5% outage is lower. Therefore, the dynamic scheduler can have a 15% gain in capacity over the semi-persistent scheduler if AMR 12.2 is used with the packet bundling. When bundling is used with lower rate codecs, the control channel capacity starts to limit the performance of the dynamic scheduler due to the high number of supported users at 5% outage, and hence the semi-persistent scheduler achieves 8–15% higher capacities than the dynamic scheduler.

When comparing the performances of different scheduling methods in the uplink, it is observed that (similar to downlink) the performance of the dynamic scheduler is purely control channel limited whereas the semi-persistent scheduler suffers much less from control channel limitation due to a looser control channel requirement. Therefore, the semi-persistent scheduler is able to have 40–80% capacity gains over the dynamic sched-uler with AMR 7.95 and AMR 5.9. Moreover, even with AMR 12.2 the semi-persistent scheduler achieves a 14% higher capacity than the dynamic scheduler.

The presented downlink results are obtained by using frequency dependent CQI report-ing. In practice the large number of supported VoIP users in LTE may necessitate the use of wideband CQI to keep the overhead from CQI feedback at a reasonable level. This would mean a lower uplink signaling overhead from CQI feedback at the cost of reduced capacity, as the frequency domain scheduling gains will be lost. To keep the capacity reduction as small as possible, the impact of lost frequency domain scheduling gains to performance should be compensated with more efficient use of frequency diversity. Therefore the use of the semi-persistent scheduler becomes even more attractive for VoIP traffic in LTE, as the performance of the semi-persistent scheduler is less dependent on the frequency domain scheduling gains than the dynamic scheduler. This is illustrated in Table 13.2, which presents the relative losses in capacity with AMR 12.2 when increasing the size of CQI reporting sub-band in the frequency domain. Relative losses are calculated against the presented capacity numbers, which were obtained assuming narrowband CQI reporting [7].

According to the simulations, the capacity of the dynamic scheduler is reduced by 15% due to the use of wideband CQI, whereas for the semi-persistent scheduler the corresponding loss is only 7%. Hence with wideband CQI the semi-persistent scheduler provides a similar performance to the dynamic scheduler for AMR 12.2, and for lower rate codecs the gains in capacity over the dynamic scheduler are increased further compared to the gains achieved with narrowband CQI.

Finally, in the light of the above results analysis, it seems that semi-persistent scheduling is the most attractive scheduling method for VoIP traffic in LTE. Nevertheless, as the used persistent allocation is indicated to the user via PDCCH signaling, some sort of

Table 13.2 Relative loss in capacity (%) due to increased CQI granularity

	CQI reporting sub-band size	
	4 PRBs	Wideband CQI
Semi-persistent scheduling	2%	7%
Dynamic without bundling	0%	3%
Dynamic with bundling	7%	15%

combination of dynamic and semi-persistent scheduling methods seems to be the best option for VoIP in LTE. Furthermore, when comparing the performances of downlink and uplink, it is concluded that the VoIP performance in LTE is uplink limited. Hence downlink capacity can accommodate additional asymmetric data traffic, e.g. web browsing.

13.7 Voice Capacity Evolution

This section presents the evolution of the voice spectral efficiency from GSM to WCDMA/HSPA and to LTE. Most of the global mobile voice traffic is carried with GSM EFR or AMR coding. The GSM spectral efficiency can typically be measured with Effective Frequency Load (EFL), which represents the percentage of the time slots of full rate users that can be occupied in case of frequency reuse one. For example, EFL = 8% corresponds to 8% × 8 slots/200 kHz = 3.2 users/MHz. The simulations and the network measurements show that GSM EFR can achieve EFL 8% and GSM AMR can achieve 20% assuming all terminals are AMR capable and the network is optimized. The GSM spectral efficiency can be further pushed by the network feature Dynamic Frequency and Channel Allocation and by using Single Antenna Interference Cancellation (SAIC), known also as Downlink Advanced Receiver Performance (DARP). We assume up to EFL 35% with those enhancements. For further information see [8] and [9]. The overhead from the Broadcast Control Channel (BCCH) is not included in the calculations. BCCH requires higher reuse than the hopping carriers.

WCDMA voice capacity is assumed to be 64 users with AMR 12.2 kbps, and 100 users with AMR 5.9 kbps on a 5 MHz carrier. HSPA voice capacity is assumed to be 123 users with AMR 12.2 kbps, and 184 users with AMR 5.9 kbps. For HSPA capacity evolution, see Chapter 17.

Capacity evolution is illustrated in Figure 13.13. LTE VoIP 12.2 kbps can provide efficiency which is 15× more than GSM EFR. The high efficiency can squeeze the

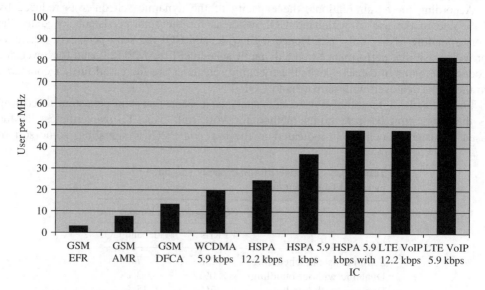

Figure 13.13 Voice spectral efficiency evolution (IC = Interference Cancellation)

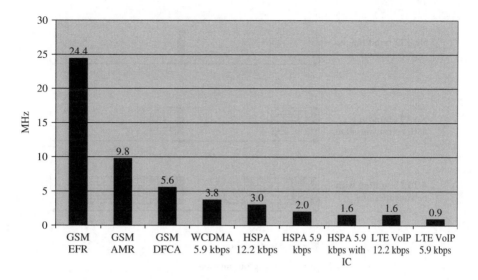

Figure 13.14　Spectrum required for supporting 1500 subscribers per sector at 40 mErl

voice traffic into a smaller spectrum. An example calculation is shown in Figure 13.14 assuming 1500 subscribers per sector and 40 mErl traffic per subscriber. GSM EFR would take 25 MHz of spectrum while LTE will be able to carry that voice traffic in less than 2 MHz. LTE can then free up more spectrum for data usage.

13.8　Uplink Coverage

Uplink coverage can be maximized when UE transmits continuously with maximum power. Since VoIP packets are small (< 40 bytes), they can easily fit into 1 ms TTI. The AMR packets arrive every 20 ms leading to only 5% activity in the uplink. The uplink coverage could be improved by increasing the UE transmission time by using retransmissions and TTI bundling. These solutions are illustrated in Figure 13.15. The number of retransmissions must be limited for VoIP to remain within the maximum delay budget. The maximum number of retransmissions is six assuming a maximum 50 ms total delay, since the retransmission delay is 8 ms in LTE.

TTI bundling can repeat the same data in multiple (up to four) TTIs [4, 10]. TTI bundling effectively increases the TTI length allowing the UE to transmit for a longer time. A single transport block is coded and transmitted in a set of consecutive TTIs. The same hybrid ARQ process number is used in each of the bundled TTIs. The bundled TTIs are treated as a single resource where only a single grant and a single acknowledgement are required. TTI bundling can be activated with a higher layer signaling per UE. The trigger could be, for example, UE reporting its transmit power is getting close to the maximum value.

The resulting eNodeB sensitivity values with retransmission and TTI bundling are shown in Table 13.3. The eNodeB sensitivity can be calculated as follows:

$$eNodeB\_sensitivity[dBm] = -174 + 10 \cdot \log_{10}(Bandwidth) + Noise\_figure + SNR$$

$$(13.2)$$

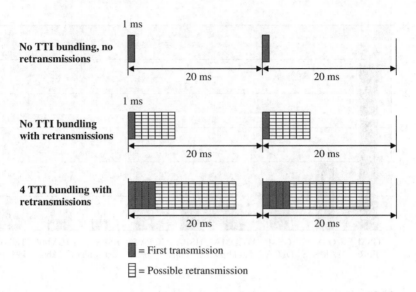

Figure 13.15   TTI bundling for enhancing VoIP coverage in uplink

Table 13.3   Uplink VoIP sensitivity with TTI bundling [11]

Number of TTIs bundled	1	4
Transmission bandwidth	360 kHz (2 resource blocks)	360 kHz (2 resource blocks)
Number of retransmissions	6	3
Required SNR	−4.2 dB	−8.0 dB
Receiver sensitivity	−120.6 dBm	−124.4 dBm

The eNodeB receiver noise figure is assumed to be 2 dB, two resource blocks are used for voice and no interference margin is included. The bundling can improve the uplink voice coverage by 4 dB.

The WCDMA NodeB sensitivity can be estimated by assuming Eb/N0 = 4.5 dB [12], which gives −126.6 dBm. To get similar voice coverage in LTE as in WCDMA we need to apply TTI bundling with a sufficient number of retransmissions. In terms of Hybrid-ARQ (HARQ) acknowledgement and retransmission timing, the TTI bundling method illustrated in Figure 13.16 is adopted for the LTE specifications [4, 10]. According to this method four (4) subframes are in one bundle for the Frequency Division Duplex (FDD) system. Within a bundle, the operation is like autonomous retransmission by the UE in consecutive subframes without waiting for ACK/NACK feedback. The Redundancy Version (RV) on each autonomous retransmission in consecutive subframes changes in a pre-determined manner.

HARQ acknowledgement is generated after receiving the last subframe in the bundle. The timing relation between the last subframe in the bundle and the transmission instant of the HARQ acknowledgement is identical to the case of no bundling. If the last subframe in a bundle is subframe N then the acknowledgement is transmitted in subframe $N + 4$. If the first subframe in a bundle is subframe k then any HARQ retransmissions begin in subframe k + 2 × HARQ RTT.

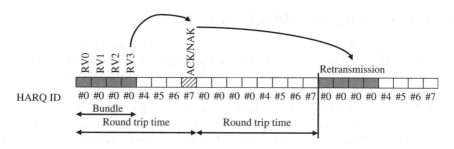

Figure 13.16 TTI bundling combining with HARQ

Table 13.4 Impact of TTI bundling on performance

	No bundling	TTI bundling
Maximal collected TTI per packet	7	12
VoIP capacity with TTI bundling at 5MHz	<120	144

In the following, the impact of four (4) TTI bundling on capacity is studied via system level simulations in a bad coverage limited macro cell scenario 3 [2]. Main system simulation parameters are aligned with [7]. Table 13.4 presents both the energy accumulation and obtained capacity for AMR 12.2 when dynamic scheduling with 50 ms packet delay budget is used. By using TTI bundling, 2.34 dB energy accumulation gain can be achieved, so VoIP capacity can be improved by 20% at least by using TTI bundling compared to no bundling.

Moreover, as the results in Table 13.5 show, the air interface delay has a clear impact on the achieved VoIP capacity and also on the gains that can be obtained from TTI bundling, i.e., with a looser delay budget capacity gains from TTI bundling are increased. This is evident as the longer air interface delay allows more energy to be accumulated from bundled TTIs, e.g. for 50 ms air interface delay, the packet will cumulate the energy from at most 12 TTIs assuming four (4) TTI bundling. Furthermore, for 70 ms air interface delay the corresponding number is increased to 20 TTIs, which would mean that 66% more energy could be accumulated. Hence the longer air interface delay implies bigger combining gains for the received signal from the bundled TTIs, and therefore the coverage is increased as the air interface delay is increased. This is again mapped to capacity gains. Besides, a longer delay budget means that the realized time domain scheduling gains are increased as each packet has more opportunities to be scheduled, and this further improves the coverage and hence improves the capacity. All in all, due to this reasoning the capacity with a 70 ms delay bound is approximately 13% higher than the capacity with a 50 ms delay bound.

Table 13.5 Performance comparison for different delay budgets

Delay budget (ms)	50	60	70
Maximal collected TTI per packet	12	16	20
VoIP capacity with TTI bundling at 5MHz	144	155	163

13.9 Circuit Switched Fallback for LTE

Voice service with LTE terminals can also be offered before high quality VoIP support is included into LTE radio and before IP Multimedia Subsystem (IMS) is deployed. The first phase can use so-called Circuit Switched (CS) fallback for LTE where the LTE terminal will be moved to GSM, WCDMA or CDMA network to provide the same services that exists in CS networks today, for example voice calls, video call or Location Services [13]. The CS fallback procedures require that the Mobility Management Entity (MME) as well as Mobile Switching Center (MSC)/Visiting Location Register (VLR) network elements are upgraded to support procedures described in this section. The CS fallback is described in [13].

When LTE UE executes the attach procedure towards the LTE core network, the LTE core network will also execute a location update towards the serving CS core network to announce the presence of the terminal to the CS core network via the LTE network in addition to executing the normal attach procedures. The UE sends the attach request together with specific 'CS Fallback Indicator' to the MME, which starts the location update procedure towards MSC/VLR via an IP based SGs interface. The new Location Area Identity (LAI) is determined in the MME based on mapping from the Tracking area. A mapped LAI could be either GERAN or UTRAN based on the operator configuration. Additionally GERAN and UTRAN can be served by different MSC/VLR in the network architecture, which causes execution of a roaming retry for the CS fallback procedure as described later. The VLR creates an association with the MME. This procedure is similar to the combined LA/RA update supported in GERAN by using Network Mode of Operation 1 (NMO1) with Gs interface.

If MME supports the connection to the multiple core network nodes similarly as SGSN does in GERAN/UTRAN, then a single MME can be connected to multiple MSC/VLR network elements in the CS core network. The benefit is the increased overall network resiliency compared to the situation when the MME is connected to only a single MSC/VLR.

For a mobile terminated call, MSC/VLR sends a paging message via SGs interface to the correct MME based on the location update information. eNodeB triggers the packet handover or network assisted cell change from LTE to the target system. The UE sends the paging response in the target system and proceeds with the CS call setup.

An overview of the mobile terminated fallback handover is shown in Figure 13.17.

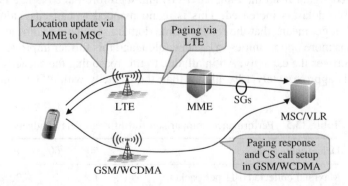

Figure 13.17 Circuit switched fallback handover – mobile terminated call

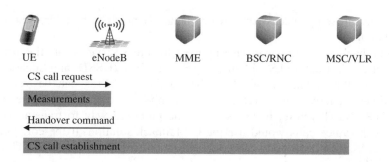

Figure 13.18   Circuit switched fallback for LTE – mobile originated call

An overview of the mobile originated call is shown in Figure 13.18. The UE sends a CS call request to the eNodeB, which may ask the UE to measure the target cell. eNodeB triggers the handover to the target system. the UE starts the normal CS call establishment procedure in the target cell. Once the CS call ends, the UE again reselects LTE to get access to high data rate capabilities.

If the service areas covered by LTE Tracking Area (TA) and GERAN/UTRAN Location Area (LA) are not similar, the serving MSC/VLR of LTE UE is different from MSC/VLR that serves the target LA. The target MSC/VLR is not aware of the UE and is not expecting a response to the paging message. This case is handled with a procedure called as Roaming Retry for CS fallback. When LTE UE has received the paging request, it executes an additional location update procedure to the target MSC/VLR to make the target MSC/VLR aware of the existence of the UE. Based on this location update the HLR of the LTE UE will send a cancel message to the MSC/VLR that was serving the LTE UE prior to the CS fallback attempt. This cancel message will stop the paging process in that MSC/VLR and will trigger the resume call handling procedure. This procedure is similar to the Mobile Terminating Roaming Retry Call procedure used in GERAN/UTRAN to request a call to be re-routed to a new MSC/VLR from a gateway MSC. The gateway MSC will re-execute the mobile terminating call procedure by executing a HLR enquiry resulting in the call being routed to the correct MSC/VLR.

If there is an active packet session in LTE and if the target system supports simultaneous voice and data connection, the packet session is relocated from LTE to the target system. If the target system does not support simultaneous voice and data, the data connection is suspended during the CS voice call. This happens when LTE UE is moved to a GERAN that does not support simultaneous voice and data connection, called Dual Transfer Mode (DTM).

Voice is one of the most demanding services in terms of delay requirements and in terms of call reliability. Many optimized GERAN (GSM) and UTRAN (WCDMA) networks today can provide call drop rates below 0.5% or even below 0.3%. The advantage of the fallback handover is that there is no urgent need to implement QoS support in LTE or SAE. The network optimization may also be less stringent if the voice is not supported initially. Also, there is no need to deploy IMS just because of the voice service. From the end user point of view, the voice service in GSM or in WCDMA is as good as voice service in LTE, except that the highest data rates are not available during the voice call. On the other hand, the CS fallback for LTE adds some delay to the call setup process as the UE must search for the target cell. With high mobility, the CS fallback for LTE may also affect the call setup success rate, which naturally needs to be addressed in practical implementation.

Similar fallback procedures can also be used for other services familiar from today's network such as Unstructured Supplementary Service Data (USSD) and Location Services (LCS). Due to the popularity of SMS, the number of SMSs can be very high and the CS fallback handover may create a large number of reselections between LTE and GERAN/UTRAN. Therefore, it was seen as the preferred way to transmit SMSs over LTE even if the voice calls would use the CS fallback solution. If the core network has support for IMS, then SMS body can be transferred over SIP messages in LTE as defined by 3GPP [14].

13.10   Single Radio Voice Call Continuity (SR-VCC)

Once the VoIP call is supported in LTE, there may still be a need to make a handover from LTE VoIP to a GSM, WCDMA or CDMA CS voice network when running out of LTE coverage. The handover functionality from VoIP to the CS domain is referred to as Single Radio Voice Call Continuity (SR-VCC). The solution does not require UE capability to simultaneously signal on two different radio access technologies – therefore it is called a Single Radio Solution. SR-VCC is illustrated in Figure 13.19. SR-VCC is defined in [15]. The Dual radio Voice call continuity was already defined in 3GPP Release 7 for voice call continuation between WLAN and GSM/WCDMA networks [16].

The selection of the domain or radio access is under the network control in SR-VCC and SR-VCC enhanced MSC Server (called 'S-IWF' in this chapter) deployed into the CS core network. This architecture has been defined to enable re-use of already deployed CS core network assets to provide the necessary functionality to assist in SR-VCC. This architecture option is illustrated in Figure 13.20.

S-IWF, co-located in the MSC Server, uses a new GTP based Sv interface towards MME function in the LTE domain to trigger the SR-VCC procedure. Architecturally the S-IWF acts similarly to an anchor MSC Server towards target 2G/3G CS domain and is also responsible for preparing the needed target radio access network resources jointly with target GERAN or UTRAN. S-IWF also connects the speech path from the target CS domain towards the other subscriber in the voice call and together with the VCC anchor also hides mobility between LTE VoIP and 2G/3G CS domain from the other side of the call. The VCC anchor is located at the IMS application server and based on the same concept that was defined by 3GPP for Release 7 WLAN Voice Call Continuity.

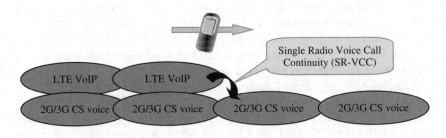

Figure 13.19   Single radio voice call continuity (SR-VCC)

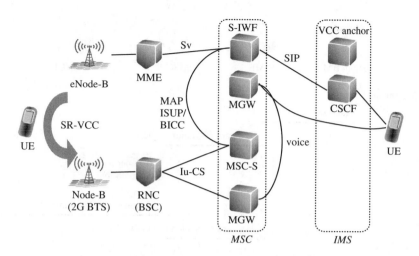

Figure 13.20 Architecture for SR-VCC. CSCF, call session control function; MGW, media gateway; MAP, mobile application port; ISUP, ISDN user part; BICC, bearer independent call control protocol

S-IWF implementation has been specified so that it may be deployed with IMS Centralized Service (ICS) architecture. S-IWF then uses Mw interface (acting as the MSC Server enhanced for ICS) towards IMS instead of, for instance, the ISUP interface. In the former situation, handover causes registration prior to a new call towards IMS on behalf of the served subscriber whereas in the latter situation no registration is needed. This kind of support for SR-VCC with both ICS and non-ICS architectures is essential to achieve the required deployment flexibility.

During the LTE attach procedure, the MME will receive the required SR-VCC Domain Transfer Number from HSS that is then given to S-IWF via the Sv interface. S-IWF uses this number to establish a connection to the VCC anchor during the SR-VCC procedure. When the VoIP session is established, it will be anchored within IMS in the VCC anchor to use SR-VCC in case it is needed later in the session. This anchoring occurs in both the originating and terminating sessions based on IMS subscription configuration.

The SR-VCC procedure is presented in Figure 13.21. LTE eNodeB first starts inter-system measurements of the target CS system. eNodeB sends the handover request to the MME, which then triggers the SR-VCC procedure via the Sv interface to the MSC-Server S-IWF functionality with Forward Relocation Request. S-IWF in the MSC-Server initiates the session transfer procedure towards IMS by establishing a new session towards the VCC anchor that originally anchored the session. The session is established by S-IWF using the SR-VCC number provided by MME. S-IWF also coordinates the resource reservation in the target cell together with the target radio access network. The MSC Server then sends a Forward Relocation Response to MME, which includes the necessary information for the UE to access the target GERAN/UTRAN cell.

After the SR-VCC procedure has been completed successfully the VoIP connection is present from Media Gateway that is controlled by S-IWF towards the other side of an ongoing session. The CS connection exists towards the radio access network to which the UE was moved during the procedure.

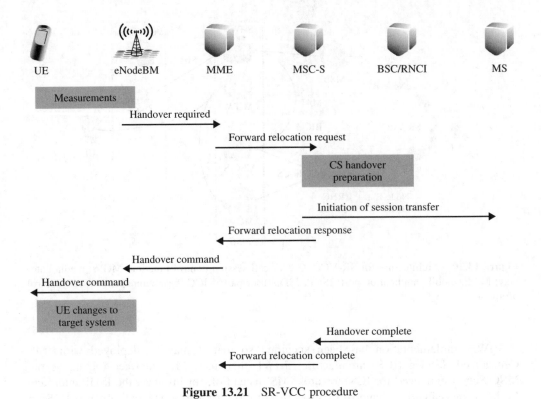

Figure 13.21   SR-VCC procedure

S-IWF functionality is not involved in a normal LTE VoIP session if the handover to CS domain is not required.

For a simultaneous voice and non-voice data connection, the handling of the non-voice bearer is carried out by the bearer splitting function in the MME. The MME may suppress the handover of a non-voice PS bearer during the SR-VCC procedure. This may happen if the target GERAN system does not support simultaneous voice and data functionality (DTM). If the non-voice bearer is also handed over, the process is done in the same way as the normal inter-system handover for packet services. The MME is responsible for coordinating the Forward Relocation Response from the SR-VCC and the packet data handover procedure.

For roaming, the Visited Public Land Mobile Network (PLMN) controls the radio access and domain change while taking into account any related Home PLMN policies.

13.11 Summary

LTE radio is not only optimized for high data rates and high data capacities, but also for high quality voice with high capacity. The system simulations show that LTE can support 50–80 simultaneous voice users per MHz per cell in the macro cellular environment with AMR data rates between 12.2 and 5.9 kbps. The lower AMR rates offer the highest capacity. The highest capacity is achieved using semi-persistent packet scheduling where the first transmission has pre-allocated resources and only retransmissions are scheduled.

The capacity with fully dynamic scheduling is limited by control channel capacity since each voice packet needs to be scheduled separately.

Voice quality depends on the voice codec and on the end-to-end delay. It is assumed that in the first phase, VoIP in LTE uses already standardized 3GPP speech codec AMR narrowband and AMR wideband due to backward compatibility with the legacy network. For end-to-end voice delay, LTE can provide values below 200 ms offering a similar or lower delay than the existing CS networks.

LTE uplink voice coverage can be optimized by using TTI bundling, which allows the combination of multiple 1-ms TTIs to increase the uplink transmission time and the average transmitted energy. The LTE voice coverage with TTI bundling is similar to the existing 2G/3G voice service.

As LTE supports only packet based services, then the voice service is also based on Voice over IP (VoIP). LTE specifications include definitions for inter-working with the existing Circuit Switched (CS) networks. One option is to use CS fallback for LTE where the UE moves to the 2G/3G network to make the voice call using a CS connection and returns to LTE when the voice call is over. Another option is to use VoIP in LTE and then allow inter-system handover to the 2G/3G CS network when LTE coverage is running out. This handover is called Single Radio Voice Call Continuity (SR-VCC). The CS fallback handover has been assumed to enable re-use of existing deployed CS core network investments as well as to provide a similar look and feel for the end users who are accustomed to current mobile network services. In some cases this kind of solution can be assumed to provide the first phase solution, but the IMS based core network architecture with support for VoIP and Multimedia Telephony is considered to be the long term solution.

Both CS fallback for LTE as well as the LTE based VoIP with SR-VCC solution can be deployed either in a phased manner or simultaneously depending on terminal capabilities. Additionally it is also possible to deploy IMS for non-VoIP services such as video sharing, presence or instant messaging, even primary voice service, by using CS fallback for LTE procedures. This kind of architecture ensures maximum flexibility for the operator to plan and deploy LTE as well as IMS solutions.

References

[1] 3GPP Technical Specifications 23.053 'Tandem Free Operation (TFO); Service description; Stage 2', V.8.3.0.
[2] ITU-T Recommendation G.114 'One way transmission time', May 2003.
[3] 3GPP Technical Specification 23.203 'Policy and charging control architecture', V.8.3.1.
[4] 3GPP Technical Specifications 36.321 'Medium Access Control (MAC) protocol specification', V.8.3.0.
[5] 3GPP TS36.300, 'Evolved Universal Terrestrial Radio Access (E-UTRA) and Evolved Universal Terrestrial Radio Access Network (E-UTRAN); Overall description; Stage 2 (Release 8)'.
[6] 3GPP TR 25.814 v7.0.0, 'Physical layer aspect for evolved Universal Terrestrial Radio Access (UTRA)'.
[7] 'Next Generation Mobile Networks (NGMN) Radio Access Performance Evaluation Methodology', A White paper by the NGMN Alliance, January 2008.
[8] T. Halonen, J. Romero, J. Melero, 'GSM, GPRS and EDGE Performance', 2nd Edition, John Wiley & Sons, 2003.
[9] A. Barreto, L. Garcia, E. Souza, 'GERAN Evolution for Increased Speech Capacity', Vehicular Technology Conference, 2007. VTC2007-Spring. IEEE 65th, April 2007.
[10] 3GPP R2-082871, 'LS on TTI Bundling', RAN1#62, Kansas City, USA, May 2008.
[11] R1-081856, 'Coverage Comparison between PUSCH, PUCCH and RACH', Nokia/Nokia Siemens Networks, Catt, RAN1#54, Kansas City, USA, May 2008.

[12] H. Holma, A. Toskala, 'WCDMA for UMTS', 4th Edition, John Wiley & Sons, 2007.
[13] 3GPP Technical Specifications 23.272 'Circuit Switched (CS) fallback in Evolved Packet System (EPS)', V. 8.1.0.
[14] 3GPP Technical Specifications 23.204 'Support of Short Message Service (SMS) over generic 3GPP Internet Protocol (IP) access', V.8.3.0.
[15] 3GPP Technical Specifications 23.216 'Single Radio Voice Call Continuity (SRVCC)', V. 8.1.0.
[16] 3GPP Technical Specifications 23.206 'Voice Call Continuity (VCC) between Circuit Switched (CS) and IP Multimedia Subsystem (IMS)', V.7.5.0.

14

Performance Requirements

Andrea Ancora, Iwajlo Angelow, Dominique Brunel, Chris Callender,
Harri Holma, Peter Muszynski, Earl Mc Cune and Laurent Noël

14.1 Introduction

3GPP defines minimum RF performance requirements for terminals (UE) and for base stations (eNodeB). These performance requirements are an essential part of the LTE standard as they facilitate a consistent and predictable system performance in a multi-vendor environment. The relevant specifications are [1–3] covering both duplex modes of LTE: frequency division duplex (FDD) and time division duplex (TDD).

Some of the RF performance requirements, for example limits for unwanted emissions, facilitate the mutual co-existence of LTE with adjacent LTE systems or adjacent 2G/3G systems run by different operators without coordination. The corresponding requirements may have been derived from either regulatory requirements or from co-existence studies within 3GPP – see [4].

Other performance requirements, for example modulation accuracy and baseband demodulation performance requirements, provide the operator assurance of sufficient performance within his deployed LTE carrier.

[2] is the basis for the eNodeB test specification [5], which defines the necessary tests for type approval. Correspondingly, [1] is the basis on the UE side for the test specifications [6] and [7].

This chapter presents the most important LTE minimum performance requirements, the rationale underlying these requirements and the implications for system performance and equipment design. Both Radio Frequency (RF) and baseband requirements are considered. First, the eNodeB requirements are considered and then the UE requirements.

14.2 Frequency Bands and Channel Arrangements

14.2.1 Frequency Bands

Table 14.1 lists the currently defined E-UTRA frequency bands, together with the corresponding duplex mode (FDD or TDD). There are currently 22 bands defined for FDD and

LTE for UMTS: Evolution to LTE-Advanced, Second Edition. Edited by Harri Holma and Antti Toskala.
© 2011 John Wiley & Sons, Ltd. Published 2011 by John Wiley & Sons, Ltd.

Table 14.1   E-UTRA frequency bands

E-UTRA Band	Uplink eNode B receive UE transmit			Downlink eNode B transmit UE receive			Duplex mode
1	1920 MHz	–	1980 MHz	2110 MHz	–	2170 MHz	FDD
2	1850 MHz	–	1910 MHz	1930 MHz	–	1990 MHz	FDD
3	1710 MHz	–	1785 MHz	1805 MHz	–	1880 MHz	FDD
4	1710 MHz	–	1755 MHz	2110 MHz	–	2155 MHz	FDD
5	824 MHz	–	849 MHz	869 MHz	–	894 MHz	FDD
6	830 MHz	–	840 MHz	875 MHz	–	885 MHz	FDD
7	2500 MHz	–	2570 MHz	2620 MHz	–	2690 MHz	FDD
8	880 MHz	–	915 MHz	925 MHz	–	960 MHz	FDD
9	1749.9 MHz	–	1784.9 MHz	1844.9 MHz	–	1879.9 MHz	FDD
10	1710 MHz	–	1770 MHz	2110 MHz	–	2170 MHz	FDD
11	1427.9 MHz	–	1452.9 MHz	1475.9 MHz	–	1500.9 MHz	FDD
12	698 MHz	–	716 MHz	728 MHz	–	746 MHz	FDD
13	777 MHz	–	787 MHz	746 MHz	–	756 MHz	FDD
14	788 MHz	–	798 MHz	758 MHz	–	768 MHz	FDD
17	704 MHz	–	716 MHz	734 MHz	–	746 MHz	FDD
18	815 MHz	–	830 MHz	860 MHz	–	875 MHz	FDD
19	830 MHz	–	845 MHz	875 MHz	–	890 MHz	FDD
20	832 MHz	–	862 MHz	791 MHz	–	821 MHz	FDD
21	1447.9 MHz	–	1462.9 MHz	1495.9 MHz	–	1510.9 MHz	FDD
22	3410 MHz	–	3500 MHz	3510 MHz	–	3600 MHz	FDD
23	2000 MHz	–	2020 MHz	2180 MHz	–	2200 MHz	FDD
24	1626.5 MHz	–	1660.5 MHz	1525 MHz	–	1559 MHz	FDD
...							
33	1900 MHz	–	1920 MHz	1900 MHz	–	1920 MHz	TDD
34	2010 MHz	–	2025 MHz	2010 MHz	–	2025 MHz	TDD
35	1850 MHz	–	1910 MHz	1850 MHz	–	1910 MHz	TDD
36	1930 MHz	–	1990 MHz	1930 MHz	–	1990 MHz	TDD
37	1910 MHz	–	1930 MHz	1910 MHz	–	1930 MHz	TDD
38	2570 MHz	–	2620 MHz	2570 MHz	–	2620 MHz	TDD
39	1880 MHz	–	1920 MHz	1880 MHz	–	1920 MHz	TDD
40	2300 MHz	–	2400 MHz	2300 MHz	–	2400 MHz	TDD
41	2496 MHz	–	2690 MHz	2496 MHz	–	2690 MHz	TDD

nine bands for TDD. Whenever possible, the RF requirements for FDD and TDD have been kept identical in order to maximize the commonality between the duplex modes.

While the physical layer specification and many RF requirements are identical for these frequency bands, there are some exceptions to this rule for the UE RF specifications, as will be discussed further in UE-related sections in this chapter. On the other hand, the eNodeB RF requirements are defined in a frequency-band agnostic manner as there are less implementation constraints for base stations. If the need arises, further LTE frequency bands can be easily added affecting only isolated parts of the RF specifications. Furthermore, the specified LTE frequency variants are independent of the underlying LTE release feature content (Release 8, 9, and so forth). Those frequency variants that were added during Release 9 or 10 timeframes can still be implemented using just Release 8 features.

Table 14.2 Usage of the frequency variants with the world's regions

	Europe	Asia	Japan	Americas
Band 1	x	x	x	
Band 2				x
Band 3	x	x		
Band 4				x
Band 5		x		x
Band 6			x	
Band 7	x	x		
Band 8	x	x		
Band 9			x	
Band 10				x
Band 11			x	
Band 12				x
Band 13				x
Band 14				x
Band 17				x
Band 18			x	
Band 19			x	
Band 20	x			
Band 21			x	
Band 22	x	x	x	x
Band 23				x
Band 24				x

The band numbers 15 and 16 are skipped because those numbers were used in ETSI specifications.

Not all of these bands are available in each of the world's regions. Table 14.2 illustrates the typical deployment areas for the different FDD frequency variants.

The most relevant FDD bands in Europe are as follows:

- Band 7 is the new 2.6 GHz band. The 2.6 GHz auctions have been running in a number of countries since 2007, and continue further during 2011.
- Band 8 is currently used mostly by GSM. The band is attractive from a coverage point of view due to the lower propagation losses. The band can be reused for LTE or for HSPA. The first commercial HSPA900 network started in Finland in November 2007. LTE refarming is considered in Chapter 10.
- Band 3 is also used by GSM, but in many cases Band 3 is not as heavily utilized by GSM as Band 8. That makes refarming for LTE simpler. Band 3 is also important in Asia where Band 7 is not generally available.
- Band 20 will be used in Europe for LTE. The first auction took place in Germany in May 2010.

Correspondingly, the new bands in the USA are Bands 4, 12, 13, 14, and 17. The LTE refarming can be carried out into Bands 2 and 5. The LTE deployment in Japan will utilize

Bands 1, 9, 11 and 18/19. In summary, LTE deployments globally will use a number of different frequency bands from the beginning.

Further frequencies for IMT (International Mobile Telephony) have been identified in ITU WRC 2007 conference – see Chapter 1.

14.2.2   Channel Bandwidth

The width of a LTE carrier is defined by the concepts of *channel bandwidth* ($BW_{Channel}$) and *transmission bandwidth configuration* (N_{RB}), see Figure 14.1. Their mutual relationship is specified in Table 14.3.

The transmission bandwidth configuration, N_{RB}, is defined as the maximum number of *resource blocks* (RB) that can be allocated within a LTE RF channel. A RB comprises 12 subcarriers and can thus be understood to possess a nominal bandwidth of 180 kHz. While the physical layer specifications allow N_{RB} to assume any value within the range $6 \leq N_{RB} \leq 110$, all RF requirements (and thus a complete LTE specification) are only defined for the values in Table 14.3. However, this flexibility within the LTE specifications supports the addition of further options for transmission bandwidth configurations (channel bandwidths), should a need arise.

The channel bandwidth is the relevant RF related parameter for defining *out-of-band* (OOB) emission requirements (for example, a spectrum emission mask, ACLR). The RF channel edges are defined as the lowest and highest frequencies of the carrier separated by the channel bandwidth – at $F_C +/- BW_{Channel}/2$.

It may be noted from Table 14.3 that the transmission bandwidth configuration measures only 90% of the channel bandwidth in case of 3, 5, 10, 15, 20 MHz LTE and less so

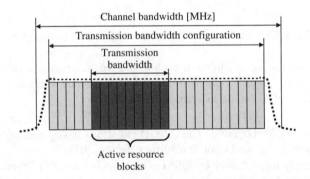

Figure 14.1 Definition of channel bandwidth and transmission bandwidth configuration for one E-UTRA carrier

Table 14.3   Transmission bandwidth configuration N_{RB} in E-UTRA channel bandwidths

Channel bandwidth $BW_{Channel}$ (MHz)	1.4	3	5	10	15	20
Transmission bandwidth configuration N_{RB}	6	15	25	50	75	100

for 1.4 MHz LTE. For example, for 5 MHz LTE one obtains a transmission bandwidth of 25*180 kHz = 4.5 MHz. In fact, the basic OFDM spectrum comprises only slowly decaying sidelobes, with a rolloff as described by the *sinc()* function. Therefore, efficient usage of the spectrum requires the use of filtering to effectively confine the OOB emissions. However, such filters require a certain amount of transition bandwidth in order (a) to be practical and (b) to consume only a small amount of the cyclic prefix duration due to the inevitable time dispersion they will cause. The values in Table 14.3, in particular for the LTE channel bandwidth of 1.4 MHz, is therefore a compromise between in- and out-of channel distortions and were extensively studied in 3GPP. The transmission bandwidth is 77% of the channel bandwidth in case of LTE 1.4 MHz.

The 1.4 and 3 MHz channel bandwidth options have been chosen to facilitate a multitude of cdma2000 and GSM migration scenarios within Bands 2, 3, 5 and 8.

Not all combinations of E-UTRA frequency bands and channel bandwidth options are meaningful and Table 14.4 provides the combinations supported by the standard. These combinations were based on inputs from the operators. It may be noted from Table 14.4, that the options with a wider channel bandwidth are typically supported for bands with a larger amount of available spectrum: e.g. 20 MHz LTE is supported in Bands 1, 2, 3, 4 but not in Bands 5, 8, and so forth. Conversely, LTE channel bandwidths below 5 MHz are typically supported in frequency bands with either smaller amount of available spectrum (for example, Bands 5, 8, and so forth) or bands exhibiting 2G migration scenarios (for instance, Bands 2, 5 and 8).

Furthermore [1] defines two levels of UE sensitivity requirements depending on the combination of bandwidths and operating band; these are also listed in Table 14.4. This is due to the fact that there will be increased UE self-interference in those frequency bands which possess a small duplex separation and/or gap. In case of large bandwidth and small duplex separation, certain relaxations of the UE performance are allowed or UE functionality is limited. For other cases, the UE needs to meet the baseline RF performance requirements.

14.2.3 Channel Arrangements

The channel raster is 100 kHz for all E-UTRA frequency bands, which means that the carrier center frequency must be an integer multiple of 100 kHz. For comparison, UMTS uses a 200 kHz channel raster and required additional RF channels for some frequency bands, offset by 100 kHz relative to the baseline raster, in order to effectively support the deployment within a 5 MHz spectrum allocation. The channel raster of 100 kHz for LTE will support a wide range of migration cases.

The spacing between carriers will depend on the deployment scenario, the size of the frequency block available and the channel bandwidths. The *nominal channel spacing* between two adjacent E-UTRA carriers is defined as follows:

$$Nominal\ Channel\ spacing = (BW_{Channel(1)} + BW_{Channel(2)})/2 \qquad (14.1)$$

where $BW_{Channel(1)}$ and $BW_{Channel(2)}$ are the channel bandwidths of the two respective E-UTRA carriers. The LTE – LTE co-existence studies within 3GPP [4] assumed this channel spacing.

However, the nominal channel spacing can be adjusted in order to optimize performance in a particular deployment scenario, for example in case of coordination between adjacent LTE carriers.

Table 14.4 Supported transmission bandwidths with normal sensitivity ('X') and relaxed sensitivity ('O')

E-UTRA Band	E-UTRA band/channel bandwidth					
	1.4 MHz	3 MHz	5 MHz	10 MHz	15 MHz	20 MHz
1			X	X	X	X
2	X	X	X	X	O	O
3	X	X	X	X	O	O
4	X	X	X	X	X	X
5	X	X	X	O		
6			X	O		
7			X	X	X	O
8	X	X	X	O		
9			X	X	O	O
10			X	X	X	X
11			X	O	O	O
12						
13	X	X	O	O		
14	X	X	O	O		
17						
18			X	O	O	
19			X	O	O	
20			X	O	O	O
21			X	O	O	
...						
33			X	X	X	X
34			X	X	X	
35	X	X	X	X	X	X
36	X	X	X	X	X	X
37			X	X	X	X
38			X	X		
39			X	X	X	X
40				X	X	X
41			X	X	X	X

14.3 eNodeB RF Transmitter

The LTE eNode-B transmitter RF requirements are defined in [2] and the corresponding test specification in [5]. In this section we discuss two of the most important transmitter requirements:

- Unwanted emissions, both inside and outside the operating band. The corresponding requirements will ensure the RF compatibility of the LTE downlink with systems operating in adjacent (or other) frequency bands.
- Transmitted signal (modulation) quality, or also known as *error vector magnitude* (EVM) requirements. These requirements will determine the in-channel performance for the transmit portion of the downlink.

Coexistence of LTE with other systems, in particular between the LTE FDD and TDD modes, will also be discussed, both in terms of the 3GPP specifications as well as within the emerging regulative framework within Europe for Band 7.

14.3.1 Operating Band Unwanted Emissions

For UMTS the *spurious emissions requirements* as recommended in ITU-R SM.329 are applicable for frequencies that are greater than 12.5 MHz away from the carrier center frequency. The 12.5 MHz is derived as 250% of the *necessary bandwidth* (5 MHz for UMTS) as per ITU-R SM.329. The frequency range within 250% of the necessary bandwidth around the carrier center may be referred to as the *out-of-band* (OOB) domain. Transmitter intermodulation distortions manifest themselves predominantly within the OOB domain and therefore more relaxed emission requirements, such as Adjacent Channel Leakage Ratio (ACLR), are typically applied within the OOB domain.

In LTE the channel bandwidth can range from 1.4 to 20 MHz. A similar scaling by 250% of the channel bandwidth would result in a large OOB domain for LTE: for the 20 MHz LTE channel bandwidth option the OOB domain would extend to a large frequency range of up to +/−50 MHz around the carrier centre frequency. In order to protect services in adjacent bands in a more predictable manner and to align with UMTS, the *LTE spurious domain* is defined to start from 10 MHz below the lowest frequency of the eNodeB transmitter operating band and from 10 MHz above the highest frequency of the eNodeB transmitter operating band, as shown in Figure 14.2. The operating band plus 10 MHz on each side are covered by the *LTE operating band unwanted emissions*.

LTE spurious domain emission limits in [2] are defined as per ITU-R SM.329 and are divided into several categories, where Categories A and B are applied as regional requirements. Within Europe the Category B limit of −30 dBm/MHz is required in the frequency range 1 GHz ↔ 12.75 GHz while the corresponding Category A limit applied in the Americas and in Japan is −13 dBm/MHz.

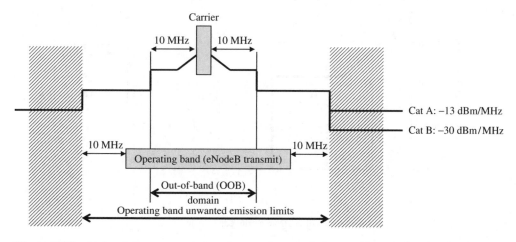

Figure 14.2 Defined frequency ranges for spurious emissions and operating band unwanted emissions

In addition to the ITU-R SM.329 spurious emission limits, [2] defines more stringent limits across the operating bands of a number of wireless systems, including UMTS, GSM, PHS (Personal Handyphone System).

The *operating band unwanted emission limits* are defined as *absolute* limits by a mask that stretches from 10 MHz below the lowest frequency of the eNodeB transmitter operating band up to 10 MHz above the highest frequency of the eNodeB transmitter operating band, as shown in Figure 14.2.

This mask depends on the LTE channel bandwidth and is shown in Figure 14.3 for E-UTRA bands >1 GHz and the limit of −25 dBm in 100 kHz as the lower bound for the unwanted emission limits. The limit of −25 dBm in 100 kHz is consistent with the level used for UTRA as spurious emission limit inside the operating band (−15 dBm/MHz). The measurement bandwidth of 100 kHz was chosen to be of similar granularity as the bandwidth of any victim system's smallest radio resource allocation (LTE 1 RB each 180 kHz; GSM 200 kHz carrier).

The operating band unwanted emission limits must also be consistent with the limits according to ITU-R SM.329 for all LTE channel bandwidth options. This means that outside 250% of the necessary bandwidth from the carrier center frequency the corresponding Category B limit of −25 dBm/100 kHz must be reached, see Figure 14.3. Even though the frequency range in which transmitter intermodulation distortions occur scales with the channel bandwidth, it was found to be feasible to define a common mask for the 5, 10, 15 and 20 MHz LTE options, which meets the SM.329 limits at 10 MHz offset from the channel edge. However, this −25 dBm/100 kHz limit must be reached already with offsets of 2.8 MHz, respectively 6 MHz from the channel edge for the LTE 1.4, respectively 3 MHz option and this necessitated definition of separate masks. For the 1.4, 3, 5 MHz

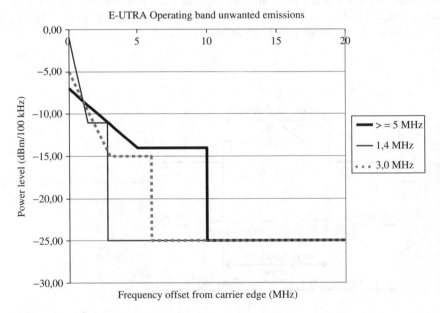

Figure 14.3 Operating band unwanted emission requirements levels relative to channel edge (E-UTRA bands >1 GHz)

LTE the same total eNodeB transmission power of 46 dBm was assumed resulting in a higher power spectral density for the smaller LTE bandwidth options, hence the mask permits higher emission levels at the channel edge.

For the North-American E-UTRA Bands (Bands 2, 4, 5, 10, 12, 13, 14) an additional unwanted emission limit is derived in [2] from FCC Title 47 Parts 22, 24 and 27. These requirements are interpreted as −13 dBm in a measurement bandwidth defined as 1% of the '−26 dB modulation bandwidth' within the first MHz from the channel edge and −13 dBm/MHz elsewhere.

14.3.2 Co-existence with Other Systems on Adjacent Carriers Within the Same Operating Band

The RAN4 specifications include also ACLR requirements of 45 dBc for the first and second adjacent channels of (a) the same LTE bandwidth and (b) 5 MHz UTRA, see Figure 14.4.

[4] contains simulation results for downlink co-existence of LTE with adjacent LTE, UTRA or GSM carriers. The required Adjacent Channel Interference Ratio (ACIR) to ensure ≤ 5% cell edge throughput loss for the victim system was found to be in the order of 30 dB in all of these cases. With an assumed UE Adjacent Channel Selectivity (ACS) of 33 dB for LTE and UTRA, the choice of 45 dB Adjacent Channel Leakage Ratio (ACLR) ensures minimal impact from the eNodeB transmit path and also aligns with the UTRA ACLR1 minimum performance requirements.

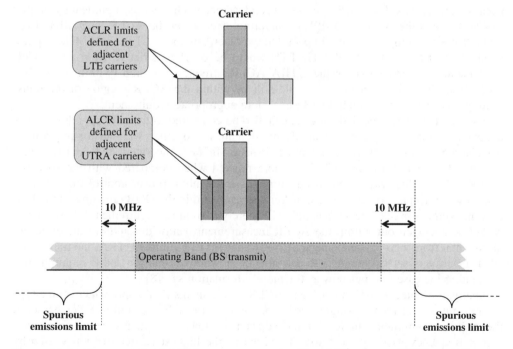

Figure 14.4 The two defined ACLR measures, one for 1st and 2nd adjacent E-UTRA carriers and one for 1st and 2nd adjacent UTRA carrier

For LTE-LTE co-existence the ACLR1, ACLR2 are only specified for the same LTE bandwidth, that is mixed cases such as, for example, 5 MHz LTE ↔ 20 MHz LTE are not covered by explicit ACLR requirements according to all possible values for the victim carrier bandwidth. However, analysis in [4] has shown that these cases are sufficiently covered by the 45 dBc ACLR requirements measured within the same bandwidth as the aggressing carrier. A large matrix of ACLR requirements for all LTE bandwidth options and, additionally, for multiple other radio technologies would have led to a large eNodeB conformance testing effort without materially adding a more robust protection.

Furthermore, the unwanted emission mask specified for LTE provides a baseline protection for any victim system. For this reason no ACLR requirements were specified, for instance for GSM victim carriers – the corresponding ACLR values can be obtained by integrating the unwanted emission mask accordingly and are well above the required ~30 dB suggested by system simulations [4]. However, ACLR requirements for the first and second UTRA channel have been added to explicitly protect adjacent 'legacy' UTRA carriers by the same level from aggressing LTE systems as specified for UTRA as LTE migration within UTRA bands was considered an important deployment scenario.

The ACLR2/3.84 MHz for an UTRA victim is specified to be the same value as the ACLR1/3.84 MHz (45 dBc), not 50 dBc as for UTRA. This is reasonable, as the second adjacent channel interference contributes only little to overall ACIR:

$$ACIR = \cfrac{1}{\cfrac{1}{ACLR1} + \cfrac{1}{ACS1} + \cfrac{1}{ACLR2} + \cfrac{1}{ACS2}} \tag{14.2}$$

because the UTRA UE ACS2 in the second adjacent channel is significantly higher (~42 dBc) than the ACS1 (~33 dBc). Therefore, decreasing the ACLR2/3.84 MHz from 50 to 45 dBc has only negligible impact. On the other hand, an ACLR2/3.84 MHz requirement of 50 dBc for 10, 15, 20 MHz LTE would be overly restrictive from an eNodeB implementation perspective as the UTRA ACLR2 frequency region (within 5–10 MHz offset from the LTE carrier edge) would still be within the ACLR1 region of the transmitting LTE carrier for which 45 dBc is a more appropriate requirement.

The ACLR values specified for LTE can also be compared to the corresponding values obtained by integrating the unwanted emission mask, together with an assumption on the eNodeB output power, for example 46 dBm. In fact, the integrated ACLR values obtained from the mask are ~2–5 dB more relaxed when compared with the specified ACLR values. This was specified on purpose, as downlink power control on RBs may lead to some 'ripples' of the leakage power spectrum within the OOB domain and the LTE unwanted emission mask was designed to be merely a 'roofing' requirement. These ripples would be averaged out within the ACLR measurement bandwidth and are therefore not detrimental from a co-existence perspective. Hence, in LTE, unlike UTRA, the transmitter (PA) linearity is set by the ACLR and not the unwanted emission mask requirements, facilitating increased output power, for more information see [8].

Finally, unwanted emission mask and ACLR requirements shall apply whatever the type of transmitter considered (single carrier or multi-carrier). For a multi-carrier eNodeB, the ACLR requirement applies for the adjacent channel frequencies below the lowest carrier frequency used by the eNodeB and above the highest carrier frequency used by the eNodeB, i.e. not within the transmit bandwidth supported by the eNodeB, which is assumed to belong to the same operator.

14.3.3 Co-existence with Other Systems in Adjacent Operating Bands

[2] contains additional spurious emission requirements for the protection of UE and/or eNodeB of other wireless systems operating in other frequency bands within the same geographical area. The system operating in the other frequency band may be GSM 850/900/1800/1900, Personal Handyphone System (PHS), Public Safety systems within the US 700 MHz bands, UTRA FDD/TDD and/or E-UTRA FDD/TDD. Most of these requirements have some regulatory background and are therefore mandatory within their respective region (Europe, Japan, North America). The spurious emission limits for the protection of 3GPP wireless systems assume typically ~67 dB isolation (Minimum Coupling Loss, MCL) between the aggressor and victim antenna systems, including antenna gain and cable losses.

To facilitate co-location (assuming only 30 dB MCL) with other 3GPP systems' base stations, an additional set of optional spurious emission requirements has been defined in [2].

However, as shown in Figure 14.2, spurious emissions do not apply for the 10 MHz frequency range immediately outside the eNodeB transmit frequency range of an operating band. This is also the case when the transmit frequency range is adjacent to the victim band, hence the above co-existence requirements do not apply for FDD/TDD transition frequencies at, for example 1920, 2570, 2620 MHz. However, the deployment of FDD and TDD technologies in adjacent frequency blocks can cause interference between the systems as illustrated in Figure 14.5. There may be interference between FDD and TDD base stations and there may be interference between the terminals as well.

From an eNodeB implementation perspective it is challenging to define stringent transmitter and receiver filter requirements in order to minimize the size of the guard band required around the FDD/TDD transition frequencies. Furthermore, different regions in the world may impose local regulatory requirements to facilitate FDD/TDD deployment in adjacent bands. It was therefore seen as not practical in 3GPP to make generic assumptions on the available guard bands and to subsequently define general FDD/TDD protection requirements, which would need to be in line with the various regulatory requirements throughout the world. Hence, additional FDD/TDD protection emission limits for the excluded 10 MHz frequency range immediately outside the eNodeB transmit frequency range of an operating band are left to local or regional requirements.

As an example for such local requirements the remainder of this section will discuss the rules defined in [9] for using FDD and TDD within the European 2.6 GHz band allocation (3GPP Band 7). The cornerstones of the emission requirements in [9] are as follows:

- In-block and out-of-block Equivalent Isotropic Radiated Power (EIRP) emission masks are defined without any reference to a specific wireless system's eNodeB standard – the limits can be considered as technology neutral. The in-block EIRP limits provide a safeguard against blocking of adjacent system's eNodeB receive path. The out-of-block masks, also called *Block Edge Masks* (BEM), impose radiated unwanted emissions limits across adjacent license block uplink/downlink receive band frequency ranges.
- The BEMs should facilitate FDD/TDD co-existence without any detailed coordination and cooperation arrangements between operators in neighboring spectrum blocks, for base station separations greater than 100 m. In this, the assumed MCL has been >53 dB and the interference impact relative to eNodeB thermal noise, I/N, has been assumed as −6 dB, with a noise figure of 5 dB. Antenna isolations measurements show that

50 dB MCL can be achieved easily on the same rooftop by using separate, vertically and/or horizontally displaced antennas for FDD and TDD respectively.

- Some of the BEMs have been derived from the UTRA eNodeB spectrum emission mask (SEM), by converting the UTRA SEM (i.e. a conducted emission requirement to be met at the eNodeB antenna connector) into an EIRP mask by making assumptions on antenna gain and cable losses (17 dBi).
- EIRP limits are requirements for *radiated* (not *conducted*) unwanted emissions and they apply within or outside a license block. Hence, these are not testable eNodeB equipment requirements as such, but they involve the whole RF site installation, including antennas and cable losses. Furthermore, a BEM requirement does not necessarily start at the *RF channel (carrier)* edge, but at the edge of the *licensee's block*, which may differ from the channel edge depending on the actual deployment scenario. Finally, the BEM must be met for the sum of all radiated out-of-block emissions, for example from multiple RF carriers or even multiple eNodeBs present at a site, not only for those pertaining to a single RF channel as is the case for the UTRA SEM.

These requirements will be discussed in the following in more detail assuming the Band 7 frequency arrangements with 70 MHz FDD uplink, 50 MHz TDD, 70 MHz FDD downlink.

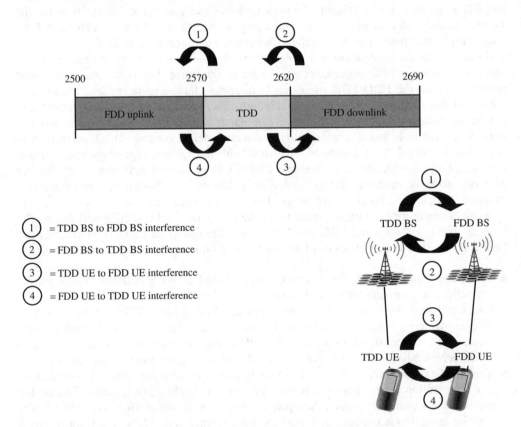

Figure 14.5 Interference scenarios between FDD and TDD (example for Band 7)

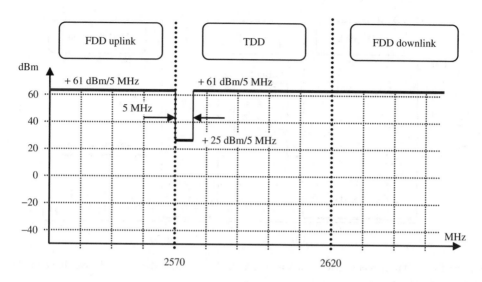

Figure 14.6   Maximum allowed in-block EIRP including restricted block

Figure 14.6 shows the maximum allowed in-block EIRP limits. The nominal EIRP limit of 61 dBm/5 MHz can be relaxed by local regulation up to 68 dBm/5 MHz for specific deployments, for example in areas of low population density where higher antenna gain and/or eNodeB output power are desirable.

The lower 5 MHz part of the TDD allocation is a *restricted block* with maximum allowed base station EIRP of 25 dBm/5 MHz for cases where antennas are placed indoors or where the antenna height is below a certain height to be specified by the regulator. The lower transmission power in the restricted 5 MHz TDD block serves to relax the FDD eNodeB selectivity and blocking requirements just below 2570 MHz.

Figure 14.7 shows the resulting out-of-block EIRP emission mask (BEM) for an unrestricted (high power) 20 MHz TDD block allocation starting 5 MHz above the FDD uplink band allocation. It can be noted that

- The whole FDD uplink band allocation is protected by an −45 dBm/MHz EIRP limit, which is chosen as a baseline requirement for protection of the uplink in the eNodeB transmitter → eNodeB receiver interference scenario. Given a MCL> 53 dB, there will be minimal impact on the FDD eNodeB uplink.
- Possible unsynchronized TDD operation is also protected by the −45 dBm/MHz EIRP limit from 5 MHz above the TDD block edge onwards.
- As the relevant interference mechanism across the FDD downlink band will be TDD eNodeB → FDD UE, a more relaxed limit of +4 dBm is sufficient. This limit is also stipulated as out-of-block EIRP limit for a FDD eNodeB across this part of the band, see Figure 14.7.
- Not shown in Figure 14.7 is the possible use of a TDD restricted block within 2570 − 2575 MHz, which has a tighter BEM requirement across the first 5 MHz just below the 2570 MHz transition frequency in order to protect the FDD uplink.
- Neither shown is the detailed slope of the BEM within the first MHz from either side of the TDD block edge, which follows essentially the shape of the UTRA SEM.

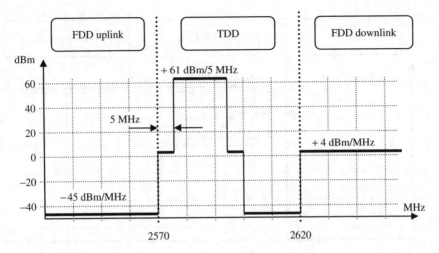

Figure 14.7 Out-of-block EIRP emission mask (BEM) for a 20 MHz TDD block above the FDD uplink band allocation

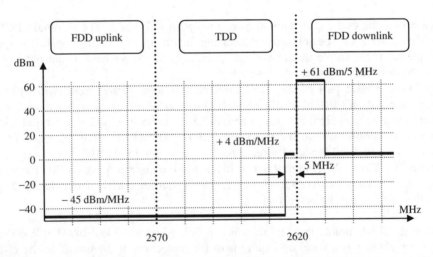

Figure 14.8 Out-of-block EIRP emission mask (BEM) for a 10 MHz FDD block above the TDD band allocation

Figure 14.8 shows the resulting out-of-block EIRP emission mask (BEM) for 10 MHz FDD block allocation starting just above the TDD band allocation. It can be noted that:

- the first 5 MHz TDD block just below the FDD downlink block may receive a higher level of interference due to the more relaxed BEM EIRP limit of +4 dBm. This is in order to facilitate FDD eNodeB transmit filters; no restricted power blocks are foreseen for the FDD blocks;
- the rest of the TDD band and FDD uplink band allocation is protected by the −45 dBm/MHz EIRP baseline requirement.

14.3.4   Transmitted Signal Quality

[2] contains the modulation-specific Error Vector Magnitude (EVM) requirements to ensure sufficiently high quality of the transmitted base station signal, see Table 14.5.

Typical impairments of the transmitted signal modulation accuracy are analog RF distortions (frequency offset, local oscillator phase noise, amplitude/phase ripple from analog filters) as well as distortions created within the digital domain such as inter-symbol interference from digital filters used for spectrum shaping, finite word-length effects, and, most important for conditions near maximum transmit power, the clipping noise from peak-to-average radio reduction schemes.

The EVM requirement will ensure that the downlink throughput due to the base-station non-ideal waveform is only marginally reduced, typically by 5%, assuming ideal reception in the UE. The required EVM must be fulfilled for all transmit configurations and across the whole power dynamic range power levels used by the base station.

14.3.4.1   Definition of the EVM

More precisely, the EVM is a measure of the difference between the ideal constellation points and the measured constellation points obtained after equalization by a defined 'reference receiver'. Unlike UTRA, the EVM is not measured on the transmitted composite time-domain signal waveform, but within the frequency domain, after the FFT, by analyzing the modulation accuracy of the constellation points on each subcarrier. Figure 14.9 shows the reference point for the EVM measurement.

The OFDM signal processing blocks prior to the EVM reference point, in particular the *constrained Zero-Forcing (ZF) equalizer*, are fundamentally not different from those

Table 14.5   EVM requirements

Modulation scheme for PDSCH	Required EVM (%)
QPSK	17.5
16QAM	12.5
64QAM	8

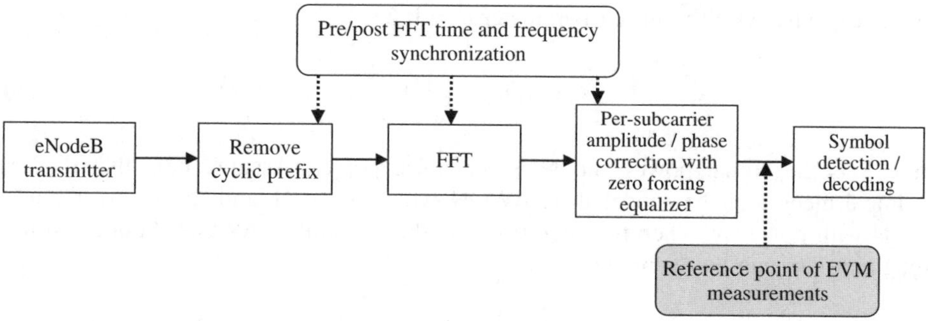

Figure 14.9   Reference point for EVM measurement

of an actual UE. The rationale for this rather complex EVM definition has been that only those transmitter impairments should 'pollute' the EVM that would not be automatically be removed by the UE reception anyway. For example, clipping noise from peak-to-average ratio reduction has AWGN-like characteristics and cannot be removed by UE equalizers, whereas a non-linear phase response from analog filter distortion can be estimated from the reference symbols and subsequently be removed by the zero forcing (ZF)-equalizer. The qualifier *constrained* within the ZF-equalization process refers to a defined frequency averaging process of the raw channel estimates used to compute the ZF-equalizer weights. Some form of frequency averaging of the raw channel estimates at the reference symbol locations will be present within an actual UE receiver implementation, and this will therefore limit the removal of some of the impairments such as filter amplitude (or phase) ripple with a spatial frequency in the order of the subcarriers spacing. Therefore an equivalent of this frequency averaging process has been defined for the EVM measurement process.

Yet another feature of the EVM definition is the chosen time synchronization point for the FFT processing. The EVM is deliberately measured not at the ideal time synchronization instant, which would be approximately the center of the cyclic prefix, but rather at two points shortly after the beginning and before the end of the cyclic prefix, respectively. This ensures that in the presence of precursors/postcursors from transmit spectrum shaping filters, the UE will see sufficiently low inter-symbol interference even under non-ideal time-tracking conditions.

The details of this EVM measurement procedure, including all these aspects are defined within [5], Annex F.

14.3.4.2 Derivation of the EVM Requirement

Next we provide a rationale for the EVM requirements based on an analytic approach presented in [10]. Let us calculate the required EVM for 5% throughput loss for each instantaneous C/I value, corresponding to a certain selected Modulation and Coding Scheme (MCS).

To start with, we consider the MCS throughput curves and approximating MCS envelope in Figure 14.10. The MCS curves were generated from link level simulations under AWGN channel conditions for a 1×1 antenna configuration (1 transmit and 1 receive antenna) using ideal channel estimation within the receiver.

We approximate the set of these throughput curves by the expression for Shannon's channel capacity with the fitting parameter $\alpha = 0.65$.

$$C = \alpha \log 2 \left(1 + \frac{S}{N} \right)$$ (14.3)

where S is the signal power and N is the noise power. Assume next that the transmit impairments can be modeled as AWGN with power M and the receiver noise as AWGN with power N. Then the condition that the transmitter AWGN should lead to 5% throughput loss can be expressed as:

$$\alpha \log 2 \left(1 + \frac{S}{M + N} \right) = 0.95 \cdot \alpha \log 2 \left(1 + \frac{S}{N} \right)$$ (14.4)

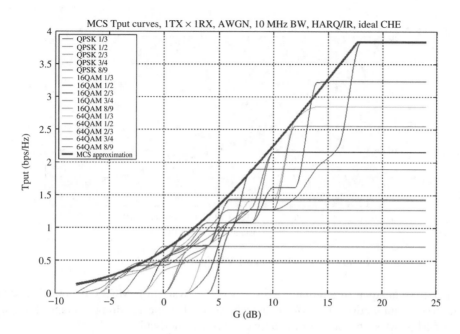

Figure 14.10   MCS throughput curves and approximating MCS envelope

Solving this condition for $EVM_{req} \equiv \frac{M}{S}$ we obtain

$$EVM_{req} \equiv \frac{M}{S} = \left[\left(1 + \frac{S}{N} \right)^{0.95} - 1 \right]^{-1} - \frac{N}{S} \qquad (14.5)$$

The required EVM is plotted in Figure 14.11 for the C/I range of the approximating envelope MCS throughput curve of Figure 14.10. As can been seen ~6.3% EVM would be required for the *highest* throughput MCS (64QAM 8/9, operating S/N ~17.7 dB). However, when assuming a single EVM requirement for all 64QAM modulated MCSs this would be too stringent as 64QAM may be chosen from a C/I range from 12 . . . 17.7 dB according to the chosen MCS set. This would indicate a required EVM in the range of 10%–6.3%. Looking at the midpoint S/N of ~15 dB for 64QAM MCS selection one obtains a 7.9% EVM requirement. Similarly, one obtains for

- 16QAM: C/I range from 6 . . . 12 dB, midpoint C/I of ~9 dB with an EVM requirement of 12.9%;
- QPSK: C/I range from −8–6 dB, midpoint C/I of ~−1 dB with EVM requirements of 29.6% and 16.3% respectively (for 6 dB C/I).

However, the above derivation assumed a smooth approximating MCS envelope. In reality, the MCS envelope has in AWGN a 'waterfall' shape as shown in Figure 14.10 and may exhibit either a larger impact from EVM in regions with steep slopes or a smaller impact in regions with flat response. In an actual system scenario the C/I distribution will

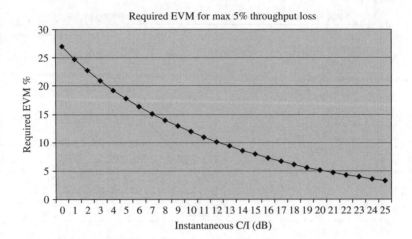

Figure 14.11 Required EVM for 5% throughput loss

average these unequal throughput losses across the 'waterfall' MCS envelope. The quasi-static system simulations in [10] do indeed verify that the resulting average throughput loss for 64QAM is in line with the above simplified derivation.

14.4 eNodeB RF Receiver

The purpose of the eNodeB RF receiver requirements is to verify different RF impairments, which have an impact on the network performance. These impairments include noise figure, receiver EVM, selectivity on different frequencies, including adjacent channel, and so forth. The following base station RF receiver requirements are described: reference sensitivity level, dynamic range, in-channel selectivity, Adjacent Channel Selectivity (ACS), blocking, receiver spurious emissions and receiver intermodulation.

Reference sensitivity level
The reference sensitivity level is the minimum mean power received at the antenna connector, at which a throughput requirement is fulfilled. The throughput is equal to or higher than 95% of the maximum throughput for a specified reference measurement channel.

The purpose of this requirement is to verify the receiver noise figure. Other receiver impairments such as receiver EVM are included within the receiver demodulation performance requirements at high SNR points. Therefore, the maximum throughput is defined at low SNR points for the sensitivity case. The reference measurement channel is based on a QPSK modulation and 1/3 coding rate.

For channel bandwidths lower than or equal to 5 MHz, the reference measurement channels are defined on the basis of all resource blocks allocated to this channel bandwidth. For channel bandwidths higher than 5 MHz, sensitivity is measured using consecutive blocks consisting of 25 RBs.

The reference sensitivity level calculation is given by (14.6). Noise Figure (NF) is equal to 5 dB and implementation margin (IM) is equal to 2 dB.

$$P_{REFSENS}\,(dBm) = -174\,(dBm/Hz) + 10\log(N_{RB} \cdot 180k) + NF + SNR + IM \quad (14.6)$$

For example, for 10 MHz channel bandwidth ($N_{RB} = 25$), the reference sensitivity level is equal to -101.5 dBm. The simulated SNR for 95% of the maximum throughput is equal to -1.0 dB.

The eNodeB noise figure is relevant for the coverage area. The uplink link budget calculation in Chapter 10 assumes base station noise figure of 2 dB, so clearly better than the minimum performance requirement defined in 3GPP specifications. The typical eNodeB has better performance than the minimum requirement since the optimized performance is one of the selling arguments for the network vendors.

Dynamic range

The dynamic range requirement is a measure of the capability of the receiver to receive a wanted signal in the presence of an interfering signal inside the received frequency channel, at which a throughput requirement is fulfilled. The throughput is equal to or higher than 95% of the maximum throughput for a specified reference measurement channel.

The intention of this requirement is to ensure that the base station can receive a high throughput in the presence of increased interference and high wanted signal levels. Such a high interference may come from neighboring cells in case of small cells and high system loading. This requirement measures the effects of receiver impairments such as receiver EVM and is performed at high SNR points. The mean power of the interfering signal (AWGN) is equal to the receiver noise floor increased by 20 dB, in order to mask the receiver's own noise floor. The maximum throughput is defined at high SNR points thus the reference measurement channel is based on 16QAM modulation and 2/3 coding rate.

The wanted signal mean power calculation is given by (14.7). NF is equal to 5 dB and IM is equal to 2.5 dB.

$$P_{wanted}[dBm] = -174[dBm/Hz] + 10log(N_{RB} \cdot 180k) + 20 + NF + SNR + IM \quad (14.7)$$

For example, for 10 MHz channel bandwidth ($N_{RB} = 25$), the wanted signal mean power is equal to -70.2 dBm. The simulated SNR for 95% of the maximum throughput is equal to 9.8 dB. The interfering signal mean power is set equal to -79.5 dBm. The dynamic range measurement for the 5 MHz case is illustrated in Figure 14.12.

In-channel selectivity

The in-channel selectivity requirement is a measure of the receiver's ability to receive a wanted signal at its assigned resource block locations in the presence of an interfering

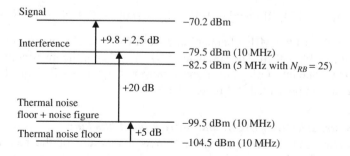

Figure 14.12 Dynamic range measurement for 10 MHz LTE

Table 14.6 Number of resource blocks
allocated for wanted and interfering signal

Channel bandwidth [MHz]	Wanted signal	Interfering signal
1.4	3	3
3	9	6
5	15	10
10	25	25
15	25	25
20	25	25

signal received at a larger power spectral density, at which a throughput requirement is fulfilled. The throughput is equal to or higher than 95% of the maximum throughput for a specified reference measurement channel.

The intention of this requirement is to address in-band adjacent resource block selectivity – the reception of simultaneous user signals at largely different power spectral density levels due to used modulation format, power control inaccuracies, other-cell interference levels, and so forth. The uplink signal is created by two signals, where one is the required QPSK modulated signal and the other one the interfering 16QAM modulated signal at elevated power. Table 14.6 presents the allocation of resource blocks for wanted and interfering signals, for different channel bandwidths. The high power level difference may happen if the interfering user is close to the base station and can use high signal-to-noise ratio while the wanted user is far from the base station and can only achieve low signal-to-noise ratio.

For channel bandwidths equal to 10, 15 and 20 MHz, the 25 resource block allocations of the wanted and interfering signals are adjacently around Direct Current (DC), in order to be sensitive to the RF impairments of the receiver image leakage, EVM, third order inter-modulation (IMD3) and the local oscillator phase noise.

The mean power of the interfering signal is equal to the receiver noise floor increased by 9.5 dB (required SNR) and additionally by 16 dB (assumed interference over thermal noise). The desensitization of the wanted resource block allocation, in the presence of the interfering resource block allocation, is equal to 3 dB.

The wanted signal mean power calculation is given by (14.8). *NF* is equal to 5 dB and *IM* is equal to 2 dB.

$$P_{wanted}[dBm] = -174[dBm/Hz] + 10\log(N_{RB} \cdot 180k) + 3 + NF + SNR + IM \quad (14.8)$$

For example, for 10 the MHz channel bandwidth ($N_{RB} = 25$), the wanted signal mean power is equal to -98.5 dBm. The simulated SNR for 95% of the maximum throughput is equal to -1.0 dB. The interfering signal mean power is equal to -77 dBm. The in-channel measurement case is presented in Figure 14.13.

Adjacent channel selectivity (ACS) and narrow-band blocking
The ACS (narrow-band blocking) requirement is a measure of the receiver ability to receive a wanted signal at its assigned channel frequency in the presence of an inter-fering adjacent channel signal, at which a throughput requirement shall be fulfilled. The

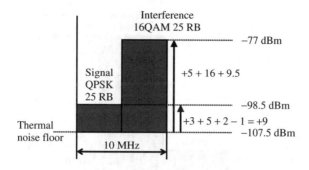

Figure 14.13 In-channel selectivity measurement for 5 MHz LTE

throughput shall be equal or higher than 95% of the maximum throughput, for a specified reference measurement channel.

The intention of this requirement is to verify the selectivity on the adjacent channel. The selectivity and the narrowband blocking are important to avoid interference between operators. The adjacent operator's mobile may use high transmission power level if its own base station is far away. If such a mobile happens to be close to our base station, it can cause high interference level on the adjacent channel.

Table 14.7 presents the ACS requirement relationship between E-UTRA interfering signal channel bandwidth, wanted signal channel bandwidth, wanted signal mean power and interfering signal centre frequency offset to the channel edge of the wanted signal. The interfering signal mean power is equal to -52 dBm.

Table 14.8 shows how the adjacent channel selectivity can be calculated from the performance requirements for 10 MHz channel bandwidth with 25 resource blocks. The

Table 14.7 Derivation of ACS requirement

Wanted signal channel bandwidth [MHz]	Wanted signal mean power [dBm]	Offset [MHz]	Interfering signal channel bandwidth [MHz]
1.4	$P_{REFSENS} + 11$	0.7	1.4
3	$P_{REFSENS} + 8$	1.5	3
5, 10, 15, 20	$P_{REFSENS} + 6$	2.5	5

Table 14.8 Calculation of the ACS requirement for 10 MHz channel bandwidth

Interfering signal mean power (A)	-52 dBm
Base station noise floor with 5 dB noise figure (B)	-102.5 dBm
Allowed desensitization (C)	6 dB
Total base station noise (D = B + C in dB)	-96.5 dBm
Allowed base station interference (E = D − B in absolute value)	-98 dBm
Base station adjacent channel selectivity (F = A − E)	46 dB

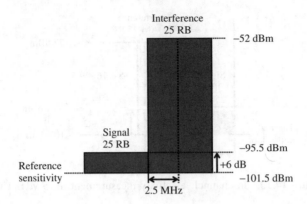

Figure 14.14   Adjacent channel selectivity measurement for 10 MHz LTE

adjacent channel selectivity test case is illustrated in Figure 14.14. Base station noise floor is given by (14.9). NF is equal to 5 dB and $N_{RB} = 25$.

$$D[dBm] = -174[dBm/Hz] + 10\log(N_{RB} \cdot 180k) + NF \qquad (14.9)$$

The narrow-band blocking measurement is illustrated in Figure 14.15. The wanted signal mean power is equal to $P_{REFSENS} + 6\,dB$. The interfering 1RB E-UTRA signal mean power is equal to $-49\,dBm$. The interfering signal is located in the first five worst-case resource block allocations. For the 3 MHz channel bandwidth, the interfering signal is located in every third resource block allocation. For 5, 10, 15 and 20 MHz channel bandwidths, the interfering signal is located additionally in every fifth resource block allocation. Such a location of the interfering signal verifies different possible impairments on the receiver performance. In case of GSM band refarming, the blocker can be narrowband GSM signal. There are no specific requirements with GSM signal used as the blocker. The GSM signal, however, is sufficiently similar to 1 RB LTE blocker for practical performance purposes.

Blocking

The blocking requirement is a measure of the receiver's ability to receive a wanted signal at its assigned channel frequency in the presence of an unwanted interferer, where a

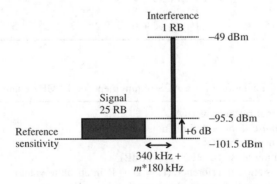

Figure 14.15   Narrowband blocking measurement for 10 MHz LTE

throughput requirement is fulfilled. The throughput should be equal or higher than 95% of the maximum throughput for a specified reference measurement channel.

The intention of this requirement is to verify the selectivity on different frequencies, excluding adjacent channel. The in-band blocking can be also called as adjacent channel selectivity for the second adjacent channel.

For in-band blocking the unwanted interferer is an E-UTRA signal. For example, for Operating Band 1 (1920 – 1980 MHz), the in-band blocking refers to the center frequencies of the interfering signal from 1900 MHz to 2000 MHz. For out-of-band blocking, the unwanted interferer is a continuous wave (CW) signal. For Operating Band 1, the out-of-band blocking refers to the center frequencies of the interfering signal from 1 MHz to 1900 MHz and from 2000 MHz to 12750 MHz. The measurements are shown in Figure 14.16.

The in-band blocking requirement for 10 MHz LTE is illustrated in Figure 14.17 and out-of-band in Figure 14.18. The wanted signal mean power is equal to $P_{REFSENS} + 6$ dB. The E-UTRA interfering signal mean power is equal to -43 dBm. The CW interfering signal mean power is equal to -15 dBm.

Receiver spurious emissions

The spurious emissions power is the power of emissions generated or amplified in a receiver that appear at the base station receiver antenna connector.

The frequency range between 2.5 channel bandwidth below the first carrier frequency and 2.5 channel bandwidth above the last carrier frequency, transmitted by the base station,

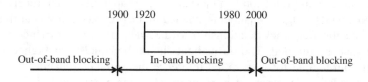

Figure 14.16 In-band and out-of-band blocking measurement for Band 1

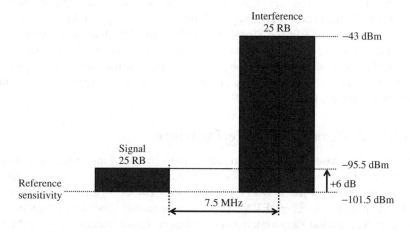

Figure 14.17 In-band blocking measurement for 10 MHz LTE

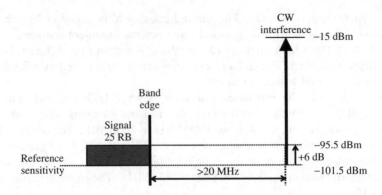

Figure 14.18 Out-of-band blocking measurement for 10 MHz LTE

may be excluded from the requirement. Frequencies that are more than 10 MHz below the lowest frequency of the base-station transmitter operating band or more than 10 MHz above the highest frequency of the base station transmitter operating band shall not be excluded from the requirement.

Additionally, the power of any spurious emission should not exceed the levels specified for coexistence with other systems in the same geographical area and for protection of the E-UTRA FDD base station receiver of own or different base station.

Receiver intermodulation

Intermodulation response rejection is a measure of the capability of the receiver to receive a wanted signal on its assigned channel frequency in the presence of two interfering signals that have a specific frequency relationship to the wanted signal, at which a throughput requirement shall be fulfilled. The throughput should be equal to or higher than 95% of the maximum throughput for a specified reference measurement channel.

The intermodulation requirement is relevant when there are two terminals from other operators transmitting at high power level close to our base station.

The wanted signal mean power is equal to $P_{REFSENS} + 6$ dB. The interfering signal mean power (both E-UTRA and CW) is equal to -52 dBm. The offset between the CW interfering signal centre frequency and the channel edge of the wanted signal is equal to 1.5 E-UTRA interfering signal channel bandwidth. The offset between the E-UTRA interfering signal center frequency and the channel edge of the wanted signal is specified on the basis of the worst-case scenario – the intermodulation products fall on the edge resource blocks of an operating channel bandwidth. The receiver intermodulation requirements are shown in Figure 14.19.

14.5   eNodeB Demodulation Performance

The purpose of the base station demodulation performance requirements is to estimate how the network is performing in practice and to verify different base station impairments which have an impact on the network performance. These impairments include RF and baseband impairments, receiver EVM, time and frequency tracking, frequency estimation, etc. The base station demodulation performance requirements are described for the following uplink channels: PUSCH, PUCCH and PRACH.

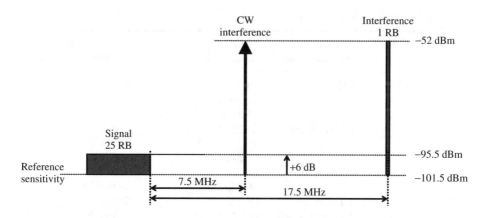

Figure 14.19 Intermodulation measurement for 5 MHz LTE

PUSCH

The PUSCH demodulation performance requirements are determined by a minimum required throughput for a given SNR. The throughput should be equal to or higher than 30% or 70% of the maximum throughput for a specified reference measurement channel, which contains data and reference symbols only.

The PUSCH demodulation performance requirements are specified for all E-UTRA channel bandwidths. For each channel bandwidth, the following various network related parameters are selected in order to match different radio system configurations:

- number of receive antennas – two or four;
- modulation and coding scheme – QPSK 1/3, 16QAM 3/4 or 64QAM 5/6;
- channel model – EPA5, EVA5, EVA70, ETU70 or ETU300, where the number indicated the Doppler shift. PA is Pedestrian A, VA is Vehicular A and TU is Typical Urban channel model;
- cyclic prefix type – normal or extended;
- number of resource blocks allocated for channel bandwidth – single or all possible.

Each channel model is described by Doppler frequency which corresponds to various velocities depending on the frequency band. For example, EPA5 corresponds to velocities equal to 7.7 km/h, 2.7 km/h and 2.1 km/h for 0.7 GHz (Band 12), 2 GHz (Band 1) and 2.6 GHz (Band 7), respectively.

Incremental redundancy HARQ, allowing up to three retransmissions is used. Table 14.9 presents a set of PUSCH base station tests for each channel bandwidth and for each configuration of receiver antennas.

For base station supporting multiple channel bandwidths, only tests for the highest and the smallest channel bandwidths are applicable.

The SNR requirements were specified on the basis of average link level simulation results with implementation margin presented by various companies during 3GPP meetings.

Below is an example of relationship between the net data rate and required SNR. 10 MHz channel bandwidth, normal cyclic prefix and QSPK 1/3 modulation and coding scheme were taken into account.

Table 14.9   Set of PUSCH base station tests

Cyclic prefix	Channel model, RB allocation	Modulation and coding scheme	Fraction of maximum throughput (%)
Normal	EPA5, all	QPSK 1/3	30 and 70
		16QAM 3/4	70
		64QAM 5/6	70
	EVA5, single	QPSK 1/3	30 and 70
		16QAM 3/4	30 and 70
		64QAM 5/6	70
	EVA70, all	QPSK 1/3	30 and 70
		16QAM 3/4	30 and 70
	ETU70, single	QPSK 1/3	30 and 70
	ETU300, single	QPSK 1/3	30 and 70
Extended	ETU70, single	16QAM 3/4	30 and 70

Table 14.10   SNR requirements for 30% of the maximum throughput (fixed reference channel A3-5)

Number of receive antennas	Channel model	SNR requirement (dB)
2	EPA5	−4.2
	EVA70	−4.1
4	EPA5	−6.8
	EVA70	−6.7

For fixed reference channel A3-5 [2], the SNR requirements were specified for full resource block allocation (50 resource blocks), for two and four receiver antennas, for EPA5 and EVA70 channel models and for 30% and 70% of the maximum throughput. Table 14.10 presents SNR requirements for 30% of the maximum throughput.

For fixed reference channel A3-5 the payload size is equal to 5160 bits and corresponds to 5.16 Mbps instantaneous net data rate. For one resource block it is 103.2 kbps, accordingly.

Thirty per cent of the maximum throughput corresponds to 30.9 kbps and 61.9 kbps for one and two resource blocks, respectively.

The link budget in Chapter 10 assumed SNR equal to −7 dB with 64 kbps and two resource blocks corresponding to 32 kbps and one resource block. The link budget assumes better eNodeB receiver performance because of more HARQ retransmissions assumed and also because typical eNodeB performance is closer to the theoretical limits than the 3GPP minimum requirement.

The uplink timing adjustment requirement for PUSCH was specified in [2]. The rationale for this requirement is to check if the base station sends timing advance commands with correct frequency and if the base station estimates appropriate uplink transmission timing. The uplink timing adjustment requirements are determined by a minimum required throughput for a given SNR and are specified for moving propagation conditions

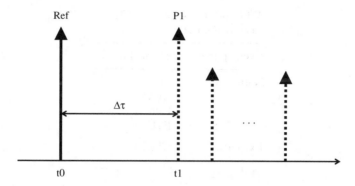

Figure 14.20   Moving propagation conditions

as shown in Figure 14.20. The time difference between the reference timing and the first tap is described by (14.10), where A $= 10\,\mu s$.

$$\Delta\tau = \frac{A}{2} \cdot \sin(\Delta\omega \cdot t) \qquad (14.10)$$

The uplink timing adjustment requirement is specified for normal and extreme conditions. In case of normal condition the ETU channel model and UE speed of 120 km/h is considered ($\Delta\omega = 0.04\,s^{-1}$). Uplink timing adjustment in extreme conditions is an optional requirement and corresponds to AWGN channel model and UE speed of 350 km/h ($\Delta\omega = 0.13\,s^{-1}$). [2] also includes eNodeB decoding requirements for high-speed train conditions up to 350 km/h. eNodeB can experience two times higher Doppler shift in the worst case if UE synchronizes to the downlink frequency including the Doppler shift (f_d) – see Figure 14.21. The maximum Doppler shift requirement is 1750 Hz, which corresponds to 350 km/h at 2.1 GHz, assuming that eNodeB experiences double Doppler shift.

PUCCH

The PUCCH performance requirements are specified for PUCCH format 1a and for PUCCH format 2. The PUCCH format 1a performance requirements are determined by a minimum required DTX to ACK probability and ACK missed detection probability for a given SNR. The DTX to ACK probability – the probability that ACK is detected when nothing is sent – will not exceed 1%. The ACK missed detection probability – the probability that ACK is not detected properly – will not exceed 1%.

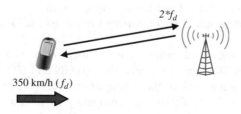

Figure 14.21   High-speed train demodulation requirement

Table 14.11 Set of PUCCH
base station tests

Cyclic prefix	Channel model
Normal	EPA5
	EVA5
	EVA70
	ETU300
Extended	ETU70

The PUCCH format 1a performance requirements are specified for all E-UTRA channel bandwidths. For each channel bandwidth, the following various network related parameters were selected in order to match different radio system configurations:

- number of receive antennas −2 or 4;
- channel model – EPA5, EVA5, EVA70, ETU70 or ETU300;
- cyclic prefix type – normal or extended.

Table 14.11 presents a set of PUCCH base station tests for each channel bandwidth and for each configuration of receive antennas.

The PUCCH format 2 performance requirements are determined by a CQI missed detection BLER probability for a given SNR. The CQI missed detection probability – the probability that CQI is not detected properly – should not exceed 1%. PUCCH format 2 performance requirements are specified for all E-UTRA channel bandwidths, normal cyclic prefix, 2 receive antennas and ETU70 channel model only.

For base station supporting multiple channel bandwidths, only tests for the highest and the smallest channel bandwidths are applicable.

PRACH

The PRACH performance requirements are specified for burst format 0, 1, 2, 3 and are determined by a minimum required total false alarm probability and missed detection probability for a given SNR. The total false alarm probability – the probability that preamble is detected (the sum of all errors from all detectors) when nothing is sent, should not exceed 0.1%. The missed detection probability should not exceed 1% and is depending on the following errors:

- preamble is not detected properly;
- different preamble is detected from the one that is sent;
- correct preamble is detected but with wrong timing estimation.

The PRACH performance requirements are specified for two and four receiver antennas and for AWGN and ETU70 channel models. For AWGN and ETU70, a timing estimation error occurs if the estimation error of the timing of the strongest path is larger than 1.04 μs and 2.08 μs, respectively. For ETU70, the strongest path for the timing estimation error refers to the strongest path in the power delay profile – the average of the delay of all paths having the same highest gain equal to 310 ns.

Table 14.12 Set of PRACH base station tests for normal mode

Channel model	f offset (Hz)
AWGN	0
ETU70	270

Table 14.13 Set of PRACH base station tests for high speed mode

Channel model	f offset (Hz)
AWGN	0
ETU70	270
ETU70	625
ETU70	1340

The PRACH performance requirements are specified for normal mode and for high-speed mode. Different frequency offsets are tested for these modes. For high speed mode, when the receiver is in demanding conditions, additional frequency offsets are specified −625 Hz and 1340 Hz. In case of a frequency offset of 625 Hz, which corresponds to half of the preamble length $(0.5 \cdot 1/0.8\,ms)$, the receiver is in difficult conditions because several correlation peaks are observed. 1340 Hz frequency offset corresponds to the velocity of 350 km/h at the 2.1 GHz frequency band. Table 14.12 and Table 14.13 present a set of PRACH base station tests for each burst format and each configuration of receive antennas, for normal mode and for high speed mode, respectively.

14.6 User Equipment Design Principles and Challenges

14.6.1 Introduction

The main requirements related to the LTE UE design are described in this section. As all LTE UEs will have to support legacy air interfaces, LTE functionality will be constructed on top of the 2G GSM/EDGE and 3.5G WCDMA/HSPA architecture. This section presents the main differences and new challenges of an LTE terminal compared to a WCDMA terminal. Both data card and phone design are discussed.

14.6.2 RF Subsystem Design Challenges

14.6.2.1 Multi-mode and Multi-band Support

LTE equipment has to provide connection to legacy air interfaces to provide customers with roaming capability in areas where LTE base stations are not yet deployed. It is essential for the acceptance of the new technology that there is continuity in the service to the user. The equipment must also support different operator, regional, and roaming requirements, which results in the need to support many RF bands. For the successful introduction of a new technology the performance of the UE must be competitive with existing technology in terms of key criteria such as cost, size, and power consumption [11].

The definition of 3GPP bands can be found Table 14.1. Although the *Phase Locked Loop* (PLL) and RF blocks of a *transceiver* (TRX) can be designed to cover almost all the bands, the designer still has to decide how many bands shall be supported simultaneously in a given phone to optimize the radio. This drives the number and frequency range of *Low Noise Amplifiers* (LNAs) and transmit buffers. The same considerations exist for the *Front-End (FE)* components in terms of the number and combination of *Power Amplifiers* (PAs), filters and thus the number of antenna switch ports. Similarly, the number of supported bands required in diversity path needs to be decided.

The support of legacy standards in combination with band support results in a complex number of use cases. These need to be studied to provide an optimum solution in terms of size and cost. Here are some of the anticipated multi-mode combinations:

- EGPRS + WCDMA + LTE FDD
- EGPRS + TD-SCDMA + LTE TDD
- EVDO + LTE FDD

The first two combinations are natural migration paths through 3GPP cellular technologies and standardization has taken care of measurement mechanisms to allow handover back and forth between each technologies without requiring operation (receive or transmit) in two modes simultaneously. This allows all three modes to be covered with a single receive or transmit path, or two receive paths when diversity is required. In the last case handover support is more complex but, again, a transmit-and-receive path combination is feasible because two receivers are available from the LTE diversity requirement.

These multi-mode requirements can be supported by one highly reconfigurable TRX *Integrated Circuit* (IC), which is often better known under the well-used term of 'Software Defined Radios'. In reality, this should be understood as software reconfigurable hardware. This re-configurability for multi-mode operation takes place mainly in the analogue baseband section of the receiver and the transmitter. The multi-band operation takes place in the RF and *Local Oscillator* (LO) section of the TRX and in the RF-FE.

The combination of the high number of bands to be supported together with the need of Multi-Mode support is driving RF subsystem architectures that optimize hardware reuse especially in the FE where size and number of components become an issue. Improvement in this area builds on the optimizations already realized in EGPRS/WCDMA terminals, driving them further to meet LTE functionality. This means that LTE functionality needs to comply with the architecture framework used for 2G and 3G but also explores new opportunities of hardware reuse:

- The LTE performance should be achieved without the use of external filters, either between LNA and Mixer or between the transmitter and the Power Amplifier, as this is already realized in some WCDMA designs. This not only removes two filters per band but also allows the design of a multi-band TRX IC to be simplified. This is especially critical for the FDD mode and in cases where large channel bandwidth is used in bands where the duplex spacing is small. These new design trade-offs are discussed in section 14.8.2. Similarly the interface to the *Base-Band* (BB) needs to multiplex every mode to a minimum number of wires. This is best achieved by a digital interface as described in section 14.6.3.

- Re-use of the same RF FE path for any given band irrespective of its operation mode. This implies using:
  - co-banding: reuse of the same receive filter for any mode, especially EGPRS (half duplex) reuses the duplexer required for FDD modes;
  - multi-mode Power amplifier: reuse of same power amplifier whatever mode or band.

The details associated with these two techniques are developed in further sections. However it is instrumental to illustrate their benefits in terms of reduced complexity and thus size and cost. Table 14.14 shows the difference in complexity between a design with none or all of the above optimizations. The use case is that of a US phone with international roaming supporting: quad-band EGPRS (bands 2/3/5/8), quad-band WCDMA (bands 1/2/4/5), and triple band LTE (bands 4/7/13). The fully optimized solution block diagram is illustrated in Figure 14.22.

The two scenarios show a difference of almost a factor of two in number of components. In addition the lower component count also significantly simplifies the TRX IC and the FE layout and size. However, a number of partially optimized solutions that fall somewhere between these scenarios also exist.

14.6.2.2 Antenna Diversity Requirement

One of the main features introduced with LTE is the MIMO operation to enhance the available data rate. The MIMO operation requires the UE to be equipped with two receive antennae and paths. The conformance testing is done through RF connectors and assumes totally uncorrelated antennae. This scenario is far from representative of real operation in the field especially for small terminals operating in the lower 700 MHz frequency band. Small terminals, like smart phones, have dimensions that only allow a few centimeters' separation between the two antennas. At low frequencies this distance results in high correlation between the two signals received at each antenna, which results in degraded diversity gain. Furthermore at these frequencies and with small terminal sizes the hand effect (modification of antennas radiation patterns due to the hand or head proximity) further degrades the diversity gain. For devices demanding higher data rates, like laptops or PC tablets, the device sizes allow proper antenna design. Also in higher frequency bands even a small terminal can provide sufficient antenna separation to grant good MIMO operation.

Table 14.14 Component count for different front end design

Implementation Blocks	Un-optimized 'side-by-side'	Fully optimized
Low Noise Amplifiers	13	10
Duplex filters	6	5.5*
Band-pass filters	17	5
Power Amplifiers	8	2
Switches	2	4
Transceivers	3	1
Total number of components	49	27

*Reuse of RX side for band 1 and 4.

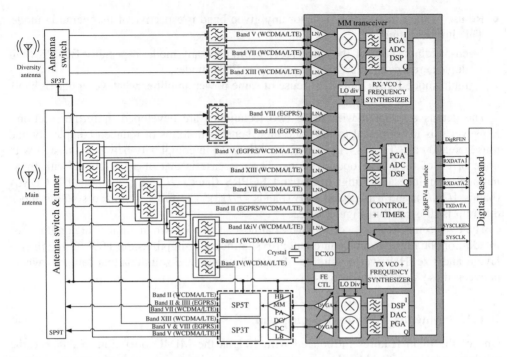

Figure 14.22 EGPRS/WCDMA/LTE optimized RF subsystem block diagram. HB = high band, LB = low band, DCXO = digital crystal oscillator, PGA = programmable gain amplifier, MM = multi-mode, FE = front-end

When the antenna design of an LTE UE is considered relative to current WCDMA phones the following design challenges are encountered:

- The overall band support: the current WCDMA antenna frequency range is from 824 to 2170 MHz whereas future LTE devices will have to cover 698 to 2690 MHz. This stretches the current state of the art in terms of antenna matching but also in terms of maintaining antenna gain throughout the larger bandwidth. This will probably drive the introduction of new technology such as active antenna tuning modules.
- Some LTE bands create new antenna-coupling issues with other systems potentially present in the UE. The other antennae to be considered are the *Global Positioning System* (GPS) antenna, the *Bluetooth* (BT) and WLAN antenna, the analog *Frequency Modulated* (FM) radio antenna, and the Digital TV (Digital Video Broadcast – Handheld, DVB-H) antenna. The related critical coexistence use cases are discussed in section 14.6.2.3.
- The support of antenna diversity: introducing an extra antenna in an already complex and relatively small UE presents a significant challenge if reasonable diversity gain is required.

The last two issues are easier to handle in data card designs where only the cellular modem is present. To some extent this is also true for the larger portable devices like PC tablets and video viewers but smart phone mechanical design may prove particularly challenging in terms of antennae placement.

14.6.2.3   New RF Co-existence Challenges

In the context of the Multi-Mode UE where multiple radio systems and multiple modems, such as BT, FM radio, GPS, WLAN and DVB-H must co-exist, the larger bandwidth, the new modulation scheme, and the new bands introduced in LTE create new coexistence challenges. In general coexistence issues are due to the transmit (TX) signal of one system (aggressor) negatively impacting another system's receiver ('RX'-victim) performance and notably its sensitivity. There are two aspects to consider: direct rise of the victim's noise floor by the aggressor's transmitter out of band noise in the receiver band and degradation of the receiver's performance due to blocking mechanisms.

Noise leakage from an aggressor TX in a victim's RX band is added to the RX noise floor further degrading its sensitivity as shown in Figure 14.23. This noise leakage is only a function of the intrinsic TX noise, TX output filter attenuation in the victim's band and antenna isolation. This desensitization mechanism depends on the Aggressor's TX design.

The desensitization of a victim due to an aggressor's transmitter out of band noise can be calculated as follows:

$$DES_{OOBN} = 10 * Log(10^{(-174+NFvictim)} + 10^{(POUTaggressor-OOBN-ANTisol)})$$
$$+ 174\text{-}NFvictim \qquad\qquad (14.11)$$

Where DES_{OOBN} is the resulting degradation of the victim's sensitivity in dB due to the aggressor's TX noise, $NFvictim$ is the victim's RX *Noise Figure (NF)* referred to the antenna in dB, $POUTaggressor$ is the aggressor's TX maximum output power in dBm, *Out Of Band Noise (OOBN)* is the aggressor's TX noise in the victim's band in dBc/Hz, and $ANTisol$ is the antenna isolation in dB.

The blocker power level present at the victim's LNA input depends on the aggressor maximum transmitted signal power, the antenna isolation and the victim's FE filter attenuation in the aggressor's TX band. Mitigation techniques can only be implemented in the victim's RX design. In both cases improved antenna isolation helps but the degrees of freedom are limited in small form factor UE especially when the aggressor and the victim operating frequencies are close to one another.

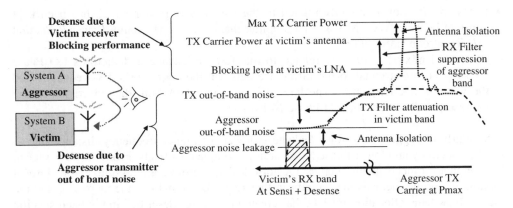

Figure 14.23 Victim/Aggressor block diagram and Aggressor transmitter leakage to victim receiver

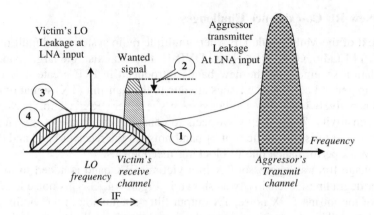

Figure 14.24 Desensitization mechanisms

The victim's desensitization due to the presence of a blocker may result from multiple mechanisms described in Figure 14.24.

Victim's RX LO phase noise being added to the wanted signal due to *reciprocal mixing* with aggressor's transmit signal leakage:

1   Victim's RX reduced gain for wanted signal due to *cross-compression* on the TX signal leakage.
2   *Second order inter-modulation distortion (IMD$_2$)* product of the TX signal leakage AM content falling at DC for *direct conversion receivers (DCR)*, which potentially overlap with the wanted signal.
3   Third order inter-modulation product of victim's LO leakage with TX signal leakage AM content falling at DC, which potentially overlaps with wanted signal (known as *cross-modulation*).

In the first two cases the actual aggressor signal characteristics and bandwidth do not play a role in the interference mechanism:

- Cross-compression is directly related to the signal peak power which has been kept constant between WCDMA and LTE uplink thanks to the *Maximum Power Reduction (MPR)* introduced in 3GPP.
- Reciprocal mixing is directly related to interference leakage and victim's LO Phase Noise at an offset equal to the difference between the aggressor's TX frequency and the victim's RX frequency. The larger distance improves both the selectivity on the aggressor leakage and the LO Phase Noise.

Similarly the aggressor's noise leakage is lower for higher frequency distance due to higher selectivity and lower Phase Noise at larger frequency offsets. From the above it can be seen that for mechanisms such as TX noise leakage, cross-compression, and reciprocal mixing, the prime factor is the vicinity in the frequency domain of the aggressor and the victim. New band allocation for LTE has created new cases described in the band specific coexistence challenge section and in section 14.8.2.

The cross-modulation and the IMD_2 products however have a spectrum bandwidth (BW) directly related to the interferer's BW. Furthermore the spectrum shape of these distortions depends on the signal statistics and modulation scheme. For these contributors it is difficult to propose a generic analysis on how the LTE interference compares to WCDMA. Analysis of which portion of the interfering spectrum is captured within the victim's channel bandwidth needs to be conducted for each LTE bandwidth and modulation combinations. A comparison of IMD2 products for WCDMA and LTE QPSK uplink modulated carriers can be found in section 14.8.2.1.

Band specific coexistence challenges

As far as the LTE out-of-band TX noise emissions are concerned, the duplexer must provide enough attenuation in the following frequency bands:

- In band 11, the TX frequency is very close to the GPS band. With the anticipated antenna isolation, more than 40 dB attenuation needs to be provided. GPS receivers may also have to improve their blocker handling capability to cope with LTE TX signal leakage.
- In bands 11/12/13/14, the TX frequencies are close to some of the TV bands, and consequently adequate attenuation of LTE TX noise is required.
- In bands 7/38/40, the TX frequencies are very close to 2.4 GHz ISM band where BT and WLAN operate. With the anticipated antenna isolation, approximately 40 dB attenuation needs to be provided. Similarly the BT and WLAN TX noise needs to be kept low in the band 7/38/40 RX frequencies. Given the blocker handling capability already required for the LTE FDD RX the BT and WLAN blockers are not anticipated to be an issue. However, BT/WLAN receivers may have to improve their blocker handling capability to cope with the LTE TX signal leakage.

As far as the LTE TX modulated carrier is concerned, with 23 dBm output power, special attention must be paid to controlling the LTE TX second harmonic power level in the following aggressor/victim combinations:

- In band 7/38, the TX second harmonic falls within the 5 GHz WLAN band, thus requiring the LTE transmitter to attenuate it to be close to −90 dBm. In comparison to the 3GPP LTE which requires a harmonic level to be lower than -40 dBm, this represents significant challenge. The WLAN RX must also be sufficiently linear to prevent re-generating this second harmonic in presence of the LTE fundamental TX signal leakage.
- In band 13, the TX second harmonic falls within the GPS band thus requiring the LTE transmitter to attenuate it to lower than −110 dBm. This level of attenuation might be hard to achieve in practice. Similarly to the above co-existence test case, the LTE TX leakage presence at the GPS RX input also imposes linearity requirements to prevent re-generating this second harmonic.

There are also other cases where a harmonic of the LTE TX falls in the 5 GHz WLAN or UWB bands. These are usually higher harmonic order that should be less of an issue.

14.6.3 RF-baseband Interface Design Challenges

In mobile handsets, the *Radio Frequency* (RF) transceiver and the *Base-Band* (BB) processor are often implemented on separate *Integrated Circuits* (IC). The transceiver IC normally contains analog signal processing, while the baseband IC is predominantly digital. Therefore *analog-to-digital* (A/D) and *digital-to-analog* (D/A) conversions are required in receive and transmit paths respectively. The location of these converters is a critical choice in wireless system design. If the converters are implemented in the RF transceiver, discrete time domain (digital) data is transferred across the interface between the BB and the TRX. On the other hand, if the converters are located in the BB IC, the interface comprises continuous time domain (analog) differential I/Q signals.

Figure 14.25 shows a typical example of an analog-based interface with receive diversity. Here this mobile is designed to support all of these radio operating options: '3G' *Wideband Code-Division Multiple Access (WCDMA)*, on one frequency band plus GSM on four frequency bands. It may readily be seen that 24 separate interconnections are required.

Despite being in mass production from many vendors, analog interface solutions face the following challenges and criticisms. The large number of interconnecting pins increases package size and cost on both sides, and complicates printed circuit board routing. This is a particular problem because the sensitive analog interconnections also require special shielding to achieve required performance. Furthermore, such interfaces are proprietary, forcing handset makers to use particular pairs of BB and RF devices. While some IC vendors prefer this forced restriction, it is a disservice to the industry by restricting

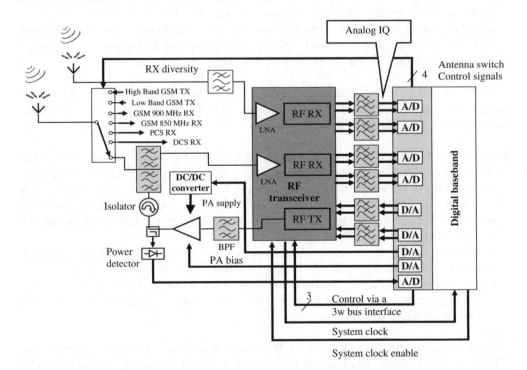

Figure 14.25 Typical HSPA monoband 3G, quad band 2G transceiver/BB block partitioning in analog I/Q interface

competition and disabling creative approaches in either the BB or RF side alone from adoption in products. BB devices must include the necessary large number of analog blocks, increasing their die size.

This last point is particularly important – much more so than the obvious economic issues involved. Clearly, larger die areas increase IC cost of manufacture. Beyond that are the facts that these analog designs do not generally shrink as well as the digital cells do with progressively smaller CMOS processes. Thus, the fraction of the BB IC that these analog designs take up is increasing in percentage. Additionally, the analog circuitry yield in production may be lower than the digital circuitry, so a perfectly fine digital baseband would be thrown away when an analog section fails. Even more of a problem is the fact that analog design information on a new CMOS process always is provided later than the digital design information – sometimes much later. When new CMOS processes become available, this forces BB designs to wait until all the necessary design information is available and sufficiently qualified. And in the newer CMOS processes, analog designs are actually getting more difficult, not easier.

Taking all of this together, it becomes clear that a purely digital interface is greatly desired, and would be a huge benefit to the mobile industry. The main organization leading the standardization effort is MIPI – the Mobile Industry Processor Interface consortium, working within IEEE-ISTO [12]. Combining the terms 'digital' and 'RF' together into the name 'DigRF^{SM}', this interface is already in its third evolutionary step as listed in Table 14.15.

Adopting these digital interfaces changes the earlier block diagrams to that of Figure 14.26. The dual objective of eliminating analog designs from the BB device and reducing pin count down to only seven pins is met. For DigRF^{SM} v4, the interface data rate necessary to just support a single antenna receive chain in a 20 MHz LTE application reaches 1248 Mbps.

One of the biggest challenges in DigRF^{SM} v4 is EMI control. EMI was not a major concern for DigRF v2 and v3, because those interface data rates are well below the mobile's radio operating frequencies. With all the bands now under consideration for an LTE mobile device use, the internal radio *low noise amplifier* (LNA) section is sensitive to energy across frequencies from 700 MHz to nearly 6 GHz. The DigRFv4^{SM} data rates and common mode spectrum emissions now exceed the radio frequencies of several mobile operating bands. This is a huge problem, clearly seen in Figure 14.29(a).

Before addressing EMI mitigation techniques, it is essential to understand how much signal isolation is available from the physical packaging of the RF IC. Adopting the vocabulary of EMI engineering, the LNA input is called the 'victim', while the interface is called the 'aggressor'. As we can see in Figure 14.27 (left), there are many possible paths for energy on the interface to couple into the 'victim' LNA. Figure 14.27 (right-plain dots) shows one example of practically available isolation.

Table 14.15 DigRF version evolutions

DigRF version	Standard	Interface bitrate (Mbps)
v2: 2G	GSM/GPRS/EDGE	26
v3: 3G	2G + HSPA	312
v4: 4G	3G + LTE	1248, 1456, 2498, 2912

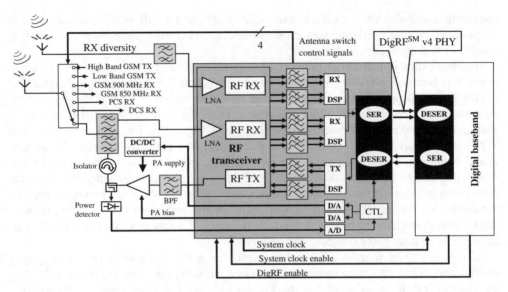

Figure 14.26 Example of digital RF-BB interface application using DigRF v4 with RX diversity. A/D = analog to digital converter, D/A = digital to analog converter, SER = serialize, DESER = deserialize. DSP functions include A/D, D/A, digital filtering and decimation (specific to RX)

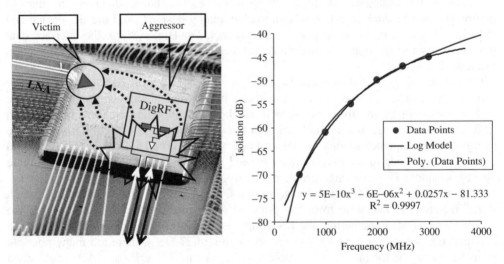

Figure 14.27 Left: examples of EMI coupling paths within an RF IC. Right: in-package isolation example at several frequencies

The upper limit to the maximum amount of aggressor noise PSD allowed at the victim's input pins is established with the set of characteristic curves shown in Figure 14.28. Considering worst cases, a cellular LNA with an intrinsic *Noise Figure* (NF) of 3 dB may be degraded to a 3.5 dB NF. According to these charts, the interfering noise must be at least 6 dB below the kT floor, i.e. at or below −180 dBm/Hz. Similarly, a more sensitive GPS LNA with an intrinsic NF of 2 dB may be degraded by at most 0.25 dB. Evaluating

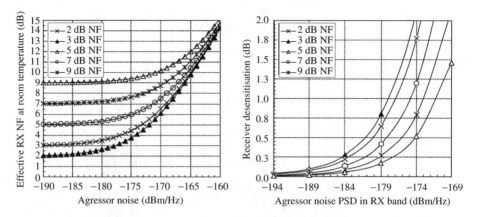

Figure 14.28 Response of a 'victim' LNA to additional noise provided at its input. Left: effective noise rise of a victim NF referred to its LNA input (room temperature). Right: corresponding victim's desense, versus aggressor noise PSD

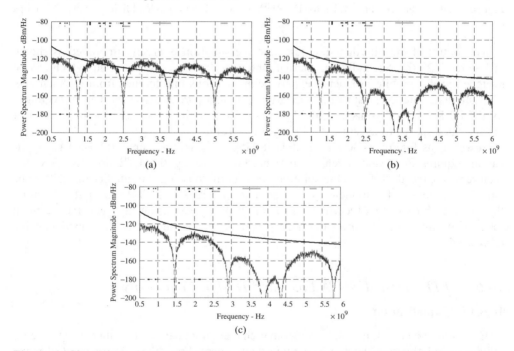

Figure 14.29 Effects from adding EMI mitigation features to the DigRF/M-PHY interface: (a) imperfect differential signaling, (b) slew rate control, and (c) alternate frequency selection of 1456 Mbit/s. Horizontal bars at top of each diagram indicate the location of telecommunication standards victim's frequency bands from 700 MHz to 6 GHz.

these two cases we conclude that the interference must be below −180 dBm/Hz and −184 dBm/Hz for a cellular and a GPS RX respectively.

The limit curve shown in Figure 14.29 (a to c) is based on combining the isolation model with the cellular LNA noise tolerance and shows violation in (a). Mitigation techniques are therefore required to ensure product success. First on the mitigation techniques list is

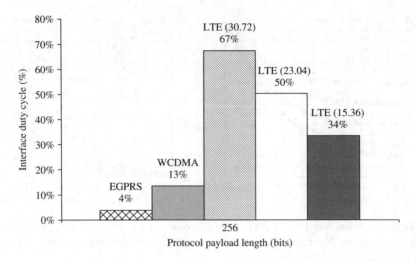

Figure 14.30 Interface duty-cycles for DigRF^{SM} v4 with 256 bit payload field size at 1.248 Gbps

to use control of the bit edge slew rate, and the situation is greatly improved as shown in Figure 14.29 (b). Secondly, an alternate interface frequency is provided, which can be switched to if a mobile product is experiencing spurious interference from the interface. This alternate interface data rate, 1456 Mbps, is also particularly useful in protecting a local GPS RX input by ensuring the LNA is now operating close to a frequency null of the interface spectrum as shown in Figure 14.29 (c).

However, as Figure 14.30 shows at 1248 Mbps, a single antenna 20 MHz LTE appli-cation requires 70% duty cycle of the interface. Adding the LTE diversity RX, the requirement exceeds 100%! One solution consists in doubling the number of 1248 Mbps lanes. Alternatively, for applications that do not have significant EMI sensitivity, doubling the interface bit-rate to 2496 Mbps solves the capacity issue, but now generates a main lobe of the interface spectrum that spans over all bands, including the super-sensitive GPS band.

14.6.4   LTE Versus HSDPA Baseband Design Complexity

14.6.4.1   Equalization

LTE aims to reduce cost per bit, among other challenging goals such as increasing spectral efficiency and allowing flexible bandwidth deployment. From the receiver point of view, one key measure is the required complexity and in particular comparison with previous releases such as WCDMA and HSDPA. Figure 14.31 shows the estimated complexity based on the baseline receiver for all transmission modes, as introduced in section 14.9, excluding channel decoding operation complexity. It can be noticed that the complexity of the LTE receiver grows linearly with respect to system bandwidth and the corresponding maximum nominal throughput. Interestingly, MIMO mode requires less than the double of SIMO mode complexity. Comparing the estimated complexity of a Release 6 optimistic HSDPA receiver assuming a low complexity chip-level equalizer Category 10 device with 13 Mbps throughput, shows LTE to be an attractive technology at least from an

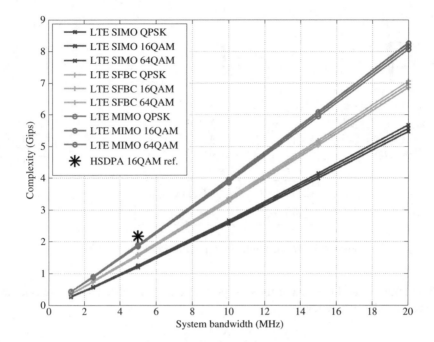

Figure 14.31   Complexity of LTE receiver

inner receiver complexity point of view because of OFDM choice. Assuming 5 MHz bandwidth and 16QAM modulation, LTE offers the same throughput with nearly half of the required complexity than HSDPA.

Nevertheless, despite this advantage in the relative comparison for the 5 MHz bandwidth case, the LTE receiver requires a considerable complexity increase compared to HSDPA since UE must support a system bandwidth of 20 MHz as minimum requirement and therefore still constitutes a challenge for mobile and chip-set manufacturers. The LTE class 4 device with 150 Mbps capability has a complexity of approximately four times higher than HSDPA category 10 device with 13 Mbps capability.

Further details in the repartition of complexity among inner receiver stages are also given in Figure 14.32. Along the supported modes, the complexity associated to FFT operations – always equal to two as the number of receiving antennas – becomes less important with respect to the overall complexity because of increased needs in channel estimation and equalization. With MIMO the operation FFT takes 47% of the calculations while channel estimation takes 34% and equalization 18%.

14.6.4.2   Turbo Decoder

In LTE, the efficiency and the complexity of the channel decoding operation grows considerably as the maximum nominal data rate increase compared to previous releases. The Turbo decoder must support rates up to 150 Mbps in Category 4 device. The Turbo code algorithm in LTE is similar to the Turbo code in HSDPA. The Turbo decoder processing delay, and hence its throughput, is roughly linearly proportional to the number of turbo

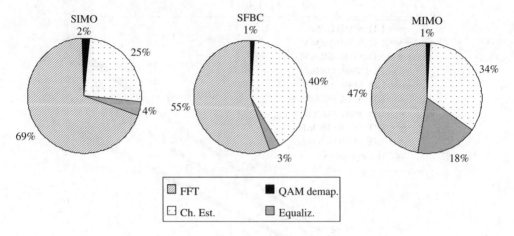

Figure 14.32   Complexity repartition in LTE receiver

decoding iterations and the size of the transport block. As a consequence, we can describe the Turbo decoder efficiency as

$$\eta = \frac{N_{it} \cdot r_{max}}{f_{clock}} \tag{14.12}$$

where N_{it} is the number iterations, r_{max} is the maximum data rate and f_{clock} is the working clock frequency of the turbo decoder. Table 14.16 and Table 14.17 show the efficiency/clock-frequency trade-off for the HSDPA and LTE cases assuming a number of iterations equal to eight and a maximum data rate of 13 Mbps and 150 Mbps respectively.

Efficiency can be seen as a measure of the parallelization required within the Turbo decoding operation. Comparing the two tables, it appears evident that to support LTE

Table 14.16   Turbo decoder efficiency/clock frequency trade-off for 13 Mbps

Efficiency η (b/s)	Clock frequency (MHz)
5.5	20
2.7	40
1.8	60
1.4	80
1.1	100
0.9	120
0.8	140
0.7	160
0.6	180

Table 14.17 Turbo decoder
efficiency/clock frequency
trade-off for 150 Mbps

Efficiency η (b/s)	Clock frequency (MHz)
12.6	100
10.5	120
9.0	140
7.9	160
7.0	180
6.3	200
5.7	220
5.2	240
4.8	260

data rate within reasonable clock frequencies – strictly dictated by power consumption and hardware technology constraints – a high level of parallelization is imposed.

It is worth mentioning that Turbo decoding parallelization was not attainable given the construction of the interleaver used for the encoding and decoding procedure of previous releases of 3GPP standard. This problem was solved for LTE specification and a contention-free interleaver has been built to allow any level of parallelization expressed as a power of two (2, 4, 8, 16...).

Nevertheless, parallelization itself does not come for free and it constitutes a challenge for manufacturers: it demands an in-depth change in the Turbo decoding algorithm, increases the surface and gate count of an amount proportional to the parallelization factor and, in principle, an operational higher clock frequency.

As a final consideration, it is worth noting that the Turbo Decoder complexity is in principle much higher than the complexity of the rest of the mobile receiver signal processing: complexity grows linearly with the target throughput and number of iterations. It can be approximated by [13].

$$C_{TD} = 200 \cdot N_{it} \cdot r_{max}(\text{MIPS}) \tag{14.13}$$

In LTE case, for a maximum data rate of 150 Mbps, the complexity approaches 240 Gips while the FFT, equalization and channel estimation complexity together are in the order of 9 Gips, as shown in Figure 14.32.

The Turbo Decoder would then appear as requiring nearly 96% of the overall complexity in the receiver but is a rather misleading conclusion. Signal processing involved in the Turbo Decoding is mainly constituted by addition operations in the max-log-MAP case. Equalization and associated functions require instead multiplication operations and, in principle, larger fixed-point sizes. Implementation choices making the difference, signal processing complexity is a valuable measure of the challenge required among standard evolutions for the same functionality but cannot allow for more general comparisons.

14.7 UE RF Transmitter

14.7.1 LTE UE Transmitter Requirement

14.7.1.1 Transmitter Output Power

The LTE specified maximum output power window is the same as in WCDMA: 23 dBm with a tolerance of $+/-2$ dB. The earlier WCDMA specifications used 24 dBm with a tolerance of $+1/-3$ dB. The SC-OFDMA [14] has higher *Peak to Average Ratio* (PAR) than the HPSK modulation of WCDMA. Figure 14.33 shows the ACLR performance of a WCDMA PA under QPSK modulated SC-OFDMA signal. The WCDMA operation assumes 24 dBm and LTE operation 23 dBm output power. The main difference is related to spectral shape and the fact that the occupied BW is slightly higher in LTE (4.5 MHz) than that of WCDMA (99% energy in 4.2 MHz), and consequently the ACLR is slightly degraded.

In a similar manner to HSDPA and HSUPA, a *Maximum Power Reduction* (MPR) has been introduced in LTE to take into account the higher PAR of 16QAM modulation and some resource block allocation. Again this ensures a proper ACLR under a complex set of TX modulations. Compared to WCDMA where the only direct interference falling in the RX band is related to spurs and the OOB noise, the LTE TX linearity should also be considered. It is particularly important in the case for 10 MHz BW in the 700 MHz bands where the duplex distance is only 30 MHz and where fifth- and seventh-order inter-modulation products of the TX overlap with the RX channel. This phenomenon is discussed in section 14.8.2.

LTE power control ranges from -40 dBm to $+23$ dBm. WCDMA transmitters offer -50 dBm to $+23$ dBm of *Transmit Power Control* (TPC) with 1 dB resolution. The same TPC techniques can be applied in LTE taking into account all MPR cases.

14.7.2 LTE Transmit Modulation Accuracy, EVM

The LTE 16QAM modulation case places stringent requirements on the TX imperfections to meet the *Error Vector Magnitude* (EVM) budget. Overall the errors should be further minimized in the case of LTE. Each contributor can be analyzed separately to meet the 17.5% EVM budget in QPSK and 12.5% in 16QAM. EVM measurements are done

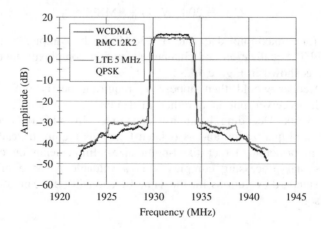

Figure 14.33 24 dBm WCDMA and 23 dBm LTE 5 MHz QPSK spectrums

after *zero-forcing* (ZF) equalization in the test equipment. For comparison, the WCDMA HPSK EVM budget is equal to 17.5%, but does not assume any equalization in the test instrument. In practice the design target is around 8%. The TX imperfections contributing to the EVM budget are considered in the next section.

- *Carrier rejection.* As the measurement is made after ZF-equalization the test equipment partially removes this contribution. LTE standardization has set a separate specification for carrier rejection to ensure that the carrier level stays within reasonable limits for the ZF algorithm. At low output power small DC offsets in the TX chain generate a high carrier leakage to a point where TPC accuracy may not be met any more. This problem is already severe in WCDMA where the TPC range is 10 dB larger at the bottom end. For this reason carrier leakage compensation is implemented in RF TRXs and this technique generally achieves 40 dB of carrier rejection throughout the TPC range making this contribution negligible in the EVM budget.
- *Even order distortion.* Even order non-linearity contributes mainly to ACLR as the main effect is to enlarge the transmitted spectrum. ACLR requirements being of 33 dB and 43 dB in adjacent and alternate channels these contributions are usually negligible in term of EVM. AM/AM in the PA can be considered similarly.
- *LO phase noise.* The induced jitter generates phase error in the modulation constellation thus contributing to EVM. As the SC-OFDMA used in transmit is single carrier the phase noise has similar contribution than for the WCDMA case.
- *PA distortion.* AM/PM distortion (together with AM/AM) has a contribution to EVM and also generates asymmetrical spectral energy in the adjacent channels. AM/PM and AM/AM distortion increases as the transmit power is increased. Hence, the LTE MPR scheme ensures that PAs dimensioned for WCDMA will not dominate the overall EVM budget.
- *Image.* The signal image generated by the quadrature imperfections in the up-mixing process can be considered as in-band noise contributing to the TX signal SNR. The use of 16QAM requires better control of such imperfections relative to WCDMA [15].
- *Group delay distortion.* Compared to WCDMA where the I/Q BB BW is 2 MHz for a minimum duplex distance of 45 MHz, LTE has an I/Q BW of 5 MHz for 30 MHz duplex distance or 10 MHz for 80 MHz duplex distance. Significant BB filtering is required to ensure that BB I/Q noise at the duplex frequency offset is very low (to allow SAW-less transmit architecture). This means that stop-band attenuation for LTE BB filter requires a more stringent filter specification than for WCDMA potentially introducing more in band distortion contributing to EVM.

Overall, the LTE EVM specification requires similar effort than for WCDMA in terms of RF imperfection but special attention needs to be taken for the higher bandwidth and smaller duplex distance. For the same reasons, the TX out of band noise is more difficult to achieve in LTE.

14.7.3   Desensitization for Band and Bandwidth Combinations (De-sense)

Although TX out-of-band noise in the RX band is not an explicit requirement in the 3GPP WCDMA specifications, reference sensitivity measurements are made in presence of the full TX power. To meet the reference sensitivity the TX noise leakage levels must stay below the thermal noise floor. Recent efforts in TX architecture have allowed the removal of the filter between the TRX IC and the PA. Interstage filters were used to clean the TRX noise before further amplification. Removal of the filter is made feasible by careful design of every noise source in the RF TRX IC.

As discussed in section 14.6 it is essential that the addition of the LTE functionality and the support of new bands do not require reintroduction of those filters. Two new challenges must be solved for this to be feasible: the smaller duplex separation of certain new band configurations (12/13/14) or the wider channel bandwidth. In some specific cases these two challenges are combined. This issue has been recognized by the standardization body as shown in Table 14.4 where the relaxed sensitivity requirements have been defined for certain combinations of band and bandwidth.

14.7.4   Transmitter Architecture

14.7.4.1   Transmit RF Modulator

Direct up conversion is the obvious choice for a 2G/3G/LTE multi-mode TX. It is the *de facto* standard for WCDMA. The large BW requirements of LTE would pose further challenges to alternative architectures such as polar modulation or other non-Cartesian approaches [16, 17]. This is especially true if the external filter requirement is to be relaxed in the FDD case. As briefly discussed above, thanks to a pragmatic approach in the LTE standardization there is a minimum number of modification needed to provide LTE transmit capability from a Cartesian transmitter already covering 2G and WCDMA. The main modifications lie in the higher bandwidth requirement on the BB DAC and filter to cover all the different BW with low out-of-band noise. In addition, extra RF bands needs to be supported which require extension of the PLLs tuning range and RF buffer bandwidths.

14.7.4.2   Multi-mode Power Amplifier

As discussed in section 14.6 one essential simplification of the world-wide RF FE is the use of a single PA line up covering multiple bands and multiple modes. The band coverage can be clustered in the following way:

- one *low band* (LB) PA line up covering all bands between 698 MHz and 915 MHz;
- one *high band* (HB) PA line up covering all bands between 1710 MHz–2025 MHz;
- one higher band PA line up covering all bands between 2300 MHz–2620 MHz.

The only band that is not covered is the Japanese band 11, which can be added for this specific phone configuration or even replace one of the other wide band PA depending on the overall band support.

Each of these line ups has to support different modulation schemes and maximum output power depending on band-mode combinations. These combinations can be found in Table 14.18, where 2 dB and 3 dB PA to antenna loss are considered for TDD and FDD modes respectively.

Taking into account the different PAR inherent to each modulation scheme and the required back off to meet ACLR requirements, a given line up has to meet a range of saturated output power (P_{outsat}) capabilities to achieve best efficiency/linearity trade offs.

Table 14.18 Modulation scheme and maximum output power per band configurations for a multi-mode PA

Sub-bands modes	Low band 698–915 MHz	High band 1710–2025 MHz	Higher band 2300–2620 MHz
GSM (GMSK)	35 dBm	32 dBm	n.a.
EDGE (8PSK)	29 dBm	28 dBm	n.a.
WCDMA (HPSK)	27 dBm	27 dBm	27 dBm
LTE (QPSK)	26 dBm	26 dBm	26 dBm

For example the LB PA has to achieve 35 dBm P_{outsat} capability for GSM, GMSK having only phase modulation the PA can be driven into saturation and achieve best efficiency. In WCDMA mode, it needs close to 31 dBm P_{outsat} capability to allow sufficient back off. A GMSK capable PA would have a very low efficiency in the WCDMA mode if nothing was done to decrease its output power capability. This would prove the Multimode PA architecture to be uncompetitive in terms of performance especially in 3G modes which are already challenging in terms of talk time. The only way to reach the best PA efficiency in all modes is to tune the output stage load line. This can be achieved in two ways:

- Tuning the output matching to transform the load impedance (usually 50 ohm) into the desired load line for every mode. This technique can be achieved for a small set of impedances and usually results in a lower Q output matching.
- Tuning the PA supply: In this case the saturated output power capability is proportional to the square of the supply voltage. If the supply is varied using a DC/DC converter then efficiency can be optimized for every mode. This technique is becoming more and more popular [18] and has the benefit to allow optimum efficiency for every mode.

14.7.4.3 Conclusion

Although this section does not provide a detailed analysis of LTE transmitter requirements, it discusses how these requirements can be achieved by simple extrapolation from GSM/EDGE/WCDMA architecture and performance. It is shown that all three modes can be supported with a single transmitter architecture with minimum of extra hardware when techniques like co-banding and DC/DC controlled multi-mode PA are introduced. This ensures easy migration towards LTE for mobile equipment manufacturers.

14.8 UE RF Receiver Requirements

The purpose of this section is to highlight the main differences between *Frequency Division Duplex* (FDD) UTRA and *Full Duplex* (FD) – FDD E-UTRA UE RF receiver (RX) system requirements. Throughout this section, WCDMA and LTE are used to refer to UTRA and E-UTRA respectively. The objective of the LTE UE RF test requirements listed in section 7 of [1] is to quantify RF impairments, which have an impact on the network performance. These impairments include *Noise Figure (NF)*, receiver *Error Vector Magnitude (EVM)*, selectivity at different frequencies, including adjacent channel, etc. The test requirements have been derived to ensure the industry can make the best possible re-use of IPs developed for WCDMA UEs. This is highlighted in the description of the UE reference sensitivity level and the *Adjacent Channel Selectivity* (ACS) system requirements. For this reason, the chapter focuses on some of the novel design challenges, which are specific to the LTE downlink modulation schemes and its associated new frequency bands. In this respect, the following challenges are discussed: RX self-desensitization, ADC design challenges, and the impact of RX EVM contributors in OFDM versus single carrier.

14.8.1 Reference Sensitivity Level

The reference sensitivity power level is the minimum mean power applied to both UE antenna ports at which a minimum throughput requirement shall be fulfilled. The throughput shall be equal or higher than 95% of the maximum throughput for a specified *Fixed Reference Channel* (FRC). Fixed Reference Channels are similar to the WCDMA reference measurement channel and in the sensitivity test case, the downlink carrier uses QPSK modulation and 1/3 coding rate.

The sensitivity requirement verifies the UE RX NF, which in the case of FDD operation, may include noise contributions due to the presence of the UE uplink modulated carrier as described in the next section. Other receiver impairments such as EVM are included within the demodulation performance requirements where a higher *signal- to-noise ratio* (SNR) is applied. Therefore, the selected FRC provides a reference maximum throughput defined for low SNR operation. Beyond this purpose, the UE reference sensitivity is of primary importance in [2] because it also serves as a baseline to set the downlink carrier power for ACS and blocker test requirements.

With LTE, the reference sensitivity requirements present several major differences compared to the WCDMA system requirements:

- LTE flexible bandwidth requires the RF receiver to implement reconfigurability of its channel select filters to support 1.4, 3, 5, 10, 15 and 20 MHz bandwidths;
- the reference sensitivity must be tested on two antenna ports: main and diversity receiver;
- new frequency bands with small *duplex gap* (DG), such as bands 5, 6, 8 and 11 introduce novel UE self-desense considerations when UE uplink transmission bandwidths are greater than 5 MHz. Bands 12, 13, 14 and the more recent band 17, all falling under the terminology UMTS 700, are also impacted by these limitations. This is not a major concern in, for example, WCDMA band I devices because the UE self-desense is primarily dominated by the UE transmitter chain noise floor at large frequency offsets, as can be seen in Figure 14.34 (left). Adding bands with small *duplex distance (DD)* and small DG, for which large BW deployment is planned, now places the UE receiver directly inside the bandwidth of the transmitter chain adjacent channel leakage shoulders as shown in Figure 14.34 (right). The test requirement therefore includes several relaxations which can be found in section 14.8.2.

The reference sensitivity level calculation is given by equation (14.14). The differences with the eNodeB equation (14.6) reside in the UE dimensioning assumptions: the NF is set to 9 dB, the *implementation margin* (IM) is equal to 2.5 dB, and a 3 dB correction factor is applied to account for the dual antenna reception gain. In addition, equation (14.14) includes a frequency band specific relaxation factor 'D_{FB}'. Note that D_{FB} is not an official 3GPP terminology and is equal to 1, 2 and 3 dB for bands in which the DD/DG ratio is

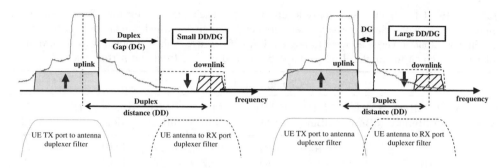

Figure 14.34 Illustration of large duplex gap (left) versus small duplex gap (right) frequency bands

greater than 1.5, 2 and 4 dB respectively, and 0 otherwise [19].

$$P_{REFSENS}[dBm] = -174[dBm/Hz] + 10\log(N_{RB} \cdot 180\,k) + NF$$
$$+ SNR + IM - 3[dB] + D_{FB} \tag{14.14}$$

D_{FB} is a metric that reflects design challenges for front-end components such as the duplexer. For a given technology resonator quality factor (Q factor), the higher the DD/DG ratio:

- The higher the *insertion loss (IL)* of each of the duplexer *band-pass filters (BPF)*. In RX, every decibel (dB) lost directly adds to the UE NF, therefore justifying a sensitivity relaxation. In TX, every dB lost causes heat dissipation in the duplexer.
- The sharper the TX BPF roll-off requirements to prevent TX noise from polluting the RX band. Due to heat dissipation, and mass production process variations, TX BPF must be designed with a slightly increased 3 dB cut-off frequency 'Fc' to ensure carriers located at the edge of the band do not suffer from too high insertion losses. In these bands the RX is more vulnerable to TX noise leakage, thereby justifying an extra relaxation.

With these assumptions listed, at the exception of the grey shaded values for which UE self-desense applies (cf. section 14.8.2), most of the FDD mode sensitivity levels listed in Table 14.19 can be computed. Several relaxations have been introduced in table 14.19 at 1.4 and 3MHz operating BW (in band II and IV for example) to account for further sensitivity degradation due to limited IIp2 and duplexer isolation performance. An introduction to UE self-desense due to IMD2 products can be found in section 14.8.2.1.

For example, Figure 14.35 shows that for a 5 MHz channel bandwidth ($N_{RB} = 25$), the reference sensitivity level is equal to −100 dBm. The simulated SNR for 95% of the maximum throughput is assumed to be equal to −1.0 dB. Link level simulations [20, 21], indicate slight variations around this nominal with SNR of −1.4 and −0.9 dB respectively (FRC A1-3). The throughput is approximately 2.1 Mbps for 5 MHz channel in the sensitivity test.

It is interesting to note that the 9 dB NF assumption used in LTE is very similar to the assumptions made for WCDMA UEs.

Depending on the balance of the link budget, the UE NF is a relevant parameter when planning cell coverage area. The fact that LTE NF requirements are similar to WCDMA commercial devices eases early delivery of UEs. In this way, the LTE standard provides a

Table 14.19 UE reference sensitivity levels applied to each antenna port for QPSK modulation. Values denoted with * in gray shaded cells are combinations for which a relaxation in N_{RB} and maximum output power is allowed to prevent UE self-desense. Values denoted with ** have been relaxed to account for limited IIp2 and/or duplexer and passive front end performance

E-UTRA Band	Channel bandwidth						Duplex Mode	DD/DG	D_{FB} (dB)
	1.4 MHz (dBm)	3 MHz (dBm)	5 MHz (dBm)	10 MHz (dBm)	15 MHz (dBm)	20 MHz (dBm)			
1	–	–	−100	−97	−95.2	−94	FDD	1.46	0
2	−102.7**	−99.7**	−98	−95	−93.2*	−92*	FDD	4	2
3	−101.7**	−98.7**	−97	−94	−92.2*	−91*	FDD	4.75	3
4	−104.7**	−101.7	−100	−97	−95.2	−94	FDD	1.13	0
5	−103.2	−100.2	−98	−95*			FDD	2.25	2
6			−100	−97*			FDD	1.29	0
7			−98	−95	−93.2*	−92*	FDD	2.4	2
8	−102.2	−99.2	−97	−94*			FDD	4.5	3
9			−99	−96	−94.2*	−93*	FDD	1.58	1
10			−100	−97	−95.2	−94	FDD	1.18	0
11			−100	−97*			FDD		
12	−101.7**	−98.7**	−97*	−94*			FDD		
13			−97*	−94*			FDD		
14		−99.2	−97*	−94*			FDD		
...									
17			−97	−94			FDD		
...									
33			−100	−97	−95.2	−94	TDD		
34	–	–	−100	−97	−95.2	−94	TDD		
35	−106.2	−102.2	−100	−97	−95.2	−94	TDD		
36	−106.2	−102.2	−100	−97	−95.2	−94	TDD		
37			−100	−97	−95.2	−94	TDD		
38	–		−100	−97	−95.2	−94	TDD		
39			−100	−97	−95.2	−94	TDD		
40			−100	−97	−95.2	−94	TDD		

NF requirement small enough to guarantee good cell coverage but not too small to allow system solution designers to deliver devices with better performance than the minimum requirement. This last point is important because sensitivity is most often a parameter used as a key selling argument. In this respect, LTE commercial devices from different vendors will deliver different sensitivity levels just like their WCDMA and GSM predecessors. An example of variability in WCDMA commercial reference sensitivity is shown in Figure 14.36 [22].

14.8.2 Introduction to UE Self-Desensitization Contributors in FDD UEs

Many of the following items are common to all FDD systems, and in particular, the reader is encouraged to refer to [11] for a detailed discussion on the impact of both TX noise

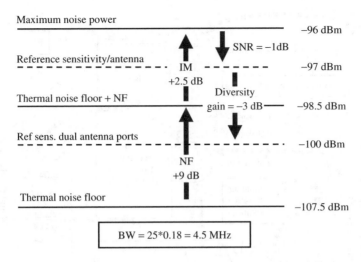

Figure 14.35 Reference sensitivity budget for LTE 5 MHz QPSK ($N_{RB} = 25$)

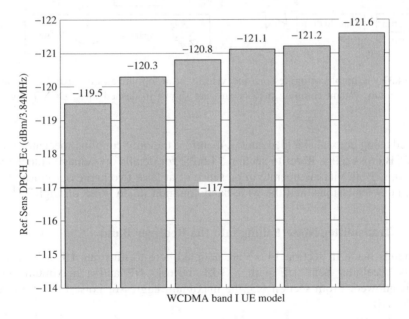

Figure 14.36 Example of class 3 WCDMA band I UE reference sensitivity performance

and TX carrier leakage in *direct conversion receivers (DCR)*. This section focuses on the key differences in LTE.

In Figure 14.37 it can be seen that the most sensitive victim is the main receiver LNA because it is only protected by the duplexer TX to RX port isolation. The diversity LNA benefits from the TX to antenna port duplexer rejection plus the extra antenna to antenna isolation. During conformance tests, this latter is fairly high because the test is

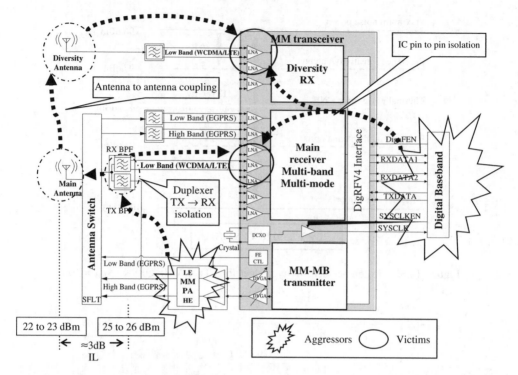

Figure 14.37 Example of aggressors and victims in an optimized quad band 2G – mono band 3G/LTE RF sub-system: coupling mechanisms are shown in dashed lines. DCXO: digital crystal oscillator

performed using coaxial shielded cables, therefore the only coupling mechanism are those due to PCB cross-talks. Refer to section 14.6.2.2 for details on radiated antenna coupling mechanisms. Both victims operate in presence of at least two aggressor noise sources: its own PA noise emissions, and the wide-band common mode noise of DigRF^{SM} v4 lines.

14.8.2.1 Transmitter Noise Falling into the Receiver Band

Assumptions made in section 14.6.3 are adopted here to illustrate UE self-desense. The victim is a cellular band UE with a 3 dB intrinsic NF and a maximum desense of 0.5 dB is allowed. From section 6.3, the maximum aggressor noise PSD must be below −180 dBm/Hz.

Large duplex distance frequency bands
The situation is identical to that experienced in existing WCDMA band I handsets. Assuming a worst-case duplexer isolation in RX of 43 dB, the maximum noise PSD falling in the RX band measured at the PA output port must be less than −180 dBm/Hz + 43 dB = −137 dBm/Hz. Most PAs tested with an ideal signal generator – a generator that provides a noise floor close to thermal noise at the duplex distance, just barely manage to meet this level [23]. This is one of the reasons why it remains a challenge to design RF IC

modulators that can deliver such low noise floors to the PA. The simplest solution consists in using an inter-stage BPF, but with the ever increasing number of bands to be supported, this solution has become unacceptable because of the associated increase in the *bill of material* (BOM). Designing filter less TX RF solutions is a subtle trade-off exercise between the amount of de-sense tolerated, the RF modulator current consumption, the BOM cost, and delivering to the customer a competitive reference sensitivity level.

Large transmission bandwidth in small duplex distance frequency bands

Although avoiding receiver de-sense is not trivial in large DG bands, solutions do exist and this leaves opportunities for innovation. In the case of the small DG, the situation is unfortunately more critical because the aggressor is no longer the out-of-band PA noise floor, but the PA ACLR shoulders as shown in Figure 14.38 (a). Therefore, a 3GPP relaxation is the only way to ensure adequate system operation.

An example of ACLR measurements performed using a commercial WCDMA band I PA with a 10 MHz ($N_{RB} = 50$) QPSK modulated carrier illustrate the issue in Figure 14.38 (b). UMTS 700 MHz front-end IL are emulated with the insertion of a 3 dB resistive attenuator at the PA output port.

As can be seen from Figure 14.38 (c), at 23 dBm output power, the desensitization reaches in a band 12 example, 16 and 10 dB for 3 and 9 dB NF respectively. To solve this issue, two mitigation techniques have been proposed in RAN4:

- *Maximum Sensitivity Degradation (MSD)* [24]: 'the victim's relaxation' technique, which consists in relaxing the reference sensitivity level by an amount similar to those of Figure 14.38 (c). The proposal maintains the UE at *maximum output power (P_{outmax})* to pass conformance test,
- Point B approach: 'the aggressor relaxation' technique [25], in which the reference sensitivity level is kept intact. This technique maintains the UE at P_{outmax} for a number *Resource Block (RB)* limited by a point called 'B'. Then, for $N_{RB} >$ point 'B', a progressive back-off of the UE output power is allowed to prevent UE self-desense. Thus, point 'B' corresponds to the maximum number of RBs at which P_{outmax} can be maintained, while point A corresponds to an output power back-off 'X' at which the maximum number of RBs can be supported as shown in Figure 14.39.

At the time of publication [26], the MSD approach has been accepted, thereby replacing the two new parameters specific to the Point B approach, with a single table listing the allowed sensitivity relaxation per band for each transmission bandwidth. Initial MSD values are proposed for certain bands [27] and must be finalized within the next RAN 4 meetings. The reference sensitivity is a topic which has triggered numerous contributions and evolves continuoulsy. For example, recent contributions suggest to remove MSD from 36.101, arguing that the reference sensitivity aims at testing the UE NF, while MSD aims at assessing the worst case performance due to other UE noise sources than pure thermal noise (NF). Finally, it is worth noting that *half duplex (HD)* – FDD operation has been accepted in RAN 1 [28]. HD-FDD is an alternative solution to the self-interference problem because the transmitter and the receiver do not operate simultaneously. In HD-FDD, the duplexer is no longer needed. This mode of operation can significantly simplify the UE RF front-end architecture as shown in [29].

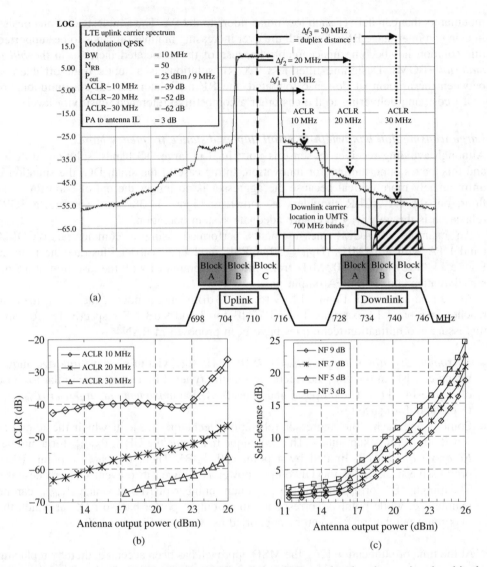

Figure 14.38 Example of LTE 10 MHz QPSK uplink ACLR overlapping the receiver band in the lower 700 MHz band: (a) ACLR spectral plot, (b) ACLR measured at 10, 20 and 30 MHz offset against antenna output power, (c) Self desensitisation versus LNA input referred NF[1]

Impact of transmitter carrier leakage presence at the receiver LNA input

In DCRs, differential structure imbalances and self-mixing are known as some of the mechanisms generating *second order intermodulation distortion (IMD$_2$)* products [30]. Self-mixing is due to finite isolation between the RF and the LO port of the down-conversion mixer. Under these conditions, the mixer behavior may be approximated as that of a squarer and therefore generates IMD$_2$ products. The squaring of CW blockers is

---

[1] Measurements in Figure 14.39 assume a Band XII application: minimum duplexer TX – RX isolation of 43 dB, maximum PA to antenna port insertion loss of 3 dB (preliminary measurements on prototypes indicate 2.5 dB maximum in duplexer, 0.5 dB in antenna switch), 23 dBm output power at antenna port.

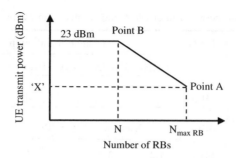

Figure 14.39 Point B approach to prevent UE self-desense

a simple DC term, which can be easily rejected using *high pass filters (HPF)*. However squaring an AM-modulated blocker generates a wideband noise-like IMD2 product, which can degrade the wanted signal SNR. In a RX filter-less architecture, the mobile's own TX leakage places the most stringent requirements on the mixer IIp2, which must receive a weak input signal (≈ -85 dBm), in presence of a TX leakage mean power of approximately -10.5 dBm at mixer input. Transmit leakage mean input power at mixer input $\approx$ PA output power ($+27$ dBm) – isolator/coupler losses (0.5 dB) - duplexer isolation (52 dB) + LNA gain (15 dB) $= -10.5$ dBm. The simplest solution consists in using an inter-stage BPF, but this is not the preferred option for the reasons explained in section 6.2. A comparison between WCDMA and LTE QPSK uplink modulated carrier IMD$_2$ products is shown in Figure 14.40.

In a WCDMA RX filter-less application, the mixer IIp2 requirements to ensure a small SNR degradation is in the range of 70 dBm [11], a figure that is extremely challenging to achieve. From Figure 14.40, it can be seen that the higher LTE IMD$_2$ PSD sets a higher mixer IIp2 requirement than for WCDMA receivers. In narrowband operation, the higher IIp2 requirements in LTE have led to reference sensitivity degradation in certain bands, such as Band II, III and Band IV.

14.8.3 ACS, Narrowband Blockers and ADC Design Challenges

Both ACS and *Narrow-Band* (NB) blocking requirements are a measure of the receiver ability to receive a wanted signal at its assigned channel frequency in presence of an interfering *adjacent channel interferer* (ACI), at which a throughput requirement shall be

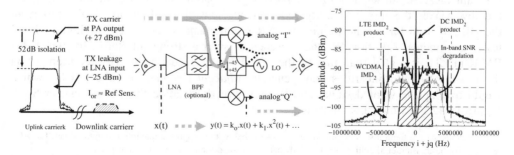

Figure 14.40 Self-mixing in direct conversion receivers – left: TX leakage at LNA input; right: I/Q spectrum observed at mixer output; dashed lines represent the wanted channel

Table 14.20 Relationship between interfering and wanted signals for ACS requirements[2]

Rx Parameter	Units	Channel bandwidth					
		1.4 MHz	3 MHz	5 MHz	10 MHz	15 MHz	20 MHz
ACS test case I							
Wanted signal mean power	dBm	Refsens + 14 dB					
$P_{Interferer}$	dBm	Refsens + 45.5 dB				Refsens + 42.5 dB	Refsens + 39.5 dB
$BW_{Interferer}$	MHz	1.4	3	5	5	5	5
$F_{Interferer}$ (offset)	MHz	1.4	3	5	7.5	10	12.5
Assumed ACS	dB	33	33	33	33	30	27
ACS test case II							
Wanted signal mean power	dBm	−56.5	−56.5	−56.5	−56.5	−53.5	−50.5
$P_{Interferer}$	dBm	−25					

fulfilled. The throughput shall be equal or higher than 95% of the maximum throughput, for a specified reference measurement channel.

The intention of this requirement is to verify the *ACI Rejection* (ACIR). Both tests are important to avoid UE dropped calls in cells where eNodeBs from adjacent operators are non-co-located. In a similar fashion to the WCDMA specifications [31], the LTE requirements are based on a 33 dB ACS budget, which has been derived through extensive co-existence simulations [4]. In order to prevent stringent selectivity requirements for the 15 and 20 MHz BW, a 3 dB and 6 dB ACS relaxation is proposed. The resulting test cases are summarized in Table 14.20.

- The ACS test case I is performed with a mean signal power set to 14 dB above reference sensitivity level, and a variable interferer power for each wanted channel bandwidth as shown in Figure 14.41.
- The ACS test case II is one the test that stresses the UE dynamic range by setting the interferer power constant to −25 dBm and variable wanted signal power so that the 33 dB ACS test condition is met. For example, in a 5 MHz BW ($N_{RB} = 25$), the *Carrier to Interferer power Ratio* (CIR) is also equal to −56.5 dBm − (−25 dBm) = −31.5 dB.

One of the UE DCR blocks that is affected by the ACS II is the LNA for which the gain must be sufficiently low to prevent overloading of the I/Q mixer inputs, and thereby relaxes the mixer linearity requirements, but also sufficiently high to prevent the UE NF from failing the minimum SNR requirements imposed by the highest MCS test requirements. Additionally, the presence of such a strong interferer, for which the PAPR in LTE is higher than that of the WCDMA case, sets a linearity requirement on the LNA. LNA non-linearities may generate *Adjacent Channel Leakage* (ACL), which would overlap the wanted signal and thereby directly degrade the wanted carrier SNR. Figure 14.42 illustrates this challenge.

---

[2] For ACS test case I, the transmitter should be set to 4 dB below the supported maximum output power. For ACS test case II, the transmitter should be set to 24 dB below the supported maximum output power.

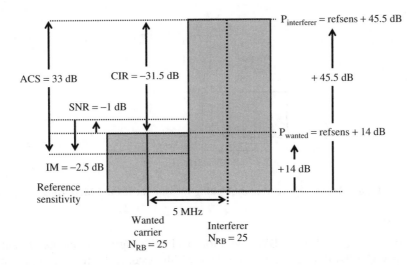

Figure 14.41 Adjacent channel selectivity test case I requirements for 5 MHz LTE

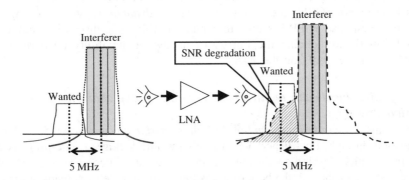

Figure 14.42 Example of LNA non-linearities in presence of the −25 dBm ACS II interferer. Left: spectrum at LNA input, right: spectrum at the LNA output in presence of LNA ACL

The NB blocking test is illustrated in Figure 14.43 for 5 MHz BW ($N_{RB} = 25$). This test ensures LTE UEs can be deployed in regions where other telecommunication standards, such as GSM/EDGE, are in service. To minimize the cellular capacity loss, it is important to minimize the guard bands. For this reason, the NB blocker test places the UE wanted channel in presence of a blocker located at small frequency offsets. This test only differs slightly from its WCDMA equivalent: in LTE, the blocker is CW vs. GMSK in 3G, the frequency offset[3] is set to 2.7075 MHz (versus 2.7 MHz in 3G) and the wanted channel benefits from 16 dB de-sense in 5 MHz BW (versus 10 dB in 3G).

---

[3] In LTE, the introduction of a small frequency offset has been proposed to ensure that the interferer does not fall in the spectral nulls of the receiver's FFT operation. The offset is an odd multiple of half the tone spacing $(2k + 1) \cdot 7.5$ kHz. Refer to Figure 14.44.

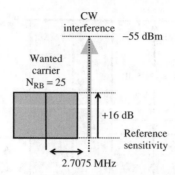

Figure 14.43   Narrowband blocking measurement for 5 MHz LTE

It can be seen that in 5 MHz BW operation, LTE filtering requirements are similar to the existing WCDMA devices. In flexible bandwidth systems the filter design strategy is a compromise articulated around two extreme scenarii: at one end, a receiver with infinite ACIR minimizes the ADC resolution and power consumption at the expense of sharp analog filter, which may distort the wanted signal. At the other extreme, a receiver that provides little or no ACIR imposes an ADC with a *dynamic range (DR)* large enough to cover the worst case 3GPP DR requirements. The following discussion highlights this trade-off by firstly introducing the impact of bandwidth flexibility on *analog channel filters* (ACF), and secondly on the ADC.

Impact of flexible bandwidth requirements onto the analog channel filter design strategy

OFDM systems overcome *inter-symbol and inter-carrier interference* (ISI and ICI) due to propagation through time-dispersive channels by introducing a *cyclic-prefix (CP)*. The CP acts in a similar fashion to time guard bands between successive OFDM symbols. Therefore, the longer the CP, the better the resilience to large delay spreads at the expense of an energy loss. The CP length must also be selected so as to avoid the signal smearing caused by the *group delay distortion* (GDD) of analog filters [33]. Yet, selecting a filter family for the worst-case delay spread foreseen in the standard (such as the ETU model for e.g.) is perhaps not the best strategy. For example, [34] suggests that in most cases the delay spread experienced by the UE is less than the CP length, and therefore the estimation of the delay spread by the BB channel estimator can be used to dynamically re-program the transfer function of a post ADC digital FIR filter. This elegant solution provides enhanced ACS performance for a given pre-ADC ACF. So, what is the limit to scaling the 3 dB cut-off frequency 'F_c' of the ACF?

Figure 14.44 illustrates the impact of scaling a baseline[4] filter's F_c optimized for WCDMA ACS and NB blocking onto the experienced GDD. The filter's F_c is either stretched or shrunk by a factor proportional to the LTE BW ratios.

---

[4] The filter used is similar to that in [34]: it has four zeros, eight poles, with a 3 dB cut-off frequency 'F_c' of 2.2 MHz in 5 MHz BW, and a notch located at 2.8 MHz offset. To reduce the group delay overshoot associated with the presence of a notch so closely located to the edge of the wanted signal, some of the poles and zeros are used to implement an analog group delay equalizer.

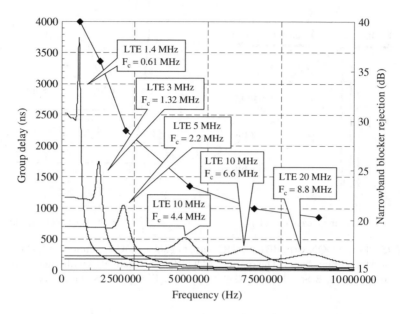

Figure 14.44 Impact of scaling 'Fc' of a 5 MHz baseline ACF optimized for WCDMA proportional to the LTE operating's BW; left *y-axis*: filter's group delay, right *y-axis*: NB blocker filter's rejection performance (diamonds)

From Figure 14.44, it can be seen that the lower the filter's 'F_c', the higher the filter's delay, GDD, and NB blocker ACIR, so much so that in the 1.4 MHz BW of operation the GDD is slightly greater than 1 µs. This latter case 'consumes' a large amount of the normal CP length of 4.7 µs and could impact ISI in large delay spread channels. From this example, it can be concluded that proportional scaling of the ACF's F_c is probably not the best trade-off. One alternative consists in taking advantage of the 3GPP relaxations to tailor the ACF ACIR to just meet the ADC DR requirements at 15 and 20 MHz operation. The ADC DR enhancements in small operating BW can then be used to relax the filter's sharpness so as to make full benefit of the CP length.

Impact of flexible bandwidth requirements onto the ADC DR
For the sake of simplicity[5], the minimum required ADC resolution is estimated by assessing the *ADC EVM (EVM_{ADC})* based upon the pseudo-quantization noise model (PQN). Figure 14.45 (left) shows the optimum *Signal to Distortion Quantization Noise Ratio (SDQNR)* at the ADC output is met for ADC clipping ratios, also denoted *ADC back-off (ADC BO)*, ranging from 10 dB to 14 dB. The resulting EVM_{ADC} is plotted in Figure 14.45 (right).

The UE EVM budget is estimated from the required SNR to achieve the highest SIMO LTE throughput, which corresponds to the 64 QAM 8/9 MCS. From section 14.3.4, 5%

---

[5] The PQN model considers the quantization error as additive noise, which has a white and a uniformly distributed spectrum and which is uncorrelated with the input signal. The reality is more subtle: work published in [35] shows that the PQN approach is not entirely sufficient to model the quantization errors due to the much larger peak-to-average power ratio of OFDM signals.

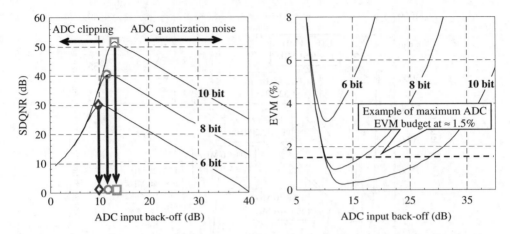

Figure 14.45 Left: SDQNR (in dB) at ADC output against ADC resolution and ADC input back-off (in dB) or ADC clipping ratio. The optimal ADC BO is respectively highlighted with a diamond, circle and square for 6, 8 and 10 bits ADC resolution. Right: corresponding ADC output EVM performance

throughput loss is met if the total composite EVM is less than 6.3%. Assuming that each EVM impairment is AWGN like, taking an example where the eNodeB EVM is equal to 4.5%, and the UE RF RX EVM performance is equal to 4%, leaves an EVM_{ADC} budget of 1.5%[6].

Let us first assume an ideal RF-BB chain, which provides an infinite ACIR, and an ideal *Analog Gain Control (AGC)* loop, so that the optimal BO is always exactly met. The situation is captured in Figure 14.46 (left) and indicates that the lowest acceptable ADC resolution is 8 bit. In a real AGC loop system, the ADC BO over the duration of a call is no longer that of a discrete point but a spread of BO statistically distributed as shown with the histogram of Figure 14.46. Taking one particular example of AGC loop inaccuracies[7] into account, it can be seen that 10 bit is the minimum ADC resolution, which provides approximately 12 dB headroom (Δ_{RF}) for RF IC imperfections, such as imperfect DC offset cancellation and ACIR (Figure 14.46 – right). This requirement is equivalent to a CW DR of 60 dB. One of the differences in LTE is that the UE AGC loop must also cope with time-varying amplitude of in-band signals due to dynamically scheduled users with varying number of active RBs transmitted at different power levels in the downlink. With 10-bit resolution, Δ_{RF} is sufficiently large to accommodate one

---

[6] Assuming AWGN-like behavior of each EVM contributor, $EVM_{ADC} = \sqrt{\dfrac{6.3\%^2 - (EVM^2_{eNodeB}}{+ EVM^2_{RFRX})}}$. Note that in conformance tests, the eNodeB emulator EVM is extremely low (typically 1% or less). Also note that the recent WiMax (802.16e) RF transceiver design in [36] can achieve $\approx 1.5\%$ RX EVM, thereby relaxing the overall EVM downlink budget.

[7] The histogram shows a recorded ADC BO distribution captured over a 10-minute-long WCDMA BLER measurement performed in a fading test case-1, 3 km/h. The AGC loop updates the analog gain at 10 ms intervals. An AGC loop for LTE is likely to operate at a faster update rate to provide better control accuracy of the ADC BO. The resulting histogram would present a smaller spread of BO, thereby relaxing the margins for RF imperfections.

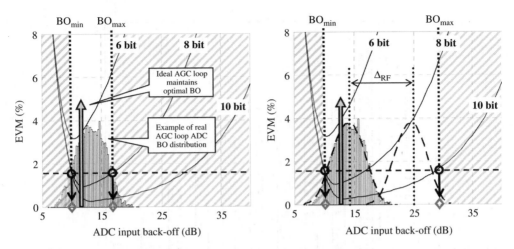

Figure 14.46 Impact of AGC loop imperfections. Left: 8 bit ADC. Right: 10-bit ADC. Dashed areas indicate regions that do not fulfill the MCS EVM$_{ADC}$ requirements. Dashed histogram envelope illustrate ADC BO margin due to RF imperfections (Δ_{RF}) present at ADC input

example of RF imperfection margins. Note that each of the assumptions[8] is listed as an example of minimum requirements as these are implementation specific.

In conclusion, the most challenging mode of operation for a multiple-standard ADC is the LTE 20 MHz operation for which a minimum of 60 dB DR must be delivered. Sigma delta ADCs represent an attractive solution to these requirements [37]. These converters shape the *Quantization Noise* (QN) floor via a high pass transfer function, thereby digging a notch into the QN PSD within the BW of the wanted carrier, and rejecting the noise out of band. The BW of the QN notch can be optimized for each BW of operation by either changing the sampling frequency and/or by reconfiguring the noise shaping filter's transfer function. An example of sigma-delta ADC flexibility is shown in Figure 14.47. It can be seen that by ensuring the DR requirements are met for LTE 20 MHz, the ADC offers a 20 dB improvement for operation at LTE 1.4 MHz. Every decibel gained in the DR can be used as an extra relaxation of the ACF filter design. In particular, the large DR performance at low operating BW relaxes significantly the ACF rejection in GSM mode for which the design of a sharp filter can be expensive in die area.

14.8.4 EVM Contributors: A Comparison between LTE and WCDMA Receivers

In section 14.3.4, the downlink LTE budget is set to approximately 8%. Compared to WCDMA, where 10 to 12% is acceptable even for HSDPA operation, this requirement

---

[8] The following margins are assumed: (1) 2 dB margin for imperfect RF receiver DC offset removal at ADC input; (2) 4 dB to account for relative power difference between a 64 QAM user and a QPSK user in the cell [2] – this is a minimum requirement; (3) Imperfect RF RX ACIR forces the AGC loop to lower the effective wanted signal target total ADC BO. With the previous assumptions 1 and 2, the DCR is allowed to be present at the ADC input a CIR of ≈4 dB (eNodeB TPC DR) + 2 dB (leaked DC offset) − 12 dB (Δ_{RF}) = −6 dB.

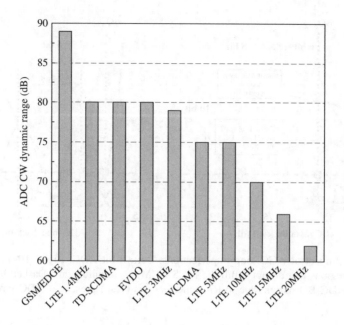

Figure 14.47   Achievable DR with sigma delta ADCs over a wide range of system's BW [38]

appears tougher. However, the novelty in LTE is that EVM measurements make use of a zero-forcing equalizer (cf. section 14.3.4). Thus, a distinction must be made between AWGN-like contributors, and contributors that can be equalized, and therefore become transparent to the UE receiver. This is an important difference from WCDMA where, for SF large enough, each EVM contributor behaves like AWGN [39]. This section aims at illustrating these differences with only a few selected impairments: I/Q gain and phase imbalance, distortions due to ACF and DCR local oscillator phase noise.

14.8.4.1   Impact of Finite Image Rejection due to I/Q Amplitude and Phase Mismatches

In real world analog/RF designs, it is nearly impossible to design DCRs with perfectly identical gain and phase response between I and Q branches. Therefore DCRs come with two natural impairments: amplitude and phase mismatches, denoted ΔA and $\Delta \Phi^9$ respectively, leading to a finite *Image Rejection (IR)*. Finite IR results in each subcarrier being overlapped with the subcarrier which is located at its frequency image mirror as shown in Figure 14.48 (c). The power ratio between a subcarrier and its image is by definition the IR. Assuming the symbols carried by the subcarriers are uncorrelated, the impact of IR on LTE is no different to that of a single carrier system and can be considered as an AWGN source [40].

---

[9] $\Delta \Phi$ may originate from either a local oscillator that does not drive each I and Q mixer in exact quadrature, or from a tiny mismatch in the 3 dB cut-off frequency of each I/Q LPF. In the latter case, $\Delta \Phi$ is not identical for all subcarriers. The net result is a frequency-dependent image rejection (IR). Note that IR can be calibrated in modern DCR design as shown in [36].

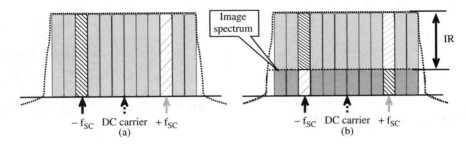

Figure 14.48 (a): transmitted OFDM carrier, (b) equivalent i + jQ baseband complex spectrum at ADC input in presence of a finite image rejection equal across all subcarriers[10]

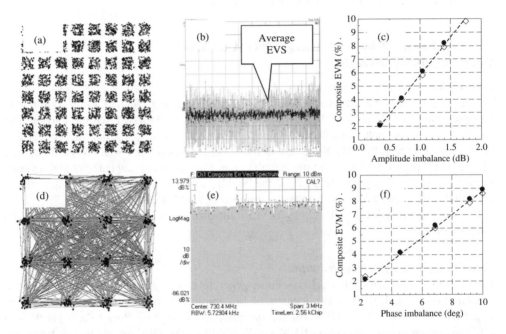

Figure 14.49 Impact of $\Delta\Phi$ and ΔA onto WCDMA and OFDM downlink carriers. Impact of $\Delta A = 1.75\,dB$ on (a) LTE 64 QAM state spreading, (b) LTE carrier EVS, (d) 16QAM HS-DSCH WCDMA, (e) WCDMA EVM spectrum. (c) and (f): comparison of LTE (plain dots) and WCDMA (empty diamonds) EVM performance against ΔA and $\Delta\Phi$

Figure 14.49 shows EVM measurements performed with an Agilent[TM] 89600 Vector Signal Analyzer, which delivers the *Error Vector Spectrum* (EVS). EVS is a tool that plots the magnitude of the error vector against each subcarrier. Figure 14.49 (b) shows the composite LTE EVS – it is an overlay of each physical channels EVS, where the darker line shows the average EVS. The EVS spectral flatness in Figure 14.49 confirms the AWGN like behavior of EVM due IR for both standards.

---

[10] In field trials, incoming subcarriers are notched due to frequency-selective fading and in practical RF IC designs, image rejection is not constant across the entire DCR reception bandwidth, resulting in frequency dependent IR.

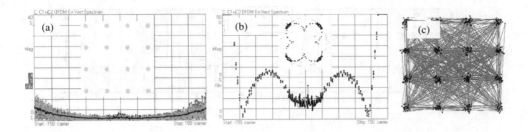

Figure 14.50 Measured[12] impact of an I/Q ACF on the composite EVM on of a 5 MHz LTE ($N_{RB} = 25$) and a WCDMA downlink carrier. (a) LTE composite EVS and 16QAM user data constellation with equalizer 'ON', (b) LTE reference symbols EVS and constellation with equalizer 'OFF', (c) WCDMA HS-DPCH constellation, EVM $\approx$ 8%

14.8.4.2   Impact of In-band Analog Filter Amplitude and Group Delay Distortions

Zero-IF low pass filter contribution

Figure 14.50 shows the measured impact of a prototype I/Q channel filter[11] similar to that presented in [34] on a 5 MHz 16 QAM LTE and a WCDMA downlink carrier. The equal spreading of each constellation points in Figure 14.50 (c) confirms the AWGN like behavior of EVM_{LPF} for WCDMA, and in this example, results in 8% EVM performance. Figure 14.50 (b) shows that, without equalization, the outermost subcarriers are severely impacted, while subcarriers located in the middle are less vulnerable. The use of the ZF-equalizer flattens EVS and brings the composite EVM down to $\approx$1.2%.

It can be concluded that LTE relaxes the LPF impairment budget compared to WCDMA modulation.

Zero-IF high pass filter contribution

IMD2 products described in earlier section generate a strong DC term, which can lead to saturation through the several decibels of I/Q amplification required to meet the ADC target BO. In WCDMA the DC term can be cancelled with *High Pass Filters* (HPF) and [39] has shown that its impact on EVM is AWGN-like. The HPF design is a compromise between EVM, capacitor size, die area, and DC settling time[13]. A passive 4.5 kHz first-order HPF with group delay distortion in excess of the CP length has been deliberately chosen to illustrate the impact on LTE. To the contrary of the LPF or BPF test case, the subcarriers located close to the center of the carrier are the most vulnerable as can be seen in Figure 14.51. For example, the *Primary Synchronization Signal* (PSS), which occupies the center of the carrier over a 1.4 MHz BW, experiences a 7.5% EVM, while the QPSK user EVM improves as the allocated BW is increased (Figure 14.51 a and c).

---

[11] Both I and Q filters are nearly perfectly matched and deliver an IR across the entire receiver bandwidth better than −40 dBc. This guarantees that the EVM performance is dominated by the LPF non-linear phase and amplitude distortions.

[12] All EVM measurements were performed with an Agilent[TM] 89600 VSA.

[13] In BiCMOS designs, it is difficult to implement analog HPF with cut-off frequencies less than 3 kHz due to cost constraints in the die area. The use of RFCMOS allows this loop to be designed partly in the digital domain with a much smaller cut-off frequency and nearly no impact on the die area.

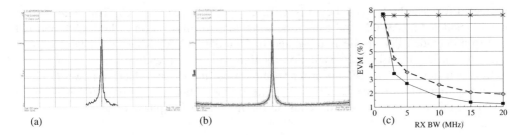

(a)   (b)   (c)
                                                                         RX BW (MHz)

Figure 14.51 Measured impact of a 4.5 kHz HPF prototype with ZF equalizer 'ON'. (a): EVS of PSS in 5 MHz (25 RB), (b) EVS of a 5 MHz (25 RB) QPSK user, (c) EVM vs. carrier's BW. Stars: average PSS EVM, diamonds: average composite EVM, squares: QPSK user average EVM

The distortion impacts so few RS that the ZF-equalizer cannot flatten the impact of the HPF. LTE therefore calls for a careful design of the DC offset compensation scheme.

14.8.4.3 Impact of Phase Locked Loop Phase Noise

If the downlink OFDM signal was just a set of un-modulated closely spaced CW tones, the resulting I or Q output of the DCR mixer would be that of each CW tone multiplied by the LO Phase Noise (PN) profile as shown in Figure 14.52 (a). Clearly, any PN exceeding the subcarrier spacing causes SNR degradation. In most PN studies [41], a distinction is made between PN contributions at high and low frequency offsets from the carrier. The close-in PN produces an identical rotation of all subcarriers and is also often referred *Common Phase Error (CPE)*. This can be estimated and therefore can be

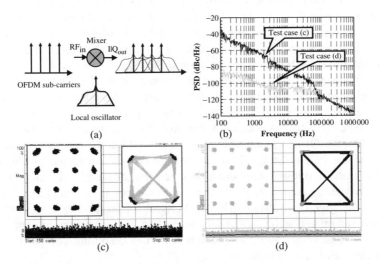

Figure 14.52 (a) Illustration of LO PN multiplication on OFDM subcarriers; (b): LO PN profiles: degraded PN in black (test case c), 'near ideal' in grey (test case d). (c) and (d): LTE 16QAM and WCDMA QPSK constellations are overlaid on the LTE EVS. (c): measurements with degraded PN, (d) measurements in near ideal PN profile

corrected for. The high-frequency offset PN components generate ICI. This contribution can be modeled as AWGN and cannot be corrected. An illustration of both CPE and ICI due to a degraded PN profile is presented in Figure 14.52 (c) where the LTE 16 QAM user data suffers from both constellation rotation and state spreading.

In conclusion, the OFDM subcarrier spacing used in LTE introduces a new LO phase noise requirement for frequency offsets ≥ 15 kHz offset. This differs from WCDMA where EVM for this impairment only depends on the integrated LO phase error.

14.9   UE Demodulation Performance

14.9.1   Transmission Modes

OFDM modulation in LTE downlink allows optimizing the transmission in time, frequency and antenna domains. The UE 3GPP standard compliancy must be ensured by satisfying the requirements covering a wide range of modes comprising Transmit/Receive diversity and spatial multiplexing. In this section we examine the principles of each mode and the corresponding receiver schemes for each case.

14.9.1.1   Single Input Multiple Output (SIMO) mode

Plain OFDM *Single Input Single Output (SISO)* transmission is not supported by LTE as UEs are required to have at least two receiving antennas. The SIMO case, then, constitutes the simplest scheme and it is a straightforward extension of the SISO case. Let us assume a transmission bandwidth of B = 20 MHz and a sampling frequency of f_s = 30.72 MHz and normal cyclic prefix mode. At the receiver side, for each receiving antenna, base-band digital signal samples after RF and analog-to-digital conversion are buffered over an OFDM symbol of duration Ts = 71.3 µs (or 71.9 µs in case of the first OFDM symbol of a slot as, in this case, the cyclic prefix samples are slightly more). The cyclic prefix samples are discarded, assuming timing offset recovery is safely done, and N = 2048 samples are FFT transformed into frequency domain equivalent signal. Due to LTE standard assumptions, at the output of the FFT, only K = 1200 samples are kept and the others are discarded. Depending on the physical channel that needs to be modulated and its specific allocation over the subcarriers, an operation of demultiplexing and re-ordering of the complex digital samples takes place.

Even in case of multi-path propagation, but provided that the channel coherence time is much larger than the OFDM symbol duration, it is well known that in OFDM the signal at the output of the FFT at each subcarrier position can be considered as affected by flat fading. In this case the optimal demodulation scheme is the simple Matched Filter and the effect of the channel is undone by multiplication by the complex conjugate of the channel coefficient at any given subcarrier. The QAM symbols are obtained by combining the derotated symbols on the same subcarriers across the two receiving antenna paths. This operation is also generally known as Maximum-Ratio-Combining and permits optimally benefi from the additional antenna diversity.

After MRC operation, the QAM symbols are demapped and the sequence of soft-bits for channel decoding operation is buffered until when the total expected amount is available after demodulation of several OFDM symbols. The soft-bits are rate-matched with respect to the parameters specific to the physical channel under consideration.

Eventually a *Hybrid-Retransmission-Request (HARQ)* combining operation is performed and the channel decoding operation is invoked. At its output, the sequence of hard decisions is verified against the transported *Cyclic-Redundancy-Check (CRC)* bits to decide whether the decoded bits are correct or not. The channel decoder to be used depend on the nature of the decoded physical channel: dedicated channels, e.g. PDSCH, are always turbo-encoded whereas channels carrying control information, such as PDCCH, are convolutionally encoded, and thus decoded by means of a Viterbi decoder.

In the case of channels supporting the HARQ protocol, the result of the redundancy check operation is fed back to the BS. The receiver performance is computed upon the success rate of the CRC in terms of throughput, which is the measure of net successfully decoded bits after the HARQ process.

14.9.1.2 Transmit Diversity Mode

Transmit diversity is implemented in LTE by means of *Spatial-Frequency-Block-Coding (SFBC)*. SFBC is nothing but the Alamouti encoding of two QAM symbols lying in neighboring subcarriers. In case of transmit-diversity transmissions, then, at receiver side, symbols at the output of the FFT require to be re-ordered in pairs accordingly to the SFBC encoding carried out at the transmitter and the Alamouti scheme is undone via linear operation.

14.9.1.3 MIMO Mode

MIMO transmission mode is the key enabler of the high LTE data rate (up to 150 Mbps for the 20 MHz bandwidth case) and allows the transmission of one or two independent data streams depending on the channel conditions experienced by the UE. In MIMO mode, the channel at each subcarrier is represented by a matrix whose size is given by the number of transmitting N_{Tx} and receiving N_{Rx} antennae. If the propagation conditions result from a rich scattering environment, the rank of the channel matrix is full and in these conditions spatial multiplexing of two data streams can be supported. In case instead the channel matrix rank is not full, ie. a rank is equal to one, only one codeword is transmitted. As for HARQ acknowledgments, MIMO closed loop mode requires continuous feed-back from the UE to the BS on a sub-frame basis. Together with the channel rank information, the UE also provides the BS with the indexes of the precoding codebook vectors to be used at the transmitter side. The closed loop MIMO precoding mechanism, at the expense of additional signaling overhead, is the method used in LTE to exploit the MIMO channel diversity effectively. This is because the precoding vector indexes, requested by the UE, are chosen such that the SNR is maximized, therefore also maximizing the throughput. The SNR is computed on the overall equivalent channel constituted by the cascade of the precoding matrix and the instantaneous propagation channel matrix.

The standard does not mandate for a particular detection scheme but instead assumes a linear *Minimum-Mean-Squared-Error* (MMSE) equalizer as a baseline detector to establish the minimum performance requirement. The MIMO transceiver scheme is shown in Figure 14.53. The equalizer coefficients are adjusted depending on the channel matrix coefficients, the precoding vector and the interference power. It can equivalently be regarded as a 2×2 matrix multiplication of the 2×1 vector constituted by the complex signal at each subcarrier at the output of the FFT of the two receiving antennae

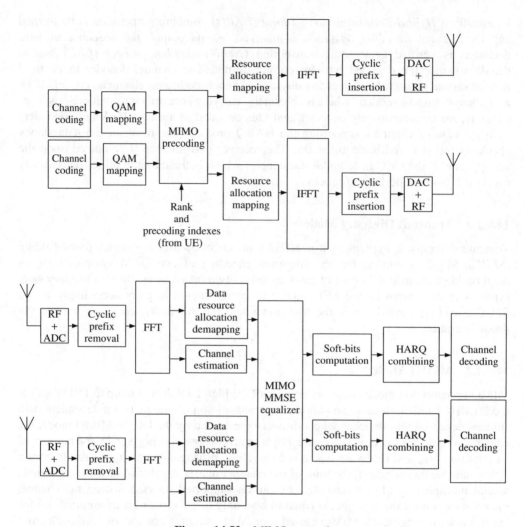

Figure 14.53   MIMO transceiver

as follows:

$$\hat{x}_i = G_{i,MMSE}\, y_i = G_{i,MMSE}(\tilde{H}_i\, x_i + n_i) = G_{i,MMSE}(H_i\, P_i\, x_i + n_i) \qquad (14.15)$$

where:

- $\hat{x}_i$ is the 2×1 detected symbol vector at subcarrier i.
- $G_{i,MMSE}$ is the MMSE equalizer 2×2 matrix at subcarrier i.
- y_i is the 2×1 received signal vector at subcarrier i.
- $\tilde{H}_i$ is the 2×2 equivalent channel matrix resulting from the cascade of the 2×2 pre-coding matrix P_i and the actual 2×2 channel matrix H_i at subcarrier i.
- n_i is the 2×1 interference vector received signal at subcarrier i.

14.9.2   Channel Modeling and Estimation

14.9.2.1   3GPP Guidelines for Channel Estimation

The coherent detection schemes mentioned earlier require the availability of a reliable channel estimate for each subcarrier, for each OFDM symbol and for each link between transmitting and receiving antennae. For this purpose, LTE system provide *Reference-Signal* (RS) whose resource elements are disposed in the time-frequency plane in a diamond-shaped uniform grid.

In case of multiple antenna transmissions, RS are interleaved and zeroed out in correspondence of the RS of other antennae in order to minimize mutual interference. Hence, thanks to this particular structure introduced specifically in LTE, the channel estimation over antennae can be performed independently for each link between each transmitting antenna and each receiving antenna. This pilot signaling scheme simplifies the channel estimation task considerably in MIMO applications.

As for the receiver scheme, the standard gives freedom in the implementation of the frequency-time channel estimation, even if some companies provided practical guidelines along the standardization discussions for the definition of the performance requirements. These guidelines proposed the channel estimation to be implemented as the cascade of two 1D filtering processes: the first dimension consists of Wiener/MMSE filtering the OFDM symbols containing RS in the frequency direction. The second dimension follows as a Wiener/MMSE filter in the time domain to obtain the full channel transfer function for all subcarriers indexes and OFDM symbols within a sub-frame.

Only the reference symbols belonging to the current sub-frame are used for time domain interpolation. The coefficients used for frequency domain filtering process are chosen from a pre-computed set as a function of signal-to-noise ratio only [42].

It is worth noting that, because channel estimation is performed after the FFT operation, the actual channel being estimated is indeed the convolution of several filters' impulse responses, amongst which the time variant component is that of the physical air interface propagation channel, and the other filters are either those of the eNodeB channel filters, or those of the UE RF front end section. In that respect, in-band distortions introduced by RF filters are naturally compensated for as long as the total delay spread does not exceed the CP length.

14.9.3   Demodulation Performance

14.9.3.1   PDSCH Fixed Reference Channels

The performance requirements in [1] include a set of Fixed Reference Channels for each transmission mode – namely SIMO, transmit diversity and MIMO modes. The FRCs so far agreed, for the SIMO case and cell-specific RS signals, fall into three categories involving a restricted set of Modulation and Coding Schemes: QPSK with coding rate 1/3, 16 QAM with coding rate $^1/_2$ and 64 QAM with coding rate $^3/_4$.

The choice of such FRC categories was made to reduce the number of tests to a minimum while having representative MCS in the overall set of 29 MCS combinations.

Within each FRC category, one test is specified for a given system bandwidth and therefore characterized by a specific transport block length, code-block segmentation parameters assuming an allocation spanning the entire available bandwidth.

An additional FRC category is also defined for single Resource Block allocation happening on the band-edge and making use of QPSK modulation.

14.9.3.2　PDSCH Demodulation Performance Requirements

The performance requirements are expressed as the required SNR $\hat{I}_{or}/N_{oc}$ in dB to obtain a given fraction of the nominal throughput of a given FRC and in given propagation conditions. The fraction of the nominal throughput is always chosen to be 30% or 70%. It is worth noting that in the case of RF section performance tests, the metric has been set to a throughput greater of equal to 95% of the maximum throughput of the reference measurement channel. Table 14.21 presents an example of the performance requirement for 64 QAM, as stated in the 3GPP standard.

The required SNR is agreed by consensus among the companies participating in the standard. The value is computed using the decoding performances of a full-floating point receiver chain where an implementation margin is taken to account for the degradation induced by fixed-point or other signal-processing non-idealities. The implementation margin is in the range of 1.5/2 dB. The principle just explained for the derivation of the performance requirement is graphically depicted in Figure 14.54.

14.9.3.3　EVM Modeling and Impact on Performance

To meet realistic performance requirements, LTE assumes that the transmitter is modeled as a non-ideal signal source characterized by transmitter Error Vector Magnitude. This non-ideality generally results from non-linearities arising from an high OFDM signal peak-to-average power ratio and limited RF transmitter amplifier dynamics.

A precise description of the EVM modeling for LTE performance requirement derivation has been provided in [43]. This model regards non-linearities as a source of additive distortion at the transmitter, as indicated by the Bussgang theorem [13] and shown in Figure 14.55.

Table 14.22 presents the EVM levels assumed in a performance simulation. The effect of the additive distortion source is to limit the attainable capacity because the effective receiver SNR cannot increase as other cell interference vanishes because of the irreducible distortion term, as Figure 14.56 shows. As an additional consequence, the MCS set is limited and upper-bounded as the general system capacity.

Table 14.21　Minimum performance 64QAM (FRC)

Bandwidth	Test Number	Reference Channel	Propagation Condition	Correlation Matrix	Reference value Fraction of Maximum Throughput (%)	Reference value SNR $\hat{I}_{or}/N_{oc}$ (dB)
10 MHz	10.1	[F34]	EVA5	Low	70	17.7
	10.2	[F34]	ETU70	Low	70	19.0
	10.4	[F34]	EVA5	High	70	19.1

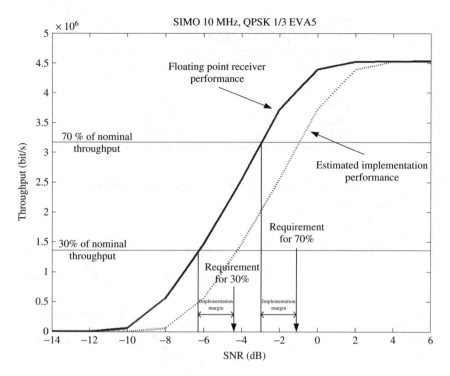

Figure 14.54 Principle for performance requirement derivation

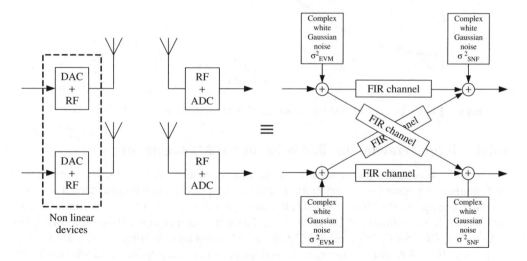

Figure 14.55 Equivalent EVM model used for performance simulations

Table 14.22 Error Vector Magnitude assumption for performance requirements

Modulation	Level	Equivalent per Tx antenna noise variance σ^2_{EVM} (dB)
QPSK	17.5%	−15.13924
16QAM	12.5%	−18.0618
64QAM	8%	−21.9382

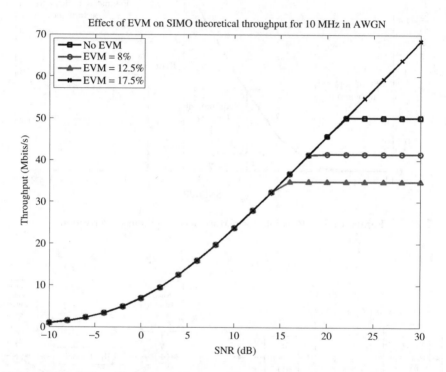

Figure 14.56 Effect of EVM on theoretical SIMO throughput for 10 MHz in AWGN

14.10 Requirements for Radio Resource Management

The mobility procedures are described in general in Chapter 7. This section focuses on the performance requirements for the mobility cases. The performance requirements for radio resource management (RRM) for E-UTRA are defined in [3] and follow a similar basic structure to those defined for UTRA in [44]. There are requirements supporting mobility in both E-UTRA RRC_Idle, and RRC_Connected states and mobility for E-UTRA intra frequency, E-UTRA inter frequency as well as performance requirements for handover and reselection to other Radio Access Technologies (RATs) including UTRA FDD, UTRA TDD and GSM. Performance is also specified for mobility to non-3GPP RATs such as CDMA2000 1x and High Rate Packet Data (HRPD), also known as CDMA EV-DO.

In the RRC_Idle state, mobility is based on UE autonomous reselection procedures, whereas in the RRC_Connected state measurement reports are made by the UE, to which the E-UTRA network may respond by issuing handover commands. Due to the autonomous nature of the idle state reselection procedures, and the relative importance of power saving in the RRC_Idle state, typical idle performance requirements allow for more implementation freedom, whereas in the RRC_Connected state, consistent UE measurement reporting is of prime importance, and so more details of measurement performance are specified to ensure that network based mobility algorithms can work with all UEs. One important difference in E-UTRA compared to UTRA is the specification of large DRX cycles in RRC_Connected state. The DRX cycle can be up to 2.56 s, for example. The large DRX cycles are similar in length to idle-state DRX cycles but measurements are reported to the E-UTRA network for handover decision making.

As well as performance requirements for supporting reselection and measurements for handover, [3] includes performance requirements for radio-related RRC procedures including RRC re-establishment and random access. It also contains requirements for UE transmit timing, and UE timer accuracy. This section gives an overview of the mobility related requirements.

14.10.1   Idle State Mobility

From a minimum performance perspective, [3] specifies how quickly the UE should be able to detect and evaluate the possibility of reselection to newly detectable target cells. In this context, an E-UTRA cell is considered detectable if it has signal-to-noise and interference ratio $SCH\_RP/I_{ot} \geq -3\,dB$. For cells that have previously been detected, [3] also specifies the minimum rate at which additional measurements should be made, and the minimum acceptable filtering and spacing of the samples used to support reselection. Maximum allowed evaluation times for cells that have previously been detected are also given. The requirements for measurement rate and maximum evaluation time have largely been copied from those used previously for UTRA because the basic needs for mobility should be similar.

Additionally, [3] specifies that reselection criteria should be re-evaluated at least every DRX cycle, because a new serving cell measurement, and probably new neighbor cell measurements, must be performed at least every DRX cycle. Once the reselection decision has been made it is expected that the UE will make a short interruption in receiving the downlink paging channels when performing reselections. For reselections to E-UTRA and UTRA target cells, this is specified as the time needed to read system information of the target cell and an additional 50 ms implementation allowance.

14.10.2   Connected State Mobility When DRX is not Active

In the RRC_Connected state, the UE continuously searches for and measures intra-frequency cells, and may additionally search for and measure inter-frequency and inter-RAT cells if certain conditions are fulfilled, including the configuration of a measurement gap sequence if one is needed by the UE. Both periodic and event-triggered reporting mechanisms are defined in RRC specifications to provide measurement results to the E-UTRA network. For different radio technologies the relevant measurement quantities are shown in Table 14.23.

In general terms, [3] defines minimum performance requirements, which indicate how quickly newly detectable cells should be identified and reported, measurement periods for

Table 14.23 Measurement quantities for different radio access technologies

Radio Technology	Applicable measurement quantities
E-UTRA	Reference Symbol Received Power (RSRP)
	Reference Symbol Received Quality (RSRQ)
UTRA FDD	CPICH Received Symbol Code Power (RSCP)
	CPICH Ec/Io
UTRA TDD	PCCPCH received symbol code power (PCCPCH RSCP)
GSM	RSSI
	Note: BSIC confirmation may be requested for GSM measurements

cells that have been detected, and the applicable accuracy requirements and measurement report mapping for each of the measurement quantities in a measurement report. Furthermore, there are requirements in Chapter 5 of [3] defining some aspects of how quickly the handover should be performed by the UE once network mobility management algorithms have made the decision to perform a handover, and have transmitted an RRC message to the UE to initiate the handover.

14.10.2.1 Cell Identification

When in RRC_Connected state, the UE continuously attempts to search for, and identify new cells. Unlike UTRA, there is no explicit neighbor cell list containing the physical identities of E-UTRA neighbor cells. For both E-UTRA FDD and E-UTRA TDD intra-frequency cell identification, there are rather similar requirements and the UE is required to identify a newly detectable cell within no more than 800 ms. This should be considered a general requirement applicable in a wide range of propagation conditions when the $SCH\_RP/I_{ot} \geq -6\,dB$ and the other side conditions given in [3] are met. The 800 ms requirement was agreed after a simulation campaign in 3GPP where cell identification was simulated by a number of companies at different SCH_RP/I_{ot} ratios.

It is also important to note that cell identification includes a 200 ms measurement period after the cell has been detected. Therefore, to comply with the requirement to identify cells within 800 ms, the UE needs to be able to detect cells internally in a faster time to allow for the measurement period.

When less than 100% of the time is available for intra-frequency cell-identification purposes – because, for example, the intra-frequency reception time is punctured by measurement gaps – the 800 ms time requirement is scaled to reflect the reduced time available.

For E-UTRA inter-frequency measurements, a similar approach is taken. Since inter-frequency cell identification is performed in measurement gaps, the basic identification time $T_{basic\_identify}$ is scaled according to the configured gap density so that

$$T_{Identify\_Inter} = T_{Basic\_Identify\_Inter} \cdot \frac{T_{Measurement\_Period\_Inter\_FDD}}{T_{Inter}}\ ms \qquad (14.16)$$

The $T_{measurement\_Period\_Inter}$ is multiplied by the number of E-UTRA carriers that the UE is monitoring (denoted as N_{freq}), which in turn means that the identification time

$T_{Identify\_Inter}$ is also multiplied by N_{freq} and so the requirement for inter-frequency cell identification becomes longer the more frequency layers are configured.

Similar requirements are also defined in [3] for UTRA cell identification when in E-UTRA RRC_Connected state.

14.10.2.2   Measurement of Identified Cells

Once cells are identified, the UE performs measurements on them, over a defined measurement period. 3GPP specifications do not define the sample rate at which the UE layer 1 is required to make measurement (or even that uniform sampling is necessary) but the measurements are specified be filtered over a standardized measurement period to ensure consistent UE behavior. The UE is also expected to meet accuracy requirements (discussed further in section 10.2.3) over the measurement period, which places constraints on how many L1 samples of the measurement need to be taken and filtered during the measurement period.

For intra-frequency cells, minimum capabilities for measurement are defined in [3]. In summary, the UE is required to have the capability to measure eight intra-frequency cells when 100% of the time is available for intra-frequency measurements. When less time is available, for example due to inter-frequency measurement gaps, then the requirement is scaled down accordingly. The period for intra-frequency measurements is specified as 200 ms, although this would be scaled up if less time is available for intra-frequency measurements.

For inter-frequency cells, there are two measurement period requirements defined in [3], one of which is a mandatory requirement, and one of which may be optionally supported by the UE. The mandatory requirement is based on a measurement bandwidth of 6 resource blocks, in which case the measurement period is $480 \times N_{freq}$ ms. When it is indicated by signaling that a bandwidth of at least 50 resource blocks is in use throughout a particular frequency layer, the UE may optionally support a measurement period of $240 \times N_{freq}$ ms. In this case, the UE knows from the signaling that it is safe to make the measurement over a wider bandwidth, and may therefore be able to achieve the required accuracy while filtering fewer measurement samples in the time domain, making possible a reduced measurement period.

For inter-frequency measurements, the minimum requirement is that the UE is capable of supporting three E-UTRA carrier frequencies (in addition to the intra-frequency layer) and on each of these three carrier frequencies it should be capable of measuring at least four cells.

14.10.2.3   Accuracy Requirements and Report Mapping

For both RSRP and RSRQ, absolute and relative accuracy requirements are defined in [3], Chapter 9. Absolute accuracy considers the difference between the actual and the ideal measurements for a single cell, whereas relative accuracy considers how much error can be expected when comparing the levels of two cells.

For RSRP, both intra-frequency and inter-frequency absolute and relative accuracy requirements are defined. For comparison of two cells measured on the same frequency, the main sources of inaccuracy can be studied in link-level simulations of the measurements, and this was the approach used to define the accuracy requirements. For absolute measurements of RSRP, uncertainty in the RF gain setting is a significant additional source of error, and this is reflected in the accuracy requirements, especially in extreme temperature and voltage conditions. When considering inter-frequency relative RSRP accuracy some of the uncertainties in RF gain setting cancel out, because they apply to both the cells that are being compared, so the performance requirement for inter-frequency RSRP relative accuracy is somewhat tighter than the inter-frequency absolute accuracy requirement, but not as tight as the intra-frequency relative accuracy requirement.

When considering the accuracy of RSRQ, it should be noted that comparison of two RSRQ measurements on the same frequency is not particularly useful. The reason is that the RSSI component of the RSRQ in both measurements will be similar because, in the intra-frequency case, both cells must be measured on the same frequency layer. For this reason, only absolute accuracy requirements are defined for intra-frequency RSRQ measurements. Since RF gain setting uncertainties to an extent affect both the measurement of the RSRP and RSSI components in RSRQ, uncertainties will cancel out somewhat, and, as such, RSRQ absolute accuracy is required to be better than RSRP absolute accuracy.

[3] also defines RSRP and RSRQ report mapping, which defines how a UE measured quantity value should be mapped to a signaled information element. This defines the range and resolution of the values that may be reported by the UE.

14.10.3   Connected State Mobility When DRX is Active

One new feature of E-UTRA compared to UTRA, which is important from a mobility aspect, is that rather large DRX cycles (for example, up to 2.56 s) may be used when the UE is in the RRC_Connected state. This means that inactive UEs may still be kept in connected state, and will use connected state mobility procedures such as handover. To allow power saving opportunities for such devices, [3] specifies different measurement performance requirements, which are applicable when large DRX cycles (80 ms and greater) are active. The basis of the large DRX measurement performance requirements is to ensure that the UE would be able to perform the mobility related measurements at or around the time when the receiver is active in the DRX cycle anyway. Power saving can be performed for the remainder of the DRX. To facilitate this, measurement periods and cell identification times are specified as a function of the configured DRX cycle length.

14.10.4   Handover Execution Performance Requirements

As with UTRA specifications, handover performance is characterized by two different delays, which are illustrated in Figure 14.57. Total handover delay is denoted by $D_{handover}$ and is the total time between the end of transmission of the handover command on the source cell and the start of UE transmissions to the target cell. In addition, maximum allowable interruption time is separately specified, and represents the time for which the UE is no longer receiving and transmitting to the source cell, and has not yet started transmission to the target cell.

RRC procedure delays are specified in [45] section 14.2 and the interruption time is given by:

$$T_{interrupt} = T_{search} + T_{IU} + 20\,ms \tag{14.17}$$

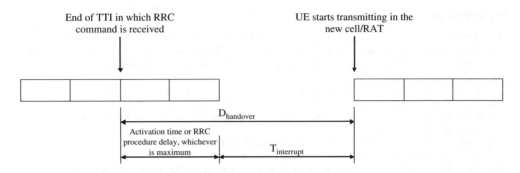

Figure 14.57   Handover delay and interruption time

where T_{search} is the time taken to search for the target cell. This will be 0 ms in the case that the target cell has previously been measured by the UE within the last 5 seconds, and is only non zero in the case of a blind handover to a cell not known to the UE. T_{IU} relates to the uncertainty in timing between the physical channel structure of the old cell. An additional 20 ms implementation margin is also included in the requirement.

As indicated in Figure 14.57, the total handover delay is either $D_{handover}$ = RRC procedure delay + $T_{interrupt}$ or activation time + $T_{interrupt}$ if an activation time is given that is greater than the RRC procedure delay in the future.

14.11   Summary

Clearly defined performance requirements are an essential part of an open well-functioning standard. The requirements are needed to provide predictable performance within an operator's own band in terms of data rates, system capacity, coverage and mobility with different terminals and different radio network vendors. The requirements are also needed to facilitate the co-existence of LTE in the presence of other systems as well as the co-existence of adjacent LTE operators in the same area. The regulatory requirements are considered when defining the emission limits. 3GPP has defined minimum RF performance requirements for LTE terminals (UE) and for base stations (eNodeB) facilitating a consistent and predictable system performance in a multi-vendor environment.

3GPP LTE frequency variants cover all the relevant cellular bands. The specifications covers 22 different variants for FDD and nine for TDD bands during the editing of the book early 2011. More frequency variants will be added in the future. The frequency variants are independent of the other content of the releases.

3GPP LTE requirements are defined in such a way that they enable efficient multimode GSM/WCDMA/LTE device implementation from the RF requirements perspective. Yet, the need to support receiver diversity required for MIMO operation, and the multiple frequency-band combinations, present a significant challenge for the front-end components optimization of such mobile devices. The SC-FDMA transmission in the uplink has similar requirements as the current HSUPA transmission. The baseband processing requirements are naturally increased due to the high data rates in LTE but the simple front end-structure in OFDMA and the Turbo decoding parallelization simplifies the practical implementations.

References

[1] 3GPP, Technical Specification 36.101, 'User Equipment (UE) Radio Transmission and Reception', v. 8.3.0.

[2] 3GPP, Technical Specification 36.104, 'Base Station (BS) RadioTransmission and Reception', v. 8.3.0.

[3] 3GPP, Technical Specification 36.133, 'Requirements for Support of RadioResource Management', v. 8.3.0.

[4] 3GPP, Technical Report 36.942, 'Radio Frequency (RF) System Scenarios', v. 8.0.0.

[5] 3GPP, Technical Specifications 36.141, 'Base Station (BS) Conformance Testing', v. 8.0.0.

[6] 3GPP, Technical Specifications 36.108, 'Common Test Environments for User Equipment (UE); Conformance testing', v. 8.0.0.

[7] 3GPP, Technical Specifications 36.114 'User Equipment (UE)/Mobile Station (MS) Over The Air (OTA) Antenna Performance; Conformance testing', v. 8.0.0.

[8] 3GPP, R4-070342, 'On E-UTRA Spectrum Emission Mask and ACLR Requirements', April 2007.

[9] European Communications Committee (ECC)/European Conference of Postal and Telecommunications Administrations (CEPT) Report 19. 'Report from CEPT to the European Commission in response to the Mandate to develop least restrictive technical conditions for frequency bands addressed in the context of WAPECS', December 2007.

[10] 3GPP, R4-070124 'System Simulation Results for Derivation of E-UTRA BS EVM Requirements', February 2007.

[11] Holma, H., Toskala, A., 'WCDMA for UMTS: HSPA Evolution and LTE', 5th Edition, Wiley, Chapter 20.

[12] The Mobile Industry Processor Interface (MIPI^{SM}) Alliance, www.mipi.org.

[13] Bannelli, P., Cacopardi, S., 'Theoretical Analysis and Performance of OFDM Signals in Nonlinear AWGN Channels', IEEE Transactions on Communications, vol. 48, no. 3, March 2000.

[14] Myung, H.G., Junsung Lim, Goodman, D.J., 'Peak-To-Average Power Ratio of Single Carrier FDMA Signals with Pulse Shaping', Personal, Indoor and Mobile Radio Communications, 2006 IEEE 17th International Symposium, September 2006: 1–5.

[15] Valkama, M., Anttila, L., Renfors, M., 'Some radio implementation challenges in 3G-LTE context', Signal Processing Advances in Wireless Communications, 2007. SPAWC 2007. IEEE 8th Workshop, 17–20 June 2007: 1–5.

[16] Priyanto, B.E., Sorensen, T.B., Jensen, O.K., Larsen, T., Kolding, T., Mogensen, P., 'Impact of polar transmitter imperfections on UTRA LTE uplink performance', Norchip, 19–20 November 2007: 1–4.

[17] Talonen, M., Lindfors, S., 'System requirements for OFDM polar transmitter', Circuit Theory and Design, 2005. Proceedings of the 2005 European Conference. Volume 3, 28 August–2 September 2005, vol. 3, page(s): III/69–III/72.

[18] Chow, Y.H., Yong, C.K., Lee, Joan, Lee, H.K., Rajendran, J., Khoo, S.H., Soo, M.L., Chan, C.F., 'A variable supply, (2.3–2.7) GHz linear power amplifier module for IEEE 802.16e and LTE applications using E-mode pHEMT technology'. Microwave Symposium Digest, 2008 IEEE MTT-S International 15–20 June 2008: 871–874.

[19] Ericsson, 'TP 36.101: REFSENS and associated requirements', TSG-RAN Working Group 4 (Radio) meeting #46bis, R4–080696, Shenzhen, China, 31 March–4 April 2008.

[20] Ericsson, 'Simulation results for reference sensitivity and dynamic range with updated TBS sizes', TSG-RAN Working Group 4 (Radio) meeting #47bis, R4-081612, Munich, 16–20 June 2008.

[21] Nokia Siemens Networks, 'Ideal simulation results for RF receiver requirements', 3GPP TSG-RAN WG4 Meeting #48, R4-081841, Jeju, Korea, 18 – 22 August 2008.

[22] STN-wireless internal WCDMA handset benchmarking test report.

[23] Anadigics AWT6241 WCDMA Power Amplifier datasheet, http://www.anadigics.com/products/handsets_datacards/wcdma_power_amplifiers/awt6241.

[24] Ericsson, 'Introduction of MSD (Maximum Sensitivity Degradation)', TSG-RAN Working Group 4 (Radio) meeting #49, R4-082972, Prague, Czech Republic, 10–14 November 2008.

[25] NTT DoCoMo, Fujitsu, Panasonic, 'Performance requirements on Self interference due to transmitter noise', TSG-RAN Working Group 4 Meeting #47, R4-080873, Kansas, USA, 5–9 May 2008.

[26] Ericsson, 'Introduction of MSD (Maximum Sensitivity Degradation)', 36.101 Change Request # 90, R4-083164, 3GPP TSG-RAN WG4 Meeting #49, Prague, Czech Republic, 10–14 November 2008.

[27] Ericsson, ST-Ericsson, Changes to REFSENS and associated MSD requirements, R4-100626, TSG-RAN Working Group 4 meeting 54, San Francisco, USA, 22–26 February 2010.

[28] 'Half Duplex FDD Operation in LTE', 3GPP TSG RAN WG4 Meeting #46, R4-080255, Sorrento (Italy), 11 to 15 February 2008.

[29] Ericsson, 'HD-FDD from a UE perspective', TSG-RAN Working Group 4 (Radio) meeting #46bis,R4-080695 Shenzhen, China, 31 March–4 April 2008.

[30] Manstretta, D., Brandolini, M., Svelto, F., 'Second-Order Intermodulation Mechanisms in CMOS Down-converters', IEEE Journal of Solid-State Circuits, Vol. 38, No. 3, March 2003, 394–406.

[31] 3GPP, Technical Specification 25.101, 'User Equipment (UE) radio transmission and reception (UTRA)', v. 8.3.0.

[32] Qualcomm Europe, Change Request 'R4–082669: UE ACS frequency offset', TSG-RAN Working Group 4 (Radio) meeting #49, Prague, CZ, 10–14 November 2008.

[33] Faulkner, M., 'The Effect of Filtering on the Performance of OFDM Systems', IEEE Transactions on Vehicular Technology, vol. 49, no. 5, September 2000: 1877–1883.

[34] Lindoff, B., Wilhelmsson, L., 'On selectivity filter design for 3G long-term evolution mobile terminals', Signal Processing Advances in Wireless Communications, 2007. SPAWC 2007. IEEE 8th Workshop on Volume, 17–20 June 2007: 1–5.

[35] Dardari, D., 'Joint Clip and Quantization Effects Characterization in OFDM Receivers', IEEE Transactions On Circuits And Systems – I: Regular Papers, vol. 53, no. 8, August 2006: 1741–1748.

[36] Locher, M. et al., 'A Versatile, Low Power, High Performance BiCMOS, MIMO/Diversity Direct Conversion Transceiver IC for Wi-Bro/WiMax (802.16e)', IEEE Journal of Solid State Circuits, vol. 43, no. 8, August 2008: 1–10.

[37] Norsworthy, S.R., Schreier, R., Temes, G.C., 'Delta-Sigma Data Converters, Theory, Design and Simulation', Wiley Interscience, 1996.

[38] Internal STN-wireless data extracted with simulation parameter modifications from: 'Ouzounov, S. van Veldhoven, R. Bastiaansen, C. Vongehr, K. van Wegberg, R. Geelen, G. Breems, L. van Roermund, A., 'A 1.2V 121-Mode $\Sigma\Delta$ Modulator for Wireless Receivers in 90nm CMOS', Solid-State Circuits Conference, 2007. ISSCC 2007. Digest of Technical Papers. IEEE International Publication Date: 11-15 Feb. 2007 On page(s): 242–600.

[39] Martel, P., Lossois, G., Danchesi, C., Brunel, D., Noël, L., 'Experimental EVM budget investigations in Zero-IF WCDMA receivers', submitted to Electronics Letters.

[40] Windisch, M., Fettweis, G., 'Performance Degradation due to I/Q Imbalance in Multi-Carrier Direct Conversion Receivers: A Theoretical Analysis', Communications, 2006. ICC apos;06. IEEE International Conference on Volume 1, June 2006: 257–262.

[41] Pollet, T., Moeneclaey, M., Jeanclaude, I., Sari, H., 'Effect of carrier phase jitter on single-carrier and multi-carrier QAM systems', Communications, 1995. ICC 95 Seattle, Gateway to Globalization, 1995 IEEE International Conference on Volume 2, 18–22 June 1995: 1046–1050.

[42] Motorola, 'Reference Channel and Noise Estimators', 3GPP E-mail reflector document for TSG-RAN Working Group 4 Meeting 46, Sorrento, Italy, 11–15 February, 2008.

[43] Ericsson, 'R4–071814: Clarification of TX EVM model', TSG-RAN Working Group 4 (Radio) meeting #44, Athens, Greece, 20–24 August 2007.

[44] 3GPP, Technical Specification 25.133 'Requirements for support of radio resource management (UTRA)', v. 8.3.0.

[45] 3GPP, Technical Specification 36.331, 'Radio Resource Control (RRC); Protocol specification', v. 8.3.0.

15

LTE TDD Mode

Che Xiangguang, Troels Kolding, Peter Skov, Wang Haiming
and Antti Toskala

15.1 Introduction

With full coverage in the 3GPP Release 8 specifications of both TDD and FDD modes
of operation, LTE can effectively be deployed in both the paired and the unpaired spectrums. LTE TDD and FDD modes have been greatly harmonized in the sense that both
modes share the same underlying framework, including radio access schemes, OFDMA in
downlink and SC-FDMA in uplink, basic sub-frame formats, and configuration protocols.
As a clear indication of harmonization, the TDD mode is included together with the FDD
mode in the same set of specifications, including the physical layer [1–4] where there
are just a few differences due to the uplink/downlink switching operation. In terms of
architecture there are no differences between FDD and TDD and the very few differences
in the MAC and higher layer protocols relate to TDD-specific physical layer parameters.
Procedures remain the same. Thus there will be high implementation synergies between
the two modes allowing for efficient support of both TDD and FDD in the same network
or user device. Coexistence would, of course, still require careful analysis.

Another key feature of the LTE TDD mode (also known as TD-LTE) is the commonality with TD-SCDMA [5]. This is an advantage, for example, in China, where the
Release 4 based TD-SCDMA (including enhancements from later Releases) has opened
for a large-scale TDD system deployment, making a way forward for further deployment of 3GPP-based LTE TDD using the available unpaired spectrum. As presented in
Chapter 14, there is a global trend to reserve significant unpaired spectrum allocations.

In this chapter, detailed aspects related to LTE TDD that differ from the FDD mode
are introduced. Further, information is given related to both link and system performance
of the LTE TDD mode of operation.

15.2 LTE TDD Fundamentals

The basic principle of TDD is to use the same frequency band for transmission and
reception but to alternate the transmission direction in time. As shown in Figure 15.1,

LTE for UMTS: Evolution to LTE-Advanced, Second Edition. Edited by Harri Holma and Antti Toskala.
© 2011 John Wiley & Sons, Ltd. Published 2011 by John Wiley & Sons, Ltd.

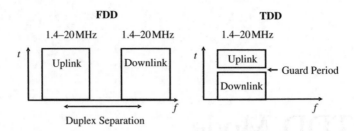

Figure 15.1 Principles of FDD and TDD modes of operation

this is a fundamental difference compared to FDD where different frequencies are used for continuous UE reception and transmission. Like FDD, LTE TDD supports bandwidths from 1.4 MHz up to 20 MHz but depending on the frequency band the number of supported bandwidths may be less than the full range. For example, for the 2.5 GHz band, it is not likely that the smallest bandwidths will be supported. Since the bandwidth is shared between uplink and downlink and the maximum bandwidth is specified to be 20 MHz in Release 8, the maximum achievable data rates are lower than in LTE FDD. This way the same receiver and transmitter processing capability can be used with both TDD and FDD modes enabling faster deployment of LTE.

The TDD system can be implemented on an unpaired band (or in two paired bands separately) while the FDD system always requires a pair of bands with a reasonable separation between uplink and downlink directions, known as the duplex separation. In an FDD UE implementation this normally requires a duplex filter when simultaneous transmission and reception is facilitated. In a TDD system the UE does not need such a duplex filter. The complexity of the duplex filter increases when the uplink and downlink frequency bands are placed in closer proximity. In some of the future spectrum allocations it is foreseen that it will be easier to find new unpaired allocations than paired allocations with sensible duplex separation, thereby increasing further the scope of applicability for TDD.

However, since uplink and downlink share the same frequency band, the signals in those two transmission directions can interfere with each other. This is illustrated in Figure 15.2 with use of TDD on the same frequency without coordination and synchronization between sites in the same coverage area. In case of uncoordinated deployment (unsynchronized) on the same frequency band, the devices connected to the cells with different timing and/or different uplink/downlink allocation may cause blocking for other users. In LTE TDD the base stations need to be synchronized to each other at frame level in the same coverage area to avoid this interference. This can be typically done by using for example satellite based solutions like GPS or Galileo or by having another external timing reference shared by the LTE TDD base stations within the same coverage area. LTE FDD does not need base-station synchronization. There is no interference between uplink and downlink in FDD due to the duplex separation of the carriers.

Two adjacent LTE TDD operators (on adjacent carriers) should preferably synchronize the base stations and allocate the same asymmetry between uplink and downlink in order to avoid potentially detrimental interference to the system reliability. If the two operators do not co-ordinate the deployments, there is instead a need for guard bands and additional filtering. Those requirements are discussed in Chapter 14.

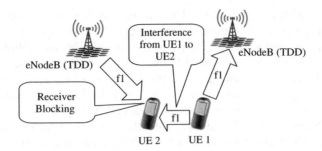

Figure 15.2   Interference from uplink to downlink in uncoordinated TDD operation

15.2.1   The LTE TDD Frame Structure

As the single frequency block is shared in the time domain between uplink and downlink (and between users as well), the transmission in LTE TDD is not continuous. While this is often also the case for data transmissions towards users in LTE FDD mode, the level of discontinuity in that case depends entirely on the scheduling function (except for half-duplex FDD terminals). For control channels, for example the PDCCH and the PHICH, the transmission for FDD is continuous. For LTE TDD, all uplink transmissions need to be on hold while any downlink resource is used and, conversely, the downlink needs to be totally silent when any of the UE is transmitting in the uplink direction. Switching between transmission directions has a small hardware delay (for both UE and eNodeB) and must be compensated. To control the resulting switching transients a *guard period* (GP) is allocated, which compensates for the maximum propagation delay of interfering components (for example, it depends on cell size and level of available cell isolation). The impact of discontinuous uplink and downlink on the link budget in LTE TDD is covered specifically in section 15.5.

To explain the exact implementation of the mechanism for switching between downlink and uplink and vice versa, consider an example setup of the LTE TDD frame structure in Figure 15.3. The sub-frames of the uplink (UL) and the downlink (DL) have a common design with LTE FDD with some minor but significant differences related to common control channels. In LTE TDD there is, at most, one DL→UL and one UL→DL transition

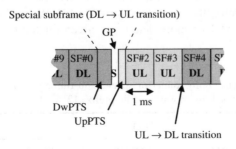

Figure 15.3   Simple illustration of the DL→UL and UL→DL transitions implemented in downlink (DL), uplink (UL), and special (S) sub-frames (SF)

per 5 ms period (half-frame). The UL→DL transition is done for all intra-cell UE by the process of time alignment. The eNodeB instructs each UE to use specific time offset so that all UE signals are aligned when they arrive at the eNodeB. Hence, uplink is synchronous as is the case for FDD. To facilitate that the UE has sufficient time to shut down its transmission and switch to listening mode, the UE does not transmit a signal during the last 10–20 microseconds of sub-frame #3 in Figure 15.3. This procedure ensures that there is no UE transmission power from the own cell that spills over into the downlink transmission. Although eNodeBs in different cells are fully synchronized, this method does not prevent UE from other cells from spilling their interference into the current sector's downlink transmission. However, in practice this is less of a problem because individual UE has limited transmission power.

While the UL→DL switching is merely an intra-cell method, the DL→UL switching method ensures that the high-power downlink transmissions from eNodeBs from other neighbor cells do not interfere when the UL eNodeB reception is ongoing in the current cell. Adopting the methodology of TD-SCDMA, LTE TDD introduces a special (S) sub-frame that is divided into three parts; the *downlink pilot time slot* (DwPTS), the *guard period* (GP), and the *uplink pilot time slot* (UpPTS). The special sub-frame replaces what would have been a normal sub-frame #1. The individual time duration in OFDM symbols of the special sub-frame parts are to some extent adjustable and the exact configuration of the special time slot will impact the performance. The GP implements the DL→UL transition and the GP has to be sufficiently long to cover the propagation delay of all critical downlink interferers on the same or adjacent carriers as well as the hardware switch-off time. Hence, the correct setting of the GP depends on network topology, antenna configurations, and so forth. To fit into the general LTE frame numerology, the total duration of DwPTS, GP, and UpPTS is always 1 ms.

DwPTS is considered to be a 'normal' downlink sub-frame and carries control information as well as data for those cases when sufficient duration is configured. High commonality is achieved by rescaling the transport block size according to its length. In this way the effective coding rate for a certain selection of payload and transmission bandwidth will stay the same. UpPTS is primarily intended for sounding reference symbol (SRS) transmissions from the UE but LTE TDD also introduces the short RACH concept so that this space may be used for access purposes as well. The flexibility for the different components of the special sub-frame is summarized in Table 15.1 for the case using the normal cyclic prefix. The GP can be up to about 700 ms thus supporting a cell range up to 100 km. For more details related to the many possible configurations including the extended cyclic prefix, the reader is referred to [2]. When discussing co-existence with TD-SCDMA in section 15.5 more details are given related to the special sub-frame configuration.

Table 15.1 Possible durations configurable for special sub-frame components – total duration of the three components is always 1 ms (normal cyclic prefix) [2]

Component	Unit	Range of duration
Downlink pilot time slot (DwPTS)	# OFDM symbols (71 μs)	3–12
Guard period (GP)	# OFDM symbols (71 μs)	1–10
Uplink pilot time slot (UpPTS)	# OFDM symbols (71 μs)	1–2

15.2.2 Asymmetric Uplink/Downlink Capacity Allocation

A key advantage to TDD is the ability to adjust the available system resources (time and frequency) to either downlink or uplink to perfectly match the uplink and downlink traffic characteristics of the cell. This is done by changing the duplex switching point and thus moving capacity from uplink to downlink, or vice versa. The LTE TDD frame structure can be adjusted to have either 5 ms or 10 ms uplink-to-downlink switch point periodically. The resulting resource split is either balanced or between the two extreme cases:

- Single 1 ms sub-frame for uplink and eight times 1 ms sub-frame for downlink per 10 ms frame.
- if we want to maximally boost the uplink capacity, then we can have three sub-frames for uplink and one for downlink per 5 ms. Thus the resulting uplink activity factor can be adjusted from 10% to 60% (if we do not consider the UpPTS).

In the case of continuous coverage across several cells, the chosen asymmetry is normally aligned between cells to avoid interference between transmission directions as described earlier. Hence, synchronization in a LTE TDD system deployed over a wide area is conducted both at frame and uplink-downlink configuration level. In practice it is expected that the uplink-downlink configuration is changed in the network only very seldom and the associated signaling of LTE TDD in Release 8 has been optimized according to this assumption. For example, a LTE TDD Release 8 cellular system would be characterized as a static or semi-static TDD system. The UE is informed about the active uplink-downlink configuration via the *system information block* (SIB-1), which is broadcasted via the *dynamic broadcast channel* (D-BCH) to the cell with an update rate of 80 ms. Knowledge about which uplink-downlink configuration is active in the cell is essential for the UE to know the location of critical control channels and timing of adaptation methods such as HARQ.

15.2.3 Co-existence with TD-SCDMA

As mentioned, there is a strong legacy between LTE TDD and TD-SCDMA. The co-existence of these systems on the same or adjacent carriers has therefore been a key discussion point during the standardization process. In the case of sharing a site with TD-SCDMA and using the same frequency band, the systems need to have an aligned uplink/downlink split to avoid interference between different base station transceivers. As TD-SCDMA slot duration does not match the LTE TDD sub-frame duration, LTE sub-frame parameterization was designed to accommodate co-existence. From knowledge of the relative uplink/downlink division, the relative timing of TD-SCDMA and LTE TDD can be adjusted to allow co-existence without BTS to BTS interference, as shown in Figure 15.4. Note the duration of fields in LTE TDD sub-frame with uplink/downlink change vary depending on the configuration thus the timings shown can take different values.

Apart from timing alignment the exact configuration of the special sub-frame in LTE TDD also plays an important role in allowing TD-SCDMA/LTE TDD co-existence. Some configurations with a good match are listed in Table 15.2 including the required detailed settings for DwPTS, GP and UpPTS. The normal cyclic prefix is assumed. For more options for configuration, the reader is referred to [2].

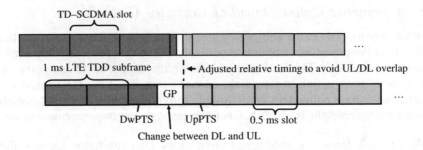

Figure 15.4 Ensuring TD-SCDMA and LTE TDD co-existence with compatible UL/DL timings

Table 15.2 Example modes for coexistence between TD-SCDMA and LTE TDD (normal cyclic prefix)

TD-SCDMA config.	LTE TDD config.	Special sub-frame config. (OFDM symbols)			
		Config.	DwPTS	GP	UpPTS
5DL-2UL	#2 (3DL-1UL)	#5	3	9	2
4DL-3UL	#1 (2DL-2UL)	#7	10	2	2
2DL-5UL	#0 (1DL-3UL)	#5	3	9	2

15.2.4 Channel Reciprocity

One interesting aspect for TDD systems is that the uplink and downlink signals travel through the same frequency band. Assuming symmetric antenna configurations and adequate RF chain calibration, there is a high correlation of the fading on the own signal between uplink and downlink, known as *channel reciprocity*. The key associated benefit is that measurements in one direction may fully or partially be used to predict the other direction, thereby enabling a reduction of the signaling overhead needed for link adaptation, packet scheduling, and use of advanced transmit antenna techniques. As already mentioned antenna configurations needs to be reciprocal in order for full reciprocity to be applicable. The baseline configuration for LTE is two antennas in both eNodeB and UE so in this case antenna configuration seems to be reciprocal. However, as the UE only has one power amplifier, it can only transmit on one antenna at a time so the only way to measure the full channel in UL is for the UE to switch the sounding transmission between the two antennas. If this is done we can say that both channel and antenna configurations are reciprocal.

An additional practical limitation towards the exploitation of reciprocity is that most adaptation schemes, including link adaptation methods and fast packet scheduling, rely on SINR or throughput assessments. As the interference level may vary significantly between uplink and downlink, the LTE TDD specifications therefore do not rely on the availability of channel reciprocity and allow for a full decoupling of uplink and downlink, thus using the very same uplink and downlink reporting methods that are available for LTE FDD. However, within the specifications there is a large degree of freedom available to optimize the total signaling overhead in cases where reciprocity has significant gain and is applicable.

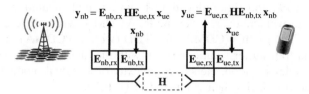

Figure 15.5 Model to illustrate the impact from RF units to channel reciprocity (capital letters indentify matrixes)

A simple linear model for the TDD channel including impact from RF-chains is shown in Figure 15.5. From this it is observed that even though the channel represented by $\mathbf{H}$ is identical for UL and DL, the channel seen by physical layer is different. In the typical application of channel state information we would like to know the channel as seen by physical layer (in DL $\mathbf{E}_{nb,tx}\mathbf{H}\mathbf{E}_{ue,rx}$) because pre-coding and decoding are all done after the signal has passed through the RF parts This information cannot be estimated from UL sounding. It is possible to improve channel reciprocity by adding calibration procedures within the Node B and UE but to calibrate the full RF chain so that $\mathbf{x}_{nb} = \mathbf{x}_{ue}$ would imply $\mathbf{y}_{nb} = \mathbf{y}_{ue}$ and this would require a standardized procedure.

Relying on channel reciprocity for pre-coding is very challenging as reciprocity of the full complex valued channel is assumed. For other purposes, such as reciprocity–based fast power control, we only rely on reciprocity of the channel attenuation and this would be less sensitive towards calibration errors [9].

15.2.5   Multiple Access Schemes

In LTE TDD the multiple access schemes are the same as those used by LTE FDD. Hence, OFDMA is used for the downlink and SC-FDMA for the uplink, as shown in Figure 15.6. The reasoning behind the selection was shared by TDD and FDD modes respectively. For a TDD device, in particular, the selection of a power-efficient uplink scheme was very important because of the more stringent uplink link budget due to the

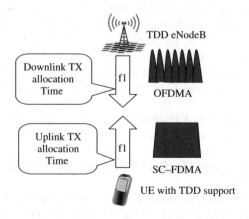

Figure 15.6   LTE TDD multiple access schemes in uplink and downlink

limited uplink transmission time allocation. From an implementation complexity view-point, this is an improvement over, for example, WCDMA, where different multiple access schemes between FDD (with WCDMA) and TDD (with TD/CDMA) modes required dif-ferent receiver and transmitter solutions (even with the identical chip rate in Release 99) in the baseband implementation. Hence, complexity of any dual mode TDD/FDD imple-mentation is reduced in LTE when compared to UTRAN.

15.3   TDD Control Design

Due to the special frame structure, the placement of critical control channels is different for LTE TDD than it is for LTE FDD. The exact placement of control channels depend on which uplink-downlink configuration is active in the cell and is thus key information for the UE to know in order to set up its connection with the network. In the following, the placement of the critical channels as well as their special meaning for LTE TDD is described. For general information about the meaning of the control channels, the reader is referred to Chapter 5, where the dynamic use of Physical Downlink Shared Control Channel (PDCCH) for uplink and downlink resource allocation is covered.

In the case of FDD operation, where downlink and uplink are in separate frequency with continuous transmission and reception on its dedicated frequency, the shared control channel design is straightforward due to one-to-one associated downlink and uplink sub-frames. However, in the case of TDD operation where downlink and uplink share the same frequency for transmission and reception but alternate the transmission direction in time, physical control channel design is rather more challenging. This was further complicated by the high flexibility in adjusting the downlink/uplink allocation.

15.3.1   Common Control Channels

To illustrate the placement of the different control channels, TDD uplink-downlink con-figuration #1 is used as reference. The critical system control channels are shown in Figure 15.6 and described below.

The *primary synchronization signal* (PSS) is placed at the third symbol in sub-frames #1 and #6, which is different from LTE FDD where the PSS is placed at the last sym-bol of first slot in sub-frames #0 and #5. The *secondary synchronization signal* (SSS) is placed at the last symbol in sub-frames #0 and #5, which is also different from LTE FDD where the SSS is placed at the second last symbol of first slot in sub-frames #0 and #5. Synchronization via the PSS is typically a robust process and the UE can sur-vive UE-UE interference of up to 100 dB for the initial cell search. This high-measuring robustness is facilitated by normalization of samples achieved over a long measuring window.

Altogether, there are five RACH preamble formats defined for LTE TDD. Four of the RACH preamble formats are common with LTE FDD, while the LTE TDD specific RACH preamble format 4 is known as *short RACH* (S-RACH) due to the short preamble sequence duration. As shown in Figure 15.7 the short RACH is transmitted on the UpPTS within the special sub-frame.

RACH needs to be protected to make access reliable and to allow for co-existence of multiple LTE TDD users. The RACH channel is fairly flexible regarding the density and placement in time and frequency. The RACH density, similarly as for LTE FDD, can

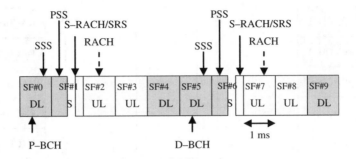

Figure 15.7 Mapping of critical control channels to TDD configuration #1

be 0.5, 1, 2, 3, 4, 5 or 6 in every 10 ms radio frame. The exact placement of RACH in LTE TDD within the available UL resources (UL sub-frames and UpPTS) in the config-ured UL/DL configuration is selected by proper network planning taking into account, for example, the requirement for co-existence with TD-SCDMA, traffic load, network topology, and so forth. Altogether, there are 58 configurations for RACH placement in time and frequency for each UL/DL configuration with the common principle that RACH channels are first distributed in the time domain among all available UL resources before being multiplexed in the frequency domain. This principle is exemplified in Figure 15.8 for UL/DL configuration #0 and the exact RACH configurations can be found in [2]. The maximum available number of RACH channels per UL sub-frame or UpPTS (for S-RACH only) is six, which is different from FDD where, at most, one RACH channel is available per UL sub-frame. This ensures that even with limited UL resources available, sufficient RACH opportunities for high traffic load can be created. As illustrated in Figure 15.9, when there is more than one RACH channel in one UL sub-frame or UpPTS, the RACH channels are distributed towards both edges of the system bandwidth for RACH in the UL sub-frame or placed from only one edge (either bottom or top) when in UpPTS. For the latter case, the frequency position is altered in the next RACH instance as shown in the figure.

The placement of the *primary broadcast channel* (P-BCH) and the *dynamic broadcast channel* (D-BCH) is the same as it is in LTE FDD. However, the detection of D-BCH is slightly different from what is done in LTE FDD. This comes from the fact that in order to detect D-BCH the size of the control channel in that DL sub-frame needs to be known to the UE. However, this raises a chicken-and-egg problem because the size of the control channel depends on the active UL/DL configuration and the UE does not know the active UL/DL configuration before it correctly detects D-BCH during the initial cell search. For all UL/DL configurations, there are three different sizes for the control channel in any

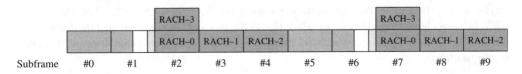

Figure 15.8 RACH distribution/placement in time and frequency in LTE TDD

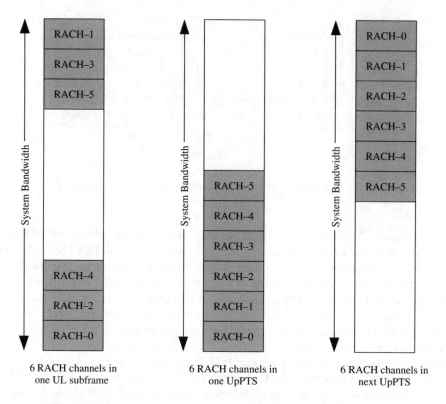

Figure 15.9 RACH placement in frequency domain in LTE TDD

DL sub-frame, thus the UE will need to perform three hypotheses on control channel size in order to detect the D-BCH. Once the UE correctly receives the D-BCH, it knows the UL/DL configuration in the cell, after which the hypothesis detection is no longer needed when detecting the D-BCH.

The paging procedure is the same for LTE TDD and FDD; however, the exact sub-frames being used for paging are slightly different in LTE TDD. For FDD, the sub-frames #0, #4, #5, and #9 can be used for paging. For TDD, sub-frames #0, #1, #5 and #6 can be used for paging.

15.3.2 Sounding Reference Signal

In LTE TDD the *sounding reference signal* (SRS) for any UE can be transmitted not only in the last symbol of one UL sub-frame as in LTE FDD but also in one or both symbols of UpPTS. Since the UpPTS is available anyway and cannot carry uplink data traffic it is expected to be the primary location for SRS in LTE TDD. The support of SRS is considered mandatory for all LTE TDD devices due to the installed bases of TD-SCDMA base stations, which in many cases have existing beam-forming antenna configurations.

15.3.3 HARQ Process and Timing

The key to designing and configuring HARQ for TDD operation is to determine the required processing time of the eNodeB and the UE for *uplink* (UL) and *downlink* (DL) respectively. The relevant processing times are as follows:

- DL UE: duration from the time when the last sample of the packet is received in downlink until an HARQ-ACK/NACK is transmitted in uplink;
- DL eNodeB: duration from the time when an HARQ-ACK/NACK is transmitted in uplink until the eNodeB can (re)-transmit data on the same HARQ process;
- UL eNodeB: duration from the time when the last sample of the packet is received in uplink until an HARQ-ACK/NACK (or a new allocation on same HARQ process) is transmitted in downlink;
- UL UE: duration from the time when a UL grant (or HARQ-ACK/NACK) is given until the UE is able to transmit the associated packet in the uplink.

In FDD the eNodeB and UE processing times for both DL and UL is fixed and assumed to be 3 ms due to invariant DL and UL sub-frame configuration and continuous DL and UL transmission and reception. In TDD, although the eNodeB and UL (minimum) processing times are the same as for FDD, the actual HARQ timing (i.e. DL UE/eNodeB, UL eNodeB/UE) varies depending on the active DL/UL configuration. This is exemplified in Figure 15.10, where it is seen, that compared to the minimum processing time of 4 ms, there is sometimes an additional delay incurred due to unavailable DL or UL sub-frame after 3 ms.

In addition to the minimum HARQ processing time, a few other decisions were made related to the special sub-frame in order to fix the required number of HARQ processes and timing in the specifications:

- There should always be a PDCCH in the DwPTS at least for UL grant or PHICH. If DwPTS spans more than three OFDM symbols it also contains a PDSCH.
- UpPTS does not contain HARQ control information or a PUSCH.

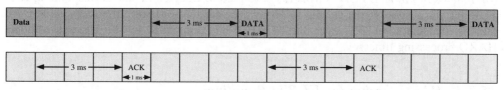

(a) Conceptual example of FDD HARQ Timing (propagation delay and timing advance is ignored)

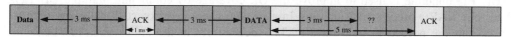

(b) Conceptual example of TDD HARQ Timing (special subframe is treated as ordinary DL subframe)

Figure 15.10 Illustration of HARQ Timing, LTE FDD (a) and LTE TDD (b)

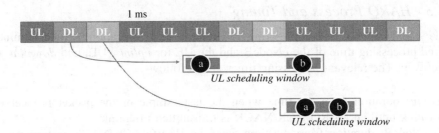

Figure 15.11 Illustration of the possible uplink multi-TTI scheduling

With the given assumptions of processing time and special sub-frame behavior, the number of DL HARQ processes and UL HARQ processes varies with TDD between four to 15 for the downlink and one to 7 for the uplink (in FDD it was always constant) As for FDD, in TDD the DL HARQ is asynchronous and the UL HARQ is synchronous. The resulting delay varies, depending on the sub-frame being used as well as on the uplink/downlink spit applied, and can take values between 4 to 7 ms for the delay k_1 between the associated UL grant (or an intended PHICH) on the PDCCH and the UL data transmission over the PUSCH. The delay k_2 between the associated UL data transmission over PUSCH and PHICH (or an UL grant for re-/new transmission) over the PDCCH also varies between 4 to 7 ms.

The multi-TTI scheduling scheme in TDD allows for efficient utilization of the downlink shared control channel resources (PDCCH) in case fewer downlink resources are available, and further decreases the UE complexity of decoding PDCCH (see Figure 15.11).

The DL HARQ timing delay k_3 between the associated DL data transmission over PDSCH and the UL HARQ-ACK transmission over PUCCH or PUSCH varies between 4 and 13 ms depending again on the uplink/downlink split applied. This is now simplified because TDD DL HARQ is operated in asynchronous mode like LTE FDD. The intended DL HARQ timing is derived by $n + k_3$, where n is the addressed sub-frame index number.

Due to the discontinuous UL/DL transmission, the TDD frame structure adds some delay to the user plane transmission when HARQ is involved. While the HARQ *round trip time* (RTT) in LTE FDD is 8 ms (as described in Chapter 14) the corresponding LTE TDD RTT is at least 10 ms and up to 16 ms. The relatively small differences come from the fact that the core RTT, even for LTE FDD, is dominated by the UE and eNodeB HARQ processing times.

15.3.4   HARQ Design for UL TTI Bundling

UL TTI bundling is used to improve the UL coverage by combining multiple UL TTIs for one HARQ process transmission. In a system where it is expected that many retransmissions are needed to transmit a packet successfully in the uplink, UL TTI bundling provides for fast automatic retransmission so that the delay penalty is minor. Although many details of UL TTI bundling are the same for LTE FDD and TDD, the HARQ timing is different due to the inherited special UL HARQ timing in LTE TDD. In terms of 3GPP discussion, only TDD UL/DL configurations #0, #1, and #6 have bundling fully defined in Release 8. To control the aspect of TTI bundling timing and HARQ process number after TTI bundling, the starting point of UL TTI bundling is limited and the number of bundled HARQ processes is fixed according to the bundle size. Thus the number of bundled HARQ processes is defined by:

$$Bundled\_HARQ\_No = \left\lfloor \frac{Original\_HARQ\_No \times Bundled\_HARQ\_RTT}{Bundled\_size \times Original\_HARQ\_RTT} \right\rfloor \quad (15.1)$$

In equation (15.1), *Original_HARQ_No* and *Original_HARQ_RTT* are fixed for each TDD configuration in the LTE uplink system, so if assuming, for instance, TDD configuration #1 and a bundle size of 4, the number of bundled HARQ processes is 2, as shown in Figure 15.12.

As to the principle of ACK/NACK timing, it is always tied to the last sub-frame in a bundle, which is exactly same as the rule in FDD. The uplink TTI bundling for LTE FDD is described in Chapter 13.

15.3.5 *UL HARQ-ACK/NACK Transmission*

In the same way as for FDD, the UL HARQ-ACK/NACK for LTE TDD is transmitted over the PHICH on PDCCH. The PHICH mapping and indexing is mostly the same for LTE FDD and TDD – the PDCCH in one DL sub-frame only contains the PHICH associated to a single UL sub-frame PUSCH. The only exception to this rule is the TDD UL/DL configuration #0 where the PHICH associated has specific exception specified.

15.3.6 *DL HARQ-ACK/NACK Transmission*

For both LTE FDD and TDD, the DL HARQ-ACK/NACK is transmitted on the PUCCH or the PUSCH depending on whether UL has simultaneous data transmission in the same UL sub-frame or not. In many cases the DL HARQ-ACK/NACK associated from more than one PDSCH, for example up to nine, will be mapped into a single UL sub-frame. This so-called multiple UL ACK/NACK transmission is, however, notably different from the FDD MIMO case in which the DL HARQ-ACK/NACK associated from a single PDSCH (for instance with two codewords) is mapped into a single UL sub-frame.

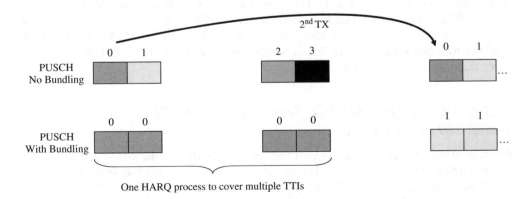

Figure 15.12 TTI bundling with TDD configuration #1

The fact that multiple downlink transmissions may need to be acknowledged within a single uplink sub-frame makes the design for good UL coverage for control channels in TDD even more challenging. A very special design arrangement has been created in order to accomplish this task while simultaneously respecting the single carrier property of the UL multiple access scheme when UE has to transmit multiple DL HARQ-ACK/NACKs. There are two DL HARQ ACK/NACK feedback modes supported in TDD operation of LTE which are configured by higher layer on a per-UE basis:

- ACK/NACK bundling feedback mode (the default mode), where a logical AND operation is performed per codeword's HARQ ACK/NACK across multiple DL sub-frames PDSCH whose associated HARQ ACK/NACK is mapped into same UL sub-frame.
- ACK/NACK multiplexing feedback mode, where a logical AND operation is performed across spatial codewords within a DL HARQ ACK/NACK process. In Release 8 LTE TDD, up to 4 bits DL HARQ ACK/NACK is supported per UL sub-frame, and some additional sub-bundling and/or scheduler limitations have been decided for UL/DL configuration #5 to generate maximally a 4 bits DL HARQ ACK/NACK.

The ACK/NACK bundling feedback mode is the most aggressive mode to relieve the coverage problem of multiple UL ACK/NACK transmission in TDD. The allocated DL resources have been decoupled from the required UL feedback channel capability; i.e. only a single DL HARQ-ACK/NACK is transmitted in a single UL sub-frame regardless of the number of associated DL sub-frames carrying PDSCH for the user. The single ACK/NACK is created by performing a logical AND operation over all associated HARQ ACK/NACK per UL sub-frame. This way, TDD has same number of HARQ ACK/NACK feedback bits and thus transmission formats on PUCCH as FDD per UL sub-frame. The ACK/NACK encoding and transmission format in PUSCH is the same as it is in FDD.

Without proper compensation in the link adaptation and packet scheduling functions, the probability of DL HARQ NACK will increase causing more unnecessary DL retransmissions when using ACK/NACK bundling. Thus control channel reliability is the key with ACK/NACK bundling. The second mode is more attractive for the case when UE has sufficient UL coverage to support multiple ACK/NACK bits on PUCCH. When in ACK/NACK multiplexing feedback mode, the status of each DL sub-frame HARQ-ACK/NACK, i.e. ACK, NACK or DTX (no data received in the DL sub-frame), one of the QPSK constellation points in certain derived PUCCH channels is selected for transmission at UE side, and the eNodeB can decode the multi-bit HARQ ACK/NACK feedback by monitoring all constellation points from all associated PUCCH channels. The exact mapping table can be found in [4].

15.3.7 DL HARQ-ACK/NACK Transmission with SRI and/or CQI over PUCCH

When in ACK/NACK bundling mode, if both bundled HARQ-ACK/NACK and SRI are to be transmitted in the same UL sub-frame, the UE will transmit the bundled ACK/NACK on its derived/assigned PUCCH ACK/NACK resource for a negative SRI transmission or transmit the bundled HARQ-ACK/NACK on its assigned SRI PUCCH resource for a positive SRI transmission. This operation is exactly the same as for FDD.

When in ACK/NACK multiplexing mode, if both multiple HARQ-ACK/NACK and SRI are transmitted in the same UL sub-frame, the UE will transmit the multiple ACK/NACK bits as described in section 0 for a negative SRI transmission, and transmit 2-bit information mapped from multiple ACK/NACK input bits on its assigned SRI PUCCH resource for a positive SR transmission using PUCCH format 1b. The mapping between multiple ACK/NACK input bits and 2-bit information depends on the number of generated HARQ-ACK among the received DL sub-frame PDSCH within the associated DL sub-frames set K. The exact mapping table can be found in [4].

15.4   Semi-persistent Scheduling

Semi-persistent scheduling (SPS) can be used with all the TDD UL/DL configurations. Many details of SPS are the same for LTE FDD and TDD, but this section details some TDD-specific aspects related to SPS. The reader is referred to Chapter 13 for more information about SPS. To match the special frame structure of TDD, the SPS resource interval must be set to equal a multiple of the UL/DL allocation period (i.e. 10 ms) to avoid the conflict of non-matching UL/DL sub-frames because the UL sub-frame and DL sub-frame do not exist simultaneously. Furthermore, LTE UL uses synchronous HARQ, and there are some problems for most UL/DL configurations in TDD because the HARQ RTT is 10 ms. When UL SPS is used for VoIP traffic (AMR codec periodicity is 20 ms), the second retransmission of a previous packet will collide with the next SPS allocation, because the period of SPS resource is two times the RTT. The collision case is shown in Figure 15.13. In the figure the numbers 1, 2 and 3 indicates different VoIP packets. For example at the 20 ms point, the second retransmission of VoIP packet #1 collides with the initial VoIP packet #2. To solve this problem, two solutions are available to the network: dynamic scheduling and two-interval SPS patterns.

Although configured for SPS, the UE will anyway to dynamic allocations on the PDCCH. Such allocations will always override an existing persistent allocation. By using

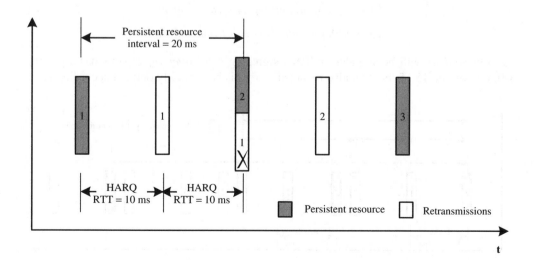

Figure 15.13   Collision between retransmissions and new transmission

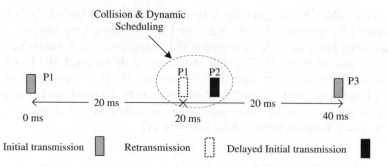

Figure 15.14 Dynamic scheduling at collision point

dynamic scheduling at the collision point it is possible to mitigate the problem as shown in Figure 15.14: If the UE is asked for a retransmission, the UE will perform a retransmission unless it has an empty buffer. With these definitions, if there are other following idle sub-frames available, the eNodeB will next schedule the retransmission in the current sub-frame, and re-schedule the initial transmission that was supposed to take place on the SPS resources in following sub-frames at the collision point.

The second solution is to use two-interval SPS pattern. Here, two-interval SPS pattern means that two periods are used for semi-persistent scheduling while only one semi-persistent scheduling period is used and allocation is persistent, based on the pre-defined period in the conventional scheme. With the two-interval SPS pattern, a resource pattern with two different intervals (T1, T2, T1, T2....) is used to avoid the main collision between the second retransmission of previous packet and the SPS allocation. The procedure is given in Figure 15.15, in which T1 is not equal to T2 and the combined set of T1 and T2 is always a multiple of 10 ms. The offset between T1 and T2 is several sub-frames, and a variable sub-frame offset is used to indicate the offset. The following formulas are used to calculate T1 and T2:

$$T1 = SPS\ periodicity + sub\text{-}frame\_offset;$$

$$T2 = SPS\ periodicity - sub\text{-}frame\_offset;$$

SPS periodicity will be signaled by RRC signaling *sub-frame_offset* (this has a positive value in Figure 15.15, but can also be negative). This is implicitly defined according to the

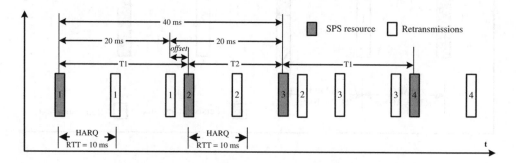

Figure 15.15 Two-interval SPS patterns

different TDD configurations and is the starting point of two-interval SPS pattern. Then T1 (the first time periodicity) and T2 (the second time periodicity) can be computed in terms of the above equations. The allocation period always starts with the first time period T1. However, even configured with two-interval SPS pattern, some residual collisions might still exist if the number of required retransmissions is large, i.e. 4. Any residual collisions can be avoided by means of dynamic scheduling as described earlier.

15.5 MIMO and Dedicated Reference Signals

LTE supports a number of different MIMO modes in DL, as described in Chapter 5, covering both closed loop schemes with UE feedback information to the Node B. Together with the information of actually adapted downlink parameters from eNodeB this sums up to significant amount of signaling to handle the DL closed loop MIMO. For TDD mode the earlier mentioned channel reciprocity can be exploited to mimic closed loop MIMO with reduced signaling overhead. Under the assumption that the channel is identical for UL and DL we can estimate the channel in UL by using sounding reference signals transmitted by the UE and then apply this channel knowledge for selecting the best DL pre-coding matrix. In this way UE feedback can be reduced or even avoided.

Going a step further, we can also eliminate the pre-coding matrix indication in DL allocation message by using *UE specific reference signals* (URS) to transfer this information. Moreover the use of URS decouples the physical transmit antenna from the UE detection complexity and system overhead resulting from having a cell-specific reference signal for each transmit antenna. URS are specified for LTE Release 8, and can be used for both FDD and TDD modes. However, this transmission mode is especially attractive when considered in a TDD setting, where channel reciprocity is available.

URS are transmitted on antenna port 5 and they are generated with same procedure as cell-specific reference signals. The only difference is that the UE RNTI affects the seed of the pseudo-random generator used to generate the code. The pattern for normal cyclic prefix and extended cyclic prefix can be seen in Figure 15.16(a) and Figure 15.16(b), respectively. There are 12 resource element used for URS per PRB per 1 ms sub-frame so the additional overhead is rather large. For the case with normal cyclic prefix, cell specific reference signals on antenna port 0 and 1, and a control channel region of three symbols antenna, the enabling URS will reduce the number of resources for data transmission by 10%. On the other hand this gives very robust performance at the cell edge or for high UE velocity.

One advantage of using URS is that pre-coding does not need to be quantified. As data and reference signals are using the same pre-coding matrix, the combined channel can be directly estimated from the UE-specific reference signals and then used for demodulating the signal.

Due to the rather large overhead of URS, this mode is mainly expected to be used when eNodeB deploys more than four antennas. The Release 8 specifications do not allow for more than four different common reference signal ports, so in this case the only way to do UE-specific pre-coding (beamforming) is to use URS.

How antennas are deployed and how pre-coding is determined when URS are used are vendor-specific issues where eNodeB implementations may differ. Two different scenarios will be discussed here. In the typical macro cell with antennas mounted above a roof-top, the azimuth spread of the channel would be low and the optimal solution would be to

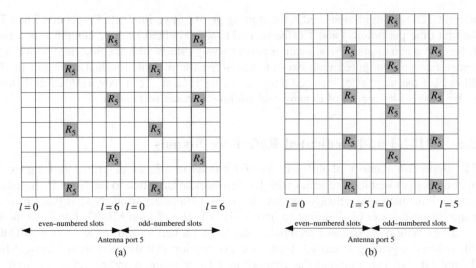

Figure 15.16   (a) URS Pattern for normal CP, and (b) URS Pattern for extended CP

have antennas with narrow spacing and use traditional angular beamforming to create a narrow beam that directs the power towards the UE while reducing the interference to other UEs. In this case, the system only relies on estimating the *direction of arrival* (DoA) from the UL transmission and this can be done without relying on channel reciprocity. Moreover, the standard allows for reuse of time and frequency resources by other users in the same sector. Sufficient angular separation should be insured to maintain performance.

In a scenario with many scatterers around the base station, the azimuth spread would be larger and angular beamforming might not work very well. Another solution could then be to rely on pre-coding determined from the eigen-vectors of the complex channel matrix determined from UL sounding. When the azimuth spread is increased, the rank of the channel could become larger than one and UE could potentially benefit from dual stream transmission. Dual stream transmission is not supported in Release 8 but has been included in Release 9, thus allowing MIMO to be used with the URS. For this purpose, transmission mode 8 is added to the physical layer LTE specifications. For more details on MIMO, beamforming and channel modeling see [6].

15.6   LTE TDD Performance

In this section the performance for LTE TDD is analyzed. For FDD mode, performance was already extensively analyzed in Chapter 10. As many of the observations provided there apply equally well to TDD mode, we try in the following to focus on the areas where TDD mode is different from FDD mode. First we look at the link performance – that is, how well the receivers in eNodeB and UE can decode the physical channels. In general this is an area where there is little difference between FDD and TDD as reference signal patterns and channel coding are duplexing mode agnostic. After the link performance, we discuss the link budget. The link budget for TDD is different from FDD because of the discontinuous transmission, so coverage for different bit rates in TDD will invariably be worse than for FDD. However there are still a number of details to pay attention to in order to evaluate the TDD link budget properly.

This section ends by discussing system performance. First we look at the best effort type of service where a fairly small number of users per sector are assumed to download large amounts of data. Assuming that we are designing a network to deliver a certain bit rate for certain number of active users, then the most important difference between networks based on FDD and TDD technologies is that UL and DL in TDD mode would need double system bandwidth compared to FDD. Data transmission would also need to be done with larger bandwidth in TDD system in order to achieve similar bit rates as FDD. Both system and transmission bandwidth affect the operation of RRM algorithms and in this the effect on system performance is analyzed.

Performance is also evaluated assuming VoIP service, and this gives a somewhat different perspective on system performance as bit rates are low and there are many users when we load the system with 100% VoIP traffic. As VoIP is a symmetric service, TDD systems can have a higher VoIP capacity than FDD because the split between UL and DL resources can be adjusted. The increased HARQ round trip time for TDD system is also shown to have some effect to the VoIP performance in cases where many UEs are coverage limited.

15.6.1 Link Performance

The two most important factors affecting the link performance for data transmission are channel estimation and channel coding. As reference signal design and channel coding are very similar for TDD and FDD, the link performance is also very similar. One source of difference is the discontinuous transmission in TDD. FDD receivers can utilize the reference signals from the previous sub-frame to estimate the channel. This is especially important for the DL link performance where the UE receiver should start decoding the control channel region as soon as it is received. In a TDD system, when a UE receiver decodes sub-frames transmitted right after UL→DL switching, it cannot rely on reference signals from previous sub-frames and this could degrade the channel estimation and thus coverage of control channels. The importance of such potential loss will depend on the UE implementation. If for example the UE could wait for the second column of reference signals then the performance degradation could be reduced.

Another potential difference in link performance between FDD and TDD is related to the special sub-frame. As explained earlier the guard period is created by adjusting the number of symbols in DwPTS. When the DwPTS length is reduced we also eliminate some reference signals as the rule is not to move them to new locations. The potential loss is minor as reference signals from the previous sub-frame could be taken into use to improve channel estimation. In the special case of UE-specific reference signals we lose one column of reference signals even with full DwPTS length. And in this case the UE cannot use reference signals from the previous sub-frame as these could have been transmitted with a different pre-coding.

Short RACH performance is clearly worse compared to 1 ms RACH preamble and thus should be used only in environments where the link budget is not seen to be an issue.

15.6.2 Link Budget and Coverage for the TDD System

The link budget calculation aims to estimate the range of different bit rates. A detailed description of how to calculate link budgets for LTE is given in Chapter 10. Here we focus on the differences between link budgets for TDD and FDD modes. The differences relate mainly to the limited maximum UE transmit power and in the following we therefore focus our attention on UL link budgets.

The TDD UE cannot transmit continuously because the transmission must be switched off during the downlink reception. The UE will thus need to transmit with larger bandwidth and lower power density to achieve similar bit rate as in a FDD system. The lower power density is due to fact that the UE transmitter is limited on total maximum power, not on power per Hz.

As a simple example, if the downlink:uplink share is, for example, 3:2, the UE transmission power density is reduced by $10 * \log 10(2/5) = -4$ dB as we need roughly 5/2 times the bandwidth for the TDD UL transmission. Another way of viewing this is that at a fixed distance from the base station the maximum achievable FDD bit rate will roughly be $2\frac{1}{2}$ times larger than the bit rate achieved with maximum UE transmit power in a TDD system. Remark that for DL, the power density can be assumed similar between FDD and TDD mode as the size of power amplifier in eNodeB can be adapted to the system bandwidth.

So from an UL coverage perspective FDD-based systems do have an advantage over TDD system due to the continuous transmission. Providing coverage with a TDD system can also be challenging because the TDD spectrum is typically situated at higher frequencies such as 2.3 GHz or 2.5 GHz. A cell range comparison for sub-urban propagation environment is shown in Figure 15.17. The best coverage is obtained by using FDD system at low frequency. The cell range for LTE900 FDD is four times larger (the cell area is 16 times larger) and LTE2500 FDD is 80% larger than LTE2500 TDD. The assumed data rate is 64 kbps and the cell range is calculated with Okumura Hata propagation model with 18 dB indoor penetration loss, 50 m base station antenna height and − 5 dB correction factor. For maximum LTE coverage, LTE TDD deployment at high frequency could be combined with LTE FDD deployment at a lower frequency.

15.6.2.1 MCS Selection and UE Transmission Bandwidth for Coverage

For a certain target bit rate different combinations of MCS and transmission bandwidth have different coverage and different spectral efficiency. From Shannon's information

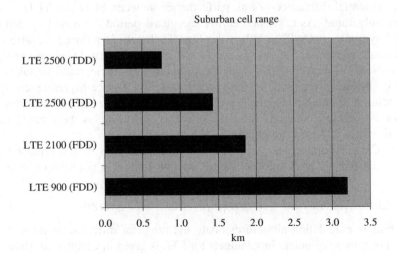

Figure 15.17 Uplink cell range for LTE FDD and TDD systems for 64 kbps

theory [7] we know that if we want to maximize coverage under fixed total transmission power constraint we should increase MCS when physical layer spectral efficiency (SE) is <1 bit/s/Hz and increase bandwidth when SE is >1 b/s/Hz. As adjusting BW does not impact SE, but increasing MCS does the link should be operated with an MCS that achieves a SE of 1 bit/s//Hz (for LTE QPSK 2/3 would do the job). Then bandwidth can be adjusted to achieve the required bit rate with optimal coverage.

In Figure 15.18 an example of this process is given. A UE at the cell edge selects MCS QPSK $\frac{3}{4}$ and sets transmission bandwidth according to the available transmit power and required SINR. Curves for both TDD and FDD UE is shown, assumptions are as given in [8]. From the figure we can see that when the UE moves towards the Node B, path loss is reduced and the link gain can be used to increase the UE transmission bandwidth. As the transmission bandwidth increases the UE bit rate also increases. At some point the UE target bit rate is reached, and if the UE path loss is reduced further we can start to increase MCS and reduce bandwidth to improve the SE while keeping the bit rate on target. At some point the maximum MCS is reached and only at this point can we start to reduce the UE total transmit power. While TDD system supports 2 Mbps in UL 300 m from the eNodeB, FDD system increases the coverage for 2 Mbps to 400 m. For more details on the interaction of UE transmit power, MCS and transmission bandwidth see Chapter 10.

15.6.2.2 Coverage for Low Bit Rates

When the data bit rate decreases, the relative overhead from header and CRC will increase. The effect of this is that even in FDD mode it does not make sense to schedule UEs in

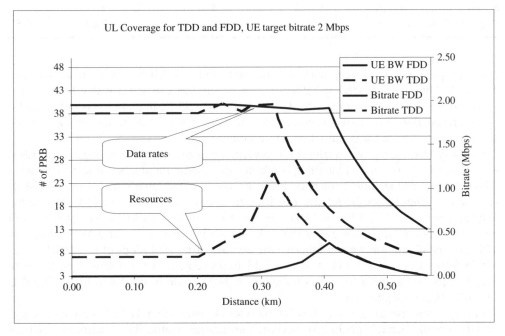

Figure 15.18 Coverage and required number of physical resource blocks for a 2 Mbps target bit rate for FDD and TDD UL (MCS in coverage limited region is QPSK $\frac{3}{4}$)

Table 15.3 Required UE transmission bandwidth to support 64 kbps for TDD and FDD

System	TDD UL with 3DL/2UL	FDD UL
Service bit rate [kbps]	64	64
MCS	QPSK 1/10	QPSK 1/10
Data bits per TTI [bits]	160	64
Header (3 byte) CRC (3 byte) [bits]	48	48
Total bits per TTI [bits]	208	112
Payload per PRB in physical layer. (2 Symbols for DM RS 1 Symbols for SRS)	27 bits	27 bits
Required # of PRB	8 PRB	5 PRB

Table 15.4 Required SINR for VoIP in TDD and FDD with and without TTI bundling. UE transmission bandwidth assumed to be 360 kHz (2 PRB)

System	TDD (3DL/2UL or 2DL/3UL)		FDD	
Bundling enabled	No	Yes	No	Yes
Number of transmissions	5	8	7	14
Required SINR	−3.3 dB	−5.3 dB	−4.7 dB	−7.7 dB

coverage problems with very narrow bandwidth and low MCS. An example for data bit rate 64 kbps is illustrated in Table 15.3. From this we can see that due to excessive overhead, UE bandwidth increase for UL in TDD 3DL/2UL configuration is only a factor 1.6, corresponding to a transmit power loss of 2 dB, not 4 dB as in the example given above where overhead was not taken into account.

One way of reducing the overhead for low bitrates is to use TTI bundling, as described in section 15.3.4. When TTI bundling is enabled the maximum number of retransmissions within a certain time limit is increased. In that way we can have more aggressive MCS selection and thus lower the relative overhead from protocol header and CRC. Due to the reduced number of UL TTIs in TDD, the potential link budget improvement from TTI bundling is not as important as for FDD. As shown in Table 15.4, for a VoIP service in 3DL/2UL configuration we can improve the number of transmissions of one VoIP packet within 50 ms from 5 to 8 TTIs that is a 2 dB improvement to the link budget. We note that the link budget gain from TTI bundling in TDD mode is similar in both 2DL/3UL and 3DL/2UL configuration.

Finally when operating at very low UL bit rates far away from the base station coverage on UL control channels could also become the limiting factor. The control channel for TDD has been designed so that if ack/nack bundling mode is selected then the required receiver sensitivity is similar between TDD and FDD.

15.6.3   System Level Performance

TDD mode is in many aspects similar to FDD and this is also valid when we analyze the systems from a performance point of view. In general, when we compare time and frequency duplexing there are some differences related to the different ways of using the spectrum, which makes it difficult to make a completely fair comparison of spectral efficiencies. Whereas a TDD mode system needs a guard period between UL and DL, a FDD system needs a large separation in frequency. Secondly if TDD systems have the same partition of UL and DL resources they can operate in adjacent bands, if not they need to be separated in similar way as UL and DL for FDD.

Another spectrum-related issue is that for a TDD system to provide a similar capacity to a FDD system the DL and UL system bandwidth needs to be double that of a FDD system. This impacts the radio resource management and users typically will need to operate with larger transmission bandwidths. That can be challenging for UL data transmission due to the limited transmission power of the UE.

One advantage for TDD RRM solution is the possibility to exploit channel reciprocity. In current RRM frame work (see Chapter 8) two parallel mechanism are available for obtaining channel state information. For DL, UE can be configured to feed back CQI, PMI and RI reports based on measurements of DL reference signals. In UL the UE can transmit sounding reference signals so that Node B can measure the radio channel. For TDD mode, in case channel reciprocity is present we ideally need only one of these mechanisms as DL channel state can be inferred from a UL channel state or vice versa. As earlier mentioned there are challenges before this could work in a practical RRM solution, such as differences in UL/DL interference levels, different UL/DL antenna configuration and lack of UL/DL radio chain calibration but, on the other hand, gains could be important. UL sounding for example can take up more than 10% of the UL system capacity. See section 15.2.4 for further discussion of channel reciprocity.

15.6.3.1   Round Trip Time for TDD Systems

Feedback control loops are used for quite a few purposes in the LTE system. As TDD systems do not have continuous transmission and reception we might expect that the round trip time for such control loops would be increased for TDD systems, potentially degrading the system performance. However, due to the need for processing time – that is, time for the UE or the Node B to decode and encode the control information – the typical round trip times between TDD and FDD are quite similar. Since the resulting delays are in the order of 10 ms for most typical cases and thus in comparison to the FDD RTT of 8 ms the impact from TDD frame structure is rather low and not likely to impact the performance of TCP/IP regardless if being run over LTE TDD or FDD.

15.6.3.2   Scheduling

One of the key LTE RRM features is the channel-aware scheduling, which is available for both UL and DL. Under the right conditions this feature can bring gains in spectral efficiency of up to 50% with even more important improvements to the coverage. To maximize the scheduling gain it is important to have frequency-selective channel knowledge and flexibility in the control signaling to allocate UEs to the optimal frequency resources.

Obtaining detailed frequency-selective channel state information and enabling flexible resource allocation in frequency domain is very costly in terms of control signaling.

For DL a number of different frequency resource allocation schemes are specified in the standard. For channel-aware scheduling the most effective allocation scheme specified gives a bit mask where each bit corresponds to a sub-band consisting of a number of continuous PRBs. To obtain the channel state for different sub-bands, the CQI report contains a similar bit mask indicating the best M sub-bands and one indication of supported MCS on these sub-bands. The number of PRBs in one sub-band and the value of M are fixed in the specification. As the size of the bit mask increases with the system bandwidth, it was decided to increase sub-band size as the system bandwidth increases. If we compare a 10 MHz FDD DL and a 20 MHz TDD DL the scheduling gain is most likely to have some loss for 20 MHz especially for channels with narrow frequency correlation bandwidth. The increase in system bandwidth could on the other hand give a TDD system an advantage due to the increased frequency diversity.

For UL due to the single carrier constraint, frequency resources have to be allocated contiguously; so here there is no difference for the resource allocation signaling between lower and higher system bandwidth. For obtaining channel state information we can enable sounding reference signal transmission from the UE. Sounding takes up resources and in contrary to the situation for downlink the larger bandwidth needs to be sounded the more resources are needed. On the other hand it is not needed to know the channel state information over full bandwidth to get channel aware scheduling gain.

The resource allocation signaling consumes a lot of DL control channel resources and as these are limited the typical number of users that can be scheduled in one TTI is 10 to 20 users (5–10 UL users and 5–10 DL users). As we assume that system bandwidth in TDD is double that of FDD the TDD UEs need to operate at double bandwidth in order to have full bandwidth utilization in the system – see Figure 15.19 The increase in UE bandwidth can impact especially UL channel-aware scheduling gain. As the frequency resources are forced to be allocated contiguously it is difficult to allocate resources according to the channel quality in the frequency domain.

The total amount of DL control channel resources varies with the UL/DL ratio as up to three symbols can be reserved in each DL sub-frame (two for DwPTS). In the

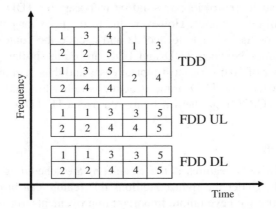

Figure 15.19   Illustration of TDD and FDD resource allocations in LTE (due to limitations on the number of users scheduled per TTI, TDD mode UEs are forced to transmit at higher bandwidth)

case of 2DL/3UL TDD potentially have fewer resources than in FDD while for other configurations there are more. The common case of 3DL/2UL has slightly more DL control channel resources than its FDD counterpart (eight symbols in 20 MHz for TDD compared to 15 symbols in 10 MHz for FDD) so here there would not be any important difference between FDD and TDD. It should be mentioned that the TDD allocation messages (DCI) have a slightly higher payload. Both due to higher system bandwidth but also due to additional bits for special features such as multi TTI and ack/nack bundling.

15.6.3.3 Uplink Power Control

The uplink power control is another very important RRM function as it has high impact on the distribution of bit rates in the sector. Moreover, fine tuning power control parameters can improve both cell spectral efficiency and coverage. As we have discussed earlier, the typical TDD UE will need to transmit at higher bandwidth compared to an FDD UE. This puts extra pressure on uplink transmit power resources, which are already limited, and can lead to poorer coverage for TDD systems. However in the case where interference limits the system performance, the performance difference is expected to be minor. In Figure 15.20 an uplink spectral efficiency comparison between a 20 MHz TDD system and a 2 × 10 MHz FDD system is shown where the TDD UEs transmit with a bandwidth of 12 PRB and FDD PRB with 6 PRB. From this figure we can see that both cell and coverage performance is very similar but slightly better for the FDD system.

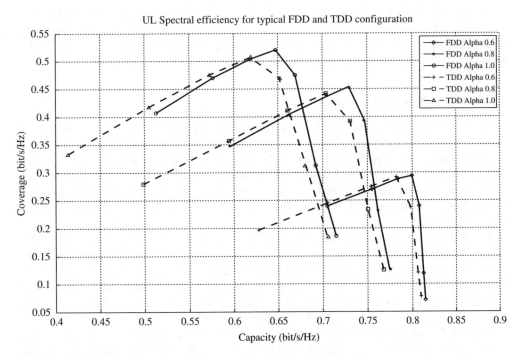

Figure 15.20 Uplink spectral efficiency for TDD and FDD system with different power control settings – see [8] for assumptions

This difference is caused by reduced channel-aware scheduling gain due to the higher bandwidth. We note that the optimal uplink interference level in the system depends on the bandwidth. For 12 PRB UE bandwidth is 3 dB lower than for 6 PRB, independent of the value of alpha. When site density decreases, the coverage performance of TDD system will decrease faster than for FDD system and we will eventually see the effect from the differences in the link budget.

15.6.3.4 Downlink HARQ Ack/Nack Bundling

Finally we mention the effect on performance of the downlink HARQ ack/nack bundling feature described in section 15.3.6. As only one ack/nack is available for multiple sub-frames, and these sub frames are from different HARQ processes at a different number of retransmissions, then there is a risk that excessive retransmission will occur. In particular, if we consider a very downlink heavy TDD configurations with aggressive MCS selection then the resulting high retransmission probability will drag down the downlink system performance.

This effect can be offset by operating HARQ with relatively low BLER for the first transmission, this on the other hand means that downlink HARQ will operate at a sub-optimal operating point. Another way to reduce excessive retransmission from ack/nack bundling is to schedule users less often but with higher bandwidth. This, on the other hand, could reduce channel-aware scheduling gain. In the most common TDD configurations with three downlink and two uplink sub-frames the impact from ACK/NACK bundling will be minor as maximum 2 ACK/NACKs will be bundled.

Simulations show that the loss in the case of four downlink/one uplink is less than 10%. However, for users in coverage problems, which rely on HARQ for the link budget, the excessive retransmission could give more important performance problems as one NACK typically will trigger retransmission of multiple TTIs.

15.6.3.5 Performance of VoIP

Voice over Internet protocol (VoIP) traffic should be supported as efficiently as possible in LTE TDD mode. However, supporting VoIP in LTE TDD faces very similar challenges as FDD mode: (1) tight delay requirements with much longer HARQ RTT; (2) various control channel restriction for different uplink/downlink configurations; and (3) even more serious coverage problem in uplink due to discontinuous transmission. Moreover, as there are multiple uplink/downlink configurations possible, this can be used to achieve higher total VoIP capacity.

The following part presents the system level performance of VoIP traffic in LTE TDD mode at 5 MHz system bandwidth. The capacity evaluation criterion is same as FDD: defined as the maximum number of per sector VoIP users that can be supported without exceeding a 5% outage level. VoIP capacity numbers are obtained from system-level simulations in Macro cell scenario 1 and Macro cell scenario 3 [9], the main system simulation parameters being aligned with [10].

Like FDD in Chapter 13, the simulation results of VoIP capacity are summarized in Table 15.5 and 15.6 for two different AMR codecs: AMR 7.95 and AMR 12.2 and for both dynamic and semi-persistent schedulers and for both Macro Case 1 and Macro Case 3 as defined in [8].

Table 15.5 VoIP capacity in LTE TDD @5 MHz for Macro Case 1

Downlink Capacity	Configuration 0 (2DL/3UL)		Configuration 1 (3DL/2UL)	
VoIP codec	AMR12.2	AMR7.95	AMR12.2	AMR7.95
Dynamic scheduler, without packet bundling	64	64	112	112
Dynamic scheduler, with packet bundling	114	122	194	206
Semi-persistent scheduler	102	140	168	220
Uplink Capacity				
VoIP codec	AMR12.2	AMR7.95	AMR12.2	AMR7.95
Dynamic scheduler	88	92	88	114
Semi-persistent scheduler	134	174	86	110

Table 15.6 VoIP capacity in LTE TDD @5 MHz for Macro Case 3

Downlink Capacity	Configuration 0 (2DL/3UL)		Configuration 1 (3DL/2UL)	
VoIP codec	AMR12.2	AMR7.95	AMR12.2	AMR7.95
Dynamic scheduler, without packet bundling	54	54	94	94
Dynamic scheduler, with packet bundling	90	94	158	168
Semi-persistent scheduler	84	114	144	194
Uplink Capacity				
VoIP codec	AMR12.2		AMR 12.2	
Dynamic scheduler, without TTI bundling	<40 (50 ms delay bound)		<40 (50 ms delay bound)	
Dynamic scheduler, with TTI bundling	<40 (50 ms delay budget) 50 (60 ms delay budget) 72 (70 ms delay budget)		<40 (50 ms delay budget) 50 (60 ms delay budget) 54 (70 ms delay budget)	

For both UL and DL simulations, the number of control symbols used in a special sub-frame is set to two, which contains six control channel elements (CCEs). In normal DL sub-frame, there are three control symbols available, which provide 10 CCEs for UL and DL control.

Regarding to the DL VoIP capacity for LTE TDD: performance of fully dynamic PS is seriously control channel limited if packet bundling is not allowed for both simulation scenarios. However with packet bundling the users having good CQI are scheduled less often and hence the capacity can be boosted up to $70 \sim 90\%$ because more control channel resources are released to be used by other users. It is no difference for packet bundling concept between FDD and TDD. The reader is referred to Chapter 13 for more information about packet bundling. And semi-persistent PS performs very well in control-channel limited situations, always outperforming dynamic PS if packet bundling is not allowed. Meanwhile, fully dynamic PS outperforms semi-persistent PS only if packet bundling is allowed and the VoIP packet payload is high enough (for example, AMR12.2 or higher) so that control channel limitations can be avoided. The performance difference between scenario1 and scenario3 is up to $15 \sim 20\%$, which is not so large because DL transmission power is not so limited even in scenario 3.

Regarding the UL VoIP capacity for LTE TDD: configuration 0 is viewed as a serious control-channel limited case, so the performance of dynamic scheduling is quite bad due to the limited number of control channels available. Performance loss reaches 40–80% compared to semi-persistent scheduling for different AMR Codec. In contrast, configuration 1 is a loose control-channel case, so the performance of dynamic scheduling is a bit better than semi-persistent scheduling due to flexible retransmission and frequency domain packet scheduling (FDPS) gain. The performance difference between macro case 1 and macro case 3 is huge – over 50% loss. The loss is very serious because macro case 3 is a very coverage-limited scenario. The UE transmit power is not large enough to provide even a proper link budget. For macro case 3, TTI bundling and longer delay bound can provide great gain on the performance. For example, by using TTI bundling in 70 ms delay budget for TDD configuration0, up to 3.0 dB energy accumulation gain can be achieved. However without TTI bundling, it is hard to achieve good VoIP capacity (i.e. < 40 users/sector) because of the deficient energy accumulation.

In order to do a fair comparison between FDD and TDD, the FDD capacity results should be scaled in terms of different TDD configurations and scheduling options. In DL, the used scaling factor for dynamic scheduler should depend on the control symbol assumptions because it is control overhead limited performance for dynamic scheduling. With two control symbols, a special sub-frame only has six CCEs available. Hence FDD results for dynamic scheduler should be scaled with the scaling factor $26/50 = 0.52$ for configuration1 and $16/50 = 0.32$ for configuration0 in DL. However, in DL the used scaling factor for SPS should depend on the number of data symbol assumptions because it is data limited performance for SPS. With two control symbols, the special sub-frame only has seven data symbols available. Hence, FDD results for SPS should be scaled with the scaling factor $27/50 = 0.54$ for configuration1 and $17/50 = 0.34$ for configuration0.

In the UL direction the special sub-frame has no data symbols, but only UL control signaling, so we just use the scaling factors 0.6 and 0.4 for configuration 0 and configuration 1 in terms of the UL data symbol ratio.

Compared with the scaled FDD results in Chapter 13, we make the following observations: In downlink macro case 1, TDD results for both configurations are at most 5% lower than the scaled FDD results. For dynamic scheduling in coverage-limited macro case 3, the average amount of transmissions per packet is increased. As scheduling flexibility in time domain is decreased for TDD and HARQ RTT is increased, the scaled TDD performance are about 10% lower than the scaled FDD results – losses are naturally higher for configuration 0 due to small number of DL sub-frames.

In UL TDD configuration 0 of macro case 1, FDD has about $7 \sim 10\%$ gain over TDD for semi-persistent scheduling due to more diversity gain in FDD moreover FDD has up to 40% gain over TDD for dynamic scheduling due to the limited control channel

Table 15.7 Impact from energy accumulation with TTI bundling

	Configuration 0 (2DL/3UL)	Configuration 1 (3DL/2UL)	FDD
No bundling	5	5	7
With bundling	8	8 (for most of packets)	12

resources in TDD. For uplink TDD Configuration1 of macro case 1, TDD has 5~20% gain for AMR 12.2 and AMR 7.95 respectively over FDD for dynamic scheduling due to loose control channel limitation. FDD has ~10% gain for AMR 12.2 and AMR 7.95 respectively over TDD for semi-persistent scheduling due to more diversity.

For macro case 3 with TTI bundling and with a slightly longer packet delay budget (70 ms), the difference between FDD and TDD reaches 15~25% for different TDD configurations due to the less energy accumulation in TDD mode. However with the shorter delay budget (50 ms), TTI bundling is not able to provide visible capacity improvement although 2 dB energy accumulation gain is achieved compared to no bundling. Table 15.7 shows the energy accumulation with 50 ms packet delay budget. With TTI bundling, 1.76 dB energy accumulation gain can be achieved in FDD compared to the value in TDD. So the improvement for VoIP capacity is obvious.

15.7 Evolution of LTE TDD

As mentioned earlier, TDD provides a very efficient way of enabling large peak data rates in unpaired spectrum conditions or when paired spectrum with significant bandwidth or duplexing distance can be found. As LTE is progressing towards LTE Advanced, introduced in Chapter 16, local area deployment and the outlook for reaching 1 GBit/s data rates will become a key driver in further developing the TDD mode. In the path towards unleashing the transmission over very wide bandwidth (for example, multiple component carriers) TDD poses some interesting challenges in terms of scalability and, on the other hand, ensuring backwards compatibility for smooth evolution.

Another important aspect related to TDD is that of co-existence. Based on Release 8 it seems that synchronization is key to control interference and provide reliable system operation in a network with full coverage for UE. However, for indoor or low-cost deployments the existing techniques for synchronization may become too expensive and hence techniques for more lean synchronization, such as *over the air* (OTA) synchronization, could be of interest.

An aspect related to TDD is that of providing dynamic switching of the uplink/downlink configuration depending on the required capacity in the network. It is yet unknown if the 3GPP LTE TDD standard will in the future provide a direct means for multiple operators to co-exist within the same geographical area or if such co-existence will be guaranteed mainly by regulation. A further step in this direction is to look at protocols for slowly or semi-statically modifying the TDD configuration according to the varying load over time. Either within a coverage area covered by a single operator or in an area with multiple operators in which case semi-static TDD, adaptation would need to be considered together with schemes for flexible spectrum usage and sharing. Finally, it is yet to be seen if fast dynamic TDD to capture an instantaneous load in a single cell will becomes possible in practice.

Given experience with first deployments, it will be interesting to follow the extent to which reciprocity can be assumed to be available taking the practical antenna aspects into account. There are several ways in which the LTE TDD standard could be further optimized towards more efficient signaling (and thus performance given fixed signaling overhead budget) in the case of reciprocity including topics coming as part of the LTE-Advanced work in upcoming 3GPP releases.

15.8   LTE TDD Summary

The LTE TDD operation is very similar to LTE FDD, including exactly the same multiple access solutions, channel encoding and so forth. From the 3GPP specifications perspective the close similarity between FDD and TDD is also demonstrated via a single set of specifications for the physical layer as well. The resulting differences are due to the uplink/downlink split inherent in the TDD operation, which cannot really be avoided as, in FDD, one does not need to change transmission direction but frames are continuous. Still, as resource allocation was with 1 ms resolution both in FDD and TDD, the slots, excluding the ones when transmission direction changes, have the same parameters and timing. The sharing of the same spectrum for uplink and downlink gives flexibility both in terms of data asymmetry and makes finding new spectrum allocations easier for IMT-Advanced needs to reach the 1 GBit/s target. TDD operation has two consequences: (a) base stations need to be synchronized to avoid uplink–downlink interference, and (b) TDD cell range is shorter than FDD due to discontinuous transmission. The LTE TDD mode (or TD-LTE) deployment has been designed to accommodate co-existence with TD-SCDMA by ensuring compatible parameterization in terms of an uplink/downlink split when sharing the same sites and antennas or for other close proximity installation. The spectrum availability for the TDD operation varies on a country basis, but the spectrum has been allocated in China and an additional TDD spectrum (on top of the older allocations like 1900 to 1920 MHz) is being allocated from 2.5 GHz in several European countries (with first licensing activities already completed). In the 3.5 GHz spectrum there have been recent allocations in certain countries with some unpaired parts as, in some countries, there are existing deployments on that band for fixed mobile broadband use. The new countries with expected TDD market potential are now India, with licensing completed in 2010 and also the US. For the US market, 3GPP is developing, in Release 10, the spectrum variant to cover the band currently being used for WIMAX in the 2.6 GHz spectrum [11]. Development of LTE TDD mode has progressed together with the LTE FDD mode, towards the LTE-Advanced in order to meet the ITU-R IMT-Advanced requirements as presented in Chapter 16.

References

[1] 3GPP, Technical Specification, TS 36.211, v. V8.4.0, September 2008.
[2] 3GPP, Technical Specification, TS 36.212, v. 8.4.0, September 2008.
[3] 3GPP, Technical Specification, TS 36.213, v. 8.4.0, September 2008.
[4] 3GPP Technical Specification, TS 36.214, v. 8.4.0, September 2008.
[5] Holma, H., Toskala, A., 'WCDMA for UMTS', 5th edition, Wiley, 2010.
[6] C. Oestges, B. Clerckx, 'MIMO Wireless Communications', 1st edition, Academic Press, 2007.
[7] Proakis, J.G, 'Digital Communications', 3rd edition, McGraw-Hill, 1995.
[8] CATT, 'Enhanced Beamforming Technique for LTE-A', 3GPP T-doc R1-082972.

[9] 3GPP Technical Report, TR 25.814, Physical Layer Aspects for Evolved Universal Terrestrial Radio Access (UTRA), 3GPP TSG RAN, September 2006.

[10] 'Next Generation Mobile Networks (NGMN) Radio Access Performance Evaluation Methodology', A White Paper by the NGMN Alliance, January 2008.

[11] 3GPP Tdoc RP-100 374, 'Work item proposal for 2600MHz LTE TDD frequency band for the US (Region 2)', Clearwire *et al*, 3GPP RAN#47, Vienna, Austria, 16–19 March 2010.

16

LTE-Advanced

Mieszko Chmiel, Mihai Enescu, Harri Holma, Tommi Koivisto, Jari Lindholm, Timo Lunttila, Klaus Pedersen, Peter Skov, Timo Roman, Antti Toskala and Yuyu Yan

16.1   Introduction

The development of LTE radio access did not stop with the first LTE Release, Release 8. As Chapter 2 indicated, Release 9 introduced important enhancements, such as position location technologies, in different areas to cover for the requirements for emergency call. With the Release 10 specifications, available from the end of 2010, 3GPP is meeting the requirements for the ITU-R IMT-Advanced process. In most cases Release 8 LTE was already fulfilling the requirements set for, in particular, the high-mobility operation of LTE-Advanced. The Release 10 version of the specification contains additions, like support for bandwidths wider than 20 MHz, in order to meet the requirements for the low-mobility operation of IMT-Advanced. This chapter presents the ITU-R IMT-Advanced process and related requirements and looks at the results obtained from the 3GPP study phase. It then presents the different specified technology components and takes a more detailed look at the key aspects of carrier aggregation, uplink and downlink spatial multiplexing enhancements, and relay operation. It concludes by briefly examining the solutions being investigated with regard to LTE Release 11.

16.2   LTE-Advanced and IMT-Advanced

3GPP is working towards LTE-Advanced within the Release 10 timeframe mentioned earlier. Further enhancements with detailed technology components will then be made in the Release 11 timeframe. It is worth noting that, whereas LTE-Advanced is the most visible radio topic in Release 10, other important topics are being worked on to be part of Release 10 LTE specifications, without a direct relationship to any of the IMT-Advanced or LTE-Advanced requirements.

From a timing perspective, the first full set of Release 10 3GPP LTE-Advanced specifications (which is embedded in the 3GPP Release LTE 10 radio specifications) was

LTE for UMTS: Evolution to LTE-Advanced, Second Edition. Edited by Harri Holma and Antti Toskala.
© 2011 John Wiley & Sons, Ltd. Published 2011 by John Wiley & Sons, Ltd.

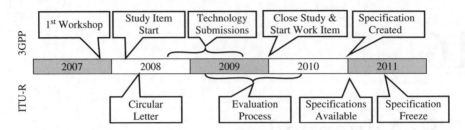

Figure 16.1   LTE-Advanced and IMT-Advanced creation timeline in 3GPP and ITU-R

available from December 2010. This allowed the ITU-R submission deadline for the IMT-Advanced process to be met. It is to be finalized during early 2011, as indicated in Figure 16.1. The process will enable later updates of the Release 10 specifications for the ITU-R to be provided in 2011 to cover for the expected corrections as part of the Release 10 specification-freezing process.

The IMT-Advanced process in ITU-R has called for proposals for radio technologies that could meet the needs of an IMT-Advanced technology, often also called the fourth-generation (4G) mobile communication system. The evaluation phase is carried out by registered evaluation groups, with 3GPP being one of the evaluation groups. The IMT-Advanced radio technology could be summarized as being a versatile wideband radio technology with support for up to 1 Gbps peak data rates.

The ITU-R defined the features of the IMT-Advanced in more detail [1] to include:

- 100 Mbps support for high mobility and up to 1 Gbps for low mobility cases;
- interworking with other technologies;
- enabling high-quality mobile services, interworking with other radio technologies and enabling devices for worldwide use.

There are further specific requirements from 3GPP for LTE-Advanced technology. Performance requirements [2] defined in 3GPP are higher in some areas than ITU-R requirements. This is partly due to the fact that Release 8 LTE is already a system based on modern design and can reach most of the targets set by ITU-R, especially for high mobility environment. The key requirements are addressed in the following section.

16.3   Requirements

The key performance requirements are summarized in Table 16.1. In addition to the system performance-related requirements in Table 16.1, there are requirements related to capability support and also related to deployment.

The velocity supported is up to 750 Hz Doppler frequency. The corresponding mobile velocity depends on the propagation frequency. A Doppler frequency at 2 GHz corresponds to 400 kph and even higher speeds at lower LTE frequencies.

16.3.1 Backwards Compatibility

Beside the natural requirement to enable handovers between earlier LTE Releases, LTE devices based on Release 8 (and naturally Release 9) are also required to operate in a

Table 16.1 Comparisons of the 3GPP LTE-Advanced and ITU-R IMT-Advanced requirements

System Performance Requirements	3GPP Requirement	ITU-R Requirement
Downlink peak spectrum efficiency	30 bits/s/Hz (max 8 antennas)	15 bits/s/Hz (max 4 antennas)
Uplink peak spectrum efficiency	15 bits/s/Hz (max 4 Tx antennas)	6.75 bits/s/Hz (max 2 TX antennas)
Uplink cell edge user spectral efficiency	0.04–0.07 bits/s/Hz	0.03 bits/s/Hz
Downlink cell edge user spectral efficiency	0.07–0.12 bits/s/Hz	0.06 bits/s/Hz
User plane latency	10 ms	10 ms

Release 10-based LTE advanced network. Naturally the performance of those devices and maximum data rates are not affected by LTE-Advanced developments but the devices should still get the necessary service. This sets requirements affecting what can be modified in terms of solutions requiring changes to common channels or, for example, the reference symbol used for mobility measurements and channel estimation. Non-backwards-compatible technologies can be considered as long as there is a significant gain and they do not prevent earlier release terminal access.

16.4   3GPP LTE-Advanced Study Phase

The 3GPP study phase on LTE-Advanced compared the performance of the proposed improvements related to 3GPP and ITU-R requirements (the former normally being the more strict ones). The study's results, from different companies, showed clearly that performance requirements can be exceeded with the technologies under consideration. An example result is presented in Figure 16.2, based on [3], studying the performance of the 4 × 4 multiuser MIMO in the downlink direction. As can be seen from the results in Figure 16.2, the 3GPP requirement for the four-antenna case of 3.7 bits/Hz/cell can be met with the MIMO methods under investigation. Following the study phase, 3GPP then started actual specification work with the first full set of specifications being available at the end of 2010. Further example results can be found in [3] for both FDD and TDD cases as well as findings on other areas of the LTE-Advanced studies reported in [4]. It is worth noting that the results in [3] are based on assumptions about the details at the time of the study phase while the results in later sections of this chapter reflect the agreed details for the actual specifications.

16.5   Carrier Aggregation

The key part of LTE-Advanced is the Carrier Aggregation (CA). The principle with CA is to extend the maximum bandwidth in the uplink or downlink (or both) directions by aggregating multiple carriers. The carriers to be aggregated are basically Release 8 carriers, which then facilitate the necessary backwards compatibility. A pre-Release 10 UE can

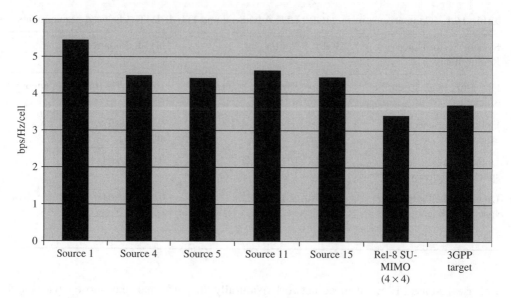

Figure 16.2 3GPP study phase performance results for the 4×4 downlink multi-user MIMO operation compared to Release 8 baseline and 3GPP requirements

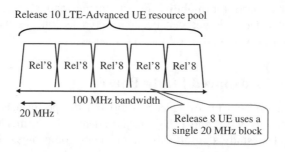

Figure 16.3 Mixing LTE-Advanced and legacy devices

access one of the component carriers while CA-capable UEs will then operate on multiple component carriers as shown in Figure 16.3. Component carriers (CCs) other than the one intended for Release 8 devices could contain performance-enhancement methods not compatible with the pre-Release 10 UE. However the methods agreed so far for Release 10 are such that Release 8 and 9 UEs can also access the carrier even if they do not benefit from the added improvements – for example, from the multi-antenna operation described in sections 16.6 and 16.7.

The LTE-Advanced CA can basically take place both within a frequency band and between frequency bands in 3GPP. It has been agreed that work will be done so that Release 10 will contain the following:

- Support for the downlink intra and inter-band CA, with the motivation that in most cases each operator does not have more than 20 MHz in a given frequency band and thus inter-band carrier aggregation in the downlink direction is more easily usable.

- In the uplink direction, carrier aggregation is not seen to be attractive because the use of two transmitters simultaneously in the UE is more challenging than two receivers. The same approach has been used with HSPA where only the downlink multiband operation has been defined in Release 9.

In Release 11, work is expected to continue on the uplink carrier aggregation between the bands. Release 10 will contain a limited set of frequency bands (and band combinations) where the downlink carrier aggregation has been defined, including UE performance and RF requirements. The bands to be considered will not be limited by the physical layer and signaling specifications but work on the necessary RF and performance studies will need to be done for a limited set of band combinations as quite a lot of work is needed for each particular frequency band combination. The principles of inter-band and intra-band downlink carrier aggregation are shown in Figure 16.4.

The Release 10 downlink CA band combinations in 3GPP are to be defined in detail during 2011. Further downlink band combinations will be added later as release-independent frequency band combinations. Detailed work on different band combinations started in 3GPP from December 2010 onwards and is expected to be finalized during 2011. For Europe, the 1800 MHz band combined with 2600 MHz will be covered first, with an expectation of adding the use of 800 MHz as one of the bands later. This reflects spectrum licensing progress – 2600 MHz has been actioned already in quite a few European countries whereas 800 MHz availability will follow somewhat later in most cases. In other regions there are more combinations. From the US, all the bands now envisaged for LTE are likely to be considered at some point in time as well as those from Region 3. Work on different band combinations will be done as separate work items, as with the example of the European 1800 MHz and 2600 MHz work item in [5].

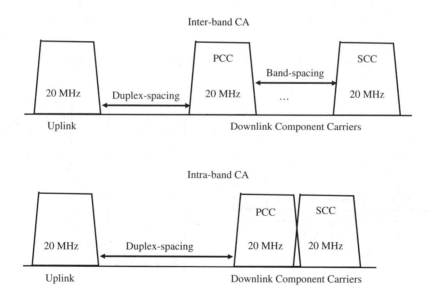

Figure 16.4 Inter-band and intra-band LTE Advanced carrier aggregation. (PCC = Primary Component Carrier, SCC = Secondary Component Carrier)

16.5.1 Impact of the Carrier Aggregation for the Higher Layer Protocol and Architecture

The use of carrier aggregation is visible in the physical layer and the MAC layer. The user plane layers above the MAC are unchanged and whether carrier aggregation is being used or not is not visible to the core network. The impact on the core is actually that it enables data rates beyond Release 8 and 9 capabilities. The MAC layer divides the data between different component carriers, as shown in Figure 16.6. Multiplexing functionality will now not limit itself to providing the data to one of the component carriers – it can do so for two or more. The example in Figure 16.5 shows the use of two downlink component carriers. This is likely to become the most common case and will be supported by RF requirements work in Release 10. Depending on the interest and frequency band availability the actual requirements for going beyond two component carriers will be then addressed in later

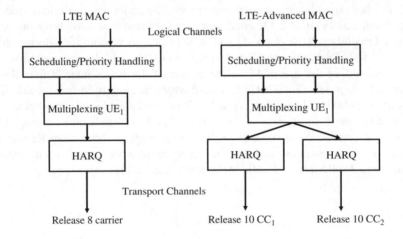

Figure 16.5   LTE and LTE-Advanced MAC layer structures

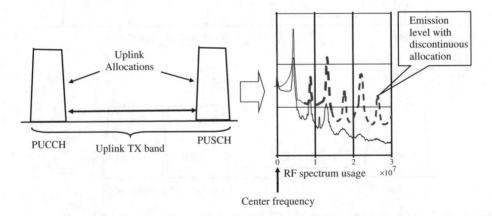

Figure 16.6   Impact of discontinuous uplink resource allocation

releases. The underlying physical layer structures and signaling is not limited to the use of only two component carriers – up to four secondary cells can be configured in addition to the primary cell.

The uplink and downlink MAC layer structures are identical but the scheduling functionality of the downlink MAC layer deals with scheduling of multiple users, as presented in Chapter 6. Different component carriers will follow the same DRX operation to enable low power consumption of the LTE-Advanced compatible devices.

16.5.2 Physical Layer Details of the Carrier Aggregation

The effects of carrier aggregation on the LTE physical layer mainly concern feedback structures for multiple carriers but there are also changes as a result of part of the single carrier uplink operation, with the introduction of discontinuous PUCCH/PUSCH and PUSCH/PUSCH allocations, as shown in Figure 16.6. There are some minor changes in procedures but in general the primary cell is the reference for UE, whether for timing or for radio link monitoring to indicate out-of-synch/in-synch status.

In the uplink direction in general it is preferable to stay with the single uplink carrier as the UE will need to use less transmitter power if multiple uplink carriers are used. The transmission power needs to be reduced as the cubic metric (CM) will increase and thus the amplifier will need to remain in the linear region to retain the necessary spectrum mask, as discussed in Chapter 4. Also the total maximum transmit power cannot be increased; thus, in the case of two uplink carriers, the maximum power per carrier would need to be reduced by 3 dB regardless of other limiting factors. The resulting CM and emission levels are also an issue within a single carrier when the uplink allocation is not continuous, as was shown in Figure 16.6. The necessary Maximum Power Reduction (MPR) could be up to 10 dB or even more to ensure the emission levels are kept under control. Thus use of multiple uplink carriers or discontinuous uplink resource allocation needs to be an option that is only used for cases where UEs are not at the cell edge, thus ensuring that the cell-edge data rate is not reduced. When a UE is closer to the eNodeB, the scheduler can decide whether the UE can use discontinuous uplink allocations based on the feedback received, without the UE encountering problems due to the required MPR. The decision can be based on the power headroom feedback from the UE.

16.5.3 Changes in the Physical Layer Uplink due to Carrier Aggregation

In the uplink direction the capability needs to exist to provide feedback for multiple carriers as well as CQI/CSI/PMI but the UE needs to be able to ACK/NACK multiple packets received in the downlink direction from multiple carriers. In the uplink, in Release 10 (or most likely later as well), one needs to be able to deal with a single uplink. The physical layer design will facilitate use of up to five carriers in total. It assumes that all carriers have the same frame structure and, in the case of TDD, the same uplink/downlink split.

The HARQ and channel state information feedback signaling need to be tailored to support up to five CCs. In Release 10 full HARQ feedback is supported in FDD mode, meaning that up to 10 ACK/NACK bits can be signaled simultaneously. The ACK/NACK codebook size is dimensioned according to the number of configured component carriers – it does not change dynamically from one sub-frame to another. There are two methods for conveying the HARQ feedback information in the uplink. Channel selection (largely similar to the TDD ACK/NACK signaling in Release 8) can be used for the payload sizes of up to 4 bits, whereas the new PUCCH format 3 based on block-spread DFT-S-OFDM can carry a larger number of bits. The basic idea in channel state information feedback is a straightforward extension of the Release-8 principles and mechanisms by copying them for multiple component carriers.

PUSCH transmissions on each uplink component carrier are power controlled independently. This means that power control commands and parameter values for the open loop part of the power control are component carrier specific. Like earlier releases, PUCCH has an independent power control. Simultaneous transmissions on multiple carriers could result in UE exceeding its maximum power capabilities. Because of this UE may need to scale down its transmit power and in this process it prioritizes control signaling so that PUCCH has the highest priority and PUSCH with uplink control information has the second highest priority. The power headroom reporting is component carrier specific.

16.5.4 Changes in the Physical Layer Downlink due to Carrier Aggregation

In the downlink, the individual component carriers operate in a similar fashion to Releases 8 and 9. The key adjustment is in the related signaling as the UE now needs to obtain information about allocations on all the component carriers that it is able to receive. The Carrier Information Field (CIF) indicates which component carrier is in the downlink scheduling. While the UE-specific PDCCH search space can be on any of the carriers, the common search space for the UE is limited to a single component carrier (Pcell). The eNodeB may provide information, using the PDCCH (UE specific search space), about any of the carriers indicating that there is data on another carrier. This is called cross-carrier scheduling.

16.5.5 Carrier Aggregation and Mobility

With carrier aggregation, there is only one RRC connection when carrier aggregation is activated, thus different carriers do not operate independently. The serving cell that handles the RRC connection establishment/re-establishment related security input becomes the Primary Serving Cell (Pcell). In the downlink it is equal to the downlink PCC and in the uplink it is equal to the uplink PCC. The eNodeB can then configure a suitable number of SCells (which consist of downlink and potentially also uplink SCCs, subject to the UE's capabilities).

Mobility is based on the PCell measurements (and PCell is used as a reference for the measurement unless the measurement is made on a frequency where a SCell is configured) and possible connection re-establishment is triggered when the PCell has a radio link failure (RLF). A RLF on a SCell does not cause re-establishment. User equipment can also provide information about the best non-serving cell on another frequency if configured to do so.

Backwards compatibility is ensured by sending the handover message towards the target cell as in Release 8. Thus an eNodeB not supporting Release 10 carrier aggregation would see the incoming handover as a Release 8 handover. This allows it to deploy CA selectively in the network, without upgrading all the eNodeBs in the coverage area.

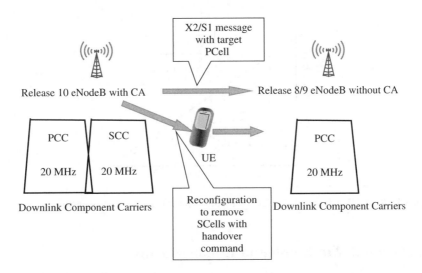

Figure 16.7 Mobility with CA to a Release 8 or 9 eNodeB

In connection with the intra-system handover, SCells can be added or removed depending on the available resources in the target cell. The source cell can provide information about which cell would be suitable as the SCell based the UE measurements on radio conditions. The target cell, together with the handover command, can then reconfigure the UE to drop the SCells if CA is not required to be used. This may be due to load reasons or when moving to a cell not supporting CA, as indicated in Figure 16.7.

16.5.6 Carrier Aggregation Performance

The use of carrier aggregation benefits system performance in two ways:

- There is an increased peak data rate when enabling the aggregation of spectra from more than single frequency band. The theoretical peak data rate from the combined use of carrier aggregation with total of 40 MHz spectrum and up to eight antennas (as presented in Section 16.6) reaches up to 1.2 Gbps in the downlink and in the uplink up to 600 Mbps with the technologies described in section 16.7 for uplink multi-antenna transmission. With 100 MHz of spectrum and five aggregated carriers the data rate would reach up to 3 Gbps in the downlink direction and 1.5 Gbps in the uplink direction.
- Improved average user throughput, especially when the number of users is not too high. Joint carrier scheduling in the eNodeB allows the optimal selection (in a dynamic fashion) of the carrier to use thus leading to better performance and optimal load balancing between the carriers. An example is illustrated in Figure 16.8, which shows the higher throughput, which then reduces as the number of users increases. With a very large number of users there are always a lot of users in good channel conditions for the scheduler to select in each of the carriers – thus the performance of the joint scheduling will then approach that of a single downlink carrier operation.

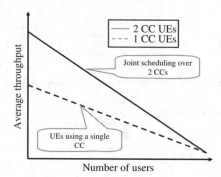

Figure 16.8   LTE-Advanced carrier aggregation performance befit over single carrier use

16.6   Downlink Multi-antenna Enhancements

Downlink MIMO is undeniably a key technology component in the LTE Release 8. Transmission modes for one, two and four eNodeB antenna ports have been specified providing peak data rates in excess of 300 Mbps. In the LTE-Advanced the natural next step was to continue with ambitious target setting to ensure its position as the leading wireless radio access technology [6]. In order to guarantee this, LTE-Advanced supports downlink transmission with up to eight transmit antenna ports. With an 8×8 antenna configuration and spatial multiplexing of up to eight layers the peak spectral efficiency increases up to 30 bps/Hz corresponding to 600 Mbps on a 20 MHz carrier. In addition to increasing the number of transmit antennas, Release 10 in the 3GPP emphasized improving the Multi-User MIMO (MU-MIMO) operation. Multi-User MIMO refers to the transmission where the parallel streams are transmitted to different UEs separated spatially while in Single-User MIMO (SU-MIMO) the parallel streams are sent to single UE.

16.6.1   Reference Symbol Structure in the Downlink

In LTE Releases 8 and 9 the MIMO operation is primarily based on cell-specific Common Reference Symbols (CRS). The reference symbol patterns are orthogonal between antenna ports and scale, depending on the configured number of transmit antenna ports. Channel state information (CSI) measurements as well as the data demodulation are typically performed using CRS. One exception is found in TDD beamforming transmission mode 7 where UE-specific reference symbols (URS) are used for demodulation. The straightforward solution in LTE-Advanced would have been to define yet another cell specific RS pattern for the case of eight transmit (TX) antennas, which would have implied usage of CRS for both CSI measurements and demodulation. That, however, would have created an issue with the backward compatibility of the Release 8 terminals, which are unaware of the existence of the new RS. The performance of the legacy terminals would inevitably be quite poor in such cases due to constant collisions between data and the new RSs. Another drawback of 8-TX CRS would have been a prohibitively high reference symbol overhead given the fact that a majority of terminals would not typically enjoy the benefits of an eight-layer transmission.

In order to cope with these challenges it was decided to adopt another reference symbol paradigm for LTE Release 10. The key idea was to decouple reference signals for CSI measurements from those needed for data demodulation as follows:

1   Channel state information reference symbols (CSI-RS) are introduced for CSI – namely CQI, PMI and RI – measurement and reporting, in the cases of 2/4/8 transmit antennas.

2   Precoded orthogonal UE-specific reference symbols are used for data demodulation with the support of up to eight spatial layers. There are mainly three key aspects justifying this choice. First, UE-specific RS allows flexible transmit precoding at eNodeB, which is seen as an enabler for competitive DL MU-MIMO. Second, the RS overhead scales according to the transmission rank and so a few high-rank-capable terminals do not penalize the whole system because of higher RS overhead as it would be the case with CRS. Third, UE-specific RS benefit from transmit precoding gain, which in turn leads to reliable channel estimation. The LTE Release 10 URS pattern for ranks 1–2 correspond to the one in LTE Release 9, while ranks 3–8 are brought up as an extension.

CSI-RS are sparse in time and frequency because requirements for CSI measurements are not as stringent as for data demodulation. CSI-RS are typically transmitted periodically (for example, every 10 ms) with very low density (1 RE/port/PRB). The periodicity of CSI-RS is configurable with duty cycle values ranging from 5 ms up to 80 ms, as DL MIMO enhancements target primarily low-mobility scenarios. This means that the impact on legacy LTE-Release 8/9 terminals is limited only to the sub-frames where the CSI-RS are transmitted and the rest of the time the legacy terminals can operate without any penalty. Furthermore, the relatively low density of the CSI-RS allows for transmission of data to Release 8 terminals in the sub-frames with CSI-RS as well, although with degraded performance. However, the MCS level needs to be downscaled accordingly to enable the UEs to cope with the additional interference.

Even though the primary driver for introduction of CSI-RS is the support for eight transmit antennas at eNodeB, the CSI-RS patterns are defined for other antenna configurations as well. Overall, specifications wise, CSI-RS and CRS may be configured independently. In Figure 16.9 the Release 10 CSI-RS patterns are depicted for eight, four, and two transmit antenna ports. The CSI-RS pattern enjoys a nested property – the pattern for low number of antenna ports is a subset of the pattern for a higher number of antenna ports. In addition to the patterns in Figure 16.9, other possible placements are enabled, and separate CSI-RS configurations are defined for normal and extended cyclic prefixes. Different patterns are also available for frame structure types 1 and 2 – that is, for FDD and TDD, with slight changes in TDD where collision with antenna port 5 may be avoided. One of the main differences of CRS is also the higher reuse factor, which, for example, in the case of two antenna ports, is 20. For comparison, the CRS reuse factor is three for two antenna ports. The higher reuse factor enables easier network planning and CSI-RS to CSI-RS collisions are avoided to a large extent, which is beneficial in case of fractional network load.

In Figure 16.10 we present two examples of URS configuration. The specification supports the use of 12/24 resource elements (REs), which can be used for URS depending

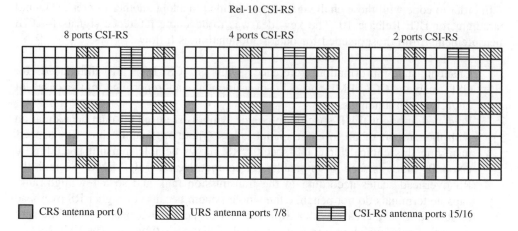

Figure 16.9   Rel-10 CSI-RS for 8, 4 and 2 antenna ports

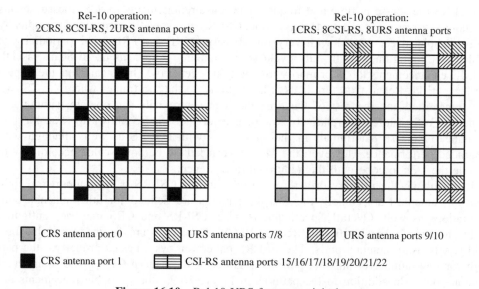

Figure 16.10   Rel-10 URS for up to eight layers

on the transmission rank. For example one and two layers are operated with 12 REs and orthogonal cover codes (OCC) of length 2 as in Release 9, whereas 24 REs with OCC length 4 are used for three to eight layers. The use of frequency division multiplexing coupled with variable OCC length allows scaling the RS overhead efficiently according to the transmission rank. We note that CSI-RS provide the opportunity for efficient CRS overhead reduction through antenna virtualization – e.g. down to one CRS port – as seen from the right part of Figure 16.10, while the UE has still has access to up to eight antenna ports via CSI-RS.

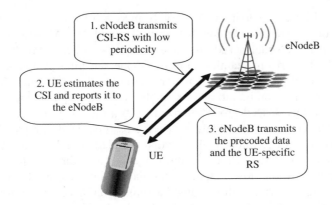

Figure 16.11   The basic procedure of CSI reporting over CSI-RS and data transmission utilizing URS on LTE Release-10

The basic principle of the system operation with CSI-RS is presented in Figure 16.11. The terminal estimates the CSI based upon the CSI-RS and transmits the CSI feedback to the eNodeB, which in turn can use the CSI in the selection of the precoder and Modulation and Coding Scheme (MCS) for the data. The data is transmitted together with user-specific (dedicated) demodulation reference symbols (URS also known as DM-RS), spanning the same physical resource blocks as the data. The same transmit precoding is applied to the data layers and associated DM-RS ports. This allows for the use of any precoding by the eNodeB, as the precoding applied remains transparent to the terminals and does not need to be signaled to the users contrary to LTE Release-8.

16.6.2   Codebook Design

The introduction of eight transmit antenna ports creates a need for a new codebook in support of the downlink MIMO operation. Furthermore, the utilization of user specific DM-RS allows the precoder selection to be very flexible at the eNodeB. In order to realize the potential benefits and gains from this, very accurate channel-state information needs to be provided to the eNodeB. In the case of TDD, channel reciprocity and sounding reference symbols can be used to some extent for obtaining the CSI by estimating it from the uplink transmission. With FDD, however, besides long-term DoA beamforming, it is not practically possible to exploit the short-term channel state information and hence a need to specify more elaborate CSI feedback mechanisms emerged in the 3GPP. The straightforward solution would be to simply specify a new codebook with larger number of elements and hence higher accuracy. That option is unfortunately impractical from the uplink signaling overhead as well as from the perspective of PMI selection complexity at the UE.

A natural approach to codebook design is to focus on scenarios of interest in conjunction with a selection of antenna configurations. In terms of use cases, the LTE Release-10 CSI feedback should support both downlink SU-MIMO and MU-MIMO, where SU-MIMO is predominant for less correlated scenarios with higher channel azimuth (i.e. angular) spread while MU-MIMO is typical for highly correlated scenarios with a small azimuth spread. Therefore the optimum operation point for SU-/MU-MIMO is very closely related to the deployment scenarios. On the other hand, traffic conditions and available multi-user diversity may vary from one TTI to another, mandating for the possibility of dynamic switching between SU- and MU-MIMO. From the latter it becomes clear that eNodeB needs to have both SU- and MU-MIMO CSI feedback available in order to perform seamless transmit mode selection. Antenna configurations and associated channel modeling play a major role in codebook design. 3GPP prioritized most practical antenna configurations for eight antenna deployment at eNodeB by focusing on (1) closely spaced cross-polarized (XP) arrays, (2) closely spaced uniform linear arrays (ULA), and (3) widely spaced cross-polarized arrays. The first two configurations are known to imply higher spatial correlation and thereby favor low-rank SU-/MU-MIMO transmission while the third one is more directed towards high-rank SU-MIMO because of the lower spatial correlation induced by large inter-element spacing. Prioritizing scenarios with low angular spread combined with close inter-element spacing at eNodeB transmit array indicate that UE feedback is primarily targeted for low rank SU- and MU-MIMO operations so that:

- Feedback in support of MU-MIMO is optimized with fine spatial granularity and UE separation in the spatial domain building on long-term channel wideband correlation properties.
- Feedback in support of SU-MIMO focuses on short-term narrow-band channel properties.

Long-term channel properties do not change very rapidly from one CSI measurement instance to another. Hence it makes sense to separate the channel state properties into a long-term and/or wideband part, which stays relatively stable over larger period of time (for example, rank and wideband beam direction), and a short-term and/or narrowband part targeted at beam selection and co-phasing over uncorrelated channel dimensions (for example, two different polarizations).

Taking advantage of these facts, efficient feedback signaling compression is achieved by decoupling the long- and short-term CSI components, and it is complemented by very competitive performance in targeted scenarios. This led the 3GPP to the adoption of a double codebook structure for LTE Release-10 CSI feedback in support of downlink MIMO with eight transmit antennas. The key principle is that a precoder W for a sub-band is composed of two matrices belonging to two distinct codebooks: W1 targets the long-term wideband channel properties while W2 aims at short-term frequency selective CSI. The resulting precoder W for each sub-band stems as the matrix multiplication of W1 and W2, i.e. W = W1 × W2. This principle is illustrated in Figure 16.12. The feedback rate for the W1 and W2 can be different, allowing for minimized uplink signaling overhead. The codebook elements themselves are based on grid-of-beams components, which are well known to provide good performance for MU-MIMO while for SU-MIMO the ability for frequency-selective precoding is maintained by allowing beam selection at sub-band level. While up to eight spatial layers are considered for 8 × 8 downlink single-user MIMO, the double codebook concept is mainly considered attractive for lower transmission ranks, namely ranks 1−2 and also ranks 3−4 to some extent, depending on the spatial correlation.

Release 10 introduced flexible support for 2, 4 and 8 CSI-RS antenna ports while URS are used for demodulation. In this environment only the 8-TX antenna codebook was newly defined. For 2- and 4-TX antennas the LTE Release 10 codebooks remain unchanged and are the corresponding codebooks from LTE Release 8, as they proved to be competitive enough.

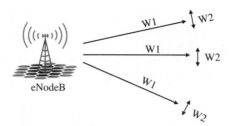

Figure 16.12 The principle of double-codebook feedback operation in LTE-Release10 – the precoder consist of two parts: W1, which targets long-term/wideband channel properties, and W2, which provides further refinement and information about short- term/narrowband properties of the channel

16.6.3 System Performance of Downlink Multi-antenna Enhancements

The main capacity benefit provided by Release 10 specifications compared to Release 8 is the Multiuser-MIMO functionality over UE-specific RS together with CSI-RS, principally in the case of four transmit antennas, that is. 4×2 and 4×4. The Release 10 solution brings no real benefit with 2×2 because of the very limited transmit precoding gain, which does not compensate for the UE-specific RS overhead. In general, the CRS overhead itself may be reduced by means of antenna virtualization. In practice one will typically configure one CRS port per polarization at the eNodeB transmit array.

The average downlink spectral efficiency is shown in Figure 16.13. Assuming uniform linear arrays at eNodeB, Release 10 MU-MIMO with 4×2 can push the capacity by +40% compared to Release 8 2×2. MU-MIMO with 4×4 can increase the capacity by +100% compared to Release 8. With 4-TX cross-polarized arrays, the gain figures are slightly reduced. The MU-MIMO benefit for the cell edge data rates is even higher than for the average data rates, as shown in Figure 16.14: 4×4 gives +150% higher cell edge data rates compared to Release 8 2×2. However, it is worth noting that cell-edge,

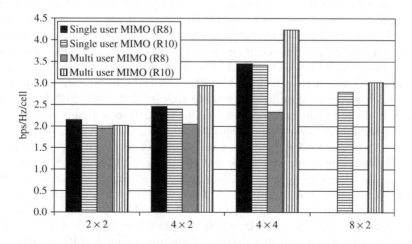

Figure 16.13 Average spectral efficiency in downlink

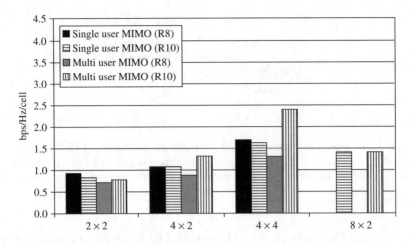

Figure 16.14   Cell edge spectral efficiency in downlink

peak and average throughput figures may be traded off one another depending on the scheduling and multi-user pairing strategy. Choosing to preserve fairness by utilizing proportional-fair frequency domain scheduling for the second spatial layer proves to lead to remarkable gains in coverage performance as shown here. On the other hand, opting for a maximum C/I type of frequency domain scheduling for the second user will boost peak throughput and thereby average throughput gains at the expense of coverage.

16.7   Uplink Multi-antenna Techniques

In order to meet the targets set for IMT-Advanced technologies, significant improvements on the uplink are required in terms of cell-edge user throughput as well as in peak and average spectral efficiency when compared to LTE Release 8. The inclusion of uplink MIMO into LTE was discussed during Release 8 standardization. However, due to time constraints and the fact that the vast majority of early terminals would not be equipped with multiple transmit chains led to uplink single-user multi-antenna techniques being left out of Release 8. The only exception is closed-loop antenna selection, as it does not actually require multiple full transmit chains but only antennas. Even this feature remains optional to the UEs and it seems unlikely that many early terminals would adopt it.

In addition to antenna switching, Multi-User MIMO is another multiple antenna technique relevant for Release 8 terminals. In Release 8 the MU-MIMO is completely transparent to the terminals and specification: the eNodeB may simply schedule multiple UEs on the same resources without the users knowing it. Provided that the reference symbols are orthogonal, the network can distinguish between the transmissions of multiple UEs and hence improve the spectral efficiency of the system. The uplink MIMO in LTE Release 10 supports two and four transmit antennas and provides peak data rates up to four times that of Release 8. Wideband (non-frequency selective) closed-loop precoding is supported for PUSCH in both TDD and FDD, meaning that the UEs may not autonomously select the precoder but it is signaled to them by the eNodeB. In the following we present some further details on various uplink MIMO related aspects in LTE Release 10.

16.7.1 Uplink Multi-antenna Reference Signal Structure

A prerequisite for MIMO operation is that orthogonal reference symbols are available for demodulation of the data. Uplink MIMO in Release 10 uses two complementary methods to provide orthogonal resources for the RSs: cyclic shifts (similar to Release 8) and orthogonal cover codes (OCC) between the two slots of the sub-frame.

The uplink reference signals in LTE Release 10 are precoded in the same way as the data. This means that the number of orthogonal resources required for the RS equals the rank of the PUSCH data transmission. The reference signals for different layers are primarily separated with cyclic shifts. During LTE Release 10 standardization it was observed that this alone does not always guarantee sufficient orthogonality between the layers, and hence improvements were studied. As a result, orthogonal cover codes were introduced to be applied between the RS blocks of the two slots of the sub-frame as shown in Figure 16.15 The orthogonal cover codes (OCC) are length-2 Walsh codes and provide an additional dimension for separating the RS in a reliable manner.

Naturally, the Sounding Reference Signals need to be tailored to support multi-antenna operation as well. The SRS transmitted from different antennas are separated using the same methods that were already available in LTE Release 8 – different cyclic shifts and transmission combs. Unlike the demodulation RS, SRS are not precoded but are antenna specific. This is necessary because the precoder selection is typically performed based on the SRS.

Another SRS enhancement in Release 10, although not only related to uplink MIMO, is the introduction of dynamic aperiodic sounding. In dynamic aperiodic sounding the eNodeB may request the UE to send SRS at any time by indicating it in the PDCCH. This helps in optimizing the SRS resource usage and in making the most of the available SRS capacity. Dynamic aperiodic sounding is useful especially with uplink MIMO because periodic SRS transmission from multiple antennas for many UEs would increase the SRS resource consumption drastically.

16.7.2 Uplink MIMO for PUSCH

Although the Uplink MIMO operation in LTE Release 10 largely resembles the downlink MIMO in Release 8, there are a few key differences. In the uplink, only codebook-based

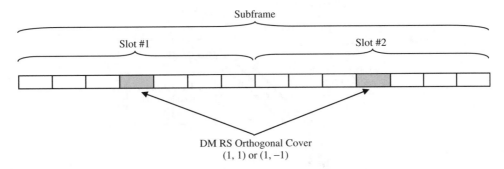

Figure 16.15 Orthogonal cover codes applied between the RS blocks of the two slots help in improving orthogonality of the reference signals

closed-loop precoding is supported. The need for open-loop transmit diversity was also discussed but the conclusion was that it does not provide any clear additional benefits given that its performance is slightly worse than what can be achieved with closed-loop precoding. Furthermore, any Release 10 terminals can be configured into single-antenna mode as a fallback providing similar coverage and data rates as a Release 8 UE when reliable CSI is not available for precoder selection.

Figure 16.16 shows the basic block diagram of the main uplink MIMO functionalities. First of all, because the eNodeB is in charge of both scheduling and the channel estimation, there is no need for CQI/PMI feedback signaling. Instead, the network configures the UE to send Sounding Reference Symbols, based on which the network can obtain uplink channel-state information (CSI) to be used for precoder and MCS selection. The eNodeB then signals the selected precoder to the UE in the PDCCH uplink grant. After that the UE transmits the precoded PUSCH data and the demodulation reference symbols.

The codebook design for the Uplink reflects the key differences between uplink and downlink multi-antenna operation. Similarly, as in any uplink considerations, in the case of uplink MIMO the low cubic metric plays a significant role as it optimizes the UE power consumption and allows for efficient usage of all power amplifiers. Due to this design criterion it was not possible, for example, simply to reuse the downlink four-transmit branch Householder codebook, but a new design was required. This is why the uplink codebooks for both two and four transmit antennas are *cubic metric preserving* – only one signal is transmitted from each of the antenna ports at the time. This guarantees that the waveform each antenna sends is pure single-carrier transmission and beneficial low peak-to-average power properties are maintained.

Another uplink specific property in the uplink codebooks is the antenna selection/turn-off elements. These can be used to switch off some of the terminal antennas to save UE battery power, for example. when a user's hand happens to block some of the antennas causing severe antenna imbalance.

16.7.3 Uplink MIMO for Control Channels

In addition to PUSCH, the multi-antenna transmission is also defined for the physical uplink control channel PUCCH. Since, in the case of PUCCH, there are no uplink

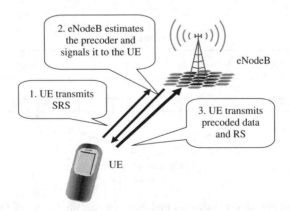

Figure 16.16 The basic principle of uplink MIMO operation in LTE-Advanced – the eNodeB estimates the CSI based on uplink SRS and signals the precoder to the UE

grants available for carrying the information on the precoder, the uplink MIMO operation needs to rely on open-loop transmission. For PUCCH formats 1/1a/1b and 2/2a/2b open-loop transmit diversity is realized using Space-Orthogonal Resource Transmit Diversity (SORTD). The basic idea in SORTD is to allocate separate orthogonal resources for each antenna port. Orthogonal resources are separated with cyclic shifts and/or orthogonal cover codes. The benefits of SORTD include improved performance compared to single-antenna transmission and simple implementation at the UE and the eNodeB. The main disadvantage is the increased resource consumption and related uplink overhead. For this reason only two-way SORTD has been standardized – the same specified diversity method with two orthogonal resources is used with both two or four transmit antennas.

16.7.4 Uplink Multi-user MIMO

The multi-user MIMO operation in LTE Release 8 is completely transparent to the UE: the eNodeB may simply schedule two UEs on the same resources without the UEs being aware of it. However, due to the uplink reference signal properties, orthogonality between the RS of different UEs can only be guaranteed when the UEs have exactly the same resource allocation – in other words, they share the same PRBs. This requirement complicates uplink MU-MIMO scheduling significantly, because the same PRBs are typically not optimal for different UEs from the frequency domain scheduling point of view. As a result the probability of finding UE that can be paired in MU-MIMO becomes sufficiently low, leaving the MU-MIMO gains rather moderate in the Release 8 uplink.

The Orthogonal Cover Codes introduced to LTE in Release 10 help to make the uplink MU-MIMO user pairing considerably more flexible. By assigning different OCC for the UEs paired in MU-MIMO the stringent requirement on exactly the same resource allocation is relaxed, and the UEs can be flexibly scheduled on the optimal frequency resources. This, in turn, helps to increase the probability of finding UEs for the MU-MIMO operation as show in Figure 16.17, and hence also improves the uplink average cell throughput by up to 15% [7].

16.7.5 System Performance of Uplink Multi-antenna Enhancements

In the uplink direction the key difference is use of more than one transmit antenna in the uplink direction. The use of two-antenna MIMO transmission brings a 16% benefit over the single antenna transmission with RX-diversity in the receiver while the use of up to four transmit-and-receive antennas can then double the capacity compared to using

Release 8 UL MU-MIMO

PRB#	0	1	2	3	4	5	6	7	8	9	10	11	12	13	14	15	16	17	18	19	20	21	22	23	24
1st MU-MIMO layer	UE1								UE2					UE3						UE4		UE5			
2nd MU-MIMO layer	unused								UE6					unused						UE7		unused			

Release 10 UL MU-MIMO

PRB#	0	1	2	3	4	5	6	7	8	9	10	11	12	13	14	15	16	17	18	19	20	21	22	23	24
1st MU-MIMO layer	UE1								UE2					UE3						UE4		UE5			
2nd MU-MIMO layer	UE6								UE7					UE8						UE9		UE10			

Figure 16.17 User pairing in LTE Release 8 and Release 10 uplink MU-MIMO

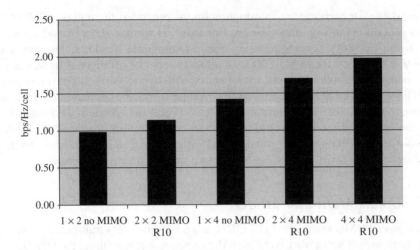

Figure 16.18   Average spectral efficiency in the uplink

a single antenna with two-antenna RX diversity, as shown in Figure 16.18. The results shown assume non-correlated antennas.

16.8   Heterogeneous Networks

The need to provide a greater amount of data for a large number of subscribers is shaping network developments as the traditional macro-cellular networks are not always adequate to deal with the increased capacity demands. With the limit in the increase in the capacity per Hz, the dimension to explore is that of cell size. As the demand for the capacity is not expected to be uniform, the networks of today are expected to evolve increasingly in the direction where the cell sizes vary drastically. In areas of less demand, improvements in macro-cell capacity can cope with the increased demands, but in densely populated areas there is a need to enhance capacity with smaller cells, from macro cell level down to micro and pico cells, and in some cases even to femto cells. The relay nodes form an additional dimension as discussed in section 16.9. Figure 16.19 shows one example of heterogeneous deployment, with different types of cells being used in the same network and with the UE being able to connect different types of cells from the link budget perspective.

As shown in Figure 16.19, from a link budget perspective, the UE could connect to different type of cells when in a high traffic area being served by micro, pico and femto cells. When the system is able to allocate different frequencies for different types of cell, there are normally no interference problems unless the power differences become such that the attenuation of the interference caused to the adjacent channel is not enough. However with the increased capacity demand, there is unlikely to be room to leave a separate carrier, up to 20 MHz, just to be used for small-cell traffic. Thus the same frequency needs to be used with different types of cells. This will lead to interference issues between different cell types and has caused a need to define mechanisms to deal with the situation. In 3GPP the work item on the Enhanced Inter-Cell Interference Coordination (eICICI) is addressing whether some of the interference could be mitigated, for example, by more effect power control or dividing the resources partly with the time domain element between different

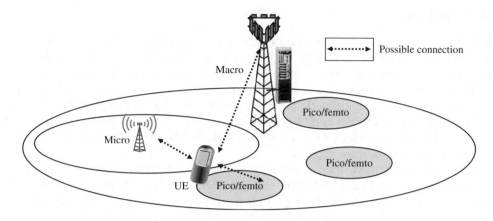

Figure 16.19   Heterogeneous deployment

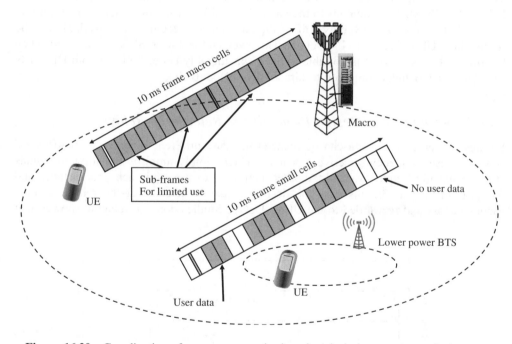

Figure 16.20   Coordination of resource usage in time domain in heterogeneous deployment

types of cell, as shown in Figure 16.20 between a macro and a small power cell. In the example case given in Figure 16.20 the small power cell will use resources (coordinated with macro cell) while leaving some of the resources empty from user data (control information and reference signals will still be transmitted – for such a sub-frame the term Almost Blank Subframe (ABS) is used in X2 signaling) and the macro cell will then avoid using the resource active in the small cell. Instead the macro cell should schedule the users in other sub-frames. To make this work there needs to a common timing reference between the sites as well.

16.9   Relays

One of the new technology components, that is standardized in release 10, is relaying. The objective of relaying is to improve performance of the LTE network by adding new nodes in areas where there are coverage problems. Relays (or Relay Nodes, RNs) have smaller transmit power than macro-eNodeBs and backhaul is realized wirelessly, so the deployment of RNs is significantly easier than deployment of macro-eNodeBs. Because of this, relays can be used to build LTE network to areas where it is hard to get wire line backhaul. The RN is connected to a Donor eNodeB (DeNB), which takes care of the data connection towards the core network.

When relaying is used, transmitted packets need to be sent two times over the air interface. Transmission between DeNB – RN and RN – UE is needed so the consumption of air interface resources can be higher than with direct eNodeB – UE transmission. However typically RN is deployed so that the link between DeNB – RN is of good quality and also the link between RN – UE is good because only nearby UEs are served by the RN. In Figure 16.21 it can be seen that, at high SNR, the spectral efficiency of UE – RN – DeNB link saturates to the value that is 50% of the maximum value of direct UE – eNodeB link, because only half of the air interface resources are available for the single link (UE – RN or RN – DeNB). However at the lower SNR values, and when location of RN is good, it is possible to have considerably better SNR on both UE – RN and RN – DeNB links compared to direct UE – eNodeB link.

16.9.1 Architecture (Design Principles of Release 10 Relays)

Wireless coverage extension can be realized in many different ways. A straightforward way is to use frequency-selective repeaters, which simply amplify and forward signals at some specific frequency band. Another method is to use equipment that decodes and forwards the signal. In this case the desired signal is detected at the relay and then encoded again and forwarded to the UE or DeNB. Studies done on relaying showed that

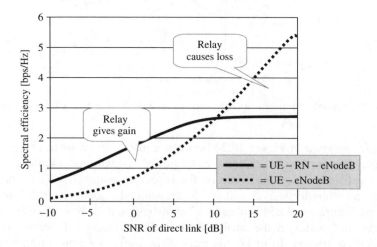

Figure 16.21   Relay performance as function of direct link quality

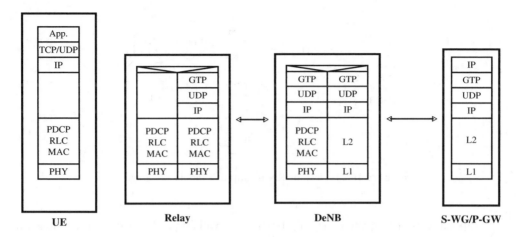

Figure 16.22   Relay architecture

decode and forward type of relaying has benefits compared to amplify and forward type of approach so this type L2/L3 relaying was selected. The next step was to decide which layers of the protocol stack are realized in the RN and which are done at the DeNB. The standardized solution in Release 10 is that RN creates an eNodeB of its own. Functions like scheduling are performed at the RN. The description of protocol stacks of the relaying system is presented in Figure 16.22. The DeNB acts as a proxy and hides the existence of the RN from the rest of the network, which sees the RN as a cell in the DeNB.

The use of repeaters is, of course, possible in the LTE networks and from a specification point of view nothing new is needed on top of Release 8 to support relaying. However, because of the benefits of higher layer relaying, the standardized solution for relaying has been carried out.

The starting point in Release 10 relaying specification work has been that RN is compatible with Release 8 terminals. New functionality that would require changes to the LTE Release 8 terminals is not specified but from a Release 8 UE point of view RN looks like any other cell.

Another starting point for relay operation is that the DeNB is a normal eNodeB so that besides serving RNs it also serves UEs – the air interface resources of the DeNB are shared between macro UEs (UEs attached to DeNB) and RNs

Both inband and outband operations for relays are supported by the specification. For inband operations, the link between DeNB – RN and the link between RN – UE use the same carrier frequency f1 and in the case of outband relaying different carrier frequencies f1 and f2 are used as depicted in Figure 16.23. From a physical layer point of view, the important thing is that backhaul and access links can be used at the same time. This is possible if isolation between tx and rx is good enough so that RN can simultaneously transmit and receive UL and DL signals on both Un and Uu interfaces. This can be realized by having sufficient frequency separation between access and backhaul carriers or by having good enough isolation of antennas for backhaul and access links. If it is possible to operate Un and Uu interfaces simultaneously then backhaul can be realized like a normal eNB – UE link from physical layer point of view.

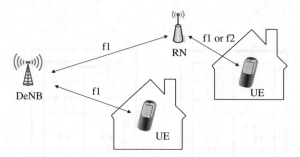

Figure 16.23   Use of RN to improve indoor coverage

Additional carrier frequency may not be always available and isolation of antennas for backhaul and access would lead to complicated and expensive implementation so most of the physical layer standardization effort of Release 10 relays is spent on specification of time domain multiplexing of access and backhaul transmission.

16.9.2   DeNB – RN Link Design

The TDM-based approach requires that RN must stop its DL transmission for a moment to receive signals from DeNB. This kind of transmission gap of DL signals must be done so that it is compatible with Release 8 UEs, so the method that is used is to configure some of the access links sub-frames as MBSFN sub-frames. In this case relay only transmits one or two PDCCH symbols in the access link and can then switch to receive signal from DeNB. In the UL direction the time domain multiplexing of access and backhaul can be realized by scheduling restrictions. RN must avoid scheduling UL transmissions from UEs that it is serving when it is transmitting to the DeNB.

The relay needs some switching time when it changes from access to backhaul transmission and back. This switching time is assumed to be about 10–20 µs so the switching cannot be done within the cyclic prefix duration. This means that one of the OFDMA symbols in the downlink and one of the SC-FDMA symbols in the uplink direction must be spent on switching between tx and rx.

Because the TDM of access and backhaul in the DL direction is realized by MBSFN configuration, it means that relay is transmitting PDCCH in the access link at the same time when DeNB is also transmitting PDCCH. This means that relay cannot receive PDCCH from the DeNB and the control information for the relays in the backhaul needs to be transmitted in PDSCH symbols. Because of this, the new R-PDCCH has been created. It has been agreed that DL assignments are transmitted in the first slot of the sub-frame and UL grants are in the second slot of the sub-frame, as shown in Figure 16.24.

In case of TDD, time domain multiplexing (or duplexing) is already used between uplink and downlink transmission and when TDM is also applied to the multiplexing of access and backhaul links it means that sub-frames for DL or UL transmission need to be further divided between access and backhaul. This division is problematic for some of the TDD uplink-downlink configurations: in configuration #5 there is only one UL sub-frame available per radio frame and in configuration #0 DL sub-frames are in the sub-frames that can not be configured as MBSFN sub-frames, so it has been agreed that

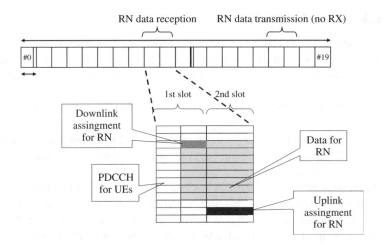

Figure 16.24   Backhaul sub-frame structure

those two configurations are not supported in Release 10 relaying. The RN operation otherwise is covered in the existing specifications but for the physical layer there is a separate specification created in Release 10 [8].

16.9.3   Relay Deployment

Both access and backhaul links need to be better than a direct link from eNodeB to the UE in order for it to make sense to use relays. Typically only a fraction of the UE is served by the relays so resources for backhaul are limited. This means that only one or two sub-frames per radio frame can normally be used for backhaul and this then means that the capacity of the backhaul is usually smaller than the capacity of the access link.

It can be assumed that RNs are not usually deployed randomly in the macro cell but they are located close to the area where coverage is limited and also the location of the RN is selected in such a way that the link to the DeNB is relatively good. It can be assumed that many times it is possible to find a location for RN where there is a line-of-sight connection to DeNB or at least strong shadowing can be avoided. In the Release 10 standardization, relays are assumed to be stationary. This is one additional thing to assure that the backhaul link is stable and of good quality.

In the typical scenario used in relay simulations there are couple of candidate locations for relay placement and the best location is selected. RNs are located outside so the penetration loss is not taken into account in the backhaul link. Because RN location is selected among a few candidates and because RN is at the height of 5–10 m, the probability of having LOS connection between RN and DeNB is higher than with randomly deployed UEs. In the access link distance between RN and UE is quite short. Penetration loss is taken into account there but it can be assumed that, besides the attenuation caused by the wall between UE and RN, the channel is often comparable to the LOS channel. Simulations with this kind of assumption show that with a small number of relays (for example, four) located close to the cell edge, both the cell edge performance and average aggregate cell throughput can be improved.

16.10   Release 11 Outlook

In Release 11, work will continue on the topics identified as part of the LTE-Advanced studies as well as on other topics beyond those of Release 10. Release 11 in 3GPP is scheduled to be finalized at the end of 2012, 18 months after closing Release 10 in mid-2011.

At the time of writing, topics for further work on LTE-Advanced are known to include:

- Carrier aggregation, which is expected to focus on adding new downlink band combinations as well as working on the uplink carrier aggregation with support for the multi-band uplink case.
- Multi-antenna enhancements, where it has been agreed that further studies will be done on Cooperative Multipoint Transmission (CoMP), which is to be a separate study item with the study conclusions expected to be reached during 2011. The CoMP idea simply aims to turn the interference signal into a useful component of the received signal or alternatively aims to use the spatial dimension to direct the signal in such a way that interference is minimized. The CoMP schemes can be divided in different ways but the classification chosen here divides them into:
  - Joint scheduling/beamforming, where the actual transmission is only coming from a single cell/sector and performance benefits are sought through coordination of the other cells. In case of co-ordinated beamforming one would aim to direct the beams scheduled on overlapping time/frequency resources in such a way that transmission could take place so that interference would be avoided in spatial domain, as illustrated in Figure 16.25. The challenges here are related to the availability of accurate information about the environment to enable intelligent decisions at the eNodeB.
  - Joint processing CoMP is based on the idea of transmission from multiple cells with active interference cancellation at the receiving end. The challenges here depend on the detailed method chosen but are mostly related to the enabling the real time joint

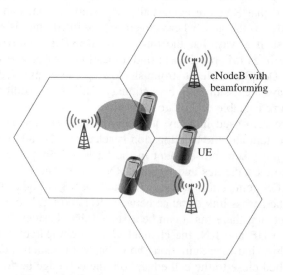

Figure 16.25   CoMP with joint beamforming

encoding and scheduling decisions from multiple eNodeBs when dealing with the inter-site operation as well as the resulting backhaul requirements as well as the resulting receiver complexity. From the performance point of view the challenge will be getting fast enough connections between different elements involved as well as realistic accuracy of the needed measurements to facilitate the CoMP operation.

Like Release 10, Release 11 will address many topics not directly linked to LTE-Advanced. Other topics identified in 3GPP discussions so far are:

- Further enhancements on the LTE MBMS (eMBMS), following the basic frame work in Release 9 (and agreed to be enhanced slightly in Release 10).
- Expected continuation of the work based on Release 10 work items or study items on topics like Self-Organizing Networks (SON), Minimization of Drive Tests (MDT) or Machine-to-Machine (M2M) connection optimization.

16.11 Conclusions

LTE-Advanced in Release 10 adds several enhancements to the LTE system. These improve LTE's peak and average data rates, especially with fragmented spectrum allocation and the use of carrier aggregation. The other factor improving peak and average data rates is the increased number of antennas and advanced antenna technologies. Coupled with important improvements in Release 10 for LTE operability as well as support for relay use and better interference management with heterogeneous networks, the added improvements together with the foreseen Release 11 solutions ensure that the LTE system will be a state-of-the-art mobile communication solution for the decade ahead. Release 10 LTE fulfills and exceeds the requirements for IMT-Advanced systems. Release 10 LTE has various UE capabilities – the highest UE capability class can achieve peak data rates as high as 3 Gbps for the downlink and 1.5 Gbps in the uplink direction with 100 MHz allocation, thus enabling it to meet end-user requirements in the years to come.

References

[1] ITU-R report, M.2134, 'Requirements Related to Technical Performance for IMT-Advanced Radio Interface(s)'.
[2] 3GPP, Technical Report, TR 36.913, 'Requirements for Further Advancements for Evolved Universal Terrestrial Radio Access (E-UTRA) (LTE-Advanced)', v. 8.0.1, March 2009.
[3] 3GPP, Technical Report, TR 36.814, 'Feasibility Study for Further Advancements for E-UTRA (LTE-Advanced)'.
[4] 3GPP, Technical Report, TR 36.912, 'Feasibility Study for Further Advancements for E-UTRA (LTE-Advanced)', v. 9.2.0, March 2010.
[5] 3GPP, Tdoc RP-100668, 'Work Item Description: LTE-Advanced Carrier Aggregation of Band 3 and Band 7', TeliasSonera, June 2010.
[6] Lunttila, T., Kiiski, M., Hooli, K., Pajukoski, K., Skov, P, Toskala, A., 'Multi-Antenna Techniques for LTE-Advanced', Proceedings of WPMC 2009, Sendai, Japan.
[7] 3GPP, Tdoc R1-094651, 'Performance of Uplink MU-MIMO with Enhanced Demodulation Reference Signal Structure', Nokia Siemens Networks, Nokia, November 2009.
[8] 3GPP, Technical Specification, TS 36.216 'Physical Layer for Relaying Operation', v. 10.0.0, September 2010.

17

HSPA Evolution

Harri Holma, Karri Ranta-aho and Antti Toskala

17.1 Introduction

High-speed packet access (HSPA) was included in the Third Generation Partnership
Project (3GPP) Releases 5 and 6 for downlink and for uplink. The 3GPP Releases 7, 8,
9 and 10 have brought a number of HSPA enhancements providing major improvements
to end-user performance and to network efficiency. The work will continue further in
future releases.

The HSPA evolution work has progressed in parallel to LTE work in 3GPP. HSPA
evolution deployments in practice take place in parallel to LTE deployments. Many of
the technical solutions in HSPA evolution and LTE are also similar. A number of 3GPP
work items are common to LTE and HSPA including home base stations (femto cells),
self-optimized networks and minimization of drive testing. The overview of the HSPA
evolution and LTE roles is illustrated in Figure 17.1. HSPA evolution is optimized for
co-existence with the WCDMA/HSPA supporting legacy Release 99 UEs on the same
carrier and is designed for simple upgrading on top of HSPA. The HSPA evolution aims
to improve performance for the end user by lower latency, lower power consumption and
higher data rates. The features of HSPA evolution are introduced in this chapter. The
HSPA evolution is also known as HSPA+.

HSPA evolution includes interworking with LTE, which enables both packet handovers
and voice handovers from LTE voice over IP (VoIP) to HSPA circuit-switched voice. The
handovers are covered in Chapter 7 and the voice call continuity in Chapter 13.

17.2 Discontinuous Transmission and Reception (DTX/DRX)

The evolution of the technology in general helps to reduce the mobile terminal power consumption. Fast and accurate power control in WCDMA helps to minimize the transmitted
power levels. The challenge in 3GPP from Release 99 to Release 6 is still the continuous reception and transmission when the mobile terminal is using HSDPA/HSUPA in
Cell_DCH state. HSPA evolution introduces a few improvements to HSDPA/HSUPA that
help to reduce the power consumption for CS voice calls and for all packet services.

LTE for UMTS: Evolution to LTE-Advanced, Second Edition. Edited by Harri Holma and Antti Toskala.
© 2011 John Wiley & Sons, Ltd. Published 2011 by John Wiley & Sons, Ltd.

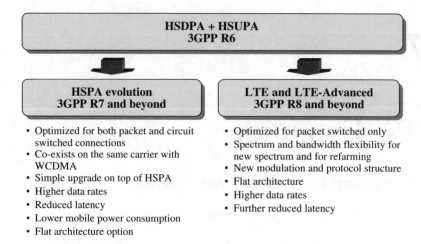

Similar technical solutions applied both in HSPA evolution and in LTE

Figure 17.1   Overview of HSPA evolution and LTE roles

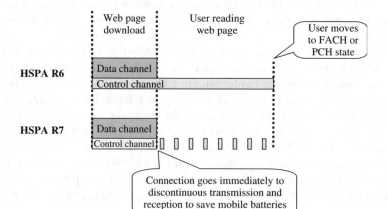

Figure 17.2   Discontinuous transmission and reception with continuous packet connectivity

3GPP Release 6 UE keeps transmitting the physical control channel even if there is no data channel transmission. The control channel transmission and reception continues until the network commands the UE to Cell_FACH or Cell_PCH state. The Release 7 UE can cut off the control channel transmission as soon as there is no data channel transmission allowing it to shut down the transmitter completely. This solution is called discontinuous uplink transmission and it brings clear savings in the transmitter power consumption.

A similar concept is introduced in the downlink where the UE needs to wake up only occasionally to check if the downlink data transmission is starting again. The UE can use a power-saving mode during other parts of the frame if there were no data to be received. This solution is called downlink discontinuous reception. The discontinuous transmission concept is illustrated in Figure 17.2 for web browsing. As soon as the web page is downloaded, the connection enters discontinuous transmission and reception.

The Release 99 FACH solution requires continuous reception by the UE, which is challenging from the power consumption point of view, especially for always-on applications transmitting frequent keep-alive messages. Each keep-alive message forces the UE to move to the Cell_FACH state and stay there until the network inactivity timer expires. The discontinuous reception is introduced also for the Cell_FACH state in HSPA evolution helping stand-by times when these types of applications are used.

Discontinuous transmission and reception was included in LTE from the beginning in Release 8 specifications. The power-saving potential in LTE is even higher than in HSPA because the TTI size is shorter than in HSPA (1 ms versus 2 ms) and because there is no need for fast power-control-related signaling.

17.3   Circuit Switched Voice on HSPA

Voice has remained as an important service for mobile operators. WCDMA Release 99 supports Circuit-switched (CS) voice on Dedicated Channel (DCH) with quite high spectral efficiency. Efficient VoIP capability on top of HSPA was defined in Release 7, but VoIP mass market has not yet started. Therefore, CS voice over HSPA was defined in HSPA evolution. CS voice over HSPA is part of Release 8 but because the capability indication for the UE support of the feature was introduced to Release 7, it is possible to implement this feature before other Release 8 features. There are two main benefits when running voice on HSPA: UE power consumption is reduced because of DTX/DRX and the spectral efficiency is improved with HSPA features.

The different 3G voice options are illustrated in Figure 17.3. CS voice over DCH is currently used in commercial networks. Dedicated Release 99 channel is used in Layer 1 and Transparent mode RLC in Layer 2. From the radio point of view, CS voice over HSPA and VoIP over HSPA use exactly the same Layer 1 including unacknowledged mode RLC on Layer 2. IP header compression is not needed for CS voice. From the core network point of view, there is again no difference between CS voice over DCH and CS voice over HSPA. In fact, the CS core network is not aware if the radio maps CS voice on DCH or on HSPA. CS voice over HSPA could be described as CS voice from the core point of view and VoIP from the radio point of view.

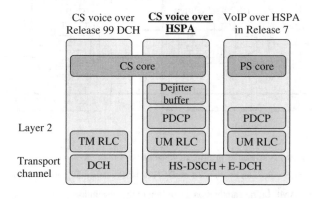

Figure 17.3   Voice options in WCDMA/HSPA

CS voice over HSPA brought the following changes to 3GPP specifications:

- Iu interface = no changes;
- Physical Layer = no changes;
- MAC layer = no changes;
- RLC layer = forwarding RLC-UM sequence numbers to upper layers;
- PDCP Layer = modification of header to identify and timestamp the CS AMR frames + interfacing/inclusion of Jitter Buffer Management;
- New de-jitter buffer = UE/RNC implementation dependent entity that absorbs the radio jitter created by HSPA operation so that the CS AMR frames can be delivered in a timely constant fashion to upper layers. The algorithm for the de-jitter buffer is not standardized in 3GPP.

CS voice over HSPA concept is presented in Figure 17.4. The CS voice connection can be mapped on DCH or on HSPA depending on UE capability and RNC algorithms. The AMR data rate adaptation can be controlled by RNC depending on the system loading. When CS voice over HSPA is used, there is a clear need for QoS differentiation in HSPA scheduling to guarantee low delays for voice packets also during the high packet data traffic load.

Since the packet scheduling and the prioritization are similar for VoIP and for CS voice over HSPA, it will be simple from the radio perspective to add VoIP support later on top of CS voice over HSPA, so CS voice over HSPA is paving the way for future VoIP introduction. The similar radio solutions make also the handover simpler between VoIP and Circuit switched domains.

CS voice on HSPA can benefit from IP protocol and packet transmissions in all interfaces: air interface carried by HSPA, Iub over IP, Iu-CS over IP and the backbone between Media Gateways (MGW) using IP. The call control signaling is still based on circuit-switched protocols [1]. The use of the interfaces is shown in Figure 17.5. CS voice

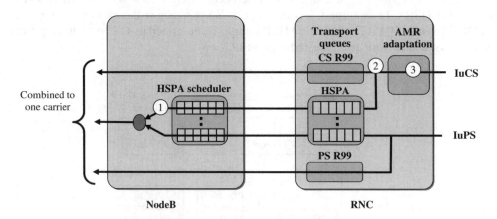

① = HSPA scheduler prioritizes voice packets

② = CS voice can be mapped on DCH or HSPA depending on UE capability

③ = AMR bit rate adaptation according to the system load

Figure 17.4   CS voice over HSPA overview

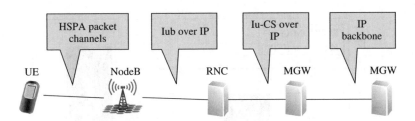

Figure 17.5   CS voice in different interfaces

over HSPA can benefit from the packet network performance and cost while maintaining the existing end-to-end protocols and ecosystem. No changes are required to charging, emergency calls or to roaming.

CS voice over HSPA increases the spectral efficiency compared to voice over DCH because voice can also benefit from the HSPA physical layer enhancements:

- UE equalizer increases downlink capacity. Equalizer is included in practice in all HSPA terminals.
- Optimized L1 control channel in HSPA reduces control-channel overheads. The downlink solution is Fractional DPCH with Discontinuous Reception and the uplink solution is Discontinuous Transmission. HS-SCCHless transmission can also be used on downlink.
- L1 retransmissions can be used also for voice on HSPA since the retransmission delay is only 14 ms.
- HSDPA optimized scheduling allows the capacity to be improved even if the tough delay requirements for voice limit the scheduling freedom compared to best effort data.

The gain in spectral efficiency with CS voice over HSPA is estimated at 50–100% compared to CS voice over DCH. The voice capacity is illustrated in Figure 17.6. Voice capacity evolution including LTE is covered in more detail in Chapter 10.

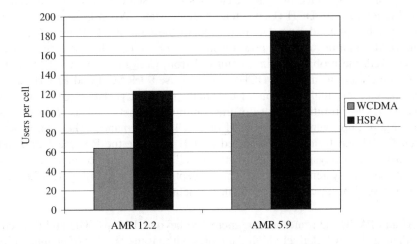

Figure 17.6   Circuit switched voice spectral efficiency with WCDMA and HSPA [2]

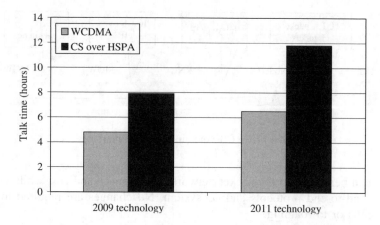

Figure 17.7   Talk time improvements with CS over HSPA and with technology evolution [3]

The voice codec with CS over HSPA can be Narrowband AMR (Adaptive Multirate Codec) or Wideband AMR.

CS voice over HSPA can benefit from DTX/DRX functionality for cutting the terminal power consumption and for providing longer talk times. The talk time benefit is analyzed in detail in [3] and summarized in Figure 17.7 assuming 900 mAh battery. The current consumption with CS voice over HSPA can be reduced below 100 mA, which enables 12 hours of talk time.

17.4   Enhanced FACH and RACH

WCDMA network data rate and latency are improved with the introduction of Release 5 HSDPA and Release 6 HSUPA. End user performance can be further improved by minimizing the packet call setup time and the channel allocation time. The expected packet call setup time with Release 7 will be below 1 s when the connection setup signaling runs on top of HSPA. Once the packet call has been established, user data can flow on HSDPA/HSUPA in Cell_DCH (Dedicated Channel) state. When the data transmission is inactive for a few seconds, the UE is moved to the Cell_PCH (Paging Channel) state to minimize the mobile terminal power consumption. When there are more data to be sent or received, the mobile terminal is moved from Cell_PCH to Cell_FACH (Forward Access Channel) and to the Cell_DCH state. Release 99 RACH and FACH can be used for signaling and for small amounts of user data. The RACH data rate is very low, typically below 10 kbps, limiting the use of the common channels. Release 5 or Release 6 do not provide any improvements in RACH or FACH performance. The idea in Release 7 Enhanced FACH and Release 8 Enhanced RACH is to use the Release 5 and Release 6 HSPA transport and physical channels in the Cell_FACH state to improve performance for the end user and system efficiency. The concept is illustrated in Figure 17.8. Enhanced FACH and RACH bring a few performance benefits:

- RACH and FACH data rates can be increased beyond 1 Mbps. The end user could get immediate access to relatively high data rates without the latency of channel allocation.

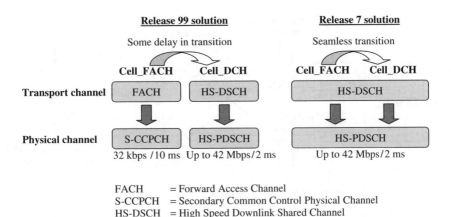

Figure 17.8 Enhanced FACH concept [4]

- The state transition from Cell_FACH to Cell_DCH would be practically seamless. Once the network resources for the channel allocation are available, a seamless transition can take place to Cell_DCH as the physical channel is not changed. Benefiting from fast power control during transmission, Enhanced RACH UE is also allowed to start the RACH preamble transmission without first waiting for a new uplink interference information on System Information Block 7 (SIB7) to be broadcast to the cell. That further reduces the state transition latency.
- Unnecessary state transitions to Cell_DCH can be avoided because more data can be transmitted in the Cell_FACH state. Many applications create some background traffic that is today carried on Cell_DCH. Therefore, Enhanced RACH and FACH can improve the network efficiency especially for the smartphones.
- Discontinuous reception could be used in Cell_FACH to reduce the power consumption. The discontinuous reception can be implemented since Enhanced FACH uses short 2 ms Transmission time interval instead of Release 99 10 ms. The discontinuous reception in Cell_FACH state is introduced in 3GPP Release 8.

The existing physical channels are used in Enhanced FACH so there are only minor changes in layer 1 specifications, which allow fast implementation of the feature. Enhanced FACH can co-exist with Release 99 and with HSDPA/HSUPA on the same carrier. No new power allocation is required for Enhanced FACH since the same HSDPA power allocation is used as for the existing HSDPA.

17.5 Downlink MIMO and 64QAM

The downlink peak data rate with Release 6 HSDPA is 10.8 Mbps with $\frac{3}{4}$ coding and 14.4 Mbps without any channel coding. There are a number of ways, in theory, to push the peak data rate higher: larger bandwidth, higher order modulation or multi-antenna transmission with Multiple Input Multiple Output (MIMO). All those solutions are part of HSPA evolution. MIMO and higher order modulation are included into HSPA evolution in Release 7 and Dual cell HSDPA in Release 8.

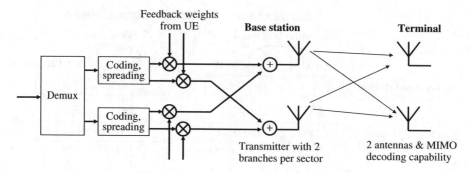

Figure 17.9 2 × 2 MIMO transmission concept

The 3GPP MIMO concept employs two transmit antennas in the base station and two receive antennas in the terminal and uses a closed-loop feedback from the terminal for adjusting the transmit antenna weighting. The diagram of the MIMO transmission is shown in Figure 17.9.

Higher order modulation allows a higher peak bit rate without increasing the transmission bandwidth. Release 6 supported QPSK (Quadrature Phase Shift Keying) and 16QAM (Quadrature Amplitude Modulation) transmission in the downlink and dual-BPSK (Binary Phase Shift Keying) in the uplink. Dual-channel BPSK modulation is similar to QPSK. The Release 7 introduces 64QAM transmission for the downlink and 16QAM for the uplink. 16QAM can double the bit rate compared to QPSK by transmitting four bits instead of two bits per symbol. 64QAM can increase the peak bit rate by 50% compared to 16QAM because 64QAM transmits six bits with a single symbol. On the other hand, the constellation points are closer to each other for the higher order modulation and the required signal-to-noise ratio for correct reception is higher. The difference in the required signal-to-noise ratio is approximately 6 dB between 16QAM and QPSK and also between 64QAM and 16QAM. Therefore, downlink 64QAM and uplink 16QAM can be utilized only when the channel conditions are favorable.

The system simulation results with 64QAM in macro cells are illustrated in Figure 17.10. The 64QAM modulation improves the user data rate with 10–25% probability depending on the scheduling (RR = round robin, PF = proportional fair). The rest of the time the channel conditions are not good enough to enable the reception of 64QAM modulation. The typical capacity gain from 64QAM is less than 10%.

64QAM and MIMO together improve the peak rate by 200% from 14 Mbps to 42 Mbps but the average cell capacity is only improved by 20% because those high data-rate features are useful only for part of the cell area. 64QAM activation can improve the cell capacity by approximately 5% and MIMO by 10%. The evolution of the peak and average rates are illustrated in Figure 17.11. The average rate corresponds to fully loaded macro cells. The practical networks are not always fully loaded, and therefore, the gains of 64QAM and MIMO features can be higher in real networks than in fully loaded simulations.

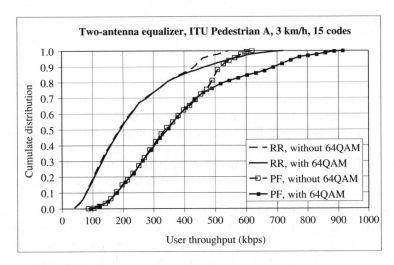

Figure 17.10 Macro cell data rates per user with 64 QAM with 20 active users in a cell [5]

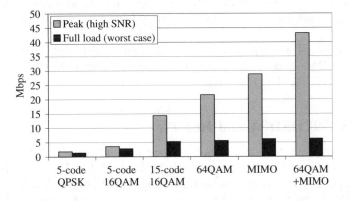

Figure 17.11 Downlink peak rate and full load cell capacity evolution with HSDPA features

17.5.1 MIMO Workaround Solutions

The original Release 7 MIMO had a few performance issues:

- Non-MIMO equalizer UEs suffer when Space Time Transmit Diversity (STTD) is activated. The equalizer receiver estimates the multi-path channel and removes the multi-path interference. When the transmission happens with STTD, there are two times more multi-path components and the equalization becomes more difficult. Also, equalization of two channels simultaneously is more challenging.
- The code multiplexing degrades performance when the precoding vectors are different for code multiplexed UEs because different precoding vectors make the signal structure suboptimal.

Further improvements were required in 3GPP specifications to improve the performance of MIMO and non-MIMO terminals. These improvements are referred to as MIMO
workaround solutions.

- Virtual antenna mapping is applied in NodeB transmission. The non-MIMO channels
  are transmitted only via one branch – virtual antenna 1. The transmission powers are
  not balanced because only one transmission branch is used. The virtual antenna mapping solution takes the signal from one virtual antenna, rotates the signal and sums to
  the other antenna. The resulting transmission powers in the physical antennas are now
  balanced and the two RF power amplifiers can also be fully utilized for non-MIMO
  and non-transmit diversity transmissions. The virtual antenna mapping does not need
  any 3GPP support.
- The original Release 7 MIMO had four precoding vectors for the feedback from UE to
  NodeB. That causes problems with virtual antenna mapping because some combinations
  of the precoding feedback weights make the powers to be unbalanced in the two power
  amplifiers. There is a need to limit the number of precoding vectors from four to two.
  The network informs UE how many precoding vectors UE can use in the feedback
  reporting. This feature was a late inclusion to Release 7 specifications during 2010.
- The relative power of Secondary-CPICH can be reduced compared to Primary-CPICH
  to reduce the interference caused by Secondary-CPICH to non-MIMO UEs.

Release 7 and 8 MIMO deployments have been only small-scale trials mainly because
of these early MIMO issues and because the same data rate of 42 Mbps can be achieved
with Dual Cell HSDPA (DC-HSDPA).

17.6   Dual Cell HSDPA and HSUPA

The LTE radio improves the data rates compared to HSPA because LTE can use a transmission bandwidth up to 20 MHz compared to 5 MHz in HSPA. The dual cell (dual carrier)
HSDPA was specified as part of Release 8 enabling HSDPA to take benefit of two adjacent
HSDPA carriers in the transmission to a single terminal using a total 10 MHz downlink
bandwidth. The uplink solution in Release 8 is still using a single 5 MHz carrier. The
concept is illustrated in Figure 17.12.

The benefit of the DC-HSDPA for the user data rate is illustrated in Figure 17.13.
DC-HSDPA can double the user data rate at low loading because the user can access
the capacity of two carriers instead of just one. The relative benefit decreases when the
loading increases. There is still some capacity benefit at high load due to frequency
domain scheduling and due to dynamic balancing of the load if both carriers are not
100% loaded all the time. NodeB scheduling can optimize the transmission between the
two carriers based on the CQI reporting – obtaining partly similar gains as in LTE with
frequency domain scheduling. The frequency domain scheduling gain in DC-HSDPA is
smaller than with LTE because the scheduling in DC-HSDPA occurs in 5 MHz blocks
while LTE scheduling occurs with 180 kHz resolution [6].

Both DC-HSDPA and MIMO can boost HSDPA data rates. Those two solutions are
compared in Table 17.1. Both solutions can provide the same peak rate of 42 Mbps
with 64QAM modulation. MIMO can improve spectral efficiency due to two antenna
transmission, while the DC-HSDPA brings some improvement to the high loaded case

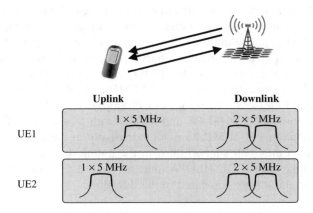

Figure 17.12   Downlink dual cell HSDPA concept

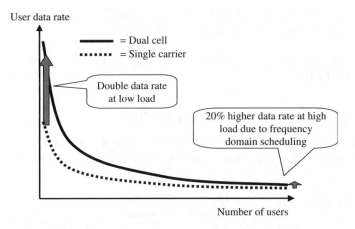

Figure 17.13   Data rate benefit of the dual carrier HSDPA

Table 17.1   Benchmarking of DC-HSDPA and MIMO

	DC-HSDPA	MIMO
Peak bit rate	42 Mbps	42 Mbps
Full load spectral efficiency improvement	20% due to frequency domain scheduling and larger trunking gain	10% due to two antenna transmissions
Data rate gain	Similar gain over the whole cell area	Largest gain close to NodeB where dual stream transmission is feasible
NodeB RF requirements	Single power amplifier and Duplex filter per sector	Two power amplifiers and Duplex filters per sector.
UE RF requirements	Possible with one antenna terminal	Two antennas required

with frequency domain scheduling and larger trunking gain. The DC-HSDPA solution looks attractive because the data rate improvement is available over the whole cell area equally whereas MIMO mostly improves the data rates close to NodeB. Also, DC-HSDPA tends to be easier to upgrade the network because it can be implemented with a single 10 MHz power amplifier per sector while MIMO requires two separate power amplifiers.

3GPP Release 8 does not define the use of MIMO and DC-HSDPA at the same time. The peak data rate in Release 8 is therefore still 42 Mbps even with DC-HSDPA. Release 9 allows the combination of DC-HSDPA and MIMO with a peak rate of 84 Mbps, which can motivate the deployment of MIMO more extensively in the networks.

Release 9 included dual cell support, also in the uplink direction. The motivation for DC-HSUPA is similar as for DC-HSDPA, but there are a few differences between the dual cell benefits in the uplink compared to the downlink:

- UE transmission power is much lower than the NodeB transmission power. The uplink data rate is more likely to be limited by transmission power than the downlink data rate.
- Frequency-selective packet scheduling is not as simple in uplink as downlink because there is no similar fast CQI reporting as with HSDPA.

Therefore, the data rate gains of DC-HSUPA are typically somewhat lower than the gains with DC-HSDPA.

17.7 Multicarrier and Multiband HSDPA

Many operators have total 15 to 20 MHz spectrum available for HSPA. Therefore, it was a natural evolution in Release 10 to support to three and four carriers. Another evolution

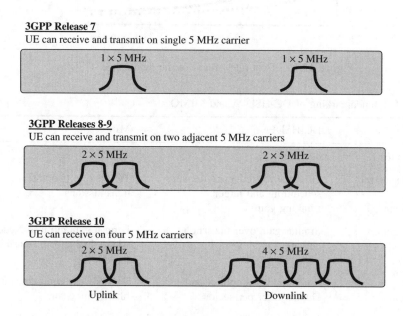

Figure 17.14 Multicarrier HSPA evolution

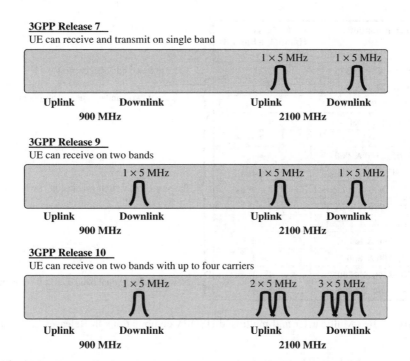

Figure 17.15 Multiband HSPA evolution with example combination of 900 MHz and 2100 MHz

was multiband HSDPA where UE can receive simultaneously on two different frequency bands to boost the data rates.

Figure 17.14 illustrates multicarrier evolution. Release 8 enabled dual cell downlink and Release 9 dual cell uplink. Release 10 enhances downlink to four carriers but does not bring any uplink enhancements beyond two carriers.

Figure 17.15 shows the multiband HSDPA evolution. The first phase of multiband solution is defined in Release 9 where the two carriers can be on different bands. Release 10 allows a total four carriers – for example, three carriers on a high frequency and one carrier on a low frequency. All multi-band combinations are shown in Figure 17.16. For example, in Europe, the typical combination would be 2100 plus 900. In the USA the combinations are 1900 plus 850 or 1900 plus 1.7/2.1. Further combinations are also defined for Japan including 2100 plus 1500, and for Asian cases 2100 plus 850.

17.8 Uplink 16QAM

HSUPA Release 6 uplink uses QPSK modulation providing 5.76 Mbps. The uplink data rate within 5 MHz can be increased by using higher order modulation or MIMO transmission. The challenge with uplink MIMO is that the UE needs to have two power amplifiers. Therefore, uplink single-user MIMO is not part of HSPA evolution or LTE in Release 8. The higher order 16QAM modulation was adopted as part of Release 7 for HSUPA, doubling the peak rate to 11.5 Mbps.

The higher order modulation improves the downlink spectral efficiency because the downlink has a limited number of orthogonal resources. The same is not true for the

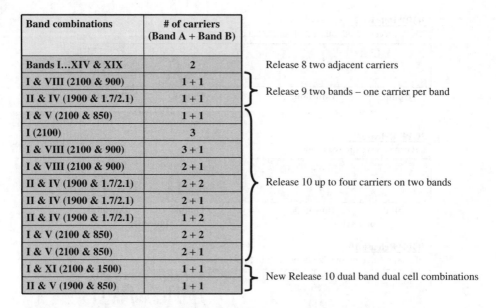

Figure 17.16 Multicarrier and multiband HSDPA combinations in Releases 8, 9 and 10

uplink because the HSUPA uplink is not orthogonal and there is a practically unlimited number of codes available in the uplink. The highest uplink spectral efficiency can be achieved by using QPSK modulation only. In other words, HSUPA 16QAM is a peak data rate feature, not a capacity feature.

The multi-path propagation affects high data rate performance. Therefore, the UE equalizer is used on HSDPA terminals. The NodeB receiver can also improve the uplink high bit rate HSUPA performance in multi-path channels by using an equalizer. Another solution is to use four-antenna reception in the uplink.

High uplink data rates require high Ec/N0. Fixed reference channel 8 with 8 Mbps and 70% throughput requires Ec/N0 = 12–16 dB [7]. The corresponding uplink noise rise will also be similar, 12–16 dB, impacting the coverage of other simultaneous users. It is therefore beneficial to use the uplink interference cancellation to subtract the high bit-rate interference from other simultaneous users. The LTE radio uses an orthogonal uplink solution avoiding the intra-cell interference.

17.9 Terminal Categories

New HSDPA terminal categories with 64QAM, MIMO and DC-HSDPA were added in Releases 7 and 8. The HSDPA categories 13 and 14 include 64QAM and categories 15–18 MIMO offering peak rates of 21.1 Mbps and 28.0 Mbps. The combination of 2×2 MIMO and 64QAM is part of categories 19 and 20 pushing the peak rate to 42.2 Mbps. The DC-HSDPA categories are 21–24. Categories 21–22 support 16QAM and categories 23–24 support 64QAM. Categories 25–28 support the combination of DC-HDSPA and MIMO. Categories 29–32 include three and four carrier HSDPA. HSDPA terminal categories are listed in Table 17.2. HSUPA terminal categories are listed in Table 17.3. Category 7 defines 16QAM and Category 8 dual cell HSUPA.

Table 17.2   HSDPA terminal categories

Cat	Codes	Modulation	MIMO	Multi-carriers	Coding	Peak	3GPP
12	5	QPSK	–	–	3/4	1.8 Mbps	Release 5
6	5	16QAM	–	–	3/4	3.6 Mbps	Release 5
8	10	16QAM	–	–	3/4	7.2 Mbps	Release 5
9	15	16QAM	–	–	3/4	10.1 Mbps	Release 5
10	15	16QAM	–	–	1/1	14.0 Mbps	Release 5
13	15	64QAM	–	–	5/6	17.6 Mbps	Release 7
14	15	64QAM	–	–	1/1	21.1 Mbps	Release 7
15	15	16QAM	2 × 2	–	5/6	23.4 Mbps	Release 7
16	15	16QAM	2 × 2	–	1/1	28.0 Mbps	Release 7
17	15	64QAM or MIMO	–	–	5/6	23.4 Mbps	Release 7
18	15	64QAM or MIMO	–	–	1/1	28.0 Mbps	Release 7
19	15	64QAM	2 × 2	–	5/6	35.3 Mbps	Release 8
20	15	64QAM	2 × 2	–	1/1	42.2 Mbps	Release 8
21	15	16QAM	–	2	5/6	23.4 Mbps	Release 8
22	15	16QAM	–	2	1/1	28.0 Mbps	Release 8
23	15	64QAM	–	2	5/6	35.3 Mbps	Release 8
24	15	64QAM	–	2	1/1	42.2 Mbps	Release 8
25	15	16QAM	2 × 2	2	5/6	46.8 Mbps	Release 9
26	15	16QAM	2 × 2	2	1/1	56.0 Mbps	Release 9
27	15	64QAM	2 × 2	2	5/6	70.6 Mbps	Release 9
28	15	64QAM	2 × 2	2	1/1	84.4 Mbps	Release 9
29	15	16QAM	–	3	1/1	63.3 Mbps	Release 10
30	15	16QAM	2 × 2	3	1/1	126.6 Mbps	Release 10
31	15	64QAM	–	4	1/1	84.4 Mbps	Release 10
32	15	64QAM	2 × 2	4	1/1	168.8 Mbps	Release 10

Table 17.3   HSUPA terminal categories

Cat	TTI	Modulation	Multi-carriers	Coding	Peak	3GPP
3	10 ms	QPSK	–	3/4	1.4 Mbps	Release 6
5	10 ms	QPSK	–	3/4	2.0 Mbps	Release 6
6	2 ms	QPSK	–	1/1	5.7 Mbps	Release 6
7	2 ms	16QAM	–	1/1	11.5 Mbps	Release 7
8	2 ms	16QAM	2	1/1	23.0 Mbps	Release 9

17.10   Layer 2 Optimization

The WCDMA Release 99 specification was based on the packet retransmissions running from Radio Network Controller (RNC) to the UE on the layer 2. The layer 2 Radio Link Control (RLC) packets had to be relatively small to avoid the retransmission of very large packets in case of transmission errors. Another reason for the relatively small RLC packet size was the need to provide sufficiently small step sizes for adjusting the data rates for Release 99 channels. The RLC packet size in Release 99 is not only small but it

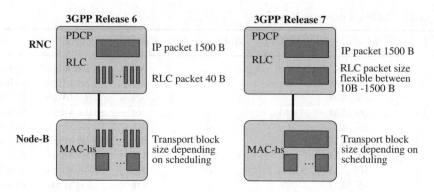

Figure 17.17   Flexible RLC concept

is also fixed for Acknowledged Mode Data and there are just a limited number of block sizes in Unacknowledged Mode Data. This limitation is due to transport channel data rate limitations in Release 99.

The RLC payload size is fixed to 40 bytes in Release 99 for Acknowledged Mode Data. The same RLC solution is applied to HSDPA Release 5 and HSUPA Release 6 as well: the 40-byte packets are transmitted from RNC to the base station in the case of HSDPA. An additional configuration option to use 80-byte RLC packet size was introduced in Release 5 to avoid extensive RLC protocol overhead, layer 2 processing and RLC transmission window stalling. With the 2 ms TTI used with HSDPA this leads to possible data rates being multiples of 160 kbps and 320 kbps respectively.

As the data rates are further increased in Release 7, increasing the RLC packet size even further would significantly affect the granularity of the data rates available for HSDPA scheduling and the possible minimum data rates.

3GPP HSDPA and HSUPA allow the optimization of the Layer 2 operation because Layer 1 retransmissions are used and the probability of Layer 2 retransmissions is very low. Also, the Release 99 transport channel limitation does not apply to HSDPA/HSUPA because the layer 2 block sizes are independent of the transport formats. It is therefore possible to use flexible and considerably larger RLC sizes and introduce segmentation to the MAC (Medium Access Control) layer in the base station.

This optimization is included for the downlink in Release 7 and for the uplink in Release 8 and it is called Flexible RLC and MAC segmentation solution. The RLC block size in flexible RLC solution can be as large as an Internet Protocol (IP) packet, which is typically 1500 bytes for download. There is no need for packet segmentation in RNC. By introducing the segmentation to the MAC, the MAC can perform the segmentation of the large RLC PDU based on physical layer requirements when needed. The flexible RLC concept in downlink is illustrated in Figure 17.17.

The flexible RLC and MAC segmentation brings a number of benefits in terms of Layer 2 efficiency and in terms of peak bit rates.

- The relative layer 2 overhead is reduced. With the RLC header of 2 bytes the RLC overhead is 5% in case of 40-byte RLC packet. When the RLC packet size increases to 1500 bytes, the RLC header overhead is reduced to below 0.2%. That reduction of the overhead can improve the effective application data throughput.

- The RLC block size can be flexibly selected according to the packet size of each application. That flexibility helps to avoid unnecessary padding which is no more needed in the flexible RLC solution. That is relevant especially for small IP packet sizes which are typical in VoIP or streaming applications.
- Less packet processing is required in RNC and in UE with an octet aligned protocol header. The number of packets to be processed is reduced since the RLC packet size is increased and octet aligned protocol headers avoids bit shifting in high data rates connections. Both reduces layer 2 processing load and makes the high bit rate implementation easier.
- Full flexibility and resolution of available data rates for the HSDPA scheduler.

17.11 Single Frequency Network (SFN) MBMS

Multimedia broadcast multicast service (MBMS) was added to 3GPP as part of Release 6. 3GPP Release 6 can use soft combining of the MBMS transmission from the adjacent cells. The soft combining considerably improves MBMS performance at the cell edge compared to receiving the signal from a single cell only. Release 6 MBMS therefore provides a very good starting point for broadcast services from the performance point of view.

Even if the soft combining can be used in Release 6, the other cell signals are still causing interference to the MBMS reception because the adjacent cells are not orthogonal due to different scrambling codes. If the same scrambling code would be used in all cells together with a terminal equalizer, the other cells transmitting the same signal in synchronized network would be seen just as single signal with time dispersion. That solution provides essentially a single frequency network with practically no neighboring cell interference. The MBMS over a Single Frequency Network (MBSFN) can enhance MBMS data rates and capacity. The single frequency can be realized with network synchronization and by using the same scrambling code for MBMS transmissions from multiple cells. The MBSFN is included into 3GPP Release 7 and extended to unpaired bands in Release 8. The unpaired band solution is called Integrated Mobile Broadcast (IMB). The MBSFN solution requires a dedicated carrier for MBMS only transmission, which makes MBSFN a less flexible solution from the spectrum point of view compared to Release 6 MBMS. Release 6 MBMS can coexist with point-to-point traffic on the same carrier.

3G MBMS has not been deployed commercially at the time of writing of this edition. Unicast video delivery works fine over HSPA and supports the general trend where the broadcast is moving from linear TV to video-on-demand, like Youtube videos.

17.12 Architecture Evolution

3GPP Release 6 has four network elements in the user and control plane: base station (NodeB), RNC (Radio Network Controller), SGSN (Serving GPRS Support Node) and GGSN (Gateway GPRS Support Node). The architecture in Release 8 LTE will have only two network elements: base station in the radio network and Access Gateway (a-GW) in the core network. The a-GW consists of control plane MME (Mobility management entity) and user plane SAE GW (System Architecture Evolution Gateway). The flat network architecture reduces the network latency and thus improves the overall performance of IP-based services. The flat model also improves both user and control plane efficiency.

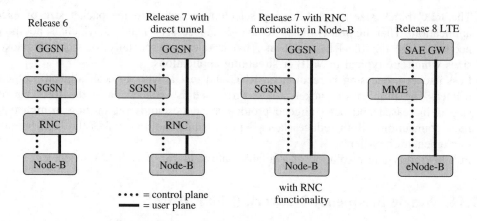

Figure 17.18   Evolution towards flat architecture

The flat architecture is also considered beneficial for HSPA and it is specified in Release 7. The HSPA flat architecture in Release 7 and LTE flat architecture in Release 8 are exactly the same: NodeB responsible for the mobility management, ciphering, all retransmissions and header compression both in HSPA and in LTE. The architecture evolution in HSPA is designed to be backwards compatible: existing terminals can operate with the new architecture and the radio and core network functional split is not changed. The architecture evolution is illustrated in Figure 17.18.

The packet core network also has flat architecture in Release 7. It is called direct tunnel solution and allows the user plane to by-pass SGSN. When having the flat architecture with all RNC functionality in the base station and using direct tunnel solution, only two nodes are needed for user data operation. This achieves flexible scalability and allows the introduction of higher data rates with HSPA evolution, with minimum impacts to the other nodes in the network. This is important for achieving low cost per bit and enabling competitive flat rate data charging offerings. As the gateway in LTE is having similar functionality as GGSN, it is foreseen to enable deployments of LTE and HSPA where both connect directly to the same core network element for user plane data handling directly from the base station.

3GPP Release 8 includes a new architecture to support small home NodeBs as well. The home NodeBs are called femto access points. The home NodeBs are installed in private homes in the same way as WLAN access points today. These home NodeBs are using operator's licensed frequency and are connected to the operator's core network. The output power level is low, typically 100 mW or below. The transport connection uses fixed DSL connections at homes. The handovers and idle mode selections between home NodeBs and the macro network are controlled by the network algorithms and parameters. The flat architecture from Figure 17.18 is not optimized for the case where there is a very large number of small base stations because the number of core network connections would be very large. Therefore, a new network element – called Home NodeB gateway – is introduced to hide the large number of home NodeBs from the core network. The gateway is located in the operator premises. The new interface between home NodeBs and the gateway is called Iuh. The interface between the gateway and the core network is the normal Iu interface. The home NodeB architecture is illustrated in Figure 17.19 [8].

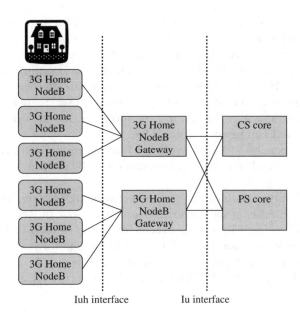

Figure 17.19   Home NodeB architecture [8]

17.13   Summary

While 3GPP defined a new radio system – LTE (Long Term Evolution) – in Release 8, it also brought quite a few improvements for HSPA. LTE deployment typically requires a new spectrum whereas HSPA evolution features can be upgraded flexibly on the existing carriers and the features can co-exist with all legacy WCDMA terminals. HSPA also supports and enhances Circuit Switched voice service, which is not available in packet-based LTE radio.

HSPA enhancements provide major improvements to the performance for the end user in increasing peak bit rates, reducing mobile terminal power consumption, and reducing the latency. The peak bit rates within 5 MHz can be tripled with 64QAM and 2×2 MIMO from 14 Mbps to 42 Mbps. The bit rate can be further increased by aggregation multiple carriers: dual cell HSPA combines two adjacent carriers together and multicarrier and multiband HSDPA allows the combination of up to four HSDPA carriers from two bands to offer peak rate up to 168 Mbps. These peak rate features also bring improvements to spectral efficiency and cell capacity.

The UE power consumption is considerably reduced in HSPA evolution due to discontinuous transmission and reception. The expected voice call talk time will improve by 50% and the usage time of bursty data applications increases even more. The practical talk times will improve further when the digital and RF technologies evolve and general power consumption is reduced.

HSPA evolution enhances basic voice service by mapping Circuit switched (CS) voice on top of HSPA channels – a concept called CS voice over HSPA. It is essentially a combination of Voice over IP (VoIP) in the radio and CS voice in the core network. CS voice over HSPA reduces UE power consumption and increases spectral efficiency compared to WCDMA voice while having no impact to the core network and end-to-end ecosystem. It will be easy to upgrade CS voice over HSPA later to VoIP over HSPA because the radio solutions are very similar.

The setup times are reduced by mapping the RACH/FACH common channels on top of HSDPA/HSUPA. It is actually all the services including signaling that can be mapped on top of HSPA in HSPA evolution. There are only a few physical layer channels left from Release 99 specifications in Release 8 – otherwise, everything is running on top of HSPA. That explains also why the end user performance and the network efficiency are considerably improved compared to Release 99.

The performance benefits of HSPA evolution features are summarized in Figure 17.20. The exact gains depend on the implementation choices and on the deployment environment.

3GPP Release 7 allows simplification of the network architecture. The number of network elements for the user plane can be reduced from four in Release 6 down to two in Release 7. Release 7 architecture is exactly the same as used in LTE in Release 8, which makes network evolution from HSPA to LTE straightforward. Most of the 3GPP Release 7 and 8 enhancements are expected to be relatively simple upgrades to the HSPA networks, as was the case with HSDPA and HSUPA in earlier releases.

		Peak rate	Average rate (capacity)	Cell edge rate	Latency gain	Talk time
Downlink	HSDPA 64QAM[1]	+50%	<10%	-	-	-
	HSDPA 2×2MIMO	+100%	<30%	<20%	-	-
	DC-HSDPA 4C-HSDPA	+100–300%	+20–300%	+20–300%	-	-
Uplink	HSUPA 10 ms (2.0 Mbps)[2]	+600%	+20–100%	<100%	Gain 20 ms	-
	HSUPA 2 ms (5.8 Mbps)	+200%	<30%	-	Gain 15 ms	-
	HSUPA 16QAM	+100%	-	-	-	-
	DC-HSUPA	+300%	+20–100%	-	-	-
	Advanced NodeB receiver	-	>30%	-	-	-
	DTX/DRX	-	-	-	-	>+50%
	HS-FACH / HS-RACH	-	-	-	Setup time <0.1 s	-
	CS voice over HSPA	-	+80% (voice)	-	-	>+50%

[1]Baseline WCDMA Release 5 downlink 14.4 Mbps
[2]Baseline WCDMA Release 99 uplink 384 kbps

= clear gain >30%
= moderate gain <30%

Figure 17.20 Summary of HSPA evolution features and their benefits

HSPA evolution continues in Release 11 and beyond. The expected topics include further carrier aggregation, uplink multi-antenna solution, coordinated multipoint transmission, and self-optimized networks.

References

[1] 3GPP, Technical Specification 24.008, 'Mobile Radio Interface Layer 3 Specification; Core Network Protocols', v.8.3.0.
[2] Holma, H., Kuusela, M., Malkamäki, E., Ranta-aho, K. and Tao C. 'VoIP over HSPA with 3GPP Release 7', PIMRC2006, September 2006.
[3] Holma, H. and Toskala, A. (eds) 'WCDMA for UMTS: HSPA Evolution and LTE', 5th edition, John Wiley & Sons, 2010.
[4] 3GPP, 'Further Discussion on Delay Enhancements in Rel7,' 3GPP R2-061189, August 2006.
[5] 3GPP, '64QAM for HSDPA', 3GPP R1-063335, November 2006.
[6] Morais de Andrade, D., Klein, A., Holma, H., Viering, I., Liebl, G. 'Performance Evaluation on Dual-Cell HSDPA Operation', VTC, September 2009.
[7] 3GPP, Technical Specifications, 25.104, 'Base Station (BS) Radio Transmission and Reception (FDD)', v. 8.3.0.
[8] 3GPP, Technical Specification, 25.467 'UTRAN Architecture for 3G Home Node B', v. 9.3.0.

Index

LTE for UMTS: Evolution to LTE-Advanced, Second Edition. Edited by Harri Holma and Antti Toskala.
© 2011 John Wiley & Sons, Ltd. Published 2011 by John Wiley & Sons, Ltd.